MODERN DIESEL TECHNOLOGY: ELECTRICITY & ELECTRONICS
2ND EDITION

Joseph A. Bell

DELMAR
CENGAGE Learning·

Australia • Brazil • Japan • Korea • Mexico • Singapore • Spain • United Kingdom • United States

Modern Diesel Technology: Electricity & Electronics, 2nd Edition
Joseph A. Bell

Vice President, Careers & Computing:
   Dave Garza

Director of Learning Solutions: Sandy Clark

Executive Editor: Dave Boelio

Director, Development–Career
   & Computing: Marah Bellegarde

Managing Editor: Larry Main

Senior Product Manager: Sharon Chambliss

Editorial Assistant: Courtney Troeger

Brand Manager: Kristin McNary

Market Development Manager: Erin Brennan

Senior Production Director: Wendy Troeger

Production Manager: Mark Bernard

Content Project Manager: Christopher Chien

Art Director: Jackie Bates/GEX

Cover Image: Courtesy of Navistar, Inc.

Cover Inset Image: © 2014 Cengage Learning;
   Photo courtesy of Joseph A. Bell

For product information and technology assistance, contact us at
Cengage Learning Customer & Sales Support, 1-800-354-9706
For permission to use material from this text or product,
submit all requests online at **www.cengage.com/permissions.**
Further permissions questions can be e-mailed to
permissionrequest@cengage.com

Library of Congress Control Number: 2012948308

ISBN-13: 978-1-133-94980-0

ISBN-10: 1-133-94980-0

Delmar
5 Maxwell Drive
Clifton Park, NY 12065-2919
USA

Cengage Learning is a leading provider of customized learning solutions with office locations around the globe, including Singapore, the United Kingdom, Australia, Mexico, Brazil, and Japan. Locate your local office at: **international.cengage.com/region**

Cengage Learning products are represented in Canada by Nelson Education, Ltd.

To learn more about Delmar, visit **www.cengage.com/delmar**

Purchase any of our products at your local college store or at our preferred online store **www.cengagebrain.com**

Notice to the Reader
Publisher does not warrant or guarantee any of the products described herein or perform any independent analysis in connection with any of the product information contained herein. Publisher does not assume, and expressly disclaims, any obligation to obtain and include information other than that provided to it by the manufacturer. The reader is expressly warned to consider and adopt all safety precautions that might be indicated by the activities described herein and to avoid all potential hazards. By following the instructions contained herein, the reader willingly assumes all risks in connection with such instructions. The publisher makes no representations or warranties of any kind, including but not limited to, the warranties of fitness for particular purpose or merchantability, nor are any such representations implied with respect to the material set forth herein, and the publisher takes no responsibility with respect to such material. The publisher shall not be liable for any special, consequential, or exemplary damages resulting, in whole or part, from the readers' use of, or reliance upon, this material.

Printed in the United States of America
1 2 3 4 5 6 7 16 15 14 13

Table of Contents

Preface for Series

The Modern Diesel Technology (MDT) series of textbooks debuted in 2007 as a means of addressing the learning requirements of schools and colleges whose syllabi used a modular approach to curricula. The initial intent was to provide comprehensive coverage of the subject matter of each title using ASE/NATEF learning outcomes and thus provide educators in programs that directly target a single certification field with a little more flexibility. In some cases, an MDT textbook exceeds the certification competency standards. An example of this is Joseph Bell's *MDT: Electricity & Electronics*, in which the approach challenges the student to attain the level of understanding needed by a technician specializing in the key areas of chassis electrical and electronics systems—in other words, higher than that required by the general service technician.

The MDT series now boasts nine textbooks, some of which are going into their second edition. As the series has evolved, it has expanded in scope with the introduction of books addressing a much broader spectrum of commercial vehicles. Titles now include *Heavy Equipment Systems*, *Mobile Equipment Hydraulics*, and *Heating, Ventilation, Air Conditioning & Refrigeration*. The latter includes a detailed examination of trailer reefer technology, subject matter that falls outside the learning objectives of a general textbook. While technicians specializing in all three areas are in demand in most areas of the country, there are as yet no national certification standards in place.

In addition, the series now includes two books that are ideal for students beginning their study of commercial vehicle technology. Thes two titles (*Preventive Maintenance and Inspection* and *Diesel Engines*) are written so that they can be used in high school programs. Each uses simple language and a no-nonsense approach suited for either classroom or self-directed study. That some high schools now option programs specializing in commercial vehicle technology is an enormous progression from the more general secondary school "shop class" which tended to lack focus. It is also a testament to the job potential of careers in the commercial vehicle technology field in a general employment climate that has stagnated for several years. Some forward-thinking high schools have developed transitional programs partnering with both colleges and industry to introduce motive power technology as early as Grade 10, an age at which many students make crucial career decisions. When a high school student graduates with credits in Diesel Technology or Preventive Maintenance Practice it can accelerate progression through college programs as well as make those responsible for hiring future technicians for commercial fleets and dealerships take notice.

As the MDT series has evolved, textbooks have been added that target specific ASE certifications, providing an invaluable study guide for certified technicians who are adding to their qualifications along with College programs that use a modular learning approach. *Electronic Diesel Engine Diagnosis* (ASE L2), *Truck Brakes, Suspension, and Steering Systems* (ASE T4 and T5) and *Light Duty Diesel Engines* (ASE A9) detail the learning outcomes required for each ASE certification test.

Because each textbook in the MDT series focuses exclusively on the competencies identified by its title, the books can be used as a review and study guide for technicians prepping for specific certification examinations. Common to all of the titles in the MDT series, the objective is to develop hands-on competency without omitting any of the conceptual building blocks that enable an expert understanding of the subject matter from the technician's perspective. The second editions of these titles not only integrate the changes in technology that have taken place over the past five years, but also blend in a wide range of instructor feedback based on actual classroom proofing. Both should combine to make these second editions more pedagogically effective.

Sean Bennett 2012

Preface

The primary purpose of this textbook is to provide an understanding of electricity and electronics to the extent required by an entry-level truck technician. The focus throughout this textbook is the fundamental principles of electricity and the application of these principles to the diagnosis of modern truck electrical systems.

AUTHOR'S BACKGROUND

My career in the automotive industry began at a service station with a garage when I was still in high school. After graduation, I joined the military where I became a truck technician. Following my military service, I worked as an automobile and truck dealership technician throughout the automotive electronics revolution of the 1980s. Like most other technicians of that era, my electrical skills were lacking, so I decided to take some college courses to increase my knowledge of electricity and electronics. After several years of part-time study while still working full-time as a technician, I was able to complete a bachelor's degree in Electrical Engineering Technology from Purdue University. I then began working for International Truck and Engine Corporation (Navistar) and became the lead electrical test engineer. Some of my projects at International include the High Performance Truck series, 2007 emissions, and ProStar models. I am currently a senior diagnostic engineer for a manufacturer of diesel engines.

REASONS FOR WRITING THIS BOOK

Trucks of the past had very simple electrical systems. Anyone who has looked at the electrical system on a modern truck knows that this is no longer true. One of my main reasons for writing this book is to provide a text designed specifically for truck technicians-in-training that stresses the importance of a strong knowledge of the fundamentals of electricity. As a former technician, I have experienced first-hand the anxiety that electrical problems can present. The goal of this book is to help alleviate the anxiety associated with troubleshooting a modern truck electrical system problem by explaining some of the mysteries of electricity and electronics.

DETAILS OF THE TEXT

The material in this textbook is presented using easy-to-understand analogies whenever possible. These analogies are comparisons of electrical concepts with concepts that are much easier for most students to understand, such as hydraulics and pneumatics. Math and non-relevant theory is kept at a minimum, but Extra for Experts sections appear at the end of most chapters where more challenging topics are addressed. The text also addresses the disassembly and testing of cranking motors (starters) and alternators. Repairing or rebuilding of cranking motors and alternators was common in truck repair facilities of the past. These days, inoperative cranking motors and alternators are usually just replaced with new or remanufactured units. Material covering the inner workings of cranking motors and alternators was included in the text because many experienced technicians have indicated that it is still important for modern truck technicians to understand how these electrical devices function. This is true even though the cranking motor or alternator is probably going to be replaced anyway. An understanding of the inner workings of cranking motors and alternators should help technicians to troubleshoot problems associated with the cranking and charging systems. This same philosophy is carried over to the coverage of electronic modules found in modern trucks. Like cranking motors and alternators, electronic modules are almost never repaired by truck technicians. Even so, the text describes some of the components that are contained within typical truck electronic modules and the manner in which these components interact with other devices in the electrical system. A basic understanding of what is occurring inside these electronic modules should help technicians to troubleshoot modern truck electrical systems.

NEW FOR THE SECOND EDITION

Enhancements for the second edition include the addition of a new chapter on electronic diesel engines with discussions on EPA 2007 and 2010 exhaust emissions including heavy-duty on-board diagnostics (HD-OBD). Hybrid electric vehicles are also introduced along with other emerging technologies. The chapter on body control modules includes updated material with enhanced coverage of the Freightliner multiplexed electrical system.

SPECIAL NOTATIONS

Throughout this book, the text contains special notations labeled Important Fact, TechTip, Caution, and Warning. **Important Fact** indicates that the information is vital for understanding a concept and is something that you should try to commit to memory. This information will usually come up again in later chapters. **TechTip** indicates that the information is something that you may find useful in the future when actually working as a technician. **Caution** is given to prevent making a mistake or error that could damage equipment or result in personal injury. **Warning** is used to emphasize that serious personal injury or injury to others could occur if the information is not heeded.

ACKNOWLEDGMENTS

I would like to thank the engineers, mechanics, and technicians who have mentored me throughout my career both as an automotive technician and as an engineer. I would also like to recognize some of my favorite professors from Indiana University–Purdue University at Fort Wayne, including Hal Broberg, Ph.D., Peter Hamburger, Ph.D., Thomas Laverghetta, MSEE, Paul I-Hai Lin, MSEE, and David Maloney, Ph.D. Thanks to Phil Christman and Jeff Calfa of Navistar for artwork permissions, including the cover art for the MDT series.

This second edition is dedicated to my grandsons Ethan, Jackson, John, and Jason.

CONTRIBUTORS

Jay Bissontz, PL
HEV Program Manager
Navistar

Piotr Dobrowolski
Product Manager
Continental Corporation

Ron Friend
Engineer
Bendix Commercial Vehicle Systems

Deborah Fogt
Senior Engineer
Cummins Inc.

Clive Harley
VP Engineering
Prestolite Electrical

Drew Harbach
Senior Product Engineer
Peterbilt Motors Company

Daniel Hilaire
Senior Master Technician
Navistar

Scott Kammeyer
Senior Engineer
Caterpillar Inc.

Edward Kelwaski
Senior Electrical Engineer
Heil Environmental

Rich Michels
Navistar Master Technician
Brattain International Trucks

Dustin Moehrman
Freightliner Technician

David Perdue
Electrical Test Engineer
Navistar

Raymond Peterson
Director – Locomotive Engineering
Union Pacific Railroad

REVIEWERS

Cynthia Bell
Physics, Chemistry, and Science Teacher
Don Bosco High School

Matt Gumbel
Senior Engineer
Navistar

John Murphy
Professor
Centennial College
School of Transportation

Aaron Dix
Madison Area Technical College

Jesus Acevedo
Western Technical College

Bill Yokley
Guilford Technical Community College

CHAPTER

1 Safety

Learning Objectives

After studying this chapter, you should be able to:

- List proper attire for working in a truck repair facility.

- Choose which type of protective eyewear should be worn for a given task.

- Describe some of the hazards associated with high voltages.

- Discuss some of the hazards associated with hybrid trucks.

- Explain the purpose of the third terminal on a North American 120V AC electrical outlet and why it is important.

- Explain personal safety precautions, including the use of MSDS.

Key Terms

electric shock

ground fault circuit interrupter (GFCI)

material safety data sheet (MSDS)

original equipment manufacturer (OEM)

INTRODUCTION

The importance of safe work practices is something that cannot be overemphasized. Many truck technicians are paid on a flat-rate pay system or similar performance-based pay system. This may tempt you to cut corners and take risks when it comes to safety in an effort to make rate. However, a work-related injury can leave you sitting at home or in the hospital and earning little or no pay (or worse).

Most shops take safety seriously because they know that an injured technician results in a loss of shop revenue and that good technicians are difficult to replace. Additionally, job-related injuries may cause governmental agencies such as the U.S. Occupational

Safety and Health Administration (OSHA) to inspect the facility and levy fines for safety violations. However, each technician is ultimately responsible for his or her own personal safety.

PROPER ATTIRE

Proper clothing for a truck technician is very important to minimize risk of injury. It is often necessary to work near moving components when troubleshooting electrical problems. Clothing should be properly fitted and worn correctly. Loose-fitting clothing or untucked shirttails can be caught in moving components, resulting in serious injury. Trucks have steps that must be climbed to enter the cab or to work at the back of cab. It may also be necessary to climb a ladder to work on a trailer electrical problem. Loose-fitting clothes can snag on ladders, causing a fall.

Jewelry should not be worn while you are working on trucks. Rings, when caught on moving components, have resulted in the loss of fingers. Chains and bracelets can easily catch on a moving component, resulting in severe injury to the technician. Jewelry is also made of metals such as gold and silver, which are excellent conductors of electricity. Jewelry can act as a short circuit between a battery positive terminal and the truck sheet metal or frame, referred to as chassis ground, resulting in serious burns.

Long hair also creates hazards when you are working around rotating components. Tie up long hair securely or tuck it into a cap.

Proper footwear is also very important for truck technicians. Truck components are typically very heavy. Sandals, athletic shoes, and similar leisure shoes have no place in a truck garage. Dropping a heavy component such as a starter (cranking) motor on your foot may cause serious injury. Work boots, especially steel-toed safety shoes, offer some level of protection. Work boots (**Figure 1-1**) should have slip-resistant soles because garage floors are often slippery.

Gloves are also very important. Truck technicians must use their hands more than just about any other body part. Trucks have many sharp edges that can cause severe cuts. Gloves designed specifically for automotive technicians have been introduced in recent years. Gloves reduce cuts and scrapes while improving grip for many general tasks. Welding, cutting, and heating using torches are common tasks for truck technicians. Specialized welding gloves should be utilized when working with a flame or when welding to prevent serious burns.

Working with chemicals, including waste oil and diesel fuel, requires the use of special chemical-resistant gloves. These not only protect against skin damage such as chemical burns but also prevent harmful chemicals from being absorbed into the body through the skin. Years of accumulating chemicals that were absorbed through the skin into the body may cause cancer or other illnesses.

Always select the correct type of glove for the job. Consult the Internet links at the end of this chapter for more information on glove selection.

Eyewear

Proper eyewear is probably the single most important piece of safety equipment for a truck technician. Squinting or looking away when grinding or cutting without wearing proper eye protection is just asking for an eye injury. Different tasks require different levels of protection. If you wear prescription glasses and need them to see while you are working, you should obtain prescription safety glasses. These should have side shields that can be attached to the frames. If you do not wear prescription glasses, clear plastic safety glasses with side shields like those shown in **Figure 1-2** should be worn at all times when working in a truck garage, even when not performing work such as metal cutting or grinding. Truck repair often requires working on a creeper underneath a truck. Road debris such as sand and salt can easily fall into your eyes.

Figure 1-1 Protective footwear.

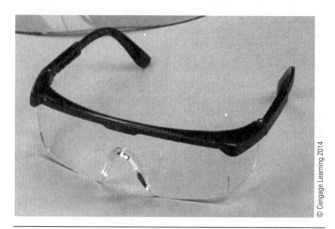

Figure 1-2 Safety glasses.

Safety glasses alone may not provide adequate eye protection when performing cutting or grinding or when working with chemicals or batteries. Safety goggles, like those shown in **Figure 1-3**, should be worn when performing these types of tasks.

A safety shield, shown in **Figure 1-3**, is not a substitute for safety glasses or safety goggles. The safety shield provides an added level of eye protection and protects the face when cutting or grinding and when working with some chemicals.

Your shop or school should have an emergency eye wash station like that shown in **Figure 1-4**. Make sure you know where it is located and test it at least monthly to ensure that it works.

Figure 1-3 Safety goggles and full face shield.

Figure 1-4 Emergency eye wash station.

Figure 1-5 Hearing protection.

Hearing Protection

Most truck shop operations are very noisy. Prolonged high noise levels can result in a substantial hearing loss, which you may not realize until it is too late. Earplugs or earmuffs should be worn when working in high-noise environments (**Figure 1-5**). This includes working near a running diesel engine.

ELECTRICAL SAFETY

Following safe practices when working with truck electrical systems can minimize your risk of injury. You should return to this section for review after studying **Chapter 2** and **Chapter 3**.

Electric Shock

The first thing that may come to mind when discussing electrical safety is **electric shock**. Most everyone has probably experienced some level of electric shock. Walking across a carpeted floor in hard-soled shoes and then touching a metal object causes you to experience a level of electric shock.

The nervous system of the body uses voltage impulses to control muscles. The amplitude of these voltages utilized by the nervous system is very low. Making physical contact with voltage sources outside of the body can cause these nervous system voltages to be overridden. This can cause muscles to contract involuntarily, such as those that control the fingers and hand. Making contact with a sufficient voltage source with the hand can cause the hand to form into a fist. It may not be possible for the person receiving the shock to release the fist, thus preventing the person from releasing grip on the voltage source.

The heart is also a muscle. Contact with a sufficient voltage source can interfere with the heart muscle impulses and cause the heart to stop beating or to beat with an irregular pattern. The diaphragm muscle utilized for breathing can be made to stop because of electric shock.

Although you may not think that you need to worry about high voltage when working on trucks, many electronic diesel fuel injection systems use pulses of 100 volts or more to control the injectors. This level of voltage can cause serious or fatal electric shock. Specific cautions provided by the truck or engine manufacturer, also known as the **original equipment manufacturer (OEM)**, must be followed when working with these high-voltage systems.

Hybrid trucks operate on voltages exceeding 500 volts, which can cause a lethal electric shock. It is critical to follow all of the OEM's instructions when servicing hybrid trucks. You should not attempt to diagnose any problems with a hybrid's high-voltage system until you have completed specialized training and have the proper personal protective equipment such as high-voltage gloves. Orange-colored wire insulation on hybrids indicates high-voltage circuits. Never cut or splice into this high-voltage wiring.

Some trucks may have DC to AC voltage converters, referred to as inverters. These devices are utilized to power refrigerators, television sets, and other 120V AC devices from the truck's batteries. The AC voltage supplied by an inverter can cause serious or fatal electric shock and should be treated with the same respect as a 120V AC wall socket.

Electric Burns

Ohm's law, which will be introduced in **Chapter 2**, indicates that when a voltage is connected across a resistance, a current will flow through the resistance. A current flowing through a resistance also generates heat. The human body has an electrical resistance. Causing a sufficient electric current to flow through the human body can cause living tissue beneath the skin to be burned. Electric welders and other shop equipment may be powered by 440V AC sources. These high voltages can cause serious burns to human tissue in addition to delivering a potentially fatal electric shock.

Severe skin burns can also result from current flow through low-resistance components such as wrenches that may bridge between the truck's 12V batteries. The area near the starter motor has a connection to the battery positive terminal. Incidental contact of the starter motor positive terminal and the grounded metal frame rail or engine block with a wrench can result in burns. The large amount of current that flows with such contact can cause pieces of hot metal to splatter, similar to arc welding.

Gold, silver, and other metals utilized to make jewelry are also excellent conductors of electricity. A ring, bracelet, or chain that makes contact with a battery positive source and ground can result in a severe burn. Always remove jewelry when working on trucks, especially when working with electricity.

Wall Socket Safety

The voltage delivered by wall sockets in North America is 120V AC. One of the conductors is referred to as the neutral conductor; the other is referred to as the hot conductor. The neutral conductor is connected to the earth ground by the electric company. Moist earth is a conductor of electricity, as is damp concrete. An AC voltmeter connected between the earth and the neutral conductor would indicate nearly zero difference in voltage. However, an AC voltmeter connected between the hot conductor and earth would indicate about 120V AC. Shoes offer some level of insulation, but perspiration and other moisture can provide a path for current flow through the shoes. Therefore, making contact between an energized or hot conductor with your hand and the earth (or damp concrete) with your shoes could result in a serious or fatal electric shock.

Many devices that are plugged into wall sockets, such as battery chargers, have electric plugs with three prongs. The two side-by-side prongs are the hot conductor and the neutral conductor. The third prong is the grounding terminal. The neutral conductor and the grounding conductor are connected together at an electrical panel. A metal rod driven into the ground (earth) is also connected to this common connection.

The grounding conductor in a metal-case device such as a battery charger is connected to the metal case. The purpose of the grounding conductor is to cause a large amount of electric current to flow should the hot conductor make contact with the metal case of the battery charger. This electric current will cause a circuit breaker to trip or a fuse to blow and open the circuit. Circuit protection devices, such as circuit breakers and fuses, will be discussed in **Chapter 4**.

Unfortunately, breaking off the grounding terminal on the power cord of a device such as a battery charger is a common practice. This is typically done because an extension cord or outlet may not have this third terminal, so someone breaks off the grounding terminal on the battery charger power cable. The device still seems to work with the grounding terminal broken off the plug, so this hazardous situation could go unnoticed for years. However, if the wiring insulation of the hot conductor wears through inside the device and makes contact with the battery charger metal case, the circuit breaker or fuse would not open. Touching the battery charger metal case would then result in a potentially lethal electric shock as the person completes a path for current flow to ground.

WARNING *Electric devices that are designed to use the grounding terminal should not be used if the grounding terminal is missing.*

Many modern electrical devices do not require the third grounding prong on the electric cord because the case of the device is made of plastic or some other nonconductive material.

Wall sockets utilized in garages may be protected by a **ground fault circuit interrupter (GFCI)**. The GFCI device may be a component of the wall socket or may be a special circuit breaker located in the electric panel. If a clamp-on inductive ammeter, introduced in **Chapter 2**, is placed around two conductors with the same amount of current flowing in opposite directions, the magnetic fields surrounding the conductors will cancel out, causing the inductive ammeter to indicate 0A of current flow. The GFCI operates on the same principle by comparing the current flow through the neutral conductor with the current flow in the hot conductor. If there is more current flow in the hot conductor than in the neutral conductor, a circuit breaker in the GFCI device will automatically trip. More current would flow through the hot conductor than through the neutral conductor if a parallel path to earth ground existed through a person's body. This current through the person would flow through the hot conductor but not the neutral conductor. This should cause the GFCI internal circuit breaker to trip to protect from shock.

These GFCI devices have a test button that causes the circuit breaker to trip. The GFCI should be tested periodically to verify that it would function when needed.

Even though you may become a highly skilled truck electrical specialist and become comfortable working with 12V DC systems, working with power line voltages is much different. Leave high-voltage electrical repair, including 120V AC, to those who are qualified.

Battery Safety

Batteries are one of the more hazardous items on a truck. Lead-acid batteries utilized in trucks contain sulfuric acid. Sulfuric acid is a hazardous substance that can cause blindness and serious chemical burns to skin. Always wear approved eye protection and protective clothing when handling, charging, servicing, or testing batteries. Battery cases can fracture if the battery is dropped, causing sulfuric acid to pour out of the battery. Keep baking soda near stored batteries to neutralize any spilled sulfuric acid.

Lead-acid batteries produce an explosive gas (hydrogen) when charging or discharging. If the gas should explode, the battery case and internal components can become shrapnel as well as spray sulfuric acid over a large area with tremendous force as shown in **Figure 1-6**. Always test or charge batteries

Figure 1-6 Lead-acid battery after a spark caused an explosion.

in a well-ventilated area to dissipate the explosive gas formed by the battery. Never create sparks or open flames or smoke near batteries, even batteries that are just being stored, as they contain explosive gas.

More battery servicing precautions are described in **Chapter 5**.

Hybrid electric trucks may use lithium-ion batteries for the high-voltage system. Lithium is a flammable metal that reacts with, and may explode when exposed to, water. Consult the OEM information before working on any hybrid system component. Additional information on lithium-ion battery safety is provided in **Chapter 5**.

OTHER PERSONAL SAFETY PRECAUTIONS

Never smoke while working on a vehicle. Not only is there a risk of causing a fire, but vapors in the air can change to more hazardous compounds when burned. Burning these vapors when drawing on a cigarette exposes the smoker to additional health hazards. Only smoke in approved areas, and do not permit others to smoke around you while you are working.

Tilting a hood on a conventional truck should be performed with care (**Figure 1-7**). Make certain that there is sufficient clearance in front of the hood in order to keep from contacting another truck or object when the hood is tilted. Springs and cables within the hood anchor assembly assist to make certain that the hood will extend only to a specific height or distance. Always make sure that these anchors are present as you start lifting because the stay cable may be

disconnected, especially when the hood has been recently removed for engine service and in other similar situations.

Heavy truck hoods use springs to assist in tilting the hood open. Dampers or struts are utilized in some applications to assist in closing the hood at a controlled rate. If these springs or dampers have been disconnected or are damaged, lowering the hood will result in the heavy hood falling closed. Always close the hood with caution because someone may have decided to look under the hood just as you are closing it.

A cab over engine (COE) truck requires additional cautions. Remove all loose objects in the cab before tilting. Truck cabs weigh up to 2,000 lb (907 kg). Work under the cab only after the cab is locked in place. Getting under a cab that is only partially lifted is like getting under a truck that is supported only by a jack without any jack stands. Either is equal to asking for a serious injury or worse.

Many truck repair shops have two or three shifts of technicians. It may be necessary to move a truck out of a stall to make room for others that are being repaired. It may not be apparent to the next shift that a truck is in a condition such that it should not be started or moved. Always place a warning tag on the steering wheel indicating that the truck should not be started or moved and the reason, such as "brakes do not function."

Always remove the ignition keys and place them in your pocket, locked toolbox drawer, or other designated location while you are working under the truck or under the hood to prevent someone from starting or moving the truck unexpectedly. Many fleets utilize a lockout bag or pouch system for key storage when trucks are undergoing repair. This is similar to the OSHA lockout/tag-out safety procedure required in industrial work environments to ensure that machines undergoing service are properly shut down and not restarted before completion of the service. See the Internet links at the end of this chapter for more information on lockout/tag-out.

Fire extinguishers are very important for truck repair shops. There are four different classes of fire extinguishers corresponding to different types of fuel, as shown in **Figure 1-8**. Make sure you know where each shop fire extinguisher is located and learn the proper operation of each type of fire extinguisher in your shop.

© Cengage Learning 2014

Figure 1-7 Tilting hood.

	Class of Fire	Typical Fuel Involved	Type of Extinguisher
Class **A** Fires (green)	**For Ordinary Combustibles** Put out a Class A fire by lowering its temperature or by coating the burning combustibles.	Wood Paper Cloth Rubber Plastics Rubbish Upholstery	Water*[1] Foam* Multipurpose dry chemical[4]
Class **B** Fires (red)	**For Flammable Liquids** Put out a Class B fire by smothering it. Use an extinguisher that gives a blanketing, flame-interrupting effect; cover whole flaming liquid surface.	Gasoline Oil Grease Paint Lighter fluid	Foam* Carbon dioxide[5] Halogenated agent[6] Standard dry chemical[2] Purple K dry chemical[3] Multipurpose dry chemical[4]
Class **C** Fires (blue)	**For Electrical Equipment** Put out a Class C fire by shutting off power as quickly as possible and by always using a nonconducting extinguishing agent to prevent electric shock.	Motors Appliances Wiring Fuse boxes Switchboards	Carbon dioxide[5] Halogenated agent[6] Standard dry chemical[2] Purple K dry chemical[3] Multipurpose dry chemical[4]
Class **D** Fires (yellow)	**For Combustible Metals** Put out a Class D fire of metal chips, turnings, or shavings by smothering or coating with a specially designed extinguishing agent.	Aluminum Magnesium Potassium Sodium Titanium Zirconium	Dry powder extinguishers and agents only

*Cartridge-operated water, foam, and soda-acid types of extinguishers are no longer manufactured. These extinguishers should be removed from service when they become due for their next hydrostatic pressure test.

Notes:

(1) Freezes in low temperatures unless treated with antifreeze solution, usually weighs over 20 pounds (9 kg), and is heavier than any other extinguisher mentioned.

(2) Also called ordinary or regular dry chemical (sodium bicarbonate).

(3) Has the greatest initial fire-stopping power of the extinguishers mentioned for Class B fires. Be sure to clean residue immediately after using the extinguisher so sprayed surfaces will not be damaged (potassium bicarbonate).

(4) The only extinguishers that fight A, B, and C classes of fires. However, they should not be used on fires in liquefied fat or oil of appreciable depth. Be sure to clean residue immediately after using the extinguisher so sprayed surfaces will not be damaged (ammonium phosphates).

(5) Use with caution in unventilated, confined spaces.

(6) May cause injury to the operator if the extinguishing agent (a gas) or the gases produced when the agent is applied to a fire are inhaled.

© Cengage Learning 2014

Figure 1-8 Fire extinguisher selection guide.

Technicians may have to work with hazardous substances. In the United States, the Hazard Communication Regulation or Right-to-Know Law is administered by OSHA. In Canada, the Canadian Centre for Occupational Health and Safety (CCOHS) acts as a resource for the Canadian Controlled Products Regulations and the Workplace Hazardous Materials Information System (WHMIS).

It is your employer's responsibility to provide you with information regarding hazardous substances utilized in the workplace. These substances must have a **material safety data sheet (MSDS)**. It is your responsibility to keep yourself informed and know how to protect yourself from hazardous materials. Your employer should make all applicable MSDS available for you to access, as shown in **Figure 1-9**. Make use of the personal protective equipment identified in the MSDS for safe handling of the material, such as a proper-fitting respirator (**Figure 1-10**).

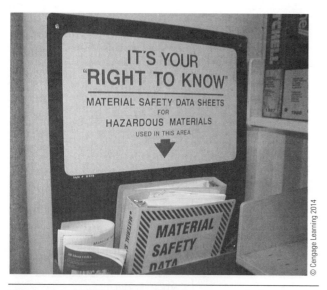

Figure 1-9 MSDS information is extremely important to your personal safety.

Figure 1-10 Chemical respirator.

Summary

- Safe work practices are the personal responsibility of each technician.

- Proper attire is very important for personal safety.

- Selection of the proper types of gloves is important to protect your hands. There are many different types of gloves. Chemical-resistant gloves should be worn to prevent harmful chemicals from being absorbed into the body through the skin.

- The proper form of protective eyewear is important to protect your vision.

- Electric shock is the flow of current through the body. Electric shock causes nerve impulses to muscles to be overridden.

- The grounding terminal present on many electrical devices can protect you from electric shock. Never use an electrical device that has had the grounding terminal removed.

- Jewelry should not be worn in the shop. Jewelry can result in severe burns or can become entangled in rotating components.

- Lead-acid batteries are very hazardous. Batteries can explode, resulting in sulfuric acid spraying over a wide area.

- Hybrid trucks operate on voltages that may exceed 500 volts and make use of lithium-ion batteries. You should never attempt to service the high-voltage system on a hybrid electric truck until you have had proper training. Orange-colored wire insulation indicates high voltage.

- Know where and what type of fire extinguishers are located in your shop. Always use the correct type of fire extinguisher for the type of fire to prevent electric shock and other hazards.

- Material safety data sheets are in the shop for your benefit. Read these data sheets to determine what type of protective equipment you need to handle the material and how to deal with emergencies that arise related to the material.

Internet Searches

Suggested Internet sites for more information on work safety:
http://www.osha.gov
http://www.ccohs.ca

Review Questions

1. Who is ultimately responsible for a technician's personal safety?

    A. Government

    B. Shop owner

    C. Service manager

    D. Each technician

2. Which of the following is a true statement?

    A. Baggy clothing should not be worn in a truck garage.

    B. Safety glasses need to be worn only when cutting or grinding metal.

    C. Hearing protection is not necessary when working around a running diesel engine unless the exhaust is disconnected.

    D. Water can always be utilized to extinguish any type of fire.

3. Which of the following is a true statement?

    A. Leather work gloves are appropriate when working with chemicals.

    B. Steel-toed work shoes are not necessary for diesel technicians.

    C. Regular prescription eyeglasses are a good substitute for safety glasses.

    D. Squinting or blinking is not a substitute for protective eyewear.

4. Technician A says that there are not any lethal voltages on any modern truck. Technician B says that 120V AC is not a high enough voltage to hurt you. Who is correct?

    A. A only

    B. B only

    C. Both A and B

    D. Neither A nor B

5. A technician needs an extension cord for a battery charger. The extension cord has only two terminals and the battery charger plug has three terminals. What should the technician do?

    A. Break off the third terminal on the battery charger cord because it is not needed on modern electrical outlets.

    B. Get an adapter that changes the three-terminal cord into two terminals and use the two-terminal extension cord.

    C. Get another extension cord with three terminals with a sufficient current rating for the battery charger.

    D. Consult the MSDS for the battery charger.

6. Technician A says that smoking while working is OK, provided you are not working with diesel fuel or other flammable liquid. Technician B says that it is OK to smoke around a battery that is being charged because the battery contains only lead and water, neither of which is flammable. Who is correct?

    A. A only

    B. B only

    C. Both A and B

    D. Neither A nor B

7. What type of information would you find on an MSDS?

    A. The types of hazards the material presents

    B. How to handle a spill of the material

    C. What to do if material is swallowed or gets in eyes

    D. All of the above

8. Technician A says that opening the hood of a heavy truck is not different from opening the hood of a large passenger car. Technician B says that on a COE truck, it is OK to work under the cab with the cab only partially tilted, provided a 2" × 4" board is propped between the cab and the frame. Who is correct?

   A. A only

   B. B only

   C. Both A and B

   D. Neither A nor B

9. Which class of fire extinguisher should be utilized on a diesel fuel fire?

   A. Class A

   B. Class B

   C. Class C

   D. Class D

10. Technician A says that orange-colored wiring insulation on a hybrid truck indicates high voltage. Technician B says that water should be added to the lithium-ion batteries on hybrid trucks. Who is correct?

    A. A only

    B. B only

    C. Both A and B

    D. Neither A nor B

CHAPTER

2 The Fundamentals of Electricity

Learning Objectives

After studying this chapter, you should be able to:

- Relate electrical terms including voltage, current, and resistance to their hydraulic or pneumatic counterpart.
- Explain the relationship among voltage, current, and resistance in an electrical circuit.
- Apply Ohm's law to simple electric circuits to determine an unknown voltage, current, or resistance.
- Perform measurement of voltage, current, and resistance using a digital multimeter (DMM).
- Describe the concept of a voltage drop across a resistance when current flows through the resistance.
- Demonstrate Kirchhoff's voltage law and current law through voltage and current measurements obtained from a series, a parallel, and a series-parallel circuit.
- Determine the total circuit resistance in series, parallel, and series-parallel electric circuits, given the individual resistor values.
- Calculate voltage drops and current draws in a series, parallel, and series-parallel electric circuit, given the individual resistor values and voltage source value.

Key Terms

alternating current (AC)	digital multimeter (DMM)	Kirchhoff's current law
ammeter	direct current (DC)	Kirchhoff's voltage law
ampere (amp)	electrical potential	ohm
analogy	electromotive force (emf)	ohmmeter
charge	electrons	Ohm's law
circuit	electron theory	parallel
conductor	equivalent resistance	parallel circuit
conventional theory	free electron	potential
coulomb	insulator	proton
current	ion	resistance

resistor   shunt   voltage

series   total resistance   voltage drop

series circuit   valence band   voltmeter

series-parallel circuit   volt

INTRODUCTION

Many truck technicians seem to have some apprehension when it comes to working with truck electrical problems. They may be very skilled with "hands-on" work such as engine, brake, and transmission repair but may not be as confident when it comes to troubleshooting an electrical problem. This apprehension about electricity seems to be caused in part by the inability to see electricity. However, most truck technicians have no problem understanding the operation of compressed air systems such as a truck air brake system. Like electricity, air is also invisible. Air is made up of particles so small that it is not possible to see them. It is possible to see smoke, dust, or water vapor suspended in air but it is not possible to see air. Since most truck technicians are familiar with compressed air, some of the similarities between compressed air and electricity will be examined.

Comparison of Electricity to Air

Even though it is not possible to see air, it is possible to observe the effects of air in motion, which is known as wind. Wind is caused by differences in atmospheric pressure at various locations on earth. At sea level, the normal atmospheric pressure is 14.7 psi (101 kPa). Because air is made up of small particles, these particles have some mass (weight). Atmospheric pressure is caused by the weight of the air above the earth's surface. The weight of all the air directly above 1 in.^{2} of the earth's surface area at sea level is 14.7 lb while the mass of all the air above an area of 1 cm^2 is 1.03 kg. This concept is shown in **Figure 2-1**. This force is exerted in all directions, not just downward. At higher altitudes, such as the top of a mountain, there is less air above the surface so the atmospheric pressure decreases as altitude increases.

Differences in air temperature and other factors can cause the atmospheric pressure to vary from one location on the earth to another location. Atmospheric pressure is also known as barometric pressure and is often expressed in units of inches of mercury or millimeters of mercury. Mercury (Hg) is a metal with a very low melting point; thus, it is in a liquid state at most temperatures found on earth. In **Figure 2-2**, a

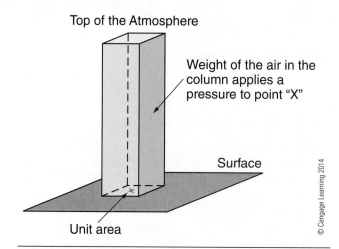

Figure 2-1 Atmospheric pressure is caused by the weight of the air above the surface of the earth.

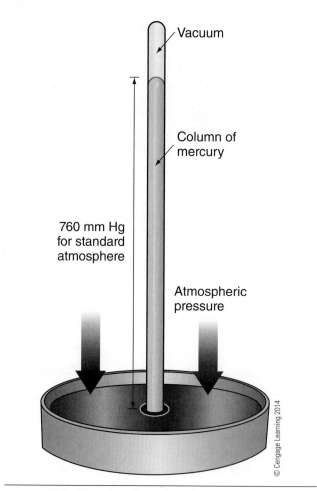

Figure 2-2 The height of a column of mercury indicates atmospheric pressure.

simplified barometer which is used to measure atmospheric pressure is shown consisting of a container of mercury and a long glass tube. All of the air has been removed from the glass tube resulting in a vacuum above the column of mercury. Increasing atmospheric pressure causes the height of the column of mercury to rise in opposition to gravity, while decreasing atmospheric pressure causes the height of the column to fall and some of the mercury to return to the container. The height of the column of mercury becomes a means of measuring atmospheric pressure. For comparison, 14.7 psi (101 kPa) is 29.9 in. Hg or 760 mm Hg, which is the typical atmospheric pressure at sea level. Note that modern barometers are electronic instruments that do not utilize mercury to measure atmospheric pressure.

Nature tends to make things equalize over time. If the atmospheric pressure is 14.5 psi (100 kPa) at one location on earth and the atmospheric pressure is 14.2 psi (98 kPa) at another nearby location, the air will move from the location with higher atmospheric pressure to the location with lower atmospheric pressure until both pressures equalize. This air movement results in wind.

Air can be compacted by an air compressor to a pressure greater than atmospheric pressure, as shown in **Figure 2-3**. The compressed air is stored in a large tank so that a reserve supply of compressed air is provided so that the electric motor will not have to run continually. Most vehicle repair shops have an air compressor that is designed to maintain the pressure of the air in the storage tank at about 120 psi (827 kPa) greater than the atmospheric pressure. Most heavy trucks also use compressed air to supply the energy to actuate the truck's brakes. This compressed air is stored in a series of air tanks. The engine-driven air compressor on trucks with air brakes is designed to maintain the pressure of air in the storage tanks at a level that is also about 120 psi (827 kPa) greater than the atmospheric pressure.

The air tank below the air compressor shown in **Figure 2-3** has a valve on the side of the tank. This valve is shown in the closed position, which keeps the compressed air trapped in the air tank. If the electric motor was switched off and the valve on the air tank was opened to the atmosphere, the compressed air in the tank would travel to the atmosphere, as shown in **Figure 2-4**. This would occur until the pressure in the air tank and the pressure of the atmosphere are equalized. In the same manner as wind, the air in the tank is moving from a location of higher pressure to a location of lower pressure. Like the wind, it is not possible to see this air in motion as it escapes the air tank.

Compressed air can be used to assist a truck technician to perform work. Instead of just opening a valve and letting the compressed air return to the atmosphere, the technician can direct the compressed air through a hose to a tool such as an air impact wrench (**Figure 2-5**). The flow of air through the impact

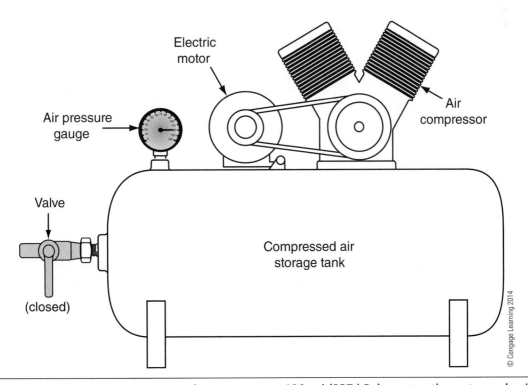

Figure 2-3 An air compressor compresses air to a pressure 120 psi (827 kPa) greater than atmospheric pressure.

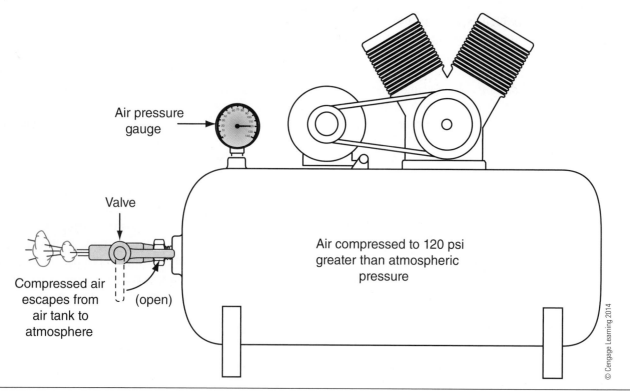

Figure 2-4 Valve on the side of an air compressor is opened, causing compressed air in the tank to escape to the atmosphere.

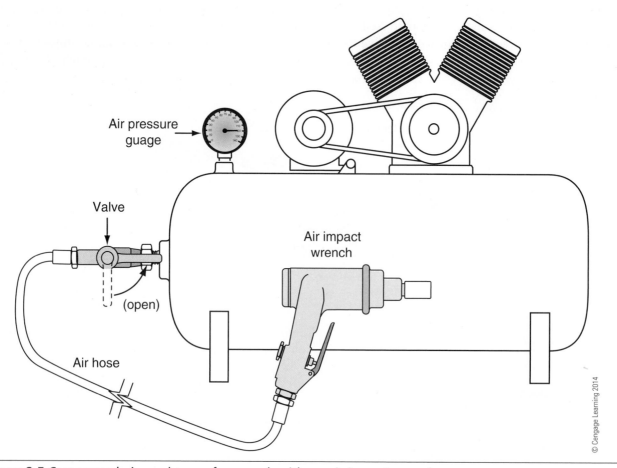

Figure 2-5 Compressed air used to perform work with an air impact wrench.

wrench results in rotary motion to loosen or tighten fasteners.

BASIC ELECTRICITY

Like compressed air, electricity is also invisible. It is possible to observe the effects of electricity, just as it is possible to observe the effects of wind, but it is not possible to see electricity. It is also possible to use electricity to perform work in much the same manner as compressed air. Electricity, like wind, results from the movement of particles so small that they cannot be seen traveling from one location to another. In the case of wind, the particles of air move due to differences in atmospheric pressure. In the case of electricity, small particles called **electrons** are drawn or attracted to other small particles known as **protons** resulting in movement of the electrons.

Charge

Electrons and protons have a fundamental physical property referred to as **charge**. Charge can be thought of as a mysterious force, like magnetism, which causes electrons and protons to be attracted to each other. Clothing that sticks together when removed from the clothes dryer is the result of charge. There are two types of charge. The charge associated with protons is a positive charge, while the charge associated with electrons is a negative charge. Particles with opposite charges are attracted to each other, while particles with like or the same charge are repelled from each other. This attraction and repulsion force is similar to the action of two magnets. You may recall from grade school science class that magnets have a north pole and a south pole. Opposite poles of magnets are attracted to one another, while like poles of magnets are repelled from one another.

If a single electron and a single proton could be suspended from strings and held at a distance from each other, the particles would be attracted to each other and would move toward each other as shown in **Figure 2-6** because of their unlike (different) charges. In the opposite way, suspending two electrons with negative charges or two protons with positive charges from strings would cause the particles to be repelled from each other because of their like (same) charge.

Atoms

Electrons and protons, along with other particles called neutrons, together form a collection of particles known as an atom. Neutrons have no charge and are not relevant to the study of basic electricity. An atom is the smallest possible piece of gold, copper, oxygen, or any of the other known elements that make up the universe.

In classical physics, an atom is described as being similar to a very small solar system. The neutrons and protons assemble to form a nucleus, which is like the

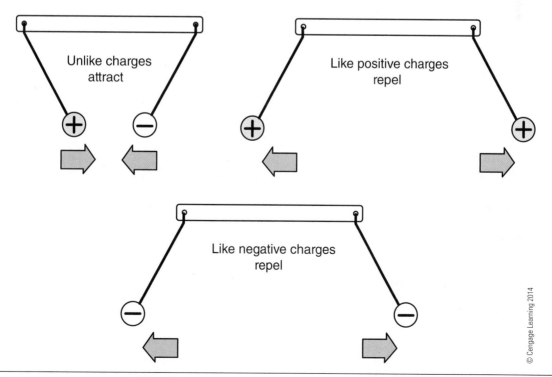

Unlike charges attract

Like positive charges repel

Like negative charges repel

Figure 2-6 Charges suspended from strings.

sun in the center of our solar system. Even though protons are typically repelled from each other because of their like positive charges, a fundamental force of the universe called the strong nuclear force acts as a glue to keep the protons and neutrons held tightly together in the nucleus.

Each of the approximately 92 different naturally occurring elements in its neutral state maintains an equal number of protons and electrons. Because electrons and protons are attracted to each other, having an equal number of protons and electrons in an atom causes the forces associated with the two different types of charge within the atom to be equalized or cancelled out.

The electrons orbit the nucleus of the atom, as shown by the copper (Cu) atom in **Figure 2-7**, in much the same way that the earth and the other planets orbit the sun. Like our solar system, the nucleus is much larger than the orbiting electrons. Unlike our solar system, there can be more than one electron in each orbital path.

The comparison of an atom to the solar system is an **analogy**. An analogy is a comparison that emphasizes the similarity between two or more different things or concepts. For example, stating that the human heart is like a simple pump is an analogy. Obviously, the heart is much more complex than a simple pump and the comparison between the two will break down once you go beyond this simple analogy. A cardiologist would require a much deeper understanding of the function of the heart beyond it being like a simple pump. The same is true for the analogy of an atom being like a small solar system. A study of quantum physics would be required to understand how science believes that an atom is actually constructed. However, such a detailed understanding of atomic theory is not necessary to diagnose any truck electrical system problem.

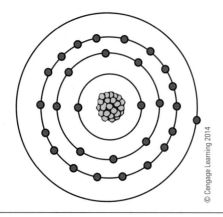

Figure 2-7 Copper atom with 29 electrons surrounding a nucleus of 29 protons and neutrons.

Displacing Electrons

In ancient times, it was discovered that rubbing two dissimilar materials together caused the materials to then be attracted to each other for a period of time. For example, rubbing a rod made of glass with a piece of silk cloth causes the two objects to be attracted to each other. The reason that the silk and the glass are attracted to each other is the difference in charge between the silk and the glass after rubbing. When certain dissimilar materials such as glass and silk are rubbed together, some of the electrons are stripped or pulled away from their outermost orbit. These stripped electrons are then forced into the outermost orbits of some of the atoms of the other material due to the rubbing. What is left after rubbing is that one of the materials is missing some of its electrons, and the other material has too many electrons. The net charge of the two materials is now different. One material has an excess of positive charge (missing some electrons), and the other material has an excess of negative charge (too many electrons). When an atom has an unequal number of protons and electrons, it is called an **ion**. An atom that has more protons than electrons is a positive ion; an atom that has fewer protons than electrons is a negative ion. The attraction between the two materials rubbed together such as glass and silk results from the imbalance in the associated positive and negative charge. This same phenomenon occurs when certain types of clothing are tossed around together in a clothes dryer. Electrons are stripped from one type of material and deposited on another type of material due to the tumbling action of the clothes in the dryer. This is the reason for "static cling." The two items of clothing stick to each other as though they were two magnets. A recently used plastic comb also illustrates the same effect when held over small pieces of paper as shown in **Figure 2-8**. The difference in charge between the paper and the comb results in the paper being attracted to the comb.

You have probably learned something about electricity when you walked across a carpet in hard-soled shoes and then touched a metal object. Walking across the carpet causes some of the carpet's electrons to be stripped out of their orbits and transferred into your atoms. When you then touch a metal object, you experience a shock and observe and hear a spark as your excess electrons travel to the metal object. No electrons or protons are created or destroyed when you walk across the carpet. You have just temporarily forced some electrons to leave their atoms. The electrons in motion cause the spark as they move from the location with too many electrons to the location with

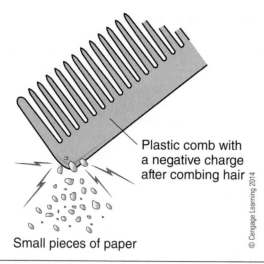

Plastic comb with
a negative charge
after combing hair

© Cengage Learning 2014

Small pieces of paper

Figure 2-8 Opposite charges on the paper and comb cause an attraction.

not enough electrons until the charge is equalized. This same equalization of charge event is the reason for lightning, but on a much larger scale.

Let us consider once again a tank of compressed air; as discussed, the particles of air moved from the air tank to the atmosphere when the valve on the tank was opened. The pressurized air in the tank flows from the tank to the lower-pressure atmosphere until the pressure in the tank and the pressure of the atmosphere are equalized. In the case of electricity, electrons flow from a location where there are atoms with too many electrons to a location where there are atoms that are missing electrons. The empty spaces in the orbits remaining when an atom loses electrons can be thought of as holes. When permitted, the excess electrons will travel to fill these holes until all the holes are filled, similar to the way that all the compressed air in a tank travels to the atmosphere when the valve on the side of the tank is opened.

Potential

Compressed air that is only being stored in a tank is not doing any work. The compressed air just has the **potential** or capability of doing something useful.

If a truck technician connects an air impact wrench to an air-line supplied with air from an air compressor, work can be performed with the air impact wrench. However, until the compressed air leaves the air tank and flows through the air impact wrench to the atmosphere, the compressed air stored in the tank just has potential. The trigger on the air impact wrench must be depressed to cause air to flow through the tool to perform any work with the compressed air stored in the air tank.

Electrical Potential. When a material with an excess of electrons (negative charge) is taken a distance from a material with an absence of electrons (positive charge or holes), an **electrical potential** exists between the two materials. Like the compressed air stored in the tank, the separation of electrons and holes just has potential until the electrons actually move to fill the holes. This electrical potential is referred to as **voltage**. Voltage is also known as the **electromotive force (emf)**. Voltage can be thought of as the electrical equivalent of pressure because both pressure and voltage describe potential.

The unit of measure of electrical potential or voltage is the **volt**. A volt is a unit of measure of electrical potential in much the same way that psi or kilopascals are units of measure of air pressure. The symbol for volts is an uppercase V. A common source of voltage is a battery. A 12-volt battery might be described as being a 12V battery. An uppercase E for emf also signifies voltage such as E = 12V.

Electron Flow

When something is described as flowing, you may think of water flowing in a river. The flow of water in a river is referred to as the current. The flow rate of a liquid, such as the flow of water through a pipe, can be measured. Liquid flow rate is measured by determining the amount of liquid, such as the number of gallons or liters, which passes a stationary point in a given period of time, such as a minute. Therefore, the flow of water through a pipe could be measured in units of gallons per minute (GPM) or liters per minute.

Like the flow of a river, the flow of electrons is also known as electric **current**. Like the flow of water through a pipe, electric current can also be measured. Electric current is measured by determining the amount of electrons that pass a stationary point in a given period of time.

Because a single electron is a very small particle, a group of electrons called a **coulomb** quantifies the number of electrons that pass a stationary point in a given period of time. More precisely, a coulomb actually indicates the charge associated with the quantity of electrons. A coulomb is defined as the charge associated with 6.25×10^{18} electrons (625 with 16 zeros behind it or 6.25 billion-billion electrons). Electric current flow is quantified by measuring the number of coulombs that pass a stationary point per second of time. Coulombs per second is similar to gallons or liters per minute because both are measuring a quantity of something passing a stationary point in a given period of time.

Instead of the term *coulombs per second* to measure electric current, the more common term for the measurement of electric current is the **ampere (amp)**. One ampere is defined as one coulomb of charge (electrons) passing a stationary point in one second of time.

The symbol for amperes is an uppercase A. The amount of current that may flow through a single truck tail lamp bulb is about 1A. Stating that 1A of current is flowing is easier than saying that 6,250,000,000,000,000,000 electrons per second are flowing through the tail lamp bulb, even though both mean the same thing. An uppercase I for the French word *intensité* also signifies electric current. Therefore, the current flowing through a tail lamp bulb could be described as I = 1A.

Direction of Current Flow. The two ends of a D cell flashlight battery are marked with a plus sign (+) and a minus sign (−). A 12V truck battery also has a plus terminal and a minus terminal. The plus sign end is referred to as the positive terminal and the minus sign end is called the negative terminal. It would seem that the plus terminal of the battery is where the excess electrons are located and the minus terminal are where the holes are located. However, Benjamin Franklin named the side of a battery with an excess of electrons negative and the side of a battery with excess holes as positive. This means that electrons actually flow from the negative end of the battery to the positive end of the battery. However, the direction that electrons actually flow is not that important because it is not possible to see electrons flow.

Most electronic symbols discussed in later chapters use arrows that indicate the direction of current flow as being from positive to negative, as do many current measurement devices which will be addressed later in this chapter. Current flow described as flowing from positive to negative is known as the **conventional theory**. Conventional theory does not describe the flow of the electrons, but rather the flow of the holes. The holes that the electrons fill can be thought of as traveling from positive to negative, in the conventional theory. Current flow described as flowing from negative to positive is known as the **electron theory**. The electron theory describes the movement of the electrons, which is actually from negative to positive. This text will use the conventional theory throughout.

One way to look at the difference between the conventional theory and the electron theory is to imagine a single line of cars waiting at a stop sign, as shown in **Figure 2-9**. Each time a car goes through the intersection, as shown by movement to the right in **Figure 2-9**, the empty space occupied by the car that moved forward is filled by the car behind it. The cars move to the right while the space is moving to the left. Electron flow is like the movement of the cars to the right, while conventional flow is like the movement of the space to the left, shown in **Figure 2-9**.

Direct Current. Almost everything in a truck electrical system is **direct current (DC)**. Direct current is current that does not change direction or amplitude. The electrical outlets in your home supply an **alternating current (AC)**. Alternating current changes

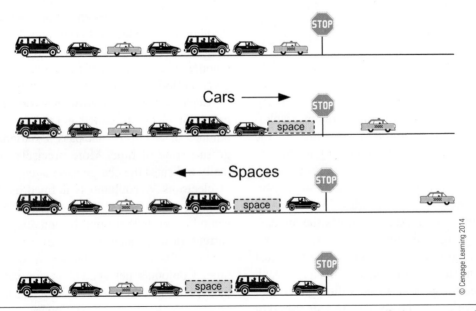

Figure 2-9 Cars waiting at a stop sign: electron flow is like the movement of the cars to the right, while conventional flow is like the movement of the spaces to the left.

amplitude and direction of flow continually. Direct current is easier to understand than alternating current, which is good news for students. Alternating current will not be addressed until later chapters.

Conductors

To make use of the compressed air stored in the air tank, it is necessary to use an air hose to direct the flow of air through a tool such as an air impact wrench. The hose provides a path for the compressed air in the tank to flow to the air impact wrench.

The electrical equivalent of an air hose is called a **conductor**. A conductor is a material that permits electrons to flow through it with very little opposition. Examples of materials that permit electrons to pass easily are metals such as copper, aluminum, and iron. Electrical wire is a conductor. Electrical wire typically found in homes is made of copper, as is most vehicle wiring.

Copper is a very good conductor of electricity. Electrons can be thought of as flowing through a copper conductor by bumping electrons in the outermost orbit or band of the copper atoms into other nearby copper atoms, which causes other easily dislodged electrons to bump into another nearby copper atom and so forth. These easily dislodged electrons are called **free electrons** because they are free to leave their atoms and bump into the orbit of adjacent atoms. The outermost orbit or band of an atom is called the **valence band**. Atomic elements that are classified as good conductors have one or two electrons in their valence band. Most metals have one or two electrons in their valence band, so they are good conductors of electricity.

To visualize electron flow through a conductor, imagine that a tube filled with marbles, like that shown in **Figure 2-10**, is like a copper conductor. Pushing an additional marble into the left side of the full tube causes a single marble to exit the right side of the tube almost instantaneously. The marble that you pushed in on the left side is not the same marble that exited the right side. However, if the marbles were all the same

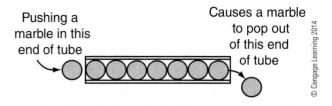

Figure 2-10 Electron flow like marbles in tube: a marble that enters the tube on the left is not the same marble that exits the tube on the right.

color and size, it would then appear that the marble you pushed in at the left side of the tube has instantly exited the right side of the tube.

Figure 2-11 shows a more accurate representation of electron flow through a conductor. The excess electrons on the right side of the conductor travel through the valence band of the atoms making up the conductor to get to the excess holes on the left side of the conductor.

Scientists believe that an energy wave is what is actually flowing through a conductor. This energy wave travels through the conductor near the speed of light (186,000 mi/s or 300,000 km/s), while the speed of the actual electrons flowing in the conductor is believed to be only a few inches per hour. An analogy of this concept of an energy wave is a line of closely spaced billiard balls being struck by the cue ball. Energy is rapidly transmitted through each ball to the next ball in the line. However, this analogy breaks down when the last ball in the line rolls into the pocket. Another analogy of an energy wave is a sports arena wave. Excited fans waiting for a sporting event to begin often stand up, raise their arms, cheer, and sit back down in a synchronized manner. To an observer, a wave can be seen traveling around the arena. However, no fans actually travel, only the energy of their wave is in motion around the arena.

To illustrate a concept, suppose one end of a thin copper wire were connected between the positive terminal of a 1.5V D cell flashlight battery and the other end of the wire was connected to the negative terminal of the battery. This would cause a large number of

Conductor

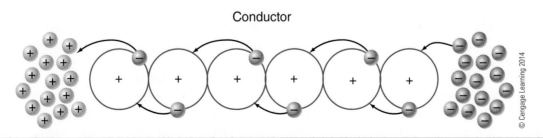

Figure 2-11 Electron flow through the valence band of conductor.

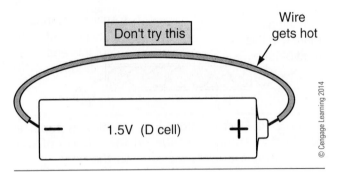

Figure 2-12 Copper wire providing an unopposed path for electrons is similar to opening a valve on a tank of compressed air and venting to the atmosphere.

electrons to leave one terminal and flow through the wire to fill the holes at the other terminal (**Figure 2-12**).

> **WARNING** *NEVER place a wire directly across the terminals of any battery, especially a large battery like a 12V truck battery. The heat generated in the wire due to the large amount of current flow may burn you, or the battery may explode.*

Placing a copper wire between the two terminals of a D cell battery is similar to opening a large valve on the side of an air tank and permitting the compressed air to escape into the atmosphere almost unimpeded. As is the case with an open valve on the side of a tank of compressed air, the electrons flowing through the wire would not perform much meaningful work other than heating the wire and the battery. The D cell battery would soon become discharged as all of the excess electrons travel through the wire to occupy the holes. This is similar to opening the valve on a tank of compressed air and the pressure of the air in the air tank equalizing with the atmospheric pressure as the tank empties.

Resistance

Consider once again a tank of compressed air. Opening a large 2-in. (51-mm) diameter valve that is installed in an air tank causes the air in the tank to flow from the tank to the atmosphere (**Figure 2-13**). The opened 2-in. (51-mm) diameter valve offers very little opposition to the flow of air out of the air tank. This will cause the air tank to empty rapidly because of the high rate of the flow of air through the open valve. The rate of flow of compressed air is often measured in standard cubic feet per minute (SCFM) or normal cubic meters per hour (NCMH). A cubic foot or cubic meter of air is the amount of compressed air at some

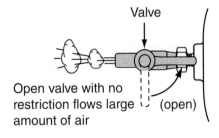

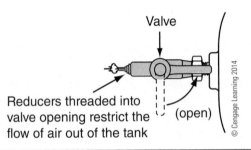

Figure 2-13 Reducers installed into the valve opening decrease the valve opening and provide opposition to the flow of air from the tank to the atmosphere.

standard pressure that will fit into a box or cube with dimensions of 1 ft or 1 m on each edge. Similar to gallons per minute or liters per minute, SCFM or NCMH is a measurement of an amount of something per unit of time.

A series of reducers or bushings have been threaded into the valve on the side of the air tank as shown in **Figure 2-13**. These reducers are typically used to permit smaller-diameter pipe to be connected to larger-diameter pipe. In this example, the 2-in. (51-mm) opening in the valve has been reduced to a 1/8-in. (3-mm) opening by the reducers. The reducers act as restrictions to the flow of air. Air will still flow from the tank to the atmosphere when the valve is opened with the reducers threaded into the valve. However, the amount of air flowing through the valve per minute will decrease substantially compared to the flow of air with no reducers threaded into the valve.

The opposition to the flow of electric current is called **resistance**. An electrical resistance restricts the flow of electric current similar to the way that reducers threaded into a valve restrict the flow of air. The unit of measurement of electrical resistance is the **ohm**. The symbol for ohms is the Greek letter omega (Ω).

Resistors. A **resistor** is an electrical component designed to have a specific value of resistance. Resistors limit current flow or cause voltage to divide. Resistors also give off heat. All of these applications will be addressed in later chapters.

Resistors are often made from carbon. Carbon is not a good conductor of electric current, but it does permit some electric current to flow. A carbon atom has four electrons in its valence band.

An uppercase R on an electrical schematic indicates that a component is a resistor. A resistor that has a value of 1000 ohms might be identified as $R = 1000\Omega$. To distinguish one resistor from another, a numeric suffix may also be added to the R. Resistors in an electrical schematic may be numbered as R_1, R_2, and so forth.

Wire Resistance.   Electric wire made of copper is a good conductor of electric current. This means that the wire offers very little restriction to the flow of electric current. Therefore, the resistance of copper wiring is very small. It is important to note that all conductors of electricity have some resistance, even though it may be a very, very small amount.

A large-diameter pipe or hose offers less opposition to flow of water than a smaller-diameter pipe or hose. In a similar manner, a large-diameter copper wire offers less opposition to electric current flow than a smaller-diameter copper wire. Therefore, the large-diameter wire has less resistance than the smaller-diameter wire.

Wire resistance also increases with wire length. A wire that is 10 m long has 10 times the resistance of a wire that is only 1 m in length provided both wires are the same diameter.

Making Use of Wire Resistance.   If you look inside a clear glass light bulb, you will observe a thin coiled metal wire called a filament, the component of the light bulb that actually glows to give off light. The filament in a light bulb is typically made of tungsten. Tungsten is a metal that is a good conductor of electricity. However, the tungsten filament has a relatively high value of resistance compared to the copper wire because of its small diameter. Electrons passing through the small-diameter filament generate friction as they collide with each other, resulting in generation of heat and light. By comparison, the relatively large-diameter copper wires that connect the battery to the light bulb have a very low resistance. The copper wires do not heat up much as current passes through them because electrons are free to pass through the large-diameter wires and little friction is generated by the moving electrons.

The conductors or wires carrying current to the light bulb are much larger in diameter than the filament (**Figure 2-14**). The wires are like a multilane express highway and electrons are like the cars on

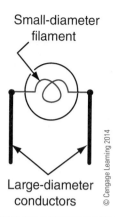

Small-diameter filament

Large-diameter conductors

© Cengage Learning 2014

Figure 2-14 Light bulb filament has a much smaller diameter than the wires that are connected to the light bulb causing an opposition to the flow of electric current.

the highway. The filament is like a construction zone on the highway, reducing four lanes of traffic in each direction to just one lane of traffic.

A D cell battery could be connected to a light bulb using wires as shown in **Figure 2-15**. As electrons pass through the light bulb's small-diameter filament, the moving electrons bump into each other, causing friction. The friction caused by electron movement results in generation of heat. This heat causes the filament to glow white-hot and give off light.

As explained previously, causing compressed air to flow through an air impact wrench permitted meaningful work to be performed by the compressed air stored in the air tank. In a similar manner, adding a light bulb to the wiring also permits meaningful work to be performed by the battery's stored potential (voltage).

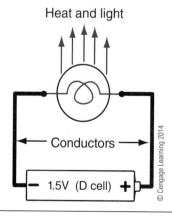

Heat and light

Conductors

1.5V (D cell)

© Cengage Learning 2014

Figure 2-15 Electrical resistance causes heat and light when electric current flows through the light bulb filament.

Insulators

A plastic or rubber material surrounds most electric wire. This material is called an **insulator**. An insulator is a material that offers a great deal of resistance to the flow of electric current. Therefore, the resistance of an insulator is said to be very high. Materials that are insulators have five or more valence electrons. Examples of insulators are plastic, rubber, and glass. These materials are all compounds made up of two or more different elements that, when bonded together, act as though they have eight valence electrons. The plastic insulation surrounding the copper conductors of an electric cord prevents the two conductors from contacting each other and prevents electric shock from occurring when an electric appliance cord is touched. In general, elements that are nonmetals are insulators or poor conductors of electric current.

OHM'S LAW

Visualize two tanks of compressed air that are the same size (capacity). One tank is maintained at a constant 120 psi (827 kPa) and the other tank is maintained at a constant 40 psi (276 kPa). The tank that is pressurized to 120 psi will fully inflate a car tire faster than the tank that is pressurized to 40 psi. This indicates that the rate of airflow is dependent upon the pressure that is causing the air to move. Given a fixed (unchanging) value of restriction, the greater the pressure of air stored in an air tank, the greater the flow of air when the air is permitted to escape the tank.

Because voltage is like electrical pressure and current is the flow of electrons, it would also seem that voltage and current are also somehow related. This leads to an important fact about electricity:

Important Fact: *Increasing* the voltage causes the electric current flowing through a fixed value of resistance to *increase*.

Returning once again to a tank of compressed air, **Figure 2-13** illustrated that adding a restriction to the valve on the side of the tank by inserting reducers or bushings into the valve, given a fixed pressure in the tank, decreased the flow of air (amount per unit of time) leaving an air tank, compared to when the valve was opened without a restriction present. The reducer acts as a restriction to the flow of air, which decreases the flow of air out of the tank.

Because electrical resistance acts as a restriction to electrical current, and electrical current is the flow of electrons, it would seem that resistance and current are also somehow related. This leads to another important fact about electricity:

Important Fact: *Increasing* the resistance while keeping the voltage source held at a fixed value causes the electric current flowing through the resistance to *decrease*.

These two principles of electricity relating voltage to current and current to resistance combine to form what is known as **Ohm's law**. If you can gain a good working understanding of Ohm's law, electricity will make a lot more sense. Unlike atomic theory, the principle of Ohm's law is something that an electrical troubleshooter will use almost daily. Ohm's law, as shown in **Equation 2-1**, states:

$$\textbf{Voltage} \div \textbf{Resistance} = \textbf{Current}$$
$$\textbf{or}$$
$$\textbf{Voltage/Resistance} = \textbf{Current}$$

The "/" sign shown in **Equation 2-1** signifies to divide the first or top number by the second or bottom number as in a fraction. Thus, 1/10 or one-tenth means to divide 1 by 10. In fact, the dots in the "÷" sign are just place holders for the two numbers in the fraction. Just replace the "/" sign with a "÷" sign if you do not like working with fractions.

The units of voltage, current, and resistance were designed so that a voltage source of 1 volt causes 1 ampere of current to flow through a resistance of 1 ohm. This is shown mathematically in **Equation 2-2**.

$$\textbf{1 volt/1 ohm = 1 ampere}$$
$$\textbf{or}$$
$$\textbf{1V/1}\Omega\textbf{ = 1A}$$

Increasing the voltage while keeping resistance at a fixed value causes the current to increase. If the voltage is increased to 2 volts and the resistance is maintained at 1 ohm, then 2 amps of current will flow through the resistance as shown in **Equation 2-3**.

$$\textbf{2 volts/1 ohm = 2 amperes}$$
$$\textbf{or}$$
$$\textbf{2V/1}\Omega\textbf{ = 2A}$$

The Ohm's law equation also indicates that if the resistance is increased, the amount of current flowing through the resistance will decrease if the voltage remains at a fixed value. If the voltage is maintained at 2 volts but the resistance is increased to 2 ohms, then 1 ampere of current will flow through the 2Ω resistance, as shown in **Equation 2-4**.

$$\text{2 volts/2 ohm} = \text{1 ampere}$$
$$\text{or}$$
$$2V/2\Omega = 1A$$

Other Forms of Ohm's Law

As indicated earlier, current is often identified as I, voltage may be identified as E, and resistance is typically identified as R. One equation for Ohm's law is shown in **Equation 2-5**.

$$E/R = I$$

If you took algebra in school, then you may remember that if any two of the variables E, I, or R are known, then it is possible to find the third variable or unknown value by manipulating this formula and solving for the unknown. If you did not take algebra or do not remember how to manipulate formulas, all three forms of Ohm's law are as shown in **Equation 2-6**.

$$I = E/R$$
$$E = I \times R$$
$$R = E/I$$

The memorization aid shown in **Figure 2-16** can be helpful for memorizing Ohm's law. Covering up the unknown value will show what two values need to be multiplied or divided to yield the unknown value. Notice that this memorization aid also shows Ohm's law using V, A, and Ω. If you prefer, use V, A, and Ω instead of E, I, and R.

The most important thing to learn about Ohm's law is the relationships among voltage, current, and resistance and what happens to the other two values when one value changes. It will probably not be too often in the real world of truck repair that a truck technician will be required to calculate what amount of current will flow through a given resistance when a specific voltage is applied. However, to become effective as an electrical troubleshooter and not just a part changer, it is important to understand the relationships implied by Ohm's law and to be able to apply them as if they were second nature. For example, a truck owner may state that his brake lamps are very dim. The two brake

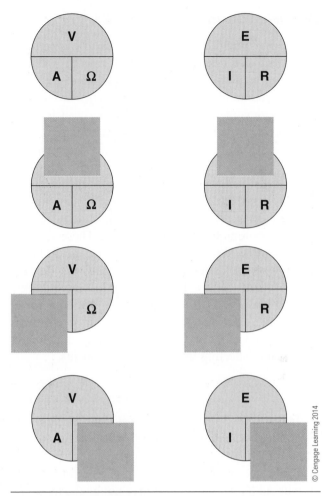

Figure 2-16 Ohm's law memorization aid.

© Cengage Learning 2014

lamps at the rear of a truck might typically draw 4A of current. If the technician measures 1A of current flow through the truck's brake lamp wiring and the battery voltage is 12V, the technician can reason that excessive resistance somewhere between the voltage source and the brake lamp filaments is the cause of the dim brake lamps. The technician does not need to calculate the exact value of the resistance to know that excessive resistance is the problem.

INTRODUCTION TO ELECTRICAL TOOLS

Most technicians in training are familiar with common hand tools like screwdrivers and wrenches. These hand tools, along with air-operated tools, are used to disassemble and reassemble mechanical components to perform repairs. Electrical tools are used mostly to diagnose or troubleshoot a problem rather than to repair the problem.

Electric Maps

In auto racing, the course that a road racing track follows is often described as being a **circuit**. The path that electric current follows is also known as a circuit. To illustrate an electric circuit, schematics are used to show the layout or relationship of the electric components. Schematics are also known as circuit diagrams or wiring diagrams. A schematic is similar to a detailed terrain map. A detailed terrain map uses symbols to indicate specific features such as rivers, bridges, roads, and so on. An electrical schematic, like a terrain map, uses symbols to indicate electrical components found in an electric circuit. These symbols will be discussed in detail in later chapters as electrical components are introduced. For now, it will be necessary to introduce a few symbols to convey some ideas. **Figure 2-17** shows common symbols for a battery and a resistor to make up a simple electrical circuit. Conductors or wires are typically drawn as a solid line connecting the various components.

Tech Tip: Trying to diagnose an electrical problem without a schematic is like driving to an unfamiliar place without using a road map. It may be possible to get to your destination without a road map, but it may not be the most efficient route.

Measuring Voltage

Because voltage is like pressure, having an understanding of a mechanical pressure gauge (gage) may assist you in understanding the concept of voltage measurement.

Pressure Gauges. A mechanical pressure gauge could be used to measure pressure, such as the pressure of compressed air stored in a tank. A mechanical pressure gauge actually measures the difference in pressure between two points and displays this pressure difference via a needle, which points at a number on the gauge, indicating the pressure. This leads to an important fact:

Important Fact: Pressure is measured with respect to some other pressure or a complete lack of pressure known as a vacuum.

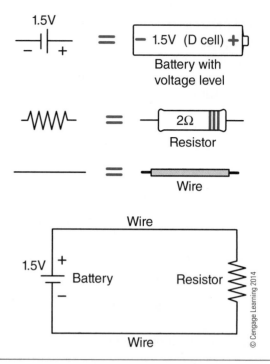

Figure 2-17 Electric symbols used to construct an electrical schematic.

Most pressure gauges display the difference between the atmospheric pressure (14.7 psi or 101 kPa at sea level) and the unknown pressure being measured. The measuring device in many pressure gauges is a Bourdon tube. A Bourdon tube is a C-shaped tube that works like a rolled-up birthday party favor noisemaker (**Figure 2-18**). Blowing into the party favor causes the flat, rolled-up paper tube to straighten and increase in length.

The pressure being measured, such as the pressure of compressed air in a tank, is routed inside the Bourdon tube. A cross-section of the Bourdon tube shown in **Figure 2-19** indicates that the tube has a flattened oval shape when the pressure in the tube is

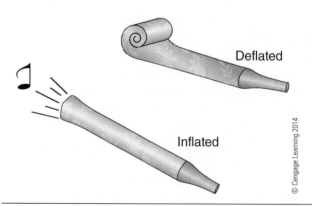

Figure 2-18 Party favor noise maker: length increases as the pressure within the paper tube increases.

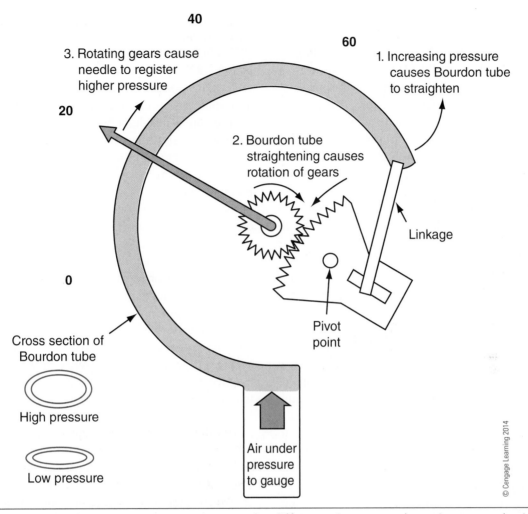

40

3. Rotating gears cause needle to register higher pressure

60

1. Increasing pressure causes Bourdon tube to straighten

20

2. Bourdon tube straightening causes rotation of gears

Linkage

0

Cross section of Bourdon tube

High pressure

Low pressure

Pivot point

Air under pressure to gauge

© Cengage Learning 2014

Figure 2-19 Bourdon tube pressure gauge measures the difference in pressure between atmospheric pressure and the pressure in the tube.

low and becomes more rounded as pressure in the tube increases. This causes the tube to unroll or straighten out as pressure is applied to the inside of the Bourdon tube. The Bourdon tube movement is translated into rotary motion by the gearing inside the gauge, which ultimately causes the indicator needle to rotate to point at a pressure value. The more the Bourdon tube is unrolled or straightened by the pressure being measured, the greater the pressure indicated by the needle.

The housing of a typical pressure gauge is vented to the atmosphere to keep the changing atmospheric pressure applied to the outside of the Bourdon tube. Atmospheric pressure tries to keep the Bourdon tube flat and rolled up, in opposition to the pressure inside the tube, which is trying to unroll the Bourdon tube. When the pressures on the inside and the outside of the Bourdon tube are identical, the pressure gauge will indicate 0 psi (0 kPa). Thus, a pressure gauge on an empty tank of compressed air will indicate 0 psi because both the pressure in the tank and the atmospheric pressure are at about the same value. The difference

between the atmospheric pressure and the empty air tank is 0 psi, as indicated by a pressure gauge on the tank.

The Bourdon tube pressure gauge installed on a full tank of compressed air may display 120 psi (827 kPa) at sea level. A complete lack of air pressure is called a vacuum. Outer space is a vacuum. If the full tank of compressed air with a Bourdon tube pressure gauge indicating 120 psi (827 kPa) at sea level with 14.7 psi (101 kPa) atmospheric pressure were taken into outer space, the pressure gauge would display 134.7 psi (928 kPa) if viewed by an astronaut. However, the pressure of air in the tank would not have really changed. The reason for the change in displayed pressure is that the "atmospheric" pressure of outer space is 14.7 psi (101 kPa) less than the atmospheric pressure at sea level on earth. Because there is zero atmospheric air pressure in outer space to oppose the pressure inside the Bourdon tube, the Bourdon tube unrolls more in outer space than it does at sea level. The pressure gauge is displaying the difference between the air pressure in the tank and the air pressure outside of the tank. Returning the tank of compressed air to sea level would cause the pressure gauge to again indicate 120 psi (827 kPa).

You may wonder that if the atmospheric pressure is 14.7 psi (101 kPa), what is atmospheric pressure measured against or referenced to? Atmospheric pressure is referenced to a vacuum, which is a complete lack of air or any other gas. **Figure 2-2** illustrated a simplified barometer, which uses a vacuum in a glass tube as a pressure reference for the atmospheric pressure. Atmospheric pressure is 14.7 psi (101 kPa) greater than a vacuum at sea level. When a vacuum is the reference pressure, the pressure is described as being an absolute pressure. Thus, atmospheric pressure at sea level is described as being 14.7 psi (101 kPa) *absolute*. Electronic fuel-injected gasoline engines and many electronically controlled diesel engines use a manifold absolute pressure (MAP) sensor to measure intake manifold pressure. This sensor measures intake manifold pressure with respect to a pure vacuum instead of atmospheric pressure. In a naturally aspirated (non-turbocharged) gasoline engine, the intake manifold pressure is always less than atmospheric pressure because of the throttle valve at the entrance of the intake manifold so the MAP sensor must be able to measure a pressure that is less than atmospheric pressure, thus the reference to a vacuum. Model year 2007 and later diesel engines with exhaust aftertreatment may have an intake air throttle that is partially closed during particulate filter regeneration resulting in an intake manifold pressure that is also less than atmospheric pressure. Sensors and diesel exhaust aftertreatment will be discussed in much greater detail in later chapters.

When a pressure is measured against or referenced to atmospheric pressure, the pressure is referred to as gauge pressure. If an air pressure gauge, which references pressure to atmospheric pressure, indicates 120 psi (827 kPa), the pressure is described as being 120 psi gauge (827 kPa gauge) or abbreviated 120 psig (827 kPag). Since most pressures are referenced to atmospheric pressure, the "g" is typically omitted and the pressure is just described as being 120 psi (827 kPa). **Figure 2-20** illustrates the relationship between gauge and absolute pressures as well as vacuum measurement.

Another type of pressure gauge is a differential pressure gauge. A differential pressure gauge has two ports instead of just one port like that found on most other pressure gauges, as shown in **Figure 2-21**. The gauge needle displays the difference in pressure between the two ports of the gauge. If one hose were left exposed to atmospheric pressure, then the gauge would just display gauge pressure like any other ordinary pressure gauge. However, if the two ports of the differential pressure gauge are connected to two different pressure sources, then the differential pressure gauge needle will indicate the difference in pressure between the two points. For example, if one port of the differential pressure gauge is connected to a 120-psi air source and the other port of the differential pressure gauge is connected to a 50-psi air source, then the differential pressure gauge will indicate 70-psi differential (120 psi − 50 psi = 70 psi). A differential pressure gauge is often used to measure the pressure drop across a filter. As a filter becomes restricted, the pressure dropped across the filter increases, indicating a need to replace the filter. Many 2007 and later diesel engines also make use of an electronic pressure sensor, which measures the differential (delta) pressure across the diesel particulate filter in the exhaust system to monitor the soot loading of the filter. This sensor will be discussed in detail in later chapters.

Pounds per square inch gauge (psig or psi)	Pounds per square inch absolute (psia)	Kilopascals gauge (kPag or kPa)	Kilopascals absolute (kPaa)	Inches of mercury vacuum (in. Hg)	Millimeters of mercury vacuum (mmHg)	Inches of mercury absolute (in. Hg)	Millimeters of mercury absolute (mmHg)
15	29.7	103.5	204.8	-	-	60.5	1536
10	24.7	68.9	170.3	-	-	50.3	1277
5	19.7	34.4	135.8	-	-	40.1	1019
0	14.7	0	101.3	0	0	29.9	760
−5	9.7	−34.4	66.9	10.2	258.6	19.7	502
−10	4.7	−68.9	32.4	20.4	517	9.6	243
−14.7	0	−101.3	0	29.9	760	0	0
*Assuming barometric pressure of 14.7 psia (101.3 kPaa)							

© Cengage Learning 2014

Figure 2-20 Comparison of gauge and absolute pressure: shaded row is atmospheric pressure.

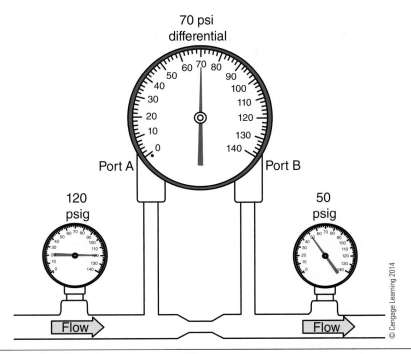

Figure 2-21 Differential pressure gauge indicates the difference in pressure between two points in the hydraulic circuit.

Voltmeter. You may wonder why all the emphasis on pressure gauges when this is a book on electricity. The reason is that voltage is very much like pressure. If you can grasp that a pressure gauge is always indicating the difference in pressure between two points, measuring voltage should make a lot more sense.

A **voltmeter** is an electrical diagnostic tool that measures voltage. Since voltage is like pressure, a voltmeter can be thought of as an electrical pressure gauge. Like a differential pressure gauge that displays the difference between two pressure sources, a voltmeter displays the difference in electrical potential (voltage) between two points. Most modern voltmeters are a component in a multimeter. A multimeter is a tool that is capable of measuring voltage, current, and resistance. Most modern multimeters are also digital devices, meaning that they have no moving needle like a mechanical pressure gauge but have a digital LCD display instead. A typical multimeter is shown in **Figure 2-22**. **Digital multimeters** are often abbreviated as **DMM**. Digital multimeters may also be known as digital volt-ohm meters (DVOM).

Like the two hoses of a differential pressure gauge, a voltmeter has provisions for connection to two wires with probes (**Figure 2-23**). Test leads are the wires with probes that connect the voltmeter to the circuit being tested. Test leads are typically colored black and red. The red lead is installed at the most positive point of the circuit; the black lead is installed at the most negative point. Two leads are necessary to measure

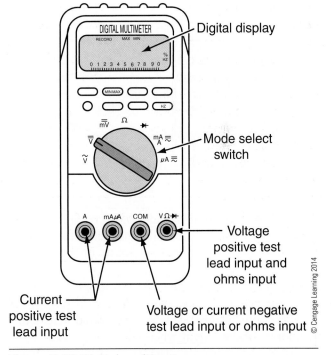

Figure 2-22 Digital multimeter.

voltage because the voltmeter displays the difference in electrical potential or electrical pressure between two points like a differential pressure gauge measures a difference in pressure between two points. A voltmeter is installed in a circuit so that the difference in electrical potential (voltage) across a device is measured, as shown in **Figure 2-24**.

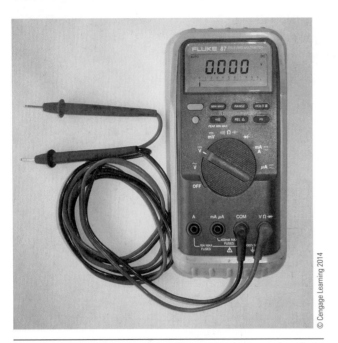

Figure 2-23 Digital multimeter with test leads attached for a voltage measurement.

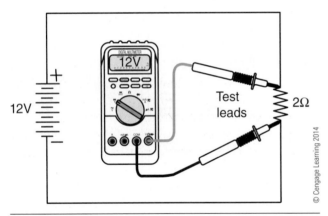

Figure 2-24 Measuring voltage: DMM is connected in parallel with the resistor and measures the difference in voltage between two points in the electric circuit.

The voltmeter test leads are connected across the two points of the circuit where the difference in voltage is being measured. The test leads must make contact with an electrically conductive material, such as a copper wire, to measure voltage. Another way of stating that a voltmeter is connected across a device is to say that a voltmeter is connected in **parallel** with a device. A parallel connection means that the voltmeter is providing a path for current flow in addition to the current that may be flowing through a device—although the amount of current flowing through the meter is negligible.

Since a voltmeter displays the difference in electrical potential between two points, placing both test leads of a voltmeter at the same point in a circuit will cause the voltmeter to indicate 0V because there is no difference in electrical potential between the two test leads. There is also no difference in electrical potential between two points where the voltmeter is not completing an electrical circuit. For example, connecting one test lead on the positive terminal of a 12V battery and holding the other test lead in the air so that it is not touching anything will also cause the voltmeter to indicate 0V because there is no difference in the electrical potential between the positive battery terminal and the air. However, connecting one test lead to the positive terminal of the battery and the other test lead to the negative terminal of the battery will cause the voltmeter to indicate a difference in electrical potential of about 12V.

Measuring Current

The flow rate of a liquid such as water is determined by measuring the amount of water passing a stationary point per unit of time. The flow rate of water or other liquid through a pipe might be measured in gallons per minute (GPM) or liters per minute. A water pipe that is flowing 5 GPM will allow enough water to flow to fill a 5-gallon bucket in 1 min.

Flow Meter. A flow meter measures the flow rate of a fluid, such as water through a pipe. One type of flow meter is shown in **Figure 2-25**. This type of flow meter uses a spring-loaded vane that rotates from vertical with no water flow toward horizontal as water flow increases. The rotation of the vane causes an indicator of some sort to display the rate of flow.

To measure the flow rate of water through a pipe, the flow meter must be installed directly in the path of flow. All the water flowing through the pipe must pass through the flow meter for the measurement to be accurate. Flow meters are sometimes used in conjunction with pressure gauges for diagnosing problems in hydraulic systems such as truck power steering systems. Often, both pressure and flow are necessary for an accurate diagnosis of a hydraulic system. In the same way, it may be necessary to measure both the voltage and current flow in a circuit to diagnose an electrical problem.

Ammeter. The flow rate of electrons (electric current) is measured using an **ammeter**. An ammeter is typically included in a DMM. Current is the measure of the number of coulombs passing a given point

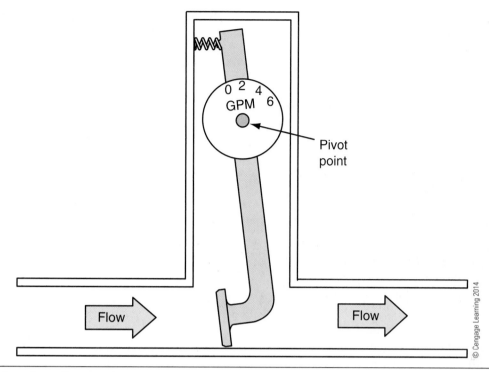

Figure 2-25 Vane-type flow meter.

per second. Therefore, the ammeter must be connected directly in the path of current flow where the current is being measured similar to a flow meter. This means that the circuit must be opened up to make the ammeter a component in the circuit so that all of the current to be measured will pass through the ammeter (**Figure 2-26**). This type of connection is called a **series** connection. *Series* in this sense means sequential or one thing after another, like the links of

a chain. Current flows through one device, then another, and so on.

To measure DC current with a DMM, the dial or knob of the DMM is placed in the DC current position. Typically, the same two DMM test leads used for obtaining voltage measurements are used to obtain current measurements as well. However, the positive lead is typically plugged into a different multimeter terminal for current measurements than it is for voltage measurements. The negative DMM test lead is typically connected to the multimeter common terminal for voltage, current, and resistance measurements.

To obtain a current measurement using an ammeter, the circuit must be opened up at the point where the current measurement is desired. The positive test lead of the ammeter is connected to the most positive point of the open circuit (the point closest to the positive battery terminal) and the negative test lead is connected to the most negative point of the open circuit (the point closest to the negative battery terminal). The ammeter then measures the current flow in the circuit, similar to the way a flow meter measures the flow of water through a pipe.

Since circuit current is passing directly through the series-connected ammeter, ammeters are limited in the amount of current that they can measure. The circuit current cannot exceed the maximum amperage rating of the ammeter or the meter may be damaged. Many modern DMMs are capable of measuring 10 amps of current or more.

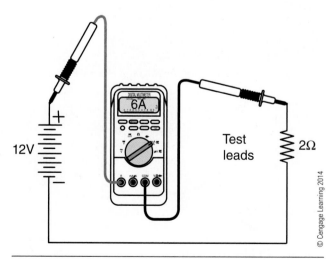

Figure 2-26 Measuring current with digital multimeter: the meter has become a component of the electrical circuit similar to a flow meter in a hydraulic circuit.

Tech Tip: Most of the readings that you will obtain with a DMM will be voltage or resistance. Both resistance and voltage measurements use the same DMM terminals on most meters. Many technicians leave the meter test leads in the meter for storage. If you ever do take amperage measurements using a DMM, return the test leads to the voltage terminals before putting the DMM away. Otherwise, the next time you use the DMM, you may accidentally connect the low internal resistance of the ammeter across a battery or other high-current source for a supposed voltage measurement. This would destroy the ammeter fuse in the DMM, at minimum.

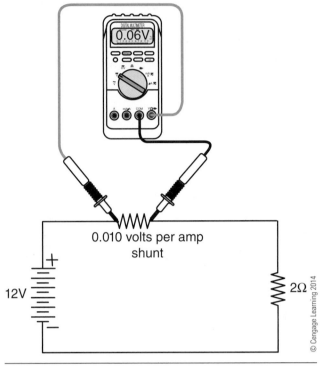

Figure 2-27 Measuring current using a shunt and voltmeter: the difference in voltage across a known value of resistance is used to calculate the current flow using Ohm's law.

Current Shunt. Measuring current with an ammeter is not as easy as measuring voltage with a voltmeter because the circuit must be opened to cause all of the circuit current to pass through the ammeter. Great care must also be taken not to damage the ammeter by exceeding the meter's design amperage or by misconnecting the meter, such as connecting the ammeter in parallel with a device instead of in series. For these reasons, a **shunt** may be utilized to measure current instead of an ammeter. A shunt is a known value of resistance with a very low value (much less than 1 ohm). The shunt is placed in series with the portion of the circuit where current is to be measured. The difference in voltage across the shunt is then measured, as shown in **Figure 2-27**. The current flow can then be calculated using Ohm's law because the value of resistance is known as is the difference in voltage across the resistance. A label on the shunt typically provides the volts per amp conversion factor. The shunt shown in **Figure 2-27** has a conversion factor of 0.010V (10 millivolts) per amp. The voltmeter displays a value of 0.06V (60 millivolts) which indicates that the current flow through the circuit is 6A. Instrument panel ammeters found in some trucks may use the shunt method to display charging system current. The charging circuit cable acts as the shunt.

Clamp-On Current Probe. Opening a circuit and installing an ammeter or shunt in series is not always possible or may be very difficult to perform. The easiest and safest way for a troubleshooter to measure electric current is to use a clamp-on current probe. Electric current flow causes a magnetic field to surround a wire through which current is flowing, as will be explained in later chapters. This magnetic field

increases as current flow increases. Clamp-on current probes permit measurement of current without having to open the circuit by measuring this magnetic field and converting the magnetic field strength to a current value. Clamp-on current probes, like those shown in **Figure 2-28**, are available, ranging from those that can

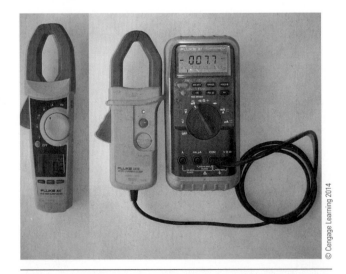

Figure 2-28 Clamp-on current probes. Probe on left is a stand-alone tool which displays current in amperes on a digital display. Probe on right is used with a DMM voltmeter. The voltmeter displays a voltage value which must be converted to current.

measure small currents much less than 1A to those that can measure the truck's 2000A initial starter motor current draw. Current probes may be a stand-alone tool complete with a digital display.

Current probes may also be designed to act as an input to a digital voltmeter. The output of this type of current probe is a voltage that is proportional to the current flow being measured. The reading displayed on the voltmeter must be converted to amperes using the current probe conversion factor. For example, the current probe shown on the right side in **Figure 2-28** provides a voltage output of 0.001V (1 millivolt) for every amp of current that is measured by the probe. Therefore, if the digital voltmeter indicated 0.025V (25 millivolts), this would indicate 25A of current flow is being measured by the current probe.

Clamp-on current probes make the troubleshooter's job a lot easier than in the past because current can now be measured without opening the circuit to install the ammeter or shunt and without fear of damaging an ammeter through an improper connection.

....

Tech Tip: It is a good idea to verify your current probe measurement setup prior to troubleshooting by measuring the value of a known current flow. The typical current flow through a single 12V low-beam headlamp is about 4A. Clamping the current probe around the wire supplying a headlamp should cause your current measurement setup to indicate about 4A. If you measure something like 40A or 0.4A instead, you may be making a math error or may be misreading the meter.

....

Measuring Resistance

Resistance can be measured using an **ohmmeter**. Most multimeters include an ohmmeter. An ohmmeter causes an electric current to flow through the device being measured to determine the resistance of the device. The battery in the ohmmeter is the voltage source for the current that flows through the device being measured. Unlike a voltmeter or ammeter, an ohmmeter is only used when there is no other voltage source in the circuit. Ohmmeters work best when measuring the resistance of a component that is completely isolated from the rest of the electrical system.

Some ohmmeters require the technician to select one of several ranges of resistance values by a rotary dial, as shown in **Figure 2-29**. For measurement of

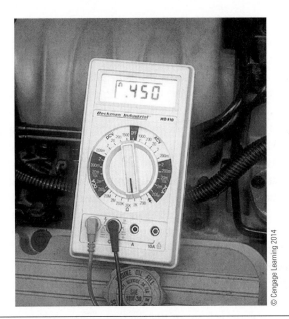

Figure 2-29 Manual range selection type of digital ohmmeter.

unknown resistance values, the highest range of resistance is selected first and the range is decreased until the display indicates the resistance. Most modern DMMs have an auto-range feature that automatically selects the correct range of resistance.

Older analog-type ohmmeters with a needle and scale require that the meter leads be connected together prior to measuring resistance and a dial rotated until the needle points at zero to calibrate the meter.

....

Tech Tip: Modern digital multimeters are fantastic tools but have fooled almost every technician at one time or another. One reason for this is that just a miniscule amount of voltage present in the circuit where resistance is being measured can result in dramatic resistance measurement errors. This miniscule voltage can be generated by engine coolant or water in contact with dissimilar metals, as will be explained in more detail in later chapters. If resistance measurements do not make sense, measure the voltage that is present between the two points of which you want to know the resistance. If the DMM displays any voltage, your resistance readings will not be accurate.

....

Negative or Positive Meter Reading. In the past, ammeters and voltmeters were analog, meaning that a moving needle pointed to a scale. Zero was typically to the left side of the scale and the needle rested at zero until a measurement was made. In the days of analog voltmeters, accidentally hooking up the volt-meter leads backward—so the positive lead was connected to a location that was more negative than the location that the meter's negative lead was connected to—caused the meter needle to peg hard against the zero stop as the needle tried to indicate a negative voltage. The same thing was true with an analog ammeter. Accidentally connecting an ammeter so that the current was flowing backward through the meter also caused the meter needle to peg against the zero stop. Either of these conditions could damage the meter.

Modern digital meters do not have these reverse polarity problems. For digital meters, connecting the test leads backward just causes the meter to display a negative sign in front of the reading. The negative sign indicates that the meter's positive lead has been connected to a point in the circuit that is more negative than the point in the circuit to where the negative test lead is connected. The leads can be reversed and the meter will indicate a positive value of voltage or current. Such negative measurements do not damage a DMM.

Although current and voltage can be negative, resistance can only be positive because it is a measure of restriction or opposition to current flow. However, some digital ohmmeters may display a negative resistance if a voltage source is still active somewhere in a circuit where resistance is being measured. The negative sign displayed when measuring resistance just lets you know that the reading is not accurate. All voltage sources must be disconnected to use an ohmmeter to measure resistance.

Additional information on digital multimeters can be found in the Extra for Experts section at the end of this chapter.

CIRCUIT ANALYSIS

Because air is compressible, the analogies between pneumatic circuits and electrical circuits begin to break down as more complicated concepts are discussed. Unlike air, a liquid is not compressible. Therefore, hydraulic circuits will be used to explain the flow of electric current through a circuit. Keep in mind for all of the following examples, the steady-state condition of the hydraulic circuit is being described. That is, all air has been purged from the system and only liquid is flowing through the circuit.

Hydraulic Circuits

If a simple hydraulic circuit were constructed consisting of a reserve of hydraulic fluid known as a sump, a pump, some tubing, and a restriction like that shown in **Figure 2-30**, it would in some ways behave in a manner very similar to an electric circuit and can be used to illustrate some electric circuit fundamentals.

The pump shown in **Figure 2-30** causes hydraulic fluid to be drawn from the tank and placed under pressure. Like compressed air, the hydraulic fluid under pressure will tend to flow from a location of higher pressure to a location of lower pressure. The zero pressure location in a hydraulic circuit is typically the storage tank or sump. In **Figure 2-30**, fluid flows from the pump through the hose to the restriction. The large-diameter hose from the pump to the restriction is like a busy four-lane highway (in each direction) that is packed bumper to bumper with cars. The restriction is like a construction zone that reduces the four lanes of traffic (in each direction) to two lanes. Traffic backs up on the four-lane highway because all cars entering the highway must pass through the two-lane construction zone, as shown in **Figure 2-31**. Beyond the construction zone, the highway returns to four lanes, but the traffic flow is very light because the two-lane restriction has limited the number of cars on the road beyond the construction zone.

Like cars on the four-lane highway before the construction zone, hydraulic fluid under pressure backs up in the hose between the pump and the restriction. The fluid under pressure is squeezed through the restriction. The fluid on the other side of the restriction is like cars on a four-lane highway beyond the two-lane construction zone. The opposition to hydraulic flow

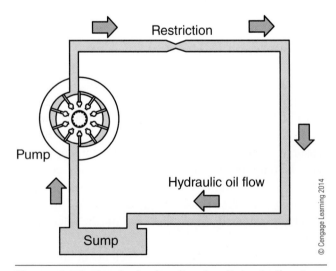

Figure 2-30 Simple hydraulic circuit is similar to a simple electric circuit.

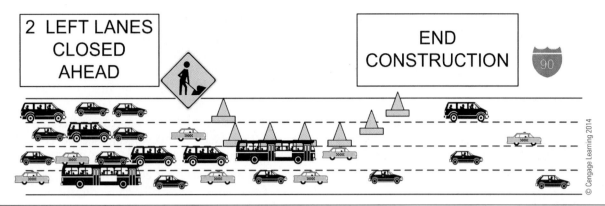

Figure 2-31 Two-lane construction zone causes traffic to back up behind the lane restriction. Traffic is light beyond the lane restriction.

beyond the restriction is minimal because of the large-diameter hose beyond the restriction. Therefore, the pressure applied to the fluid is reduced to almost zero as the fluid flows unimpeded all the way back to the sump.

The pressure applied to the fluid with respect to atmospheric pressure at the inlet to the restriction and at the pump outlet is the same value, in this case 75 psi (517 kPa) as shown by gauge A in **Figure 2-32**. From this point, the fluid flows through the restriction, resulting in a pressure drop across this restriction. The term *pressure drop* indicates that the pressure in the system referenced to atmospheric pressure is lower at

the restriction output than it was at the restriction input. The pressure beyond the restriction is shown to be about 0 psi as shown by gauge B in **Figure 2-32**. This indicates that the difference in pressure between gauge A and gauge B is 75 psi (75 psi – 0 psi). Therefore, the pressure drop across the restriction is 75 psi (517 kPa).

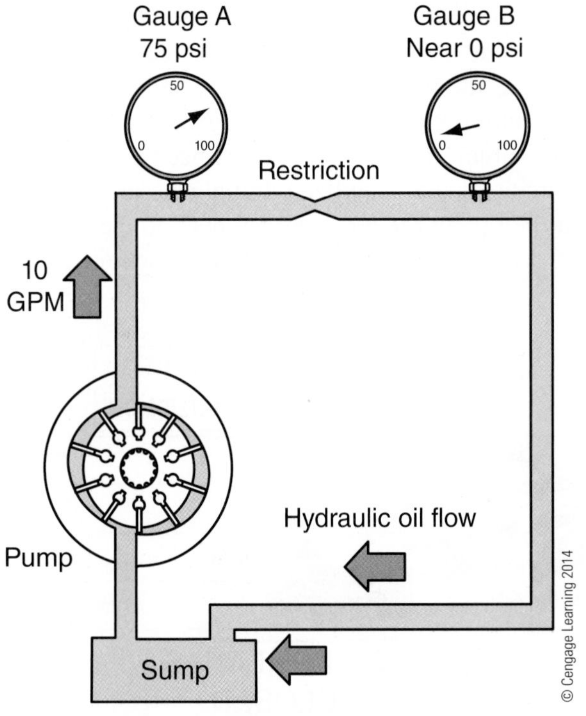

Instead of using two conventional pressure gauges to measure the pressure drop across the restriction, a single differential pressure gauge could be connected across the restriction as shown in **Figure 2-33**. This gauge will also indicate a 75-psi drop across the restriction.

If flow meters were inserted at the pump outlet, the restriction inlet, the restriction outlet, and the sump inlet, all of the flow meters would indicate the same rate of fluid flow (after all the air in the system was purged). This is a simple yet important concept because a similar thing happens in an electric circuit. There is only one path for fluid flow through this hydraulic circuit, and the fluid is a liquid that cannot be compressed. The amount of hydraulic fluid leaving the pump per minute is the same amount of fluid that returns to the tank per minute, which is the same amount flowing through the restriction and all the hoses. The only way for more fluid to leave the pump than to return to the sump is for a leak to cause the fluid to spill onto the floor. Likewise, the only way for less fluid to leave the pump than to return to the sump is for some other source of fluid outside this circuit to add to the flow.

A second restriction is added to the simple hydraulic circuit, as shown in **Figure 2-34**. This second restriction is connected in series with the first restriction and has a smaller diameter than the first restriction. Therefore, this second restriction is more limiting than the first restriction. The pump in this example is regulated to maintain 75 psi at its outlet, for illustration purposes, regardless of the amount of restriction in the circuit.

Figure 2-32 Pressure drop across a restriction in a hydraulic circuit.

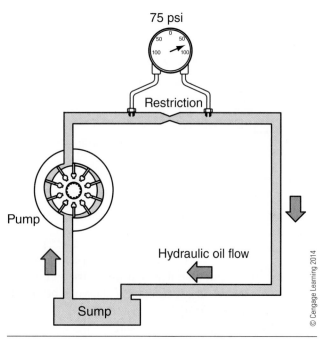

Figure 2-33 Using a differential pressure gauge.

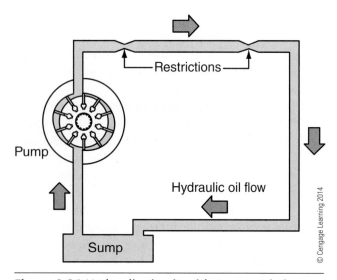

Figure 2-34 Hydraulic circuit with two restrictions.

The fluid flow in the circuit is now similar to traffic flow when there are two construction zones on a four-lane highway at rush hour (four lanes each direction) that are separated by a mile or so of unrestricted lanes, as shown in **Figure 2-35**. The first construction zone cuts the four lanes of traffic down to two lanes. Beyond the first construction zone, traffic is free to drive on all four lanes. However, the second construction zone a mile down the road is worse than the first construction zone because it cuts traffic from four lanes to one single lane. This really causes traffic to back up. Eventually, the section of four-lane highway between the two construction zones fills up with cars and traffic flow decreases to a trickle because every car must pass through the single-lane construction zone. Beyond the second restriction, the highway returns to four lanes and the few cars on the four lanes of highway beyond both construction zones can speed away.

Having two restrictions of different sizes plumbed in series with each other in a hydraulic circuit is similar to the example of two construction zones. As with the single restriction circuit, fluid under pressure builds up at the inlet of the first restriction. The fluid squeezes through the first restriction to the section of hose between the two restrictions. Unlike the single-restriction hydraulic circuit, fluid under pressure also builds up in this section of the hose beyond the first restriction because of the second restriction. The fluid then squeezes through the second restriction. The opposition to hydraulic flow beyond the second restriction is minimal. Therefore, the pressure applied to the fluid beyond the second restriction is reduced to nearly nothing as the fluid flows unimpeded all the way back to the sump. Note that this analogy describes the hydraulic circuit after all the air has been purged out of the circuit and has stabilized. Initial pressure drops that occur before all the air is purged out and the system is filled with hydraulic fluid would be much different.

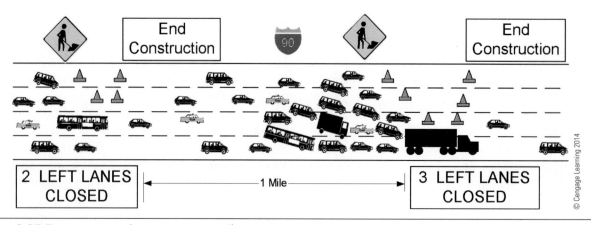

Figure 2-35 Two construction zones one mile apart.

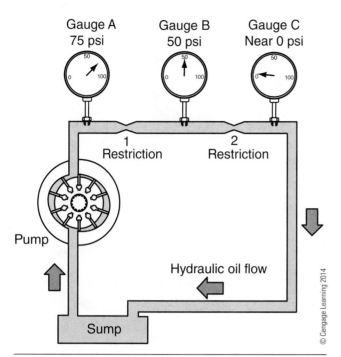

Figure 2-36 Pressure drop across each restriction.

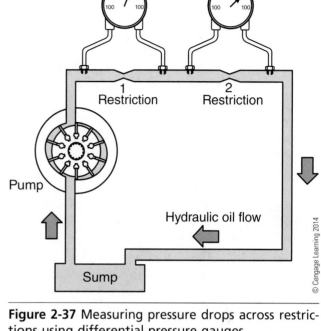

Figure 2-37 Measuring pressure drops across restrictions using differential pressure gauges.

In the example shown in **Figure 2-36**, the fluid is under 75 psi (517 kPa) of pressure at the inlet of the first restriction shown by gauge A. Unlike the single-restriction circuit, there is 50 psi (345 kPa) in the hose beyond the first restriction shown by gauge B because of the second restriction. This indicates that 25 psi (172 kPa) of pressure has been dropped across the first restriction (75 psi – 50 psi = 25 psi). At the inlet to the second restriction, the fluid is under 50 psi of pressure. Because pressure gauge C indicates 0 psi, the pressure drop across the second restriction is 50 psi. Note that adding the pressure drops of 25 psi and 50 psi together results in the 75 psi of pressure supplied by the pump.

A differential pressure gauge could also be utilized to measure the pressure drop across the restrictions, as shown in **Figure 2-37**. The gauge across the first restriction indicates that 25 psi (172 kPa) is dropped across the first restriction and 50 psi (345 kPa) is dropped across the second restriction. Instead of measuring pressures referenced to atmospheric pressure and subtracting the difference to find the pressure drop, the differential pressure gauge measures the difference in pressure between the point before the restriction and the point beyond the restriction. This permits a direct reading of the pressure dropped across the restriction.

If the two restrictions were switched around, as shown in **Figure 2-38**, the pressure drop across

each restriction would still be the same as shown in **Figure 2-38**. The pressure drop follows the restriction. The flow rate of fluid everywhere in the circuit shown in **Figure 2-38** would be the same as the circuit shown in **Figure 2-37**.

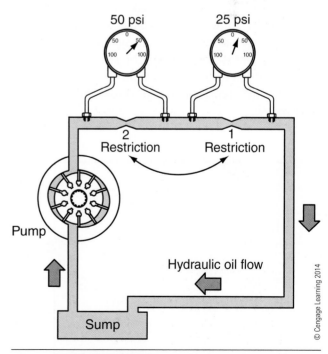

Figure 2-38 Restrictions switched around, pressure drop follows restriction.

Simple Electrical Circuit

Using the electrical symbols introduced previously, **Figure 2-39** shows a simple electric circuit. This circuit consists of a 12V voltage source (battery), wires, and a 2-ohm resistor. This circuit is similar to the single-restriction hydraulic circuit described in the previous section. If this electrical circuit is compared to the hydraulic circuit, the battery's positive terminal acts like the pressure-regulated pump, the wires are like the hoses, the resistor is like the restriction, and the battery's negative terminal is like the sump (assuming conventional current flow).

The battery shown in **Figure 2-39** has two terminals: negative (–) and positive (+). A wire connects the battery's positive terminal to the resistor. Another wire connects the other end of the resistor to the battery's negative terminal. The 12V battery causes 6 amps of current to flow through the 2-ohm resistor as predicted by Ohm's law, shown in **Equation 2-7**.

$$12V/2\Omega = 6A$$
$$\text{or}$$
$$12 \text{ volts} \div 2 \text{ ohms} = 6 \text{ amps}$$

In **Figure 2-40**, a voltmeter is measuring the difference in voltage across the resistor. This is similar to using the differential pressure gauge to measure the pressure drop across the restriction in the hydraulic circuit. The ammeter in **Figure 2-40** is measuring the current flowing through the circuit in a similar manner to the flow meter, which measured hydraulic fluid flow rate in the hydraulic circuit. Note that the DMM acting as an ammeter in **Figure 2-40** is connected in series with the circuit so that all current being measured flows through the DMM. The DMM acting as a voltmeter in **Figure 2-40** is connected in parallel with the resistor so that the difference in voltage across the resistor is being measured.

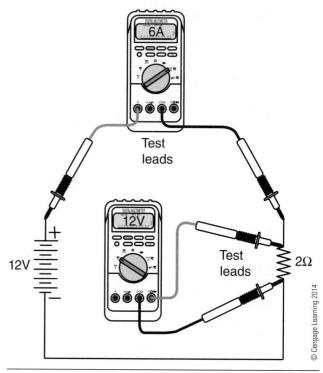

Figure 2-40 Measuring voltage drop and current flow: note the position of the test leads and selector dials on DMMs.

Returning again to the single-restriction hydraulic circuit shown in **Figure 2-33**, if the restriction shown was replaced with a restriction with a smaller diameter, the pressure-regulated pump would still maintain 75 psi output pressure, but the flow rate of hydraulic fluid through the circuit would decrease. Less fluid passes through the smaller, more limiting restriction than would flow through the original restriction with a larger diameter that it replaced.

Modifying the electric circuit in **Figure 2-39** by replacing the 2-ohm resistor with a 3-ohm resistor will change the amount of current flowing in the circuit. Increasing the resistance to 3 ohms will cause the current to decrease from 6 amps to 4 amps as predicted by Ohm's law. Note that the DMM ammeter in **Figure 2-41** now indicates 4A with the resistance increased from 2 ohms to 3 ohms. The battery voltage remains fixed at 12V, like the pressure-regulated pump maintaining 75 psi of output pressure, as shown in **Figure 2-41** by the DMM measuring the voltage across the resistor. Increasing the electrical circuit resistance decreases the electric current flow in the same way that a restriction with a smaller diameter reduces the flow rate in the hydraulic circuit.

It is important to note that the current flow is 4 amps everywhere in the circuit shown in **Figure 2-41**. Moving the ammeter to a point between the bottom of the resistor and the negative battery terminal will still

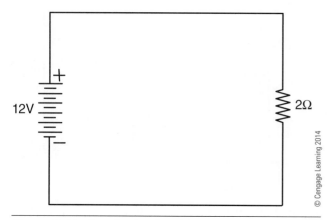

Figure 2-39 Electric circuit with one resistor.

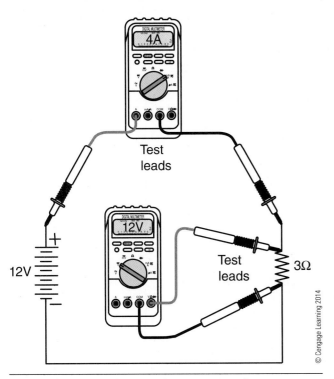

Figure 2-41 Increasing resistance to 3 ohms.

cause the ammeter to read 4 amps as shown in **Figure 2-42**. This type of circuit in which there is only one path for current flow is called a **series circuit**. In a series electrical circuit, the current is the same

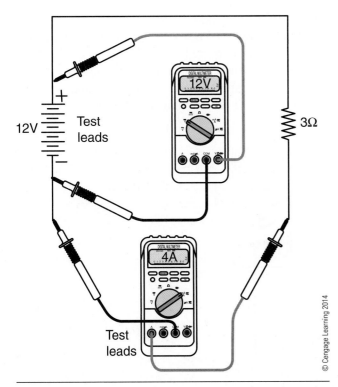

Figure 2-42 Current is the same everywhere in series circuit.

everywhere in the circuit because there is only one path for current flow. This is similar to the hydraulic circuits shown previously, which also only have one path for hydraulic fluid to flow through.

Two-Resistor Series Electrical Circuit

Figure 2-43 shows an electrical circuit with two resistors and one battery. The resistors are connected end to end with each other and with the battery terminals. This circuit is like the hydraulic circuit shown earlier, with two restrictions in series with each other. Any electric current that leaves the positive terminal of the battery must pass through both resistors before returning to the negative terminal of the battery (assuming conventional current flow). Because there is only one path for current flow, this two-resistor circuit is a series circuit.

Like the differing pressure drops across the unequal restrictions in the two-restriction hydraulic circuit shown in **Figure 2-37**, voltage will drop across the two resistors shown in **Figure 2-43** when current flows through the circuit. Because the current is the same throughout this entire series circuit and the values of resistance are different, Ohm's law would indicate that the voltage measured across each of the resistors should be different.

In **Figure 2-44**, the three DMM voltmeters indicate that the difference in voltage across the upper resistor is 4V, while the difference in voltage across the lower resistor is 8V. Another way of stating this is that 4 volts are being dropped across the top resistor and 8 volts are being dropped across the bottom resistor. **Voltage drop** refers to the difference in voltage across a device, in the same way that pressure drop refers to a difference in pressure across a device in a hydraulic circuit.

The voltmeter connected across the battery in **Figure 2-44** indicates that the battery voltage is 12V.

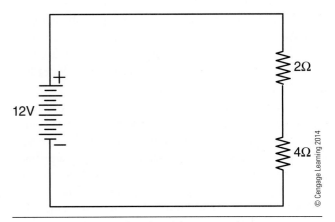

Figure 2-43 Series circuit with two resistors.

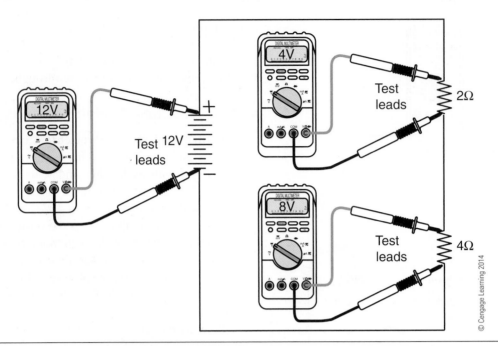

Figure 2-44 Difference in the voltage dropped across unequal series resistors.

Note that adding the 4V dropped across the top resistor to the 8V dropped across the bottom resistor results in 12V. Recall that adding the pressure drops together in the hydraulic circuit with two restrictions resulted in the value of the pressure supplied by the pump. This leads to an important principle of electricity:

Important Fact: The sum of all of the voltage drops in a series circuit is equal to the source voltage.

This important fact is taken from **Kirchhoff's voltage law**, which is also known as Kirchhoff's second law. Kirchhoff's voltage law is another tool for trouble-shooters. A skilled troubleshooter knows that if 5V were measured across a dim brake lamp bulb and the vehicle batteries are supplying 12V, then the other 7V are being dropped elsewhere in the circuit. The troubleshooter can then use a voltmeter to find where the 7V are being dropped and isolate the cause of the trouble. Kirchhoff's voltage law is discussed in further detail in the Extra for Experts section at the end of this chapter.

The concept of voltage drop is important for a troubleshooter to understand. Think of voltage drop as being like pressure drop in a hydraulic circuit. The voltmeter leads are like hoses of a differential pressure gauge that are connected to the circuit to measure the difference in electrical potential (voltage) between two points.

One analogy of Kirchhoff's voltage law is money. If you start out with $12 in your pocket and then spend $5 for a snack and $7 for gasoline, you will have $0 remaining. The money in your pocket is like the source voltage (i.e., battery voltage) and each place that money is spent is like a voltage drop. In the case of an electric circuit, all of the source voltage is always "spent" on the voltage drop(s) around the circuit leaving 0V remaining.

Total Series Circuit Resistance. If the two resistors in the circuit shown in **Figure 2-43** were removed from the circuit and connected together end to end with a piece of wire, an ohmmeter would indicate that the combined resistance of the 2Ω resistor and the 4Ω resistor would be 6Ω, as shown in **Figure 2-45**.

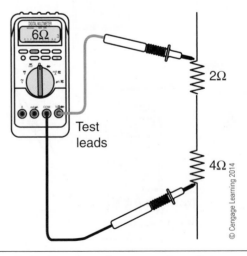

Figure 2-45 Total resistance of series-connected resistors is the sum of their resistance values.

This 6Ω is the **total resistance** value of the circuit and represents the amount of resistance "seen" by the battery. This leads to another important fact regarding series circuits:

Important Fact: In a series circuit, the sum of the values of resistors connected in series is the total circuit resistance.

Replacing the two resistors in **Figure 2-45** with a single 6Ω resistor will cause the same amount of current to flow as a circuit with a 2Ω and a 4Ω resistor connected in series. If you were a battery, you would not know if you were connected to a single 6Ω resistor or a 2Ω and a 4Ω resistor connected in series. The same amount of current flows in both circuits. The voltage dropped across a single 6Ω resistor or the voltage dropped across the combination of the 2Ω and 4Ω resistors connected in series would also be the same.

There really is no reason to physically replace the two resistors with a single resistor. However, it is very easy to find the current flow in a circuit with one resistor and one voltage source. The resistor replacement is just done on paper so that the circuit can be analyzed for total resistance, which is then used to find the circuit current.

Because circuits with more than one resistor are being introduced, Ohm's law must be applied with some care. Ohm's law can be applied either to an entire circuit or to each component in the circuit. When Ohm's law is applied to the entire circuit, the total voltage, total current, and total resistance must be used in the Ohm's law equations to find the unknown total voltage, total current, or total resistance. When Ohm's law is applied to just one component in the circuit, then the voltage drop across the component, the current flow through the component, and the resistance of the single component must be used in the Ohm's law equations to find the unknown voltage drop across, current flow through, or resistance of the single component. This should make more sense after some practice with circuit analysis.

Ohm's law can be applied to the entire series circuit to find the value of the total current flowing in the circuit after finding the total resistance. Because the current is the same everywhere in a series circuit, the total current is the same as the current flowing through each circuit component. To find total current for the circuit shown in **Figure 2-43**, the total voltage, 12V, is divided by the total resistance, 6Ω. Doing so will result in **Equation 2-8**.

$$12V/6\Omega = 2A$$

Because **Figure 2-43** is a series circuit and the current is the same everywhere in a series circuit, the total current could also be calculated by finding the current flow through either resistor. Dividing the voltage dropped across either resistor by the value of its resistance also yields 2A, as shown in **Equation 2-9**.

$$4V/2\Omega = 2A$$
$$\text{and}$$
$$8V/4\Omega = 2A$$

This proves that the total current is the same throughout the series circuit and that current can be calculated by applying Ohm's law to the entire circuit, using total resistance and total voltage in the Ohm's law equation. This also proves that current in a series circuit can be calculated by using the voltage dropped across a single resistor, along with the value of this single resistor, in the Ohm's law equation and the calculations will still arrive at the same answer for total current in the series circuit.

Tech Tip: The simpler a circuit is, the easier it is to solve for unknown values. The goal in circuit analysis is often to simplify a circuit to a single voltage source and a single resistor. From there, the current that flows through all the series-connected devices can be calculated. After the current has been calculated, then the voltage drops across the individual resistors can be calculated with Ohm's law.

The purpose of circuit analysis skills for a truck technician is to develop electrical troubleshooting skills and not just perform meaningless calculations. Sketching the portion of the electrical system that has a problem and being able to analyze this circuit will assist you in determining the source of the trouble while your co-workers are scratching their heads or avoiding dealing with a truck with an electrical problem that just does not seem to make sense.

Parallel Circuits

Parallel circuits are circuits that have more than one path for flow. Hydraulic circuits with parallel paths have some of the same characteristics as parallel electric circuits.

Parallel Hydraulic Circuit. Hydraulic circuits can also be plumbed so that two or more paths will be provided for fluid to flow through, as in **Figure 2-46**. This type of arrangement is referred to as a parallel circuit be-cause the two restrictions and associated plumbing or hoses are parallel to each other. The restrictions are labeled *restriction 3* and *restriction 4*. The two re-strictions in this circuit have different diameters, like the two restrictions connected in series shown in earlier examples. In this example, restriction 4 is more re-strictive (has a smaller diameter) than restriction 3. It should be obvious that more fluid is going to flow through restriction 3 than restriction 4 if fluid under the same pressure—50 psi (345 kPa) in the example shown in **Figure 2-46**—is applied to both restrictions.

The pressure dropped across both restrictions will be 50 psi because both gauges A and B indicate 50 psi and gauge C at the bottom indicates 0 psi. It is im-portant to note that the pressure drop across all parallel restrictions is the same even though they may have different diameters. This is different from series hy-draulic circuits in which the amount of pressure dropped across the restriction depended on the diam-eter of the restriction opening.

If a flow meter were inserted in each of the hydraulic hoses shown in **Figure 2-46**, the flow meter might in-dicate that 10 gallons of fluid leave the pump per min-ute. In this example, restriction 3 might permit 8 GPM to flow, while restriction 4 may only permit 2 GPM to flow. This is also different from the series hydraulic circuit in which the flow rate of fluid is the same ev-erywhere in the circuit regardless of restriction size. The flowing fluid recombines after flowing through the

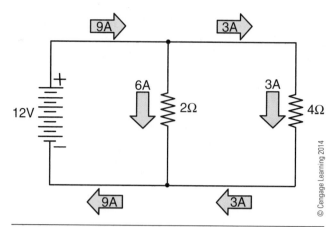

Figure 2-47 Parallel electric circuit: the voltage dropped across unequal resistances is the same, but the current flow through each resistor is different.

parallel restrictions and returns to the tank. The total flow rate of fluid that returns to the tank is 10 GPM.

Parallel Electric Circuit. A parallel electric circuit has some of the same properties as a parallel hydraulic cir-cuit. A parallel electric circuit is shown in **Figure 2-47**. This circuit has two paths for current to flow through, like the parallel hydraulic circuit. Some of the electrons will flow through the 2-ohm resistor; some will flow through the 4-ohm resistor, shown in **Figure 2-47**.

Unlike the series electrical circuit, the voltage dropped across any of the parallel-connected resistors will always be the same (12V in this case), in the same way as the equal pressure drop across parallel hy-draulic circuit restrictions. Also unlike a series circuit, the current is not the same everywhere in a parallel circuit. Ohm's law indicates that the current through the 2-ohm resistor will be 6A and the current through the 4-ohm resistor will be 3A. This is similar to the hydraulic flow being different thorough unequal re-strictions in a parallel hydraulic circuit.

Like the flow rates in the parallel hydraulic circuit shown in **Figure 2-46**, the current leaving the battery and the current returning to the battery will be the same in a parallel electrical circuit like that shown in **Figure 2-47**. Because 6A is flowing through one branch of this parallel circuit and 3A is flowing through the other branch, the total current in the circuit is 9A. This is like the 10 GPM shown in **Figure 2-46**. This leads to another important fact:

Important Fact: The sum of currents flowing in each branch of a parallel circuit is equal to the total current that is supplied by the voltage source.

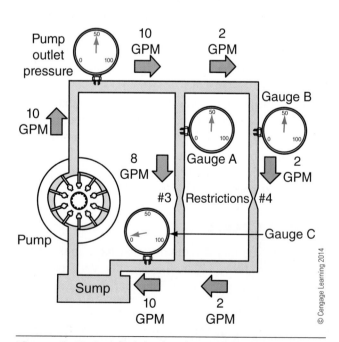

Figure 2-46 Parallel hydraulic circuit: the pressure dropped across unequal restrictions is the same, but the flow through each restriction is different.

This is taken from **Kirchhoff's current law**, which is also known as Kirchhoff's first law. Kirchhoff's current law is another of the fundamental laws of electricity, like Ohm's law. Adding the amount of current flowing in each parallel-connected resistor will equal the same amount of current that is being supplied by the battery. Additionally, the amount of current that leaves the battery-positive terminal must be the same amount of current that returns to the battery-negative terminal, assuming conventional current flow. This is like the flow of fluid in the simple hydraulic circuits in which the amount of fluid that leaves the pump is the same amount that returns to the sump. Kirchhoff's current law is discussed in further detail in the Extra for Experts section at the end of this chapter.

A parallel electric circuit with which you are probably familiar is house wiring. The wall sockets in a house are connected in parallel. In North America, most wall sockets have a voltage of about 120V AC. Thus, a device plugged into any wall socket in the house will be supplied with 120V AC.

The lights in your car such as the headlamps are also wired in parallel. The voltage measured across each headlamp is 12V. If the headlamps were wired in series instead of parallel, each headlamp would have only 6V dropped across it because the headlamps are equal resistances. This would result in very dim headlamps. In addition, if headlamps were wired in series and if one burned out, neither headlamp would work because the flow of current would be interrupted by the burned-out headlamp.

Total Parallel Circuit Resistance. Finding the total value of resistances that are connected in parallel is a little more difficult than finding the total resistance in series circuits. In a series circuit, the total circuit resistance is always *greater* than the *largest* resistor because total series resistance is found by adding all series resistors together. In a parallel circuit, total circuit resistance is always *less* than the *smallest* resistor. This may not make sense without looking at the parallel hydraulic circuit again. If additional parallel restrictions were added and the pump pressure were regulated to some value, the amount of total fluid flow would increase each time another path for fluid flow were added, regardless of the size of the restriction. This says that the total restriction to fluid flow must be decreasing (becoming less restrictive) each time another path for fluid flow is added because more fluid is leaving the pump and returning to the sump as more parallel paths for hydraulic flow are added.

This leads to an important fact regarding parallel electric circuits:

Important Fact: In a parallel electrical circuit, adding additional parallel resistors decreases the total circuit resistance resulting in an increase in the total circuit current.

If you have ever operated too many home appliances at the same time—such as a toaster, microwave, and hair dryer—you may have already discovered this fact. House wiring and vehicle wiring are protected from excessive current flow by fuses or circuit breakers. Circuit protection devices like fuses and circuit breakers will be addressed in **Chapter 4**. The higher the current flow through a wire, the higher the temperature of the wire. If you plug too many devices into wall outlets protected by the same circuit breaker or fuse, the circuit breaker or fuse opens up the electrical circuit to protect the wiring from getting hot enough to start a fire. Ohm's law would indicate that if the voltage were fixed (120V AC) and the amount of current goes up as evidenced by the open circuit breaker or fuse, then the total circuit resistance must be decreasing as additional appliances are plugged into wall outlets. Thus, plugging more devices into parallel-connected outlets protected by the same circuit breaker causes the current to increase, indicating that the total circuit resistance has decreased.

To find the total value of resistance in a parallel circuit with two parallel-connected resistors, a rule known as the product over the sum rule can be utilized. This rule says to multiply the resistors together (product) and divide this value by the sum of the two resistors. If there is a 2-ohm and a 4-ohm resistor connected in parallel, such as **Figure 2-47**, the total resistance would be calculated as shown in **Equation 2-10**.

$$(2 \times 4)/(2 + 4) = 8/6$$
or approximately 1.3 ohms

Note that the 1.3 ohms of total resistance is smaller than the smallest resistor in the parallel circuit, the 2-ohm resistor.

If both parallel-resistor values are the same value, a shortcut can be used to find total resistance by dividing the resistor value in half. For example, two 10Ω resistors connected in parallel have a total resistance of 5Ω. This shortcut can also be applied when there are more than two parallel-connected resistors of identical value. Divide the value of one of the resistors by the number of resistors to find total parallel resistance.

For example, three 12Ω resistors connected in parallel have a total resistance of 4Ω.

When there are more than two parallel resistors of unequal value, finding the total resistance is a little more difficult. The technique is addressed in the Extra for Experts section at the end of this chapter.

....

Tech Tip: You have likely heard the statement "electricity always takes the path of least resistance." This statement has resulted in much confusion for students of electricity when it comes to parallel circuits. This statement is often interpreted to mean that all the electricity (current) will flow through the smaller of two unequal-value resistors that are connected in parallel. This is not correct. If a 1Ω resistor was placed in parallel with a 1,000,000Ω resistor and the parallel combination was connected across a battery, current would flow through both resistors. The misleading statement can be restated as "the majority of the electric current always takes the path of least resistance."

....

Series-Parallel Circuits

As the name implies, **series-parallel circuits** are combinations of series and parallel circuits. There is not a lot new to learn here; it is just necessary to treat each piece of a series-parallel circuit as a separate series or parallel circuit and combine the results.

In a series circuit, the current is the same everywhere in the circuit and the voltage divides proportionally to the resistance (higher-value resistors have a greater voltage drop). In a parallel circuit, the voltage dropped across each component is the same, but the current through each parallel device depends on the resistance. The greater the resistance value, the lower the current flow through that branch of the circuit.

In the circuit shown in **Figure 2-48**, there are three 4-ohm resistors connected in a series-parallel arrangement. The task is to find the voltage drops across each resistor and the current flow through each resistor. One way to find these values is to convert this circuit to a single-resistor, single-voltage-source circuit on paper and solve for the total circuit current first. The total circuit current is the current that leaves and returns to the battery.

The first step is to find the total resistance. To find the total resistance, it is necessary to find the **equivalent resistance** that could replace all of the parallel-connected resistors and still cause the same amount of

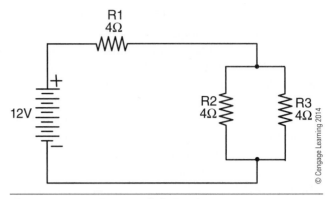

Figure 2-48 Series-parallel circuit.

current to leave and return to the battery. Equivalent resistance is a single resistor that has the same value as a group of other resistors regardless of how the group of resistors is connected. If you were a battery, you could not determine whether you were connected to a single resistor or to a group of resistors with an equivalent resistance of the same value as the single resistor.

This single resistor representing the equivalent resistance of the circuit does not physically replace the group of resistors. The equivalent resistance is only used in calculations to simplify the circuit. Since both parallel resistors identified as R2 and R3 have the same resistance value of 4Ω, the one-half rule can be used, which provides an equivalent resistance value of 2Ω for the two parallel-connected resistors. The circuit can then be redrawn so that the parallel resistors are replaced with a single 2Ω resistor, as shown in the top drawing of **Figure 2-49**.

The former series-parallel circuit now consists of a series circuit consisting of a 4Ω and a 2Ω resistor. Adding these series resistors together yields a total circuit resistance of 6Ω as shown in the lower drawing of **Figure 2-49**. The equivalent resistance of this circuit is 6Ω. The circuit has now been reduced to a single-resistor, single-voltage-source circuit.

Now that the total resistance is known, the total circuit current can be found using Ohm's law. Ohm's law indicates that if the total voltage is divided by the total resistance, the total current flow in the circuit will be provided. In this example, dividing 12V by 6Ω yields 2A of total circuit current. This 2A of total current is the value of current that leaves and returns to the battery.

Now that total current is known to be 2A, the principles of electricity for series and parallel circuits can be applied to each portion of the circuit. Because the total current is 2A, and the single 4Ω resistor identified as R1 is in series with the battery, 2A must

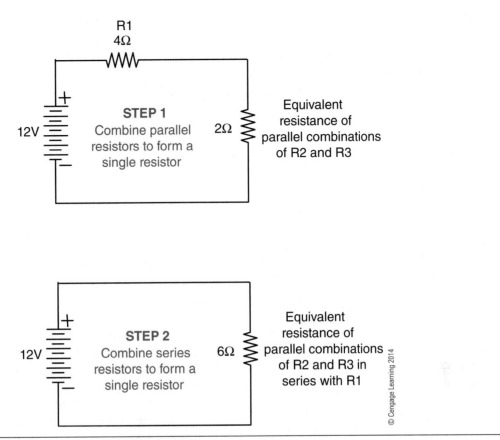

STEP 1
Combine parallel resistors to form a single resistor

Equivalent resistance of parallel combinations of R2 and R3

STEP 2
Combine series resistors to form a single resistor

Equivalent resistance of parallel combinations of R2 and R3 in series with R1

Figure 2-49 Series-parallel circuit simplification progression: goal is a single-resistor circuit.

be flowing through R1. The voltage dropped across R1 can then be found using Ohm's law. Ohm's law indicates that if the current flow through a resistor is multiplied by the resistance value, the voltage dropped across the resistor will be determined. Thus, multiplying 4Ω by 2A results in 8V being dropped across resistor R1.

The voltage dropped across R1 is 8V. This only leaves 4V remaining from the 12V supplied by the battery for the parallel combination of R2 and R3 per Kirchhoff's voltage law. Because the voltage dropped across parallel resistors is the same, the voltage dropped across both R2 and R3 will be 4V.

The current through R2 can then be found by using Ohm's law again. Ohm's law indicates that dividing the voltage dropped across a resistor by the resistance value will yield the current flow through the resistor. Therefore, dividing the 4V dropped across the 4Ω resistor R2 indicates that 1A of current is flowing through R2.

The current through R2 is 1A. The equation could be solved again for R3 to find the current flow through R3. This would yield 1A flowing through R3 as well because it has the same resistance value and the same voltage dropped as R2. The current through R3 also could have been found by applying Kirchhoff's current

law. Because 2A is flowing into the parallel-connected R2 and R3 resistors, a total of 2A must also be flowing out of these resistors. The current through R2 was calculated as 1A, so this leaves 1A of current to flow through R3 as well because they have the same resistance value (4Ω). The solved circuit is shown in **Figure 2-50**.

There are several other different ways to solve for all the voltage and currents in this circuit, but each method should provide the same answers. Otherwise, you have made a math mistake or are not following all the rules of series and parallel circuits and Ohm's law.

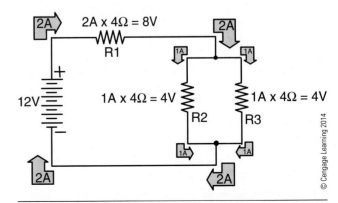

Figure 2-50 Solved series-parallel circuit.

As a troubleshooter, there will probably not be too many times that you need to calculate voltage drops for series-parallel circuits. However, the fundamentals of circuit analysis, including Ohm's law and Kirchhoff's voltage and current laws, will assist you in pinpointing an electrical problem as opposed to replacing components until the problem is solved.

METRIC PREFIXES

Metric prefixes are used extensively in electricity and electronics. A metric prefix is added to the front of a unit of measure that acts as a power of ten multiplier. For example, the prefix "kilo" added to the front of the metric unit of measure of length "meter" yields a kilometer. Kilo (k) indicates 1000 or 10 raised to the third power (10 × 10 × 10). Thus, 10,000 m would be 10 km. A computer hard drive may have 10,000,000,000 bytes or 10 billion bytes of storage space. Instead of writing all the zeros, the hard drive would be described as a 10-gigabyte hard drive.

Metric prefixes can describe very small things as well. For example, the prefix milli means 1/1000th. A millimeter is 1/1000th of a meter. One millimeter is abbreviated as 1 mm. Common metric prefixes you will probably see when working with truck electrical systems are shown in **Figure 2-51**, along with abbreviations and examples. The symbol μ (mu) is a Greek alphabet character and indicates micro or 1 millionth.

Because very large or very small values of voltage, current, and resistance are common when working with electricity, metric prefixes minimize decimal point errors and reduce the length of numbers by eliminating zeros before or after the decimal point. For example, the resistance of a typical antilock braking

system (ABS) wheel speed sensor found on a truck may be 1500Ω. This resistance may be identified in troubleshooting guides for the ABS system as 1.5kΩ. This means 1.5Ω × 1000 or 1500Ω. An ohmmeter that is measuring the resistance of a wheel speed sensor may display the answer as 1500 or 1.5 k, depending upon the meter and the meter settings. The current that flows through a single instrument panel illumination lamp may only be about 0.01 A or 1/100th of an amp. This current could be described using metric prefixes as 10 milliamps or 10 mA. The prefix *milli* indicates 10 × 1/1000th of an amp. Once again, an ammeter that is measuring this current may indicate 0.010 amps or 10 mA, depending on the meter and the meter settings.

When working with electricity, an answer that is only wrong by one zero or one decimal point position is an answer that is either 1/10th or 10 times the correct answer. Such a large measurement error would probably lead to an incorrect diagnosis of a problem. This one decimal point error becomes very evident when dealing with money. A paycheck that is wrong only by one decimal point position will make either a very happy or a very upset technician. A paycheck that is supposed to be $1041.90 becomes either $10,419.00 or $104.19 when it is wrong by just a decimal point position. Practice with metric prefixes is especially important for students who are not familiar with the metric system.

EXTRA FOR EXPERTS

This section provides additional information that you should find useful after you have mastered basic electricity.

Finding Total Parallel Resistance with Multiple Resistors

The reciprocal of a number is a math operation used to find the total value of more than two parallel resistors. The reciprocal of a number is 1 divided by the number. For example, the reciprocal of 10 is 1/10 or 0.1. To find total resistance of resistors in parallel, find the reciprocal of each resistor, using a calculator, and add these reciprocals together. This value is the sum of reciprocals. Next, the reciprocal of the sum of these reciprocals is then determined, which is the total parallel resistance (confused yet?). For example, if a 10Ω, a 20Ω, and a 30Ω resistor are all connected in parallel, as shown in **Figure 2-52**, the total resistance would be found by first finding the reciprocal of each of the three resistors. This is done by dividing 1 by 10, dividing 1 by 20, and dividing 1 by 30. These three

PREFIX	SYMBOL	RELATION TO BASIC UNIT
Mega	M	1,000,000
Kilo	k	1,000
Milli	m	0.001 or $\frac{1}{1000}$
Micro	μ	.000001 or $\frac{1}{1000000}$
Nano	n	.000000001
Pico	p	.000000000001

© Cengage Learning 2014

Figure 2-51 Common metric prefixes used in electrical system troubleshooting.

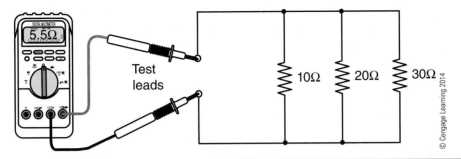

Figure 2-52 Three resistors in parallel: note the measured resistance value is less than any of the parallel-connected resistors.

answers are then added together to find the sum of the reciprocals. Total resistance is then found by dividing 1 by the total, which is the sum of the reciprocals. In this example, the total resistance for the three parallel-connected resistors is about 5.5Ω. Notice that this value is smaller than the smallest single parallel resistance of 10Ω. If your total parallel resistance answer is not smaller than the smallest parallel resistor, you have a mistake in your calculation and need to refigure the problem.

Tech Tip: The reciprocal key (1/x or x^{-1}) found on most calculators is very helpful when performing parallel circuit calculations. See the instructions for your calculator for more information on the use of the reciprocal key.

Additional Digital Multimeter Information

A differential pressure gauge that is measuring the pressure drop across a restriction does not permit any hydraulic fluid to flow through the pressure gauge. If fluid did flow through the pressure gauge, it would increase the hydraulic flow rate and would decrease the pressure drop across the restriction. A good test tool, such as a pressure gauge, does not change the system that is being tested.

A voltmeter measures the voltage (electrical potential) between two different points, similar to a differential pressure gauge in a hydraulic circuit. A DMM has a very high resistance between the meter's voltage measurement terminal and common terminal. This high resistance is actually contained inside the meter's internal circuitry and is described as being the internal resistance of the meter. The internal resistance of a typical digital voltmeter is 10 million (10 Mega) ohms or more. The high internal resistance permits only a very small amount of current to flow through the

voltmeter. This is like a differential pressure gauge permitting no fluid to flow through the gauge.

Placing a voltmeter across or in parallel with two different points in a circuit to obtain a voltage measurement does not change by very much the total current flow in the circuit or the voltage drop across the points being measured. This is because the meter's high internal resistance is placed in parallel with the points in the circuit where voltage drop is to be determined. Placing a 10-Mega-ohm resistor (the meter's internal resistance) in parallel with a 100-ohm resistor has almost no impact on the total resistance. The total resistance would still be about 100 ohms.

The high internal resistance of a voltmeter is also why a voltmeter is not typically connected in series with electrical components in the circuit. The rules of series circuits indicate that the voltage dropped across components that are connected in series is proportional to the resistance. The 10-Mega-ohm internal resistance of the voltmeter would normally be much larger than any other component in the circuit; thus, almost all the voltage is dropped across the voltmeter's internal resistance. Some of the problems associated with using a high internal resistance voltmeter will be discussed in more detail in later chapters.

Flow meters, which are used to measure the flow of some fluid, are placed directly in the path of flow. Because the flow meter is directly in the path of flow, it does cause some restriction to the flow of the fluid. However, a flow meter is designed to cause only a minimal amount of restriction. Any restriction caused by the flow meter affects the accuracy of the flow measurement.

Like a flow meter, an ammeter is also placed directly in the path of current flow. All ammeters have an internal resistance. Unlike the internal resistance of a voltmeter, the ammeter's internal resistance is designed to be as low as possible (fractions of an ohm). Circuit current passes directly through the ammeter so the internal resistance of an ammeter placed in series with the circuit reduces the flow of current in a circuit by only a small amount. This is the reason that a DMM has separate positive (+) input terminals for amperes and voltage. However, both the voltmeter and the ammeter of a DMM typically share a common negative (–) input terminal.

Because the internal resistance of an ammeter is very low, an ammeter is never used to measure the circuit current across or in parallel with a component as is performed with a voltmeter. Doing so will cause most of the current flow to bypass the device in the circuit and pass through the low-resistance ammeter instead, which will possibly damage the meter. Ammeters have a replaceable fuse so that a misconnection of the ammeter will (it is hoped) only blow the ammeter fuse before any damage is done to the ammeter or circuit. Fuses will be discussed in greater detail in **Chapter 4**. For now, a fuse can be thought of as the electrical equivalent of a safety valve.

A second input and scale on some high-quality DMMs also permits accurately measuring very small currents.

Kirchhoff's Laws

Ohm's law is useful for solving simple series and parallel circuits with one voltage source. However, in circuits that are more complex other circuit analysis tools such as Kirchhoff's laws are necessary along with Ohm's law to solve for unknown voltage drops and currents.

Kirchhoff's voltage law states that the algebraic sum of the voltage sources and the voltage drops in a closed circuit must equal zero. An algebraic sum means that negative numbers and positive numbers will be added together following the rules of addition in algebra. **Equation 2-11** illustrates an example of an algebraic sum.

$$-5 + 3 + 2 = 0$$

As indicated previously, Kirchhoff's voltage law is behind the Important Fact stating that the sum of all of the voltage drops in a series circuit is equal to the source voltage. This statement is true when the circuit has one voltage source and all resistors are connected in series.

Kirchhoff's current law states that the algebraic sum of the currents entering and leaving a point must equal zero. Kirchhoff's current law is behind the Important Fact stating that the sum of currents flowing in each branch of a parallel circuit is equal to the total current that is supplied by the voltage source.

Kirchhoff's current law is illustrated in **Figure 2-53**. The arrows indicate that 4A is flowing through R_1 and

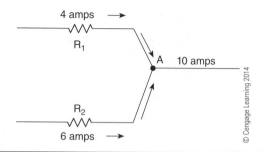

Figure 2-53 Kirchhoff's current law: the sum of the currents entering point A is the same amount of current that leaves point A.

6A is flowing through R_2. Both currents combine at point A to form 10A. Currents flowing into a point are designated as positive currents; currents flowing out of a point are designated as negative currents. The Kirchhoff's current law equation for **Figure 2-53** is shown in **Equation 2-12**.

$$+4A + 6A + (-10A) = 0A$$
$$\text{or}$$
$$4A + 6A - 10A = 0A$$

Figure 2-54 shows a more complex series-parallel circuit. To make the math simpler and eliminate decimals and fractions, a voltage of 120V is used instead of the typical 12V found in other examples in this chapter. Conventional current flow from positive to negative is illustrated in these examples. Arrows and polarities would be the opposite if electron flow were utilized. It really does not matter which theory you use though, as long as you remain consistent in your choice.

If we were to simplify this circuit to a single resistor in series with a voltage source, we would find that the total circuit resistance is 60 ohms. Therefore, the total circuit current would be 2A. You should apply what you already know about circuit analysis to verify these values. In **Figure 2-54**, 2A is shown leaving point A of the circuit. This current passes through R_1 and then splits into two currents at point B of the circuit. There is 0.8A of current flowing through the branch containing R_2 while 1.2A flows through the branch containing R_4, R_5, and R_6. If we apply Kirchhoff's current law to point B of the circuit, **Equation 2-13** shows the result.

$$+2A - 0.8A - 1.2A = 0A$$

At point E in **Figure 2-54**, 0.8A is entering point E from the R_2 branch along with 1.2A from the R_4, R_5, R_6 branch. These two currents join together to pass through R_3 and return to the battery's negative terminal.

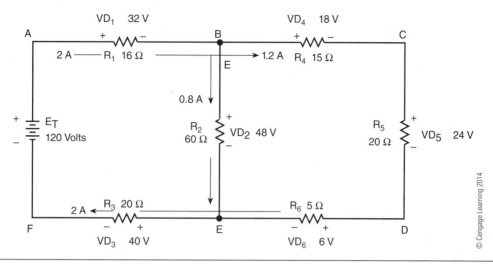

Figure 2-54 Kirchhoff's current law and voltage law combined.

Applying Kirchhoff's current law to point E results in **Equation 2-14**.

$$+0.8A + 1.2A - 2A = 0A$$

Kirchhoff's voltage law states that the algebraic sum of the voltage sources and the voltage drops in a closed circuit must equal zero. In **Figure 2-54**, we will designate the end of a resistor where conventional current flow enters as positive and the other end of the resistor as negative. The voltage drops (VD) across each resistor in **Figure 2-54** are indicated as VD_1, VD_2, etc. along with the polarity based on conventional current flow. The polarity of the voltage source designated as E_T is also shown.

The circuit shown in **Figure 2-54** has three separate closed loops. Loop ACDF contains voltage drops VD_1, VD_4, VD_5, VD_6, VD_3, and source voltage E_T. Applying Kirchhoff's voltage law to this loop results in **Equation 2-15**.

$$+VD_1 + VD_4 + VD_5 + VD_6 + VD_3, -E_T = 0$$
$$\text{or}$$
$$+32V + 18V + 24V + 6V + 40V - 120V = 0V$$

The reason that E_T is shown with a negative sign and each of the resistor voltage drops are shown with a positive sign is that as we pass through the loop using conventional current flow after exiting the battery's positive terminal, we first encounter positive signs for each of the resistors. When we make it back to the battery's negative terminal, on the way back to point A we first encounter a negative sign of the voltage source, thus the reason for the polarities indicated in **Equation 2-15**.

The Kirchhoff's voltage law equation for closed loop ABEF in **Figure 2-54** is shown in **Equation 2-16**.

$$+32V + 48V + 40V - 120V = 0V$$

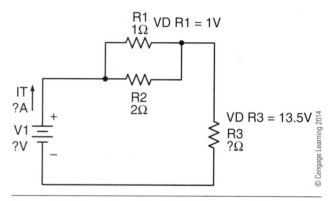

Figure 2-55 Applying Kirchhoff's laws and Ohm's law to find unknown values.

Putting Ohm's law and Kirchhoff's laws together with the rules of series and parallel circuits, you should now be able to demonstrate all that you have learned. In **Figure 2-55**, the voltage dropped across R1 is 1V and the voltage dropped across R3 is 13.5V. The source voltage V1, total current I_T, and the resistance of R3 are all unknown. Therefore, the task is to find these unknown values. This could be done in a variety of ways, but let us find I_T first. Since R1 and R2 are in parallel, the voltage dropped across R2 is also 1V. Therefore, the current through R2 is 0.5A and the current through R1 is 1A, as indicated by Ohm's law. Kirchhoff's current law indicates that the total circuit current I_T is the sum of these two branch currents, which is 1.5A.

Now that we have the total current I_T, we can find the resistance of R3 using Ohm's law. Therefore, R3 is 9Ω. For the source voltage V1, apply Kirchhoff's voltage law, which indicates that V1 is the sum of the two voltage drops, which is 14.5V.

Summary

- A difference in electric potential between two points causes electrons to flow between the two points when a conductive path is provided between them. Voltage is electrical potential and is measured in units of volts. Voltage can be thought of as being like pressure in a hydraulic or pneumatic circuit.

- The flow of electrons is called current and is measured in units of amperes (amps). An ampere describes a quantity of electrons passing a stationary point per second. Electric current is like the fluid that flows in a hydraulic circuit.

- The opposition to electric current flow is called resistance and is measured in units of ohms.

- Ohm's law mathematically describes the relationship among voltage, current, and resistance. Ohm's law indicates that dividing the value of the voltage dropped across a resistance by the value of the resistance in ohms yields the current flowing through the resistance in amperes. Ohm's law can be manipulated so that when any two quantities are known (voltage, current, or resistance), the third unknown quantity can be calculated.

- A voltmeter is like a differential pressure gauge in that the difference in electric potential between two points is being measured. A voltmeter is typically placed in parallel (across) a component to measure the voltage being dropped across the component.

- An ammeter is like a flow meter in that the quantity of electrons passing a stationary point per second is being measured. An ammeter is always placed in series with the portion of the circuit in which the current flow is to be measured. Placing an ammeter in parallel with a component can damage the ammeter.

- An ohmmeter measures resistance and is only used when there are no voltage sources in the circuit.

- In a series electrical circuit, there is only one path for current flow. The current is the same everywhere in a series circuit. The voltage dropped across components in a series circuit is proportional to the component's resistance value relative to other components in the circuit.

- In a parallel electrical circuit, there is more than one path for current flow. The current is divided among each parallel path. The greater the resistance of a path, the less current that flows through that path compared to the other paths. The voltage dropped across each component in a parallel circuit is the same, regardless of resistance values of the components.

- Series-parallel circuits have components that are connected in parallel and components that are connected in series. The voltage drops and branch currents can be found by determining the equivalent resistance of the entire circuit. The equivalent resistance is then used to determine the total circuit current. The rules of series and parallel circuits are then applied to the circuit to determine the voltage drops and branch currents.

Suggested Internet Searches

For more information on electrical tools, search the following web sites:
http://www.fluke.com
http://www.esitest.com

Review Questions

1. Which of the following would *not* be a correct way of referring to voltage?

   A. Electrical potential

   B. emf

   C. Coulombs per second

   D. Volts

2. Technician A says that the best way to measure current through a resistor is to place the ammeter in parallel with the resistor. Technician B says that an ammeter should never be placed directly in the path of current flow. Who is correct?

   A. A only

   B. B only

   C. Both A and B

   D. Neither A nor B

3. Which of the following is the correct method to measure voltage dropped across a component?

   A. Place the voltmeter in series with the component such as a resistor that the voltage is to be measured across.

   B. Place the voltmeter in parallel with the component such as a resistor that the voltage is to be measured across.

   C. Place one lead of the ammeter on the positive battery terminal and the other lead on the negative battery terminal.

   D. Use an ammeter to measure the current flowing in parallel with the component and apply Ohm's law to determine the voltage drop.

4. Match the electrical terms to their nearest hydraulic counterpart:

   A. Voltage

   B. Current

   C. Resistance

   D. Conductor

   1. Restriction

   2. Pressure

   3. Flow rate

   4. Hose

5. Technician A says that an ohmmeter should only be used in a circuit that does not have a voltage source. Technician B says that current can be measured by determining the voltage that is dropped across a known resistance and applying Ohm's law to solve for the current. Who is correct?

   A. A only

   B. B only

   C. Both A and B

   D. Neither A nor B

6. Two resistors and a 12V battery are all connected in series, resulting in a current flowing through the resistors. Which of the following statements is true?

   A. If the resistors are of equal value, then the resistor closest to the battery-positive terminal will have a larger voltage drop.

   B. If the resistors are of unequal value, more current will flow through the smaller resistor.

   C. If the resistors are of equal value, 6V will be dropped across one of the resistors.

   D. The amount of current that leaves the battery-positive terminal will be much greater than the current that returns to the battery-negative terminal.

7. A technician measures the voltage dropped across three resistors that are connected in parallel with each other and are connected across the terminals of a 12V battery. Which of the following statements is true?

   A. The largest parallel resistor has a higher voltage drop than the other two resistors.

   B. If the resistors are of unequal values, the same amount of current will flow through each resistor.

   C. The sum of all of the currents flowing through the three resistors is the same amount of current that leaves the battery-positive terminal.

   D. The total circuit resistance can be found by adding all three resistors together.

8. Technician A says that in a series circuit, adding together all of the individual voltage drops should produce a sum equal to the source voltage. Technician B says that in a circuit in which a battery is supplying current, the amount of current that leaves the battery-positive terminal must be the same as the amount of current that returns to the battery-negative terminal. Who is correct?

   A. A only

   B. B only

   C. Both A and B

   D. Neither A nor B

9. Which of the following is *not* a true statement?

    A. Current flow occurs because of a difference in electrical potential (voltage).

    B. Increasing the total resistance in a circuit without a change in the voltage supplied to the circuit causes the total current flow in the circuit to decrease.

    C. An increase in the total circuit current without a change of total circuit resistance indicates that the voltage supplied to the circuit has increased.

    D. The total resistance of a circuit that has one resistor will increase if additional resistors are connected in parallel with this resistor.

10. Which of the following is a true statement?

    A. The current is the same everywhere in a series-parallel circuit.

    B. The voltage dropped across each component is always the same in a series-parallel circuit.

    C. Truck headlamps are typically connected in parallel.

    D. Devices that are connected in parallel always have the same amount of current flow through each device.

11. Solve for the following values shown in **Figure 2-56**:

    A. Voltage dropped across R1

    B. Voltage dropped across R2

    C. Current through R1

    D. Current through R2

    E. Total circuit current

    F. Total circuit resistance

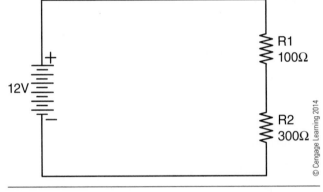

Figure 2-56

12. Solve for the following values shown in **Figure 2-57**:

    A. Voltage dropped across R1

    B. Voltage dropped across R2

    C. Current through R1

    D. Current through R2

    E. Total circuit current

    F. Total circuit resistance

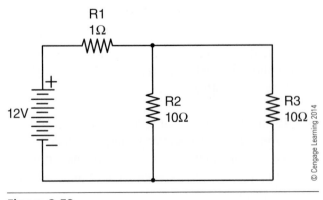

Figure 2-57

13. Solve for the following values shown in **Figure 2-58**:

    A. Voltage dropped across R1

    B. Voltage dropped across R2

    C. Voltage dropped across R3

    D. Current flow through R1

    E. Current flow through R2

    F. Current flow through R3

    G. Total circuit current

    H. Total circuit resistance

Figure 2-58

14. Find the following shown in **Figure 2-59**:

    A. Current through R1

    B. Current through R5

    C. Voltage dropped across R2

    D. Current through R2

    E. Current through R3

    F. Voltage dropped across R4

    G. Total circuit current

    H. Total circuit resistance

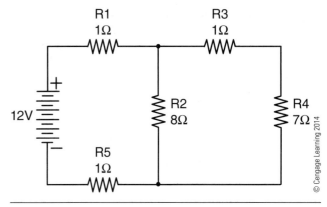

Figure 2-59

15. Referring to **Figure 2-60**, technician A says that since R2 is the path of least resistance, current will only flow through R2. Technician B says that the voltage dropped across R1 and R2 would both be 5V because R1 and R2 are in series with each other. Who is correct?

    A. A only

    B. B only

    C. Both A and B

    D. Neither A nor B

16. Find the following for **Figure 2-60**:

    A. Current through R1

    B. Current through R2

    C. Voltage dropped across R2

    D. Total circuit current

    E. Total circuit resistance to six decimal places

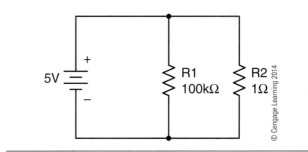

Figure 2-60

17. Find the following for the circuit shown in **Figure 2-61** (hint: redraw diagonal resistor circuits with all vertical lines if shape is confusing):

    A. Current through R1

    B. Current through R3

    C. Voltage dropped across R2

    D. Voltage dropped across R3

    E. Total circuit current

    F. Total circuit resistance

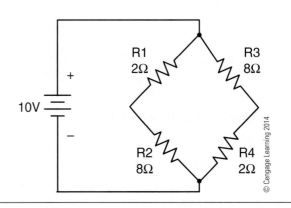

Figure 2-61

18. Referring to **Figure 2-62**, the voltage dropped across R1 is 5V. The source voltage V1 and total circuit current IT is unknown. Find the following (hint: find current flow through R1 first):

    A. Source voltage V1

    B. Total circuit current IT

    C. Voltage dropped across R2

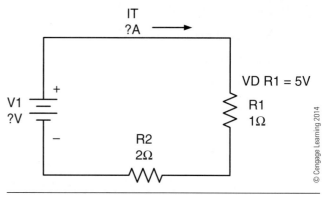

Figure 2-62

19. Referring to **Figure 2-63**, the voltage dropped across R3 is 9V. Find the following:

    A. Source voltage V1 (hint: solve for voltage dropped across R2 first and then use Kirchhoff's voltage law to find V1)

    B. Current flow through R1 using Kirchhoff's current law

    C. Resistance of R1

    D. Resistance of R3

    E. Current flow through R3

    F. Total circuit resistance to three decimal places

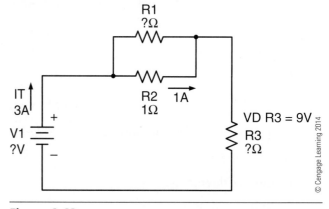

Figure 2-63

20. Referring to **Figure 2-64**, the voltage dropped across R1 is 2V and the voltage dropped across R3 is 10V and the total circuit current IT is 2A. Find the following:

    A. Source voltage V1.

    B. Current flow through R2 (hint: resistances connected in parallel have the same voltage drop)

    C. Resistance of R1 (hint: find current through R2 first and then use Kirchhoff's current law to find current through R1, and use Ohm's law to solve for R1)

    D. Current flow through R3

    E. Resistance of R3

    F. Total circuit resistance

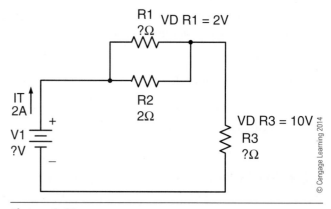

Figure 2-64

3 Physics for Electricity

Learning Objectives

After studying this chapter, you should be able to:

- Use Watt's law to solve for electric power.

- Discuss the concepts of electric fields and magnetic lines of force.

- Explain how an electromagnet is formed.

- Put into your own words how a voltage is induced in a conductor by movement through a magnetic field.

- Describe how an inductor and a capacitor store energy.

- Discuss how a negative voltage spike is generated when current flow through an inductor ceases.

- Explain the similarities between a capacitor and an accumulator used in a hydraulic system.

Key Terms

capacitor	Lenz's law	right-hand rule
counter emf (CEMF)	magnetic lines of force	self-inductance
dielectric	negative voltage spike	suppression
electric fields	nonlinear	transformer
electromagnet	oscilloscope	watt
energy	potential energy	Watt's law
induced	power	work
inductor	reluctance	

INTRODUCTION

This chapter provides an introduction to physics as it relates to truck electrical systems. The important concept of magnetism and its relationship to electric current will be discussed in this chapter. A basic understanding of magnetism will be useful for understanding the operation of devices such as electric motors, generators, and diesel fuel injectors. Electric power will also be discussed.

ELECTRIC POWER

If your car has ever run out of fuel, you may have pushed your car to the nearest service station. In doing this, you have performed **work**. The physics definition of work is force that results in movement in the direction of the applied force. Multiplying the force exerted in the direction of movement times the distance moved yields the amount of work that has been performed. Force in the English system is measured in pounds (lb) and distance is measured in feet (ft). Therefore, the unit of work in the English system is pounds-feet (lb·ft). In the metric system (SI), force is measured in newtons and distance is measured in meters. Therefore, the metric unit of work is the newton-meter. The newton-meter is also called a joule (J). One lb·ft of work is equal to approximately 1.36 J.

The definition of mechanical **power** is the rate at which work is performed; in other words, the amount of work that is performed in a given period of time. It may have taken you 15 minutes to push your car ¼ mile (0.4 km) to the service station. Once you refill the tank, the engine may be able to propel the car over the same ¼ mile in less than 15 seconds. This indicates that the engine is capable of producing substantially more power than your muscles. In the SI system, power is expressed in units of joules per second. One joule per second is called a **watt** (W). Power is also expressed using the English unit horsepower (hp). One horsepower is approximately 746 W.

Like mechanical power, electric power is also measured in units of watts. Electric power indicates the rate at which electric **energy** is being transferred. Energy is defined as the capacity or ability to perform work.

Watt's Law

Electric power is calculated by multiplying the voltage times the amperage, as shown in **Equation 3-1**.

$$\textbf{Watts} = \textbf{Volts} \times \textbf{Amps}$$

This equation is known as **Watt's law**. Watt's law can be manipulated like Ohm's law to find an unknown value when the other two values are given. The other two forms of Watt's law are shown in **Equation 3-2**.

$$\textbf{Volts} = \textbf{Watts/Amps}$$
$$\textbf{Amps} = \textbf{Watts/Volts}$$

A light bulb, like that used in a household lamp, has a wattage rating. This indicates how much heat the light bulb will produce per second. A standard 60W light bulb supplied with 120V AC will draw 0.5A (500mA) of AC current as predicted by Watt's law. A 100W light bulb will draw about 0.83A (830mA) of AC current. The heat dissipated by the 100W bulb is much greater than that given off by a 60W bulb. Household lamp fixtures typically indicate the maximum bulb size

that can be safely used. Using a higher wattage bulb than specified can result in a fire. The importance of selecting the proper wattage of bulb in truck electrical systems will be discussed in **Chapter 9**.

MAGNETISM

Magnetism and electricity are closely related. Magnetism is an important part of the study of electricity because current flow through a conductor results in the production of magnetism. Conversely, magnetism can be used to produce an electric current. Magnetism is also the basis for any electric motor. Magnetism permits the conversion of mechanical energy to electrical energy, such as a generator, and the conversion of electrical energy to mechanical energy, such as a starter (cranking) motor.

Magnetic Fields

In science fiction movies or television space shows, an invisible force field is often used to keep someone from entering an area. The actor walks into this invisible force field and responds as though he walked into a wall. A magnetic field can be thought of as being like this science fiction force field. An invisible magnetic field surrounds a magnet. Every magnet has two distinct poles or ends. These poles are named the north pole and the south pole, like the poles of the earth. You may have performed an experiment in grade school science class in which iron filings were sprinkled on a piece of paper with a bar-shaped magnet beneath the paper, as shown in **Figure 3-1**. Tapping on the paper causes the pieces of iron to align into a pattern. These lines illustrate **magnetic lines of force** or lines of magnetic flux. The pattern shows that the magnetic lines of force are most concentrated or closest together at the north and south poles of the bar magnet.

The invisible magnetic lines of force do not flow or rotate, even though they are nearly always shown with arrows, as illustrated in **Figure 3-2**. The arrows simply indicate the direction in which the north end of a magnetic compass needle would point if the compass were placed in the magnetic field. Since it is not possible to see the magnetic lines of force, visualizing flow or rotation may make it easier to understand the concept. The arrows on the magnetic lines of force always point away from the north pole of the magnet and toward the south pole of the magnet. Therefore, it may be useful to think of the magnetic lines of force as originating at the north pole and "traveling" to the south pole. The arrows on the magnetic lines of force

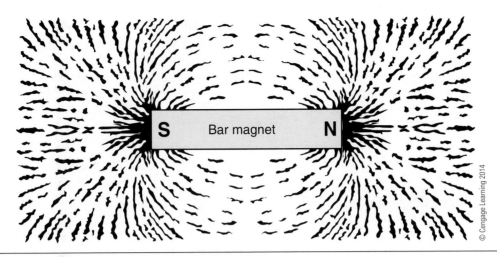

Figure 3-1 Magnetic lines of force illustrated by iron filings.

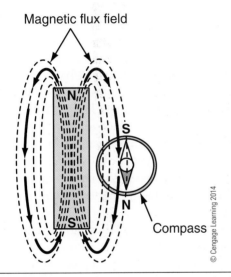

Figure 3-2 Directional arrows on magnetic lines of force indicate direction north end of a compass needle would point when placed in the magnetic field.

also illustrate what happens when magnetic fields from different sources interact with each other.

If the unlike poles of two bar magnets are placed close to each other, the unlike poles of the two magnets (north and south) are attracted to each other, as shown in **Figure 3-3**. This force of attraction increases as the magnets are brought closer and closer together. The magnetic force pulling the magnets together increases by a factor of four each time the distance between the magnets is reduced by a factor of two. If you have ever held two magnets near each other with unlike poles facing each other, you have felt this rapid increase in the force of attraction as you moved the magnets closer and closer to each other. As the magnets near each other, it may become

difficult to keep the magnets from contacting each other. If the force of attraction and the distance between the two magnets were plotted on an *X-Y* graph, the line connecting the points on the graph would be curved. This type of relationship, in which a rapid increase or decrease in one of the values takes place as change occurs at a steady rate, is referred to as a **nonlinear** relationship. The line drawn on an *X-Y* graph that connects the plotted points of a nonlinear relationship is curved. This type of plot with a curved line will be shown in several examples throughout the text.

Reversing one of the magnets so the two like poles are facing each other causes the individual magnetic fields emitted by both magnets to oppose each other, as shown in **Figure 3-3**. This interaction between the two magnetic fields produced by the two magnets causes the magnets to repel each other. This force of repulsion increases rapidly as the magnets are brought closer and closer together at a steady rate. You can feel this force of repulsion as you move like poles of two magnets closer to each other. If the magnets are strong enough, you may not be able to cause the like poles of two magnets to contact each other.

There are two important facts to remember about magnetism:

Important Facts: (1) Like magnetic poles repel each other; unlike magnetic poles attract each other.

(2) The magnetic attraction or repulsion force that exists between two magnets increases at a nonlinear rate as the distance between two magnets decreases.

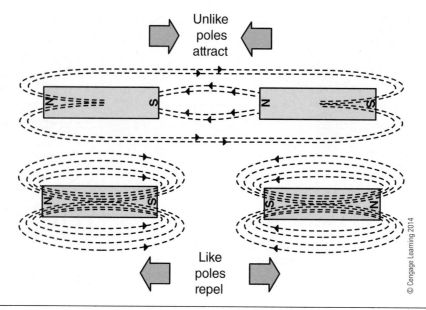

Figure 3-3 Unlike magnetic poles attract; like magnetic poles repel.

Current Flow and Magnetism

Current flowing through a conductor causes a magnetic field to surround the conductor. The magnetic field can be visualized by magnetic lines of force that encircle the conductor, as shown in **Figure 3-4**. The arrows on the magnetic lines of force appear to indicate that the magnetic lines of force are in motion. However, the magnetic lines of force do not rotate, as explained previously. Other than the circular shape, the magnetic lines of force surrounding the conductor are like the oval-shaped magnetic lines of force that are produced by a bar magnet.

If the direction of current flow were reversed in the conductor, the arrows associated with the magnetic lines of force surrounding the conductor would also reverse and point in the opposite direction. To assist in remembering the direction of arrows on magnetic lines of force surrounding a conductor that is carrying

(transporting) current, the **right-hand rule** can be used, provided you are using conventional current flow theory (positive to negative). The right-hand rule indicates that if you grasp a conductor with your right hand so that your thumb points in the direction of conventional current flow, your fingers will be pointing in the direction of the arrows on the magnetic lines of force surrounding the conductor as shown in **Figure 3-5**.

Many of the following illustrations and others in later chapters show the cross-section of a conductor (wire) as a circle. Imagine that a piece of wire has been cut and you are looking directly into the cut end of the wire. The cut end of wire would appear as a circle in two dimensions. To illustrate the direction of current flow in a conductor that is drawn in two dimensions, it is common to illustrate the direction of conventional current flow by means of a dart or arrow like that shown in **Figure 3-6**.

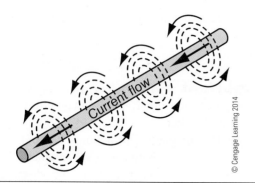

Figure 3-4 Lines of force around a current-carrying conductor.

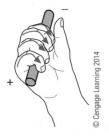

Figure 3-5 Right-hand rule: thumb points in direction of conventional current flow, fingers point in direction of magnetic lines of force surrounding the conductor.

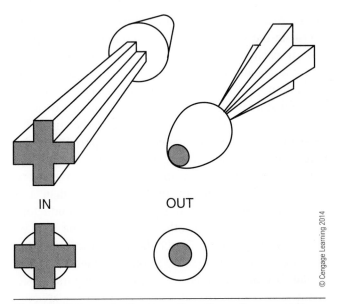

Figure 3-6 Darts indicating direction of current flow.

When a cross-section of a conductor is shown as a circle with a + sign, this indicates that the direction of current flow in the conductor is into the page or into the cross-section of the conductor. The + sign is all you observe of the fins on the dart as it is traveling away from you (**Figure 3-7**). When the cross-section of a conductor is shown with a dot in the center, the direction of conventional current flow is out of the page or out of the cross-section of the conductor. The dot is the tip of the dart traveling toward you.

The strength of the magnetic field surrounding a conductor is directly proportional to the current flow. In other words, the magnetic field gets stronger as current flow is increased. This is the basis of clamp-on ammeters introduced in **Chapter 2**. A clamp-on ammeter measures the strength of the magnetic field caused by current flow and converts this magnetic field measurement into a corresponding current value.

To illustrate a stronger magnetic field, the magnetic lines of force surrounding the conductor are shown as an increased number of lines packed more closely together and extending further from the conductor (**Figure 3-8**).

If two conductors with current flowing in the opposite directions are placed next to each other, the two conductors will be repelled from each other like the north poles of two magnets placed next to each other. The reason that this occurs is the interaction of the magnetic fields surrounding each conductor. In the space between the two conductors that are carrying

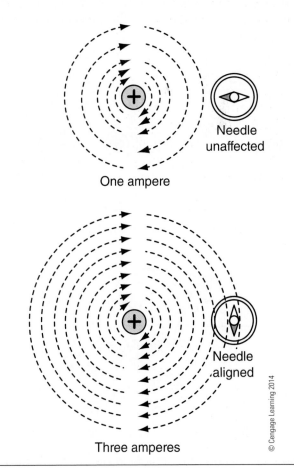

Figure 3-7 Dot and cross used to show direction of current flow looking into a cross-section of a conductor.

Figure 3-8 Magnetic field around conductor with 1 amp of current flow and 3 amps of current flow.

current in opposite directions, the arrows associated with both magnetic fields will be pointing in the same direction, as shown in **Figure 3-9**. In this example, the arrows associated with both fields are pointing in the down direction in the space between the two conductors. Because the arrows on the magnetic lines of force, caused by current flow through the conductors, are pointing in the same direction, the magnetic lines of force combine to form a stronger field in this area between the conductors than exists outside of this area.

Both conductors will be compelled (want) to move away from this strong magnetic field in the space between the conductors to a location where the magnetic field is not as strong. This leads to an important fact:

....

Important Facts: Conductors that are carrying current are compelled (want) to move out of a stronger magnetic field into a weaker magnetic field.

....

If the direction of current flow through conductors placed next to each other is the same direction in both conductors, the two conductors will be attracted to each other. This occurs because the direction of arrows on the magnetic lines of force surrounding each conductor in the space between the conductors are opposite of each other, as shown in **Figure 3-10**. The arrows associated with one conductor's magnetic field are pointing up, while the arrows associated with the other conductor's magnetic field are pointing down. In the space between the two conductors, the magnetic field of one conductor cancels out a portion of the magnetic field of the other conductor. This canceling-out effect results in a net magnetic field in the space between the conductors that is weaker than the magnetic field outside of this area. The conductors are drawn toward each other because of the weaker magnetic field in the space between the two conductors compared to the stronger magnetic field outside of this area between the conductors. Once again, the conductors are moving from an area of stronger magnetic field to an area where the magnetic field is decreased. The important principle is that the interaction of magnetic fields can cause movement of current-carrying conductors. This is the basis of electric motors, as will be discussed in greater detail in **Chapter 8**.

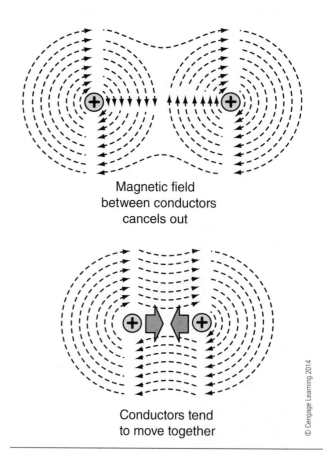

Magnetic field
between conductors
cancels out

Conductors tend
to move together

© Cengage Learning 2014

Figure 3-10 Magnetic field cancels out in the space between conductors with current flow in same direction. This causes the two conductors to move toward each other.

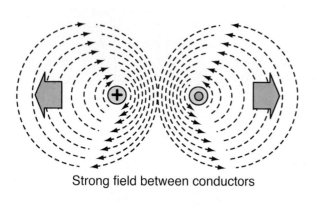

Strong field between conductors

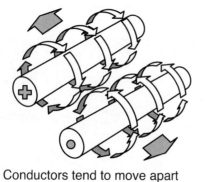

Conductors tend to move apart

© Cengage Learning 2014

Figure 3-9 Strong magnetic field produced in space between conductors with current flow in opposite directions.

Magnetic fields from two or more sources interact with each other to form a single magnetic field. For example, if both conductors shown in **Figure 3-10** were each carrying 10A with the current flow in the same direction, the strength of the magnetic field surrounding the pair of conductors would be the same as a single wire carrying 20A, as shown in **Figure 3-11**. A clamp-on ammeter clamped around two conductors each carrying 10A of current in the same direction would indicate that 20A of current was flowing through the combined conductors. Conversely, if the direction of current flow were reversed in one of the conductors so that both conductors were each carrying 10A of current but in opposite directions, a clamp-on ammeter clamped around the two conductors would indicate 0A. This is because the two magnetic fields cancel each other out, resulting in no net magnetic field.

Forming Wire into Loops

When a wire is formed into a loop and current flows through the wire, a magnetic field surrounds the wire in the same way that a magnetic field surrounds a straight piece of wire (**Figure 3-12**).

However, when several loops of wire are placed next to each other, the magnetic field caused by each loop of wire combines with adjacent loops to form a single magnetic field. Each individual magnetic field caused by each loop of wire adds up with the magnetic field created by each of the other loops of wire to form

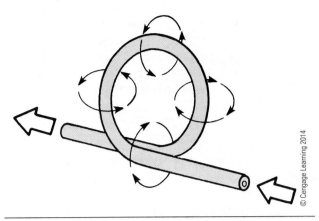

Figure 3-12 Magnetic field around one loop of a conductor.

a stronger magnetic field. The combined magnetic field surrounding a coil appears very similar to the magnetic field that surrounds a bar magnet (**Figure 3-13**). If the direction of current flow through the coiled wire shown in **Figure 3-13** were reversed, the north and south poles of this electrically produced magnet would reverse as well.

Directing Lines of Force

Magnetic lines of force pass through some materials more easily than other materials. This is similar to electric current, which also flows through some materials easier than other materials. Material that can be magnetized—such as iron, nickel, and steel—provides a path of low magnetic opposition for magnetic lines of force. Nonmagnetic materials such as aluminum, copper, and air have much greater opposition to magnetic lines of force than magnetic materials. Opposition to the setting up of or "flow" of magnetic lines of force in a material is referred to as **reluctance**. The majority of the magnetic lines of force take the path that has the lowest reluctance. Materials that permit magnetic lines of force to pass through them with little opposition are called low-reluctance materials.

If a wire is looped around a piece of iron and current is flowing through the wire, then the magnetic

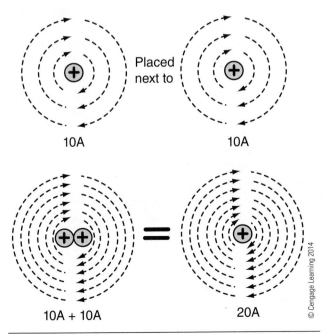

Figure 3-11 Two conductors with 10A of current flowing through each conductor has the same combined magnetic field strength as 20A flowing through a single conductor.

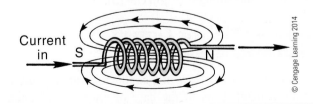

Figure 3-13 The magnetic field surrounding each loop of a conductor is added to the field from adjacent loops to produce a combined magnetic field.

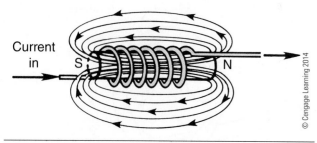

Figure 3-14 Adding an iron core to the coil to form an electromagnet.

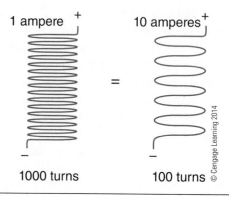

Figure 3-15 Magnetic field strength is determined by the amount of current flow and the number of loops or turns, referred to as ampere-turns.

lines of force that surround the conductor will concentrate in the piece of iron (**Figure 3-14**). The iron has a much lower reluctance than air, causing the magnetic lines of force to be directed into the iron. The magnetic lines of force will cause the piece of iron to become magnetized for as long as current is flowing through the wire. This type of magnet is known as an **electromagnet**. An electromagnet is a piece of nonmagnetized metal that is temporarily changed into a magnet by electric current flow. Most scrap yards use a crane with a large electromagnet on the end of a cable to lift scrap metal. The electromagnet can be switched off and on as needed. Many truck electrical components make use of electromagnets such as motors, relays, and solenoids.

Ampere-Turns

The magnetic field surrounding a conductor increases as the amount of current flowing through the conductor increases. The total magnetic field surrounding a coil of wire also increases with each loop that is added to the coil. The loops of wire in a coil are called turns. The ampere-turn is the unit used to indicate the strength of a magnetic field created by a coil of wire. An electromagnet that has 1 amp of current flowing through 1000 turns or loops of wire causes a magnetic field that has a magnetic strength of 1000 ampere-turns. An electromagnet that has 10 amps of current flowing through 100 turns of wire also causes a magnetic field that has a magnetic strength of 1000 ampere-turns. Both electromagnets would have magnetic fields with the same strength (**Figure 3-15**).

Creating Electric Current with Magnetism

Electric current flow in a conductor can be used to create a magnet, as in the case of an electromagnet. Conversely, a magnet can be used to produce an electric current flow in a conductor. The magnet actually produces a voltage or difference in electrical potential between the ends of the conductor, but the

voltage that is produced can result in an electric current if a path for current flow is provided.

Consider a horseshoe-shaped magnet like that shown in **Figure 3-16**. Think of a horseshoe magnet as a bar magnet that has been bent into a horseshoe shape. In the space between the poles of a horseshoe magnet, the magnetic lines of force are shown as being straight up and down because one flat section of the horseshoe magnet is a north pole and the other flat section is a south pole. If a conductor of electricity such as a piece of copper wire is moved through a magnetic field so that magnetic lines of force are "cut" by the conductor, a voltage is **induced** in the conductor. *Induced* means to produce or to cause to happen. Cutting magnetic lines of force is illustrated in **Figure 3-16**. The conductor must interrupt or cut through the magnetic lines of force to cause a voltage to be induced in the conductor. Only movement that is perpendicular (90-degree angle) to the magnetic lines of force causes a voltage to be induced in the conductor. Moving the conductor parallel to the magnetic lines of force, such as the up and down motion shown in **Figure 3-17**, does not induce any voltage in the conductor because no magnetic lines of force are being cut by the conductor.

A voltage can also be induced in a conductor by moving the magnetic lines of force and holding the conductor still. It really makes no difference which is moved—the magnetic lines of force or the conductor. Either causes a voltage to be induced in the conductor. The important fact to remember is:

Important Facts: Cutting magnetic lines of force with a conductor causes a voltage to be induced in the conductor.

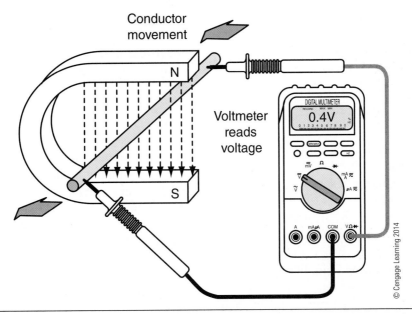

Figure 3-16 Cutting magnetic lines of force to induce a voltage in the conductor: conductor is moving from the right to the left perpendicular to the magnetic lines of force.

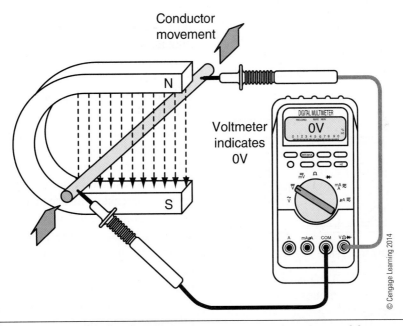

Figure 3-17 Up and down movement of conductor parallel to magnetic lines of force resulting in no induced voltage because no magnetic lines of force are being cut by the conductor.

This important fact will come up repeatedly in this text. Inducing a voltage in a conductor by cutting through magnetic lines of force is the basis for generators (alternators) like those used to keep a truck's batteries charged. Many other components in a modern truck make use of this basic principle, as will be shown in later chapters.

INDUCTORS

Wire that is formed into a spiral or coil is referred to as an **inductor**. The symbol for an inductor is a series of loops of wire, as shown in **Figure 3-18**. Inductors are rated in units known as the henry (H). The number of henries that an inductor has depends on the number of loops making up the coil, the material in

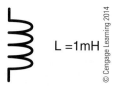

L =1mH

Figure 3-18 Inductor schematic symbol.

the middle of the coil called the core, and the dimensions of the coil. The symbol L is used to indicate that a component is an inductor, just as R indicates that a component is a resistor.

A resistor cannot store energy; resistors can only convert electrical energy into heat. Unlike resistors, inductors are capable of storing energy. Inductors store energy in the form of a magnetic field, like that shown as magnetic lines of force in previous sections of this chapter. The energy storage feature causes inductors to exhibit some seemingly strange behavior when the current flow through the inductor changes.

To describe what occurs when the current flow in an inductor changes, a hydraulic circuit will be evaluated. In the hydraulic circuit shown in **Figure 3-19**, a pump causes fluid to flow through the hoses to a turbine mounted inside a section of hose. A turbine is like a propeller or fan blade and is found in diesel engine turbochargers and jet engines. The turbine is attached

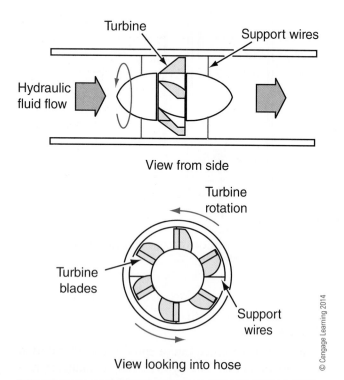

Figure 3-19 Turbine installed in a hydraulic line initially causes a severe restriction to hydraulic oil flow but becomes a means of storing energy as it begins to rotate.

with support wires such that it is free to spin inside the section of hose. Initially, when fluid first starts to flow through the hose, the turbine will be a restriction to hydraulic flow. However, as hydraulic fluid continues flowing through the turbine, the blades of the turbine will begin to spin like a fan. The spinning turbine then offers very little opposition to the steady hydraulic fluid flow as the speed of the turbine increases to match the flow rate of the hydraulic fluid.

If the hydraulic pump, which is causing hydraulic fluid flow, in **Figure 3-19** were suddenly shut off, the turbine would still be spinning for a brief period of time because of its inertia. The spinning turbine tends to keep fluid flowing in the same direction through the circuit like a pump for a brief period of time after the real pump is shut off.

Initially, the turbine acted as a hydraulic restriction until it was spinning at a rate consistent with the rate of hydraulic flow. Once the turbine was spinning, it offered little resistance to hydraulic flow. When the hydraulic pump was switched off, the inertia of the turbine resulted in the turbine acting as a second hydraulic pump until it stops spinning. The hydraulic flow in the circuit would continue flowing in the same direction at a decreasing rate until the turbine stops spinning.

A similar event occurs with an inductor when the current flow through the inductor changes. When current starts to flow through an inductor, the inductor initially acts as a restriction to the current flow, similar to a large value resistor. As the current flow through the inductor continues over time, the inductor becomes less of a restriction to current flow, similar to the turbine increasing in speed to match the flow rate in the hydraulic circuit. Eventually, the current is flowing at a steady rate and the inductor offers very little resistance to the steady current flow, similar to the turbine example where the turbine is spinning at the same rate as the hydraulic oil flow.

When the current flow through the inductor is then interrupted, the inductor tries to maintain the flow of current in the circuit in the same direction as the original current flow, similar to the spinning turbine example where the hydraulic pump was shut off. For a brief period of time, the inductor becomes a voltage source and attempts to keep current flowing in the same direction, similar to the example of the spinning turbine in the hydraulic hose acting as a pump.

The inductor initially limits the current flow because the magnetic field created by the current flow "chokes" or opposes current flow through the inductor. An inductor or coil is often referred to as a choke for this reason. Current flow through a loop of wire causes a magnetic field to be developed that surrounds the loop. The magnetic field that surrounds the loop is radiating or expanding from each loop to adjacent loops of the coil when the current first begins to flow through the inductor. The radiating magnetic field is like the waves that occur when a rock is thrown into a smooth lake. The waves radiate outwardly from the point in the water where the rock entered. This radiating movement of magnetic lines of force as the current begins flowing through each loop of the coil causes a voltage to be induced in the other loops of wire in the coil (**Figure 3-20**). The loops are not moving through the magnetic lines of force, but the expanding magnetic lines of force are moving through the stationary loops. Thus, the magnetic lines of force cut through the conductor to induce a voltage in the conductor. The concept in which the current flow through an inductor causes a voltage to be induced within the inductor is referred to as **self-inductance**. As the name implies, self-inductance is the act of an inductor inducing a voltage in itself through electromagnetic induction.

The self-induced voltage due to the expanding magnetic field causes the inductor to act as a voltage source. The self-induced voltage that is developed across the inductor has a polarity that opposes the original voltage source, such as a battery, that caused the initial current to flow through the inductor. This opposing voltage source produced by the inductor is referred to as **counter emf (CEMF)** or back emf. The concept where two voltage sources connected in series can limit the current flow in a circuit will make more

sense after studying series-opposing battery connections in later chapters. For now, just try to accept that the opposing self-induced CEMF voltage generated by the inductor acts to limit the initial current flow through the inductor. This initial opposition to current flow caused by the inductor is like the hydraulic example in which the stationary turbine blades initially acted as a restriction to hydraulic flow. The opposing self-induced CEMF voltage produced by the inductor decreases with time because the magnetic field stops expanding as the current flow through the inductor reaches a peak steady (unchanging) value. The loss of the self-induced CEMF effect after the magnetic field stops expanding causes the inductor to offer no resistance to steady current flow other than the resistance of the wire that makes up the inductor. This is like the turbine in the hydraulic circuit example reaching a speed that matches the hydraulic flow rate so it is no longer acting as much of a restriction.

The current flow through the inductor, with respect to time, when voltage is first applied at 0 millisecond is plotted in **Figure 3-21**. Because the line on this graph is curved, the relationship between initial current flow through an inductor and time is nonlinear. The actual time for the current to rise to a steady level depends on the value of the inductor (henries) and other factors, but the shape of the curve would be the same for all inductors.

A steady or unchanging magnetic field now surrounds the inductor, as shown in **Figure 3-22**, when the current flow through the inductor reaches a steady level. The inductor has stored energy in the form of a magnetic field surrounding the inductor. This is like the turbine in the hydraulic example discussed earlier in this section storing kinetic energy by the inertia of the spinning turbine. The steady-level magnetic field will remain around the inductor as long as the current flow through the inductor also remains at a steady level. The amplitude of the steady level current is determined by the resistance of the wire making up the inductor.

Interrupting Current Flow through an Inductor

If the voltage source that is causing the current to flow through a circuit that only has resistors is interrupted or switched off, the current flow through the resistors also immediately ceases. Because an inductor stores energy in the form of a magnetic field, switching off the voltage source that is causing current to flow through an inductor does not cause the current flow through the inductor to immediately cease. When the current flow through an inductor starts to decrease

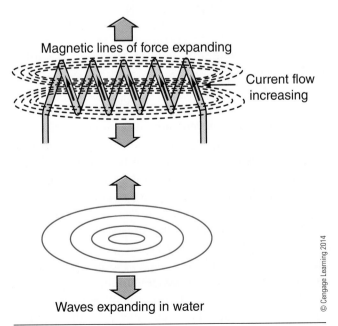

Magnetic lines of force expanding

Current flow increasing

Waves expanding in water

© Cengage Learning 2014

Figure 3-20 Self-inductance caused by an increasing magnetic field as current initially starts to flow through the inductor.

Increasing current flow through inductor

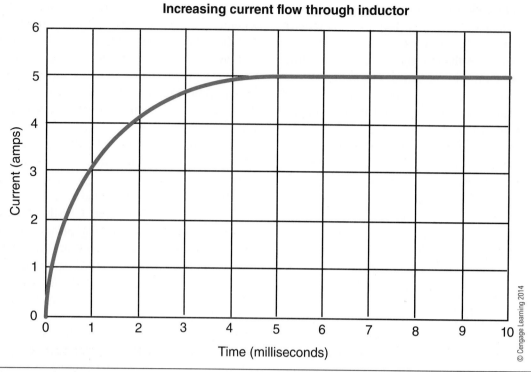

Figure 3-21 Current through the inductor increases with time to some steady value.

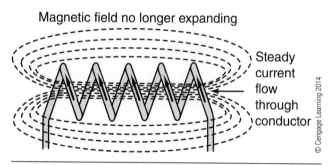

Figure 3-22 Magnetic field surrounding an inductor with no change in current flow: inductor is storing energy in the form of a magnetic field.

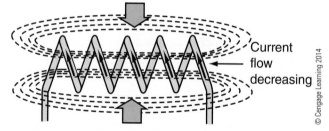

Figure 3-23 Decreasing current flow through inductor causes the magnetic field to collapse back into the conductor.

when the original voltage source is switched off, the magnetic field surrounding the inductor also decreases. The decreasing magnetic field causes the magnetic lines of force to retreat into the inductor (**Figure 3-23**).

This decrease in the strength of the magnetic field surrounding the inductor causes magnetic lines of force to once again cut through the loops of the inductor. However, the direction of the movement of the magnetic lines of force is inward towards the inductor. This is the opposite direction as when the current first started flowing through the inductor. It is like watching a movie of a rock being thrown into a smooth lake, but shown in reverse. The waves retract back into the point where the rock entered the water. This cutting of magnetic lines of force by the stationary loops of the inductor causes a voltage to be induced into the

inductor. This same self-induction CEMF phenomenon occurred when the current started flowing through the inductor except that the direction of movement of the magnetic lines of force is now in the opposite direction. Instead of traveling outward away from the inductor, the magnetic lines of force collapse inward toward the inductor when the current flow through the inductor decreases. This directional change in the movement of the magnetic lines of force as they collapse through the inductor causes the polarity of the self-induced voltage to produce current flow through the inductor that has the same direction as the original current. In the hydraulic circuit example, the turbine continued spinning in the hydraulic line for a period of time after the pump was shut off resulting in hydraulic fluid flow in the same direction for a brief period of time.

In a similar manner, the collapsing magnetic field causes the current to continue to flow in the same direction as that caused by the original voltage source (i.e., battery) for a period of time after the original voltage source is no longer supplying current to the circuit. This leads to an important fact:

....

Important Facts: The polarity of an induced voltage across an inductor due to a decreasing magnetic field has an opposite polarity of the voltage that produced the initial current flow through the inductor.

....

This principle of an opposite-polarity voltage being produced across an inductor when current flow decreases is taken from **Lenz's law**. Lenz's law states that the polarity of an induced voltage is such to oppose the change in current that produced it. Lenz's law is illustrated in the series-parallel circuit shown in **Figure 3-24**. The switch in this example permits current supplied by the battery to be interrupted. Switches are electrical devices used to control the flow of

electric current in a circuit and will be discussed in detail in **Chapter 4**. With the switch closed, conventional current flow causes the voltage dropped across each device to have a polarity, as shown in the upper drawing in **Figure 3-24**. At some point, the magnetic field surrounding the inductor L1 will stop expanding and reach a steady level. The magnetic field surrounding L1 is a form of stored energy. When the switch is opened, as shown in the lower drawing in **Figure 3-24**, the current flow from the battery is interrupted, which causes the magnetic field surrounding inductor L1 to collapse back into the inductor. This collapsing magnetic field causes a voltage to be induced in the inductor with a polarity that is opposite the polarity of the original voltage dropped across the inductor, per Lenz's law.

It is important to note that the direction of the current flow through the inductor caused by the collapsing magnetic field is still the same as the original current flow, as shown by the arrows in the lower drawing of **Figure 3-24**. Even though the polarity of the voltage measured across the inductor has reversed, the direction of current flow through the inductor remains the same. What has changed is that the inductor has become the voltage source supplying the circuit current instead of the battery. The voltage measured across the resistor R1 in **Figure 3-24** has also changed polarity, like the inductor voltage, because the direction of current flow through R1 has changed, as shown by the arrows. However, the polarity of the voltage dropped across resistor R2 in series with the inductor remains the same because the direction of current flow through R2 remains unchanged.

The current flow caused by the induced voltage produced in the inductor from the collapsing magnetic field will dwindle away in a brief period of time after all the magnetic lines of force have retreated back into the inductor. This is similar to the example of the turbine in the hydraulic line, which stops spinning shortly after the pump is switched off. The magnetic field is the source of the stored energy in the inductor that permits it to continue to act as a voltage source. This is similar to the way that the kinetic energy of the turbine (inertia) is the energy source that causes hydraulic flow to continue for a brief period of time after the pump is shut off in the turbine example.

Amplitude of Induced Voltage

The amplitude of the voltage induced across an inductor when the current flow decreases through the inductor depends on two things:

- The value of the inductor measured in henries. This value depends on factors such as the

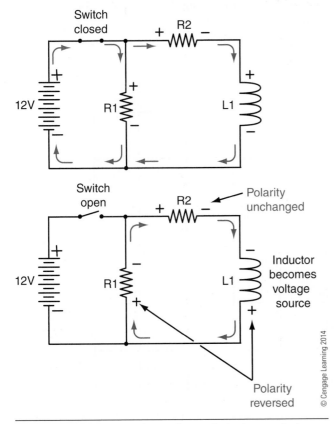

© Cengage Learning 2014

Figure 3-24 Current flow through inductor is interrupted, causing a reverse polarity voltage to be induced across the inductor as the magnetic field surrounding the inductor collapses.

number of loops (turns), the length of the coil, and the material in the center of the coil, such as iron.

- The rate of change in the decrease of the current flow through the inductor. *Rate of change* means how much the current decreases in a given period of time.

When magnetic lines of force cut through a conductor, a voltage is induced in the conductor. The amplitude of the induced voltage depends on how many magnetic lines of force are being cut by the conductor per second. The more lines of force that are cut by the conductor per second, the higher the amplitude of the voltage induced in the inductor. This concept is important for the study of generators and will be addressed in greater detail in **Chapter 7**.

When current flow rapidly decreases through an inductor, the magnetic lines of force surrounding the inductor also rapidly collapse into the inductor such as when the switch was opened in **Figure 3-24**. When the magnetic lines of force collapse back into the inductor, they rapidly move inward through the coil of wire, causing a large number of lines of force to cut through the windings of the inductor in a brief period of time. These collapsing magnetic lines of force due to the rapid decrease in current flow through the inductor can cause a very high reverse-polarity voltage to be induced across the inductor. This leads to an important fact of magnetism:

Important Facts: The more rapid the decrease in the current flow through an inductor, the higher the amplitude of the reverse-polarity voltage that is developed across the inductor.

This principle only speaks of the change in the current through the inductor, not the original level of the current. Thus, a relatively small amount of current flowing through an inductor that decreases very rapidly can cause a very high reverse-polarity voltage to be induced across the inductor. The amplitude of this reverse-polarity voltage can be much greater than the voltage source that caused the original current flow through the inductor.

Suppression

Even though the large-amplitude reverse-polarity voltage across an inductor where the voltage source was switched off may only last for a very brief time as shown in **Figure 3-25**, this **negative voltage spike** can damage electrical and electronic components and

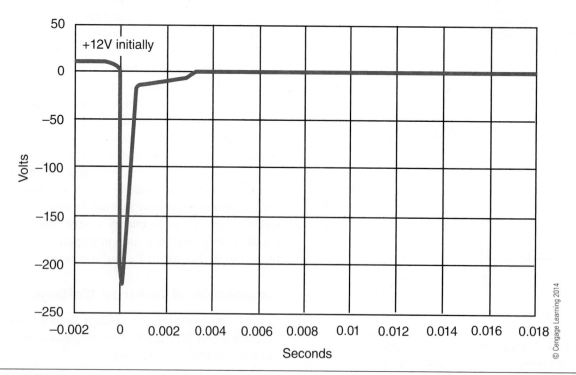

Figure 3-25 Oscilloscope display of a negative voltage spike across an unsuppressed relay coil. Interrupting the flow of current through the coil causes a −220V negative voltage spike.

cause intermittent problems or unexpected operation in the truck electrical system. **Figure 3-25** shows the voltage with respect to time across an inductor located in an electrical device called a relay. Relays are remote control switches and will be studied in **Chapter 4**. The tool used to display this negative voltage spike is an **oscilloscope**. Oscilloscopes are used to display the amplitude of a changing voltage or current with respect to time and will be discussed in later chapters.

To minimize these negative voltage spikes, most inductors have some form of **suppression**. *Suppression* means to reduce or hold down. A common method of reducing the negative voltage spike produced by a relay coil is to place a resistor in parallel with the inductor as shown in **Figure 3-26**. For many inductors used on trucks, the suppression resistor has a value of approximately 700Ω. As the negative voltage induced in the coil begins to increase rapidly when the voltage supply is switched off, the resistor in parallel with the coil acts somewhat like an electrical shortcut across the coil. The resistor prevents the negative voltage, induced in the coil due to the collapsing magnetic field, from increasing to a high level. The high voltage across the inductor never gets a chance to build up because the energy of the collapsing magnetic field is instead dissipated as heat by the parallel suppression resistor. Other types of suppression devices will be studied in later chapters after the introduction of electronics.

A high-amplitude negative voltage spike is not always a bad thing. For example, the ignition system used on gasoline (spark ignition) engines requires a voltage source that produces several thousand volts. An ignition coil is a type of **transformer** that is used to produce this high voltage. A transformer, like that shown in **Figure 3-27**, is an electrical device that can be used to increase or step-up a voltage. A typical

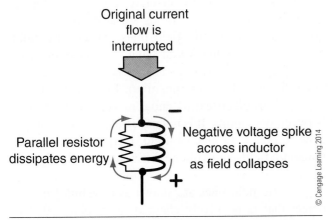

Figure 3-26 Inductor suppressed by parallel resistor limits the negative voltage spike to a low amplitude.

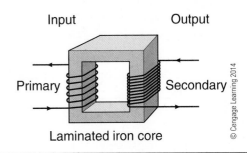

Figure 3-27 Transformer steps up a changing voltage: a transformer cannot increase a DC voltage.

transformer consists of two inductors that are wound around a laminated iron core. The ratio of number of turns or loops in the two inductors determines the increase in voltage that will occur. The two windings are referred to as the primary winding and the secondary winding. When used as a step-up transformer, the secondary winding has more turns than the primary winding. The supply voltage that is to be stepped-up is connected to the primary windings. The output or load is connected to the secondary windings. When current flows through the primary windings, the resulting magnetic lines of force are directed into the laminated iron core. These magnetic lines of force cut through the secondary windings, which causes a voltage to be induced in the secondary windings. Since there are more secondary windings than there are primary windings, the voltage induced in the secondary windings will be higher than the voltage dropped across the primary windings. The amplitude of the voltage induced in the secondary windings is proportional to the ratio of the number of turns in the secondary windings to the number of turns in the primary windings. In other words, if there are 100 turns in the primary windings and 10,000 turns in the secondary windings, then the ratio is 10,000 to 100, which can be reduced to 100 to 1. Therefore, the voltage induced in the secondary windings will be 100 times greater than the voltage dropped across the primary windings.

A transformer will only step-up a changing voltage because the voltage induced in the secondary winding depends on the magnetic lines of force cutting through the stationary secondary windings. Therefore, a transformer will not step-up a steady DC voltage. However, a DC voltage that is switched off and on is a changing voltage and causes magnetic lines of force to cut through secondary windings. This is how a gasoline engine spark ignition system operates.

An ignition coil, like that shown in **Figure 3-28**, is a special type of transformer. The primary windings are shown as being wound around the secondary windings. The primary windings are supplied with 12V

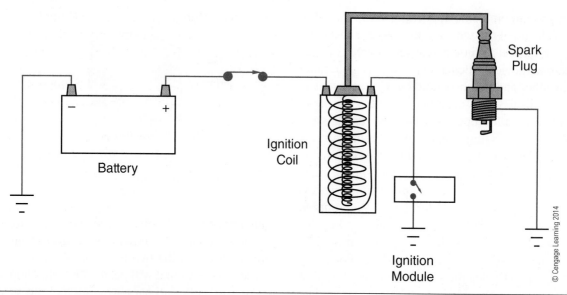

© Cengage Learning 2014

Figure 3-28 Spark-ignition engine ignition system utilizes a negative voltage spike and a type of transformer to produce thousands of volts.

from the battery through the ignition switch. The ground for the primary windings is provided through a second switch indicated as an ignition module. The 12V supplied by the battery is used to provide about 3A of current flow through the primary windings when the switch completing the path to ground is closed. Up until the early 1970s, a simple cam-operated switch known as the ignition contacts (points) located in the distributor was used to interrupt the flow of current supplied to the ignition coil. A distributor cam, typically driven off the engine camshaft, causes the ignition contacts to open at just the right moment resulting in the rapid collapse of the magnetic field surrounding the primary windings. The collapsing magnetic field causes the voltage across the primary windings to reverse polarity per Lenz's law and produce a negative voltage spike, similar to that shown in **Figure 3-25**. This causes a very high voltage (20,000V or more) to be induced in the ignition coil's secondary windings. The output of the secondary windings is connected to a spark plug, as shown in **Figure 3-28**. This high voltage induced in the secondary windings then causes current to arc across the spark plug's air gap like a small lightning bolt to ignite the fuel-air mixture.

Transformers can also step-down a voltage if the connections are reversed so that the supply voltage is connected to the winding with the greater number of turns and the load is connected to the winding with fewer turns. Electric power companies increase the voltage generated at a power plant up to 750kV AC for long-distance transmission using step-up transformers. A series of step-down transformers then decrease the 750kV AC down to 120V AC for home use in North America.

Some 2007–2010 Caterpillar diesel engines also make use of an ignition coil and spark plug placed in the exhaust system to regenerate the diesel particulate filter, as will be discussed in **Chapter 14**.

ELECTRIC FIELDS

Some clothes, such as socks and sweaters, are often found to be stuck together when they are removed from a clothes dryer. You may remember rubbing a balloon on your hair when you were a child and discovering that the balloon would then stick to the wall for a period of time. This phenomenon seems to be similar to magnetism. Nonmagnetic materials, such as socks and sweaters, are sometimes attracted to or repelled from each other in the same way as a magnet and a piece of steel.

The attraction or repulsion force of nonmagnetic materials is caused by **electric fields**. Similar to magnetic fields, electric fields are also invisible and are defined by electric field lines. An electric field exists between objects that have an excess of electrons (negative) and objects that have an excess of holes for the electrons (positive). Electric fields are the result of electrical potential or voltage difference between two points. It is not necessary to have current flow to create an electric field. Current flow creates magnetic fields and voltage difference creates electric fields.

Electric field lines are shown as traveling from an object that is positively charged to an object that is negatively charged, similar to magnetic lines of force shown as traveling from the north to the south pole of a magnet. Objects that have opposite charge (positive and negative) are attracted to each other, but objects that have the same charge are repelled from each other. This is similar to magnets as well.

CAPACITORS

A **capacitor** is an electrical device used to store energy in the form of an electric field. A capacitor is represented by one of the symbols shown in **Figure 3-29**. The two lines represent plates made of a conductive material. The area between these plates indicates a space with a very high resistance through which current cannot flow, similar to an open switch.

Capacitors are rated in farads (F). One farad is a very large unit of capacitance. Some car stereos make use of a 1-farad or larger capacitor. However, most common capacitors are in the microfarad region. The symbol C indicates that a component is a capacitor.

Hydraulic Accumulator Circuit

A hydraulic system component that behaves similar to a capacitor in DC circuits is an accumulator.

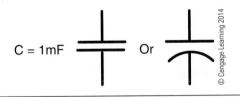

Figure 3-29 Capacitor schematic symbol.

An accumulator is used to store hydraulic fluid under pressure for later use. An accumulator is also often used to minimize pressure pulsations in a hydraulic system. Accumulators can be found in some automatic transmissions and some new generations of truck hydraulic brake systems. A simple hydraulic accumulator consists of a spring and a piston, as shown in **Figure 3-30**. Hydraulic fluid under pressure causes the piston to compress the spring. Energy is stored in the compressed spring. The spring in some accumulators is a compressed gas such as nitrogen.

A hydraulic circuit with an accumulator is shown in **Figure 3-31**. The pump shown in this example is regulated to supply a constant pressure of 50 psi

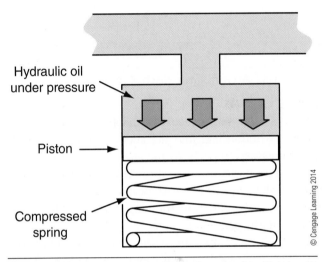

Figure 3-30 Spring accumulator in a hydraulic circuit stores energy.

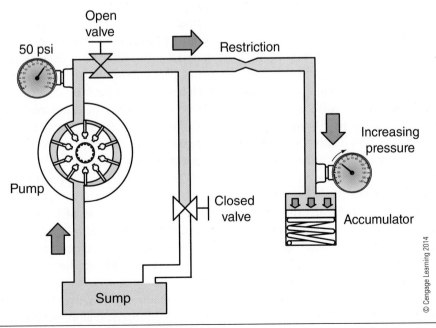

Figure 3-31 Accumulator charging: rate of charge is dependent on the size of the restriction and the size of the accumulator.

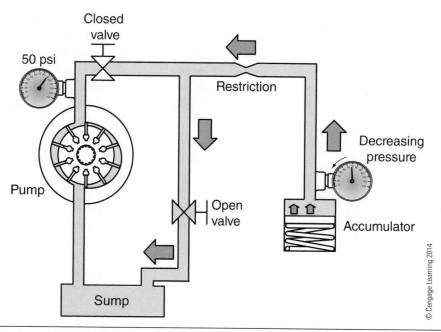

Figure 3-32 Accumulator discharging through the restriction back to the sump.

(345 kPa). The pump causes fluid to flow through a series restriction and into the accumulator. The fluid under pressure compresses the accumulator spring.

The restriction in series with the accumulator limits the rate that the fluid can flow into the accumulator. The size of the series restriction, along with the volume of the accumulator, determines how long it will take for the pressure of the fluid in the accumulator to be the same as the pump supply pressure of 50 psi (345 kPa). The smaller the diameter of the orifice of the series restriction and the larger the volume of the accumulator, the longer it will take for the accumulator pressure to reach the 50 psi pump supply pressure.

When the accumulator pressure reaches 50 psi, no additional hydraulic fluid flows in this circuit because there is no path back to sump due to the closed valve.

After the pressure in the accumulator reaches the pump supply pressure of 50 psi, the valve between the pump and the restriction shown in **Figure 3-32** is manually closed so that the flow from the pump is blocked. The accumulator remains pressurized at 50 psi because there is nowhere for fluid to flow. The valve between the restriction and the sump is then manually opened to provide a path for the fluid under pressure in the accumulator to return to the sump. The fluid under pressure in the accumulator will flow through the series restriction to the sump. The direction that the fluid flows through the restriction will be opposite to the direction of fluid flow that filled the accumulator because the accumulator is now acting like a pump. The series restriction will control how fast the accumulator emptied, similar to the way that it

controlled how fast the accumulator filled. The energy provided by the spring in the accumulator will decrease to zero as the spring moves upward. When the spring in the accumulator has reached the end of its travel, the accumulator is discharged and no additional hydraulic fluid will flow in this circuit.

Capacitor Circuit

A capacitor is similar to an accumulator in that energy is stored by the capacitor. A capacitor stores energy in the form of an electric field. A capacitor consists of two metal plates that are spaced close together. Between the two plates, a thin nonconductive material referred to as a **dielectric** separates the two metal plates. A dielectric is a material that has a very high resistance value. The dielectric between the plates of the capacitor is made of materials such as paper, porcelain, or even air. Because of the very high resistance of the dielectric material, nearly zero current flows through the capacitor.

When the plates of a capacitor are connected across a voltage source, an electric field will exist between the two plates because of their difference in polarity. This field is illustrated as electric field lines as shown in **Figure 3-33**. The capacitor stores energy in the form of an electric field, similar to an inductor storing energy in the form of a magnetic field.

A capacitor is shown in an electrical circuit in **Figure 3-34**. Some features of this electrical circuit are very similar to the hydraulic circuit shown in the previous examples. When the switch is moved from

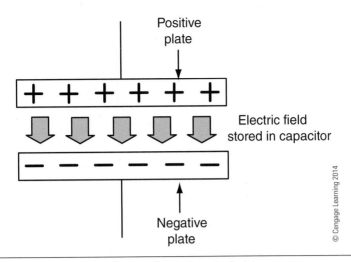

Figure 3-33 Capacitor stores energy in the form of an electric field.

position 2 to position 1, the capacitor will begin to charge, like the hydraulic accumulator. The time it takes the capacitor to fully charge to nearly the same voltage as the battery will depend on the value of the series resistance and the size (capacitance value) of the capacitor. This is similar to the size (diameter) of

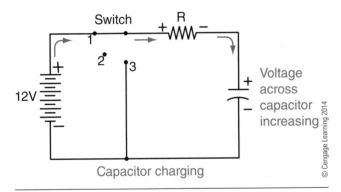

Figure 3-34 Capacitor charging: rate of charge is dependent on the value of resistance and the value of the capacitor in farads.

the series restriction in the hydraulic circuit and the volume of the accumulator determining how long it would take the accumulator to be charged to the pump pressure in the previous examples. With very large values of series resistance (mega ohms), it may take several minutes for the capacitor to charge to a level near that of the battery voltage. With very small values of series resistance, it may take only fractions of a second for the capacitor to charge to a level near that of the battery voltage. When the capacitor reaches close to full charge, the current flow in the circuit will nearly cease because the high resistance of the dielectric of the capacitor permits nearly zero current to flow through the capacitor.

When the capacitor reaches full charge (voltage across capacitor is nearly the same as the battery voltage), the switch is manually moved back to position 2 (**Figure 3-35**). This interrupts any current flow from the battery. The voltage across the capacitor will remain at almost 12V even though the connection with the battery has been opened by the switch. Since there

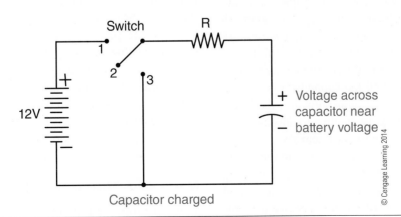

Figure 3-35 Capacitor maintaining its charge with no current flow in the circuit.

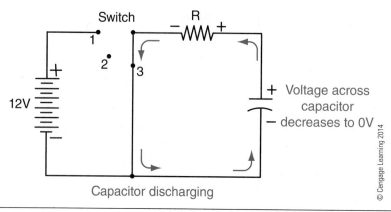

Figure 3-36 Capacitor discharging through resistor: rate of discharge is dependent upon value of resistance and value of capacitor.

is no path for current flow, the capacitor will remain charged indefinitely (in theory) with the switch in position 2. This is similar to the hydraulic accumulator circuit shown in **Figure 3-31** in which the accumulator is charged to the pump pressure of 50 psi and both valves are then closed to trap the hydraulic fluid under pressure in the accumulator.

If the switch is then manually moved to position 3, a path for current flow will be provided and the capacitor will discharge (**Figure 3-36**). The rate of discharge will be controlled by the series resistor in a manner similar to the series restriction in the hydraulic accumulator circuit example. Current will flow through the resistor in the opposite direction that it did when the capacitor was charging because the capacitor is now supplying the voltage in the circuit, not the battery. The voltage across the capacitor will decrease as the energy in the capacitor stored in the electric field decreases. When nearly 0V exist across the plates, the capacitor is fully discharged and all the current flow in the circuit effectively ceases.

Plotting the voltage measured across a capacitor over time on a graph as it charges and discharges will produce a curved line as shown in **Figure 3-37**. Because this line is curved, the charge and discharge rate of a capacitor is said to be nonlinear. The top graph in **Figure 3-37** shows that the capacitor charges to more than half the final voltage level relatively rapidly and then slowly charges to nearly the source voltage value. The same thing occurs when the capacitor discharges, except in the opposite direction as shown in the lower graph. The capacitor rapidly begins to discharge at first, and then slowly discharges to near 0V.

The shape of the curved lines on a graph showing capacitor charging and discharging voltages would remain the same if the series resistance were increased or decreased or the value of capacitance were modified. What would be different, though, would be the

specific time for the capacitor to charge or discharge. This time could be microseconds or several days, depending on the value of the series resistance and the value of the capacitance.

The value of capacitance measured in farads determines how long it takes for the capacitor to charge and discharge through a fixed value of resistance. Similarly, a large hydraulic accumulator would take longer to charge or discharge than a smaller accumulator that was placed in an otherwise identical hydraulic system. The value of capacitance indicates the energy storage capability of the capacitor.

Three factors determine the amount of capacitance (number of farads) in a capacitor:

1. The area (size) of the plates.
2. The distance between the plates.
3. The dielectric constant of the nonconductive material between the plates.

As the area (size) of the plates increases, the capacitance also increases so the capacitance is directly proportional to the area of the plates. However, the capacitance is inversely proportional to the distance between the plates of the capacitor. Therefore, as the distance between the plates decreases, the capacitance value increases. The dielectric constant value of the material between the plates is directly proportional to the capacitance value. As the dielectric constant increases, the value of capacitance also increases. The dielectric constant varies greatly from air (which has a dielectric constant of 1) to glass, which has a dielectric constant of 10. Other materials have much higher dielectric constants and are used to make special electrolytic capacitors.

Uses of Capacitors

Capacitors are used for several purposes in a truck electrical system. Like an accumulator used to smooth

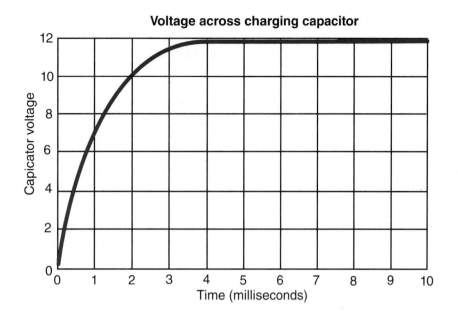

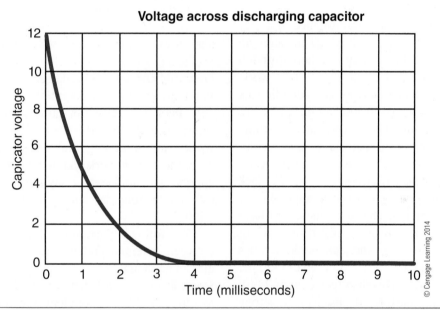

Figure 3-37 Voltage measured across a capacitor while charging (top) and while discharging (bottom).

out pressure pulsations, a capacitor can be used to smooth out voltage pulsations. The storage action of the capacitor causes it to release or absorb energy to maintain a steady DC voltage. Devices like alternators have a large capacitor in parallel with the output terminal for this purpose. Entertainment radios also contain large capacitors to smooth out the power supply voltage. Capacitors are often used in electronic circuits to filter out electrical noise.

Capacitors can also be used in timing circuits because the time it takes for a capacitor to charge to a specific level of voltage with a given value of series resistance can be calculated.

Some pressure-sensing devices also utilize variable capacitors to measure an unknown pressure. The pressure

being measured causes the distance between the plates of a capacitor within the sensor to change. The resulting change in capacitance is used to determine the unknown pressure, as will be discussed in **Chapter 11**.

EXTRA FOR EXPERTS

This section covers some additional information related to physics.

Work, Energy, and Power

Work, *energy*, and *power* are physics terms that are often interchanged in everyday conversation even though they do not have the same meanings.

Work. The word *work* is normally used to describe the place of employment where most people go to trade their time for money. However, the physics definition of *work* involves neither time nor money. In physics, work is an activity that involves force and a movement in the direction of that force, as described by **Equation 3-3.**

Work = Force × Distance

Force in the English system is measured in pounds (lb) and distance is measured in feet (ft). Therefore, the unit of work in the English system is pounds-feet (lb·ft). Do not confuse work with the English measurement of torque, which also has units of lb·ft or ft·lb. In the metric system (SI), force is measured in newtons and distance is measured in meters. Therefore, the metric unit of work is the newton-meter. The newton-meter is also called a joule (J). One lb·ft of work is equal to about 1.36 J.

Work is the product of force and distance. If a very strong truck technician were to show off his strength by lifting a 500-lb (227-kg) weight to a height of 1 ft (30.5 cm), he has performed 500 lb·ft (678 J) of work. If a physically weaker technician used a hydraulic jack to lift the same 500-lb weight to a height of 1 ft, he has performed the same amount of work as the strong technician. The difference is that the weaker technician has applied a much smaller force over a greater total distance. Each downward stroke of the jack handle might require 50 lb of downward applied force from the weaker technician. The jack handle may have to be pushed downward a distance of 2 ft (61 cm) for each stroke, so five strokes of the jack handle cover a distance of 10 ft (3 m) with a force of 50 lb (23 kg) per stroke of the jack handle. This also results in 500 lb·ft (678 J) of work being performed, just like the strong technician. The fact that it took more time for the weaker technician to perform 500 lb·ft of work than the strong technician does not matter when it comes to the physics definition of *work*.

Anything multiplied by zero is zero. Thus, if someone else trying to lift the 500 lb exerts a force and he or she is not able to cause the weight to rise any distance (0 ft), this person has not performed any work. He or she may be straining, but no work has been performed because the weight remains on the ground. No distance moved in the direction of the force exerted means that no work has been performed.

Energy. Fellow workers may complain that they do not feel like working because they just do not have enough energy. However, the physics definition of energy is the capacity or ability to perform work.

Energy indicates that the capacity or ability to perform work is present. You must have energy to be able to perform work. Energy could be thought of as the currency or money used to pay for work that has not yet been performed. Therefore, the units of energy are joules or lb·ft, like work. To perform 500 lb·ft of work, you must have 500 lb·ft (678 J) of stored energy. It is as though an amount of work, such as lifting 500 lb a distance of 1 ft, has a price tag that indicates that this task will cost 500 lb·ft (678 J) of energy. In the physics world, there are no energy credit cards. Work must be paid in full by transforming stored energy as the work is performed. Both technicians in the previous example would have to spend at least 500 lb·ft of energy to raise the 500 lb weight to a height of 1 ft.

Energy has many forms. One form of energy is the energy of motion, referred to as kinetic energy. The kinetic energy of an object depends on its mass and velocity. Heat is also a form of energy. Energy can also be stored in a stationary form for later use.

Potential Energy. Stored energy is **potential energy**. Potential energy and potential used to describe voltage are not exactly the same things. Potential just indicates that the ability to perform some amount of work is present. Potential energy indicates specifically how much work can be performed by the stored energy. Potential is like saying, "I have some money" while potential energy is like saying, "I have $500."

The following example illustrates the difference between potential and potential energy. Compressed air can be thought of as having the potential to perform work. Two air tanks that both contain air compressed to 120 psi (827 kPa) both have the potential to perform some amount of work. Which of these two air tanks can perform more work? In order to determine which can perform more work, it is also necessary to know how much compressed air each tank holds, or the volume of each tank. For example, one of the air tanks may be a small portable air tank used to inflate tires and have a volume of 2 cubic feet (0.06 cubic meter), and the other tank is part of a shop-sized air compressor that may have a volume of 20 cubic feet (0.6 cubic meter). Even though both tanks may contain air compressed to 120 psi the larger-volume air tank has much more stored potential energy than the smaller air tank.

A 6V flashlight battery has 6 volts of electrical potential. A physically larger 6V truck battery that requires two persons to lift it into the truck's battery box also has 6 volts of electrical potential. Both batteries have the same electrical potential (6V) but not the same amount of potential energy.

Power

In the earlier example of the two technicians lifting a 500-lb weight, the stronger technician lifts the weight in much less time than the technician using a hydraulic jack to lift the weight. Even though both technicians performed the same amount of work, the stronger technician has produced more power because the time required to perform the work is less.

The power produced by an electric motor such as a starter motor is often expressed in watts. A diesel engine starter motor may be rated at several thousand watts. A small electric motor, like that used to power a remote control car, has a power rating that is thousands of times less than a diesel engine starter motor. However, in theory, the remote control car motor could also be used to crank over a diesel engine by connecting the small remote control car motor to a series of gears. Through gear reduction, the torque or rotational twisting force developed by the motor could be multiplied many times over. The cost for increasing torque through gearing is a corresponding reduction of rotational speed (rpm) at the output of the final gear.

Using a remote control car electric motor to crank a diesel engine through a series of gears will not crank the engine over very fast. Using such a small motor may require several hours for the engine to be rotated one complete revolution and would require several AA-sized batteries. By comparison, a typical starter motor is capable of rotating a diesel engine at 200 rpm or more. Even though a remote control car motor and the large engine starter motor can perform the same amount of work, such as rotating a diesel engine one revolution, the rate of performing this work, or power, is substantially different between the starter motor and the remote control car motor.

Horsepower. Another measurement of power besides the watt that may be more recognizable to U.S. students is horsepower. One horsepower (hp) is approximately 746 watts. One horsepower is defined as the amount of power necessary to lift 550 lb at a rate of 1 ft/s. This definition was obtained long ago by measuring the performance of a typical workhorse. Horsepower is often used to describe the power of an engine or electric motor. An engine with a higher horsepower rating can perform a given amount of work in less time than an engine with a lower horsepower rating.

The 550 lb·ft/s measurement in the definition of horsepower does not have anything to do with torque, even though torque is commonly measured in units of lb·ft. The 550-lb·ft measurement in the definition of horsepower describes applying 550 lb of force over a distance of 1 ft, not a twisting effort like torque. Both torque and work use units of lb·ft, but torque and work are not describing the same thing. Instead, horsepower and torque are related by another formula. In the SI system, power (watts) is equal to torque (measured in newton-meters) times angular velocity (how fast something is rotating, measured in radians per second). A radian is similar to an angular measurement, like degrees. The equation for power in watts is shown in **Equation 3-4**.

Power = torque × angular velocity

The equation for power using the English units of horsepower, torque (lb·ft), and angular velocity (revolutions per minute or rpm) is shown in **Equation 3-5**.

Horsepower = (torque × rpm)/5252

Most motor sports enthusiasts are familiar with this equation. The 5252 in this equation is a conversion factor from SI units to English units.

Kinetic Energy

Kinetic energy is the energy of motion and depends on the velocity and the mass of the object. The formula for kinetic energy measured in joules using units of kilograms for mass and meters per second for velocity is shown in **Equation 3-6**.

KE = 1/2 × mass × velocity²

For example, a tractor and trailer with a combined weight of 80,000 lb traveling at 55 mph (36,287 kg at 24.59 m/s) has 8,091,643 lb·ft (10.97 MJ) of kinetic energy.

Forms of Potential Energy

One form of potential energy is gravitational potential energy (GPE). GPE depends on the mass of an object and the distance it is held above the earth. The formula for GPE measured in joules using SI units is shown in **Equation 3-7**. The acceleration due to gravity is approximately 9.8 m/s².

GPE = mass × acceleration due to gravity × height

An 80,000 lb tractor-trailer combination that is held 101 ft (30.8 m) above the ground by a very large crane has 8,091,643 lb·ft (10.97 MJ) of gravitational potential energy. Note that this is the same amount of kinetic energy of the tractor trailer traveling at 55 mph. Dropping the tractor trailer from a height of 101 ft would cause it to accelerate to 55 mph at the point when it hits the ground (ignoring the effects of air resistance). A coin dropped from 101 ft would also attain a velocity of 55 mph (again ignoring the effects of air resistance). However, the coin has much less gravitational potential energy when held at 101 ft above the ground than the tractor trailer because the coin has much less mass.

Spring potential energy is another form of potential energy and is the result of compressing or extending a spring from its free (natural) length. This spring could be the typical coil type of spring, a beam, or even an air spring. Potential energy is stored in the extended or compressed spring. The amount of potential energy depends on how far the spring is compressed or extended from its free length and how "stiff" the spring is. The spring from a ballpoint pen that is compressed from its free length by 0.25 in. (6 mm) has much less potential energy than a diesel engine valve spring that is also compressed from its free length by 0.25 in. Truck air brake systems depend on the spring potential energy of large springs to supply the energy to apply the spring brakes when parking.

Transforming Forms of Energy

Energy can be transformed from one form to another. Dropping a bowling ball from a height of 10 ft (3 m) causes the ball's gravitational potential energy to be transformed into kinetic energy, the energy of motion. If the bowling ball were dropped directly onto a large coil spring, the ball's kinetic energy would be transformed into spring potential energy as the kinetic energy of the falling ball causes the spring to compress. Once the ball stops traveling downward, it has stopped moving at an instant indicating that all of the kinetic energy has been converted (velocity = 0; therefore, KE = 0). The compressed spring will then cause the ball to accelerate upward again, converting the spring potential energy back into kinetic energy. The upward motion of the ball restores some of the ball's gravitational potential energy, which is then converted back into kinetic energy as the ball falls back on the spring again to repeat the process. Each time the process repeats, the ball will travel upward to a decreasing height. The coil spring is also compressed less and less each time that the process repeats. The reason that the maximum height of the ball decreases

each time the process repeats itself is because there is a price to be paid for converting between energy forms. The price for energy conversion is some of the energy, which causes the ball eventually to end up motionless atop the spring.

Energy Conversion Losses

No energy-to-energy conversion process is 100 percent efficient. Losses occur in any energy form conversions. These losses occur mostly in the form of heat, another form of energy. Some of these losses are caused by friction, including air (wind) resistance. Friction causes heat. Friction is the reason that the bowling ball travels to a lower height each time the spring is compressed in the previous section. The ball is traveling through the air, which causes the air to heat up slightly. Compressing the coil spring causes the spring to heat up each cycle as well. In the end, the bowling ball is stationary atop the spring and the original gravitational potential energy of the bowling ball has mostly been transformed into heat. A small amount of gravitational potential energy remains due to the ball's height above the ground, along with some spring potential energy due to the weight of the ball causing the spring to be partially compressed.

Another form of potential energy is chemical potential energy. Diesel fuel has approximate heat energy of 43.1 MJ/kg. In the process of converting diesel fuel to kinetic energy (truck in motion), only a small portion of the actual potential energy stored in the diesel fuel is converted to kinetic energy. A significant amount of the heat energy transformed from the diesel fuel's stored chemical energy exits through the exhaust or into the engine's cooling system. Friction throughout the drivetrain steals a portion of the diesel fuel's energy, as do wind resistance and other forms of friction, including the tires on the road surface. Anything that heats up as the result of the truck running down the road—including tires, oil, coolant, the air surrounding the truck, and so on—is taken from the energy stored in the diesel fuel.

Not all heat generated is due to unintentional losses. The brake system on a truck purposely converts the kinetic energy of the truck into heat energy. When the truck's brakes are overheated, such as when a truck is descending a mountain grade too fast, the brakes become ineffective at stopping the vehicle. This is when it is time to look for a runaway truck ramp so that the truck's kinetic energy can be converted back into gravitational potential energy.

Summary

- Power describes the rate at which work is being performed and has units of watts.

- Electric power measured in watts is the product of volts and amps.

- Magnetic lines of force are a means of graphically illustrating the strength and orientation of a magnetic field. Magnetic lines of force are drawn so that they appear to originate at the north pole of a magnet and end at the south pole of a magnet.

- Like magnetic poles repel each other, and unlike magnetic poles attract each other. This attraction or repulsion force increases at a rapid rate as the distance between two magnets decreases.

- Current flow through a conductor causes a magnetic field to encircle the conductor. The direction of the arrows on the magnetic lines of force can be determined using the right-hand rule.

- A conductor that is carrying current is compelled to move out of a strong magnetic field into a weaker magnetic field.

- A wire that is formed into a coil is called an inductor. Inductors store energy in the form of a magnetic field. Inductors are rated in units of henries.

- Current flow through an inductor causes a magnetic field to surround the coil much in the same way that a magnetic field surrounds a permanent magnet. Placing a piece of metal such as iron inside the inductor causes the metal to temporarily become a magnet. This temporary magnet is an electromagnet. Electromagnets are used in many truck electrical components.

- Passing a conductor through a magnetic field so that magnetic lines of force are interrupted or cut by the conductor causes a voltage to be induced in the conductor. This is the basis of generators.

- Changing the current flow (increasing or decreasing) through an inductor causes the expanding or contracting magnetic field to induce a voltage across the coil. This is known as self-inductance.

- The polarity of the voltage induced across an inductor due to a decreasing magnetic field is opposite the polarity of the voltage that produced the initial current flow through the inductor. This is described by Lenz's law.

- The negative voltage spike produced by an inductor when current flow is rapidly interrupted can damage a truck's electronic components. Suppression is typically used across the inductor to reduce the negative voltage spike.

- Capacitors store energy in the form of an electric field. Capacitors are rated in units known as farads.

- The time it takes for a capacitor to charge depends on the series resistance in the circuit. The larger the value of series resistance, the longer it takes the capacitor to charge.

Suggested Internet Searches

Try searching the following terms for more information:
Conservation of energy principle
Electromagnetic induction
Conversion between forms of energy

Review Questions

1. How much current is drawn by a light bulb that dissipates 12W of power when the voltage is 12V?

   A. 5A   C. 1A

   B. 10A   D. 2A

2. Current is flowing through a conductor. Which of the following is a true statement?

A. A magnetic field will only surround the conductor if the conductor is looped into a coil.

B. The copper wires turn into a permanent magnet.

C. A magnetic field surrounds the conductor.

D. The right-hand rule indicates that the direction of the arrows on the magnetic lines of force point in the same direction regardless of the direction of the current flow.

3. A clamp-on ammeter is placed around a conductor and the meter indicates that there is 10A of current flowing through the conductor. If the wire were wrapped into 10 loops and the clamp-on ammeter were clamped around the 10 loops, what value of current would the clamp-on ammeter indicate?

A. 10A

B. 1A

C. 0A

D. 100A

4. Two conductors with current flowing in the same direction are placed next to each other. Technician A says the two conductors will try to push away from each other. Technician B says that clamping a clamp-on ammeter around the two wires will cause the ammeter to measure a combined current of 0A. Who is correct?

A. A only

B. B only

C. Both A and B

D. Neither A nor B

5. Magnetic lines of force are cutting through a conductor. Which of the following is *not* a true statement?

A. A voltage will be induced in the conductor.

B. Increasing the number of magnetic lines of force being cut per second by the conductor will increase the voltage induced in the conductor.

C. Magnetic lines of force can be seen if smoke is directed toward the magnetic field.

D. Magnetic lines of force are used to quantify a magnetic field's strength.

6. Current flow through an inductor is interrupted rapidly by opening a switch. Technician A says that the polarity of the voltage measured across the inductor will reverse rapidly and could become a very high level of voltage. Technician B says that the direction of current flow through the inductor caused by the collapsing magnetic field when the switch is opened will remain the same and that this current will decrease to 0A. Who is correct?

A. A only

B. B only

C. Both A and B

D. Neither A nor B

7. A capacitor is charging. Which of the following is a true statement?

A. The time it takes for the capacitor to charge to a level nearly that of the battery voltage depends on the value of capacitance.

B. The time it takes the capacitor to charge to a level nearly that of the battery voltage depends on any resistance that is in series with the capacitor and battery.

C. The capacitor will remain charged for a period of time if removed from the circuit.

D. All of the above are true statements.

8. Technician A says that the magnetic field that surrounds an inductor (coil) with current flowing through it will remain in place for a long period of time when the inductor is removed from a circuit. Technician B says that the initial current flow through an inductor is a very high level that decreases with time as the magnetic field surrounding the inductor reaches a steady value. Who is correct?

    A. A only   C. Both A and B

    B. B only   D. Neither A nor B

9. Which of the following is *not* a true statement?

    A. A capacitor stores energy in the   C. The resistance between the plates of a capacitor is very
       form of an electric field.   low.

    B. An inductor stores energy in the   D. Electric current flow can be used to create a magnet; a
       form of a magnetic field.   magnet can be used to create an electric current.

10. A light bulb is connected by two wires to the positive and negative terminals of a battery, causing the bulb to illuminate. The two wires are placed next to each other so that the current flow through the wires is in opposite directions at the point in the circuit where a clamp-on ammeter is installed around both wires. If the voltage supplied to the light bulb is 12V and there is 24W of power being dissipated by the light bulb, what value of current would the clamp-on ammeter indicate?

    A. 2A   C. 20A

    B. 4A   D. 0A

11. Inductors are measured in units referred to as which of the following?

    A. farads   C. henries

    B. ohms   D. watts

12. Capacitors are measured in units referred to as which of the following?

    A. farads   C. henries

    B. ohms   D. watts

13. Which of the following controls how fast a capacitor will discharge?

    A. Series resistance   C. Reluctance of the dielectric

    B. Capacitor value   D. Both A and B

14. Which 12V truck electrical item draws the most current?

    A. 2400W grid heater   C. 70W headlamp

    B. 20W mirror heater   D. All draw the same amount of current

15. A conductor is held stationary in a magnetic field that is not changing. What will the amount of voltage induced in the conductor be?

    A. Very high voltage with reverse   C. 0V
       polarity
                                             D. Not able to determine the amount of voltage without
    B. Determined by Lenz's law   conductor resistance

CHAPTER

4 Electrical Components

Learning Objectives

After studying this chapter, you should be able to:

- Describe the components that make up a truck wiring harness.

- Repair a damaged section of wiring.

- Determine the correct wire gauge to use based on the amperage and length of the conductor.

- Describe the different types of electrical switches using terms such as *poles*, *throws*, *normally open*, and *momentary contact*.

- Use a voltmeter to determine if a switch is open, closed, or defective.

- Define the function of each terminal of a standard miniature ISO relay.

- Troubleshoot a circuit controlled by a relay and list the most likely failures of a relay.

- Explain how circuit protection devices (CPDs) operate and the importance of only replacing a CPD with the rating specified by the OEM.

- Use a voltmeter to determine if a CPD is open.

- Locate an open circuit and a shorted to ground circuit, using wiring schematics and electric test tools.

Key Terms

American Wire Gage (AWG)

armature

blown

case-based reasoning

circuit breaker

circuit protection device (CPD)

clipping

common

connectors

contacts

control circuit

earth

energized

fretting corrosion

fuse

fusible link

ground

harness

inrush current

insulation

latching

load

momentary contact

normal state

open circuit

pole

positive temperature coefficient (PTC)

relay

routing

short to ground

Society of Automotive Engineers (SAE)

terminals

throw

tripped

INTRODUCTION

This chapter provides an introduction to common electrical system components, including wiring, switches, relays, and circuit protection devices. These components are all relatively simple in their functionality but are important for the proper operation of the vehicle electrical system.

Electronics, introduced in trucks over the past several years, have revolutionized many of the concepts discussed in this chapter. Body control modules and other electronic modules now perform many of the tasks performed by some of the electromechanical devices described in this chapter. However, it is important to understand the fundamental concepts of these devices to become an effective electrical troubleshooter. Body control modules and electronics will be introduced in later chapters.

WIRING

Wiring is often taken for granted because it appears to be the electrical equivalent of hose or piping. However, wiring is probably the single largest cause of trouble in a truck electrical system. Wiring is subjected to heat, vibration, abrasion, water, and road chemicals. If the wiring system fails, the truck can become inoperative. Troubleshooting and repairing wiring problems is one of the most common electrical-related tasks required of a technician.

Wiring Harnesses

Most trucks are built on an assembly line, so most wires are not installed and routed one wire at a time between the various electrical components. To aid in the manufacturing and servicing of the truck, truck wiring is divided into **harnesses**. A wiring harness is a modularized section of wires designed to be connected to electrical components and to other wiring harnesses, as shown in **Figure 4-1**. For example, the cab or interior of a truck may have most of the wiring for the cab built into one harness, which the OEM may refer to as the cab harness. This cab harness could be considered as one component or assembly. The harness might have several **connectors**. Connectors are interlocking devices, typically made of an insulating material like plastic, that permit the wiring harness to be connected to other vehicle wiring harnesses or electrical components like switches or motors. Connectors contain **terminals**. Terminals are the electrically conductive pins (male) and sockets (female) that provide the electrical connections between two connectors.

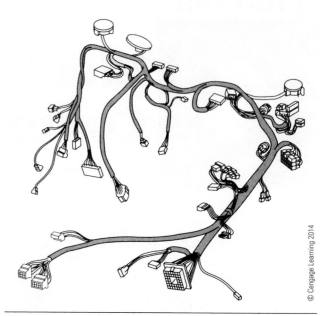

Figure 4-1 Typical wiring harness.

© Cengage Learning 2014

A truck's cab wiring harness may be connected to the engine wiring harness. The engine wiring harness may be connected to a transmission wiring harness and the front lighting wiring harness. This interconnection among major wiring harnesses permits a wire that originates in the cab wiring harness to route through several wiring harnesses to the end destination, such as a headlamp connector. Dividing the truck wiring into harnesses with interlocking connectors makes it easier to manufacture and service the truck. However, these connection points often become a cause of electrical problems due to water intrusion, vibration, or strain.

Water mixed with a small amount of road salt or other minerals causes a chemical reaction called electrolysis to occur with metals that are connected to a positive electrical potential. This chemical reaction dissolves metals. Most metals also corrode when exposed to moisture to form a white or green substance that has a high electrical resistance. This layer of corrosion acts as a series resistance. This series resistance will reduce or stop current flow in the circuit and will eventually destroy the terminals and wiring.

Tech Tip: You may find that one or more terminals may be missing in a multi-terminal connector that has been exposed to road salt mixed with water, while the other terminals may only have a small amount of corrosion. Terminals in the connector that are supplied with constant battery power can dissolve in a very brief amount of time when exposed to salt water.

Wire Details

Wire used in truck electrical systems is typically made of stranded copper. *Stranded* means that several smaller-diameter wires make up a wire's construction, as shown in **Figure 4-2**. Stranded wire is more flexible than solid wire. If solid wire is bent back and forth several times, it will eventually fatigue and break. Stranded wire can be bent back and forth with less frequent occurrence of breakage than solid wiring. Stranded wiring also has less resistance than a comparable sample of solid wire because electrons tend to mostly flow on the outer surface of a conductor.

Wire Diameter. Wire diameter in much of North America is measured by the **American Wire Gauge (AWG)** number or International Organization for Standardization (ISO) metric size. The AWG number is a standard that designates a number for wire based on the actual diameter of the wire conductor, not including the **insulation** material thickness. As the AWG number gets smaller, the wire conductor diameter gets larger. The wire gauge numbers are like shotgun gauge sizes in this respect. A 12-gauge shotgun has a larger bore diameter than a 20-gauge shotgun. A truck alternator output wire may be a 1 AWG-sized cable while a panel illumination light may use a 20 AWG wire. The 1 AWG alternator cable is much larger in diameter than the 20 AWG wire.

The word *cable* is often used instead of wire to describe large-diameter wires such as those used between the alternator and battery. Cable, such as that used as a conductor between the batteries and starter motor, may be larger than 0 AWG. For example, battery cable may be identified as 4-0 AWG. This indicates 0000 AWG or four-aught cable, and not 4 AWG. 4-0 AWG is a substantially larger diameter cable than 4 AWG, so don't confuse the two. The more zeros, the larger the diameter of the cable.

ISO metric wiring indicates the cross-sectional area of the wire in mm². The larger the metric number, the larger the wire diameter. ISO metric wire diameters along with the closest AWG size are shown in **Figure 4-3**.

Wire Resistance. The more lanes that a highway has, the more cars the highway can handle and still maintain traffic flow. The same idea is true of electrical wire. The larger the diameter of the wire, the lower the wire's resistance and the more current the wire can safely carry. Wire resistance in milliohms per foot and milliohms per meter for AWG stranded copper wire is shown in **Figure 4-4**.

The resistance of a wire increases as the length of the wire increases. The milliohms per foot or milliohms per meter information shown in **Figure 4-4** can be used to find the resistance of a length of wire. For example, to find the resistance of a section of 12 AWG wire that is 10 ft long, multiply the milliohms-per-foot value for 12 AWG wire, taken from **Figure 4-4**, by 10 ft.

Metric Size (mm²)	AWG (Gauge) Size	Ampere Capacity
0.50	20	4
0.75	18	6
1.0	16	8
1.5	14	15
2.5	12	20
4.0	10	30
10	8	40
16	6	50
25	4	60

© Cengage Learning 2014

Figure 4-3 Approximate AWG and metric ISO wire sizes.

AWG (Gauge) Size	Wire Resistance (milliohms per Foot)	Wire Resistance (milliohms per Meter)
20	10.4	34.1
18	6.5	21.4
16	4.1	13.4
14	2.6	8.5
12	1.6	5.3
10	1.0	3.3
8	0.6	2.1
6	0.4	1.3

© Cengage Learning 2014

Figure 4-4 Stranded copper wire resistance in milliohms per foot or milliohms per meter.

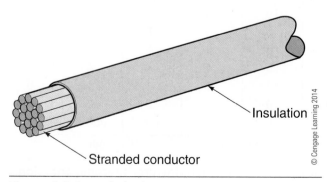

Insulation

Stranded conductor

© Cengage Learning 2014

Figure 4-2 Stranded wire.

Because wire resistance increases with wire length, it may be necessary to increase the wire gauge (smaller AWG number) for electrical devices that are located a long distance from the voltage source. Because wire has a resistance, it has a voltage drop when current flows through the wire, just like a resistor. The wire becomes a series resistor that reduces the voltage available for an electrical device such as a light. This voltage drop on the wire is often ignored but can become substantial for very long wire runs, such as those in trailer wiring. The voltage dropped on the wire also increases as the current flow through the wire increases per Ohm's law. The chart shown in **Figure 4-5** increases the recommended wire gauge to a larger diameter for a given current value as the length of the circuit increases. In general, a larger-diameter wire (smaller AWG number) can be substituted for the recommended wire gauge, provided the terminals, connectors, and seals (if applicable) can accommodate the larger wire size. However, a smaller-diameter wire (larger AWG number) should never be used in place of the recommended wire gauge. Doing so will cause an excessive voltage to be dropped on the wire and can cause the wire to overheat.

Tech Tip: Think of wiring as being a resistor that is in series with the electrical device such as a light. When looking at the wiring schematic, imagine the wiring and connectors as being low-value resistors. A voltage drop occurs across each resistor in a series circuit. Having too much voltage drop on the wiring causes the voltage left over for the electrical device to be too low, as indicated by Kirchhoff's voltage law.

Wire Insulation. There are several types of wire insulation. Most modern truck wire insulation is made of

Total Approximate Circuit Amperes	Wire Gauge (for Length in Feet)								
	3	5	7	10	15	20	25	30	40
1	18	18	18	18	18	18	18	18	16
1.5	18	18	18	18	18	18	18	16	16
2	18	18	18	18	18	18	18	16	16
3	18	18	18	18	18	18	16	16	16
4	18	18	18	18	18	16	16	16	14
5	18	18	18	16	16	16	14	14	12
6	18	18	16	16	16	14	14	14	12
7	16	16	16	16	16	14	14	12	12
8	16	16	16	16	14	14	14	12	12
10	14	14	14	14	14	14	12	12	10
11	14	14	14	14	14	12	12	10	10
12	14	14	14	14	14	12	10	10	10
15	14	14	14	14	12	12	10	10	8
18	12	12	12	12	12	12	10	8	8
20	12	12	12	12	12	10	10	8	8
22	10	10	10	10	10	10	8	8	6
24	10	10	10	10	10	8	8	8	6
30	10	10	10	10	8	8	8	6	6
40	8	8	8	8	8	6	6	6	4
50	6	6	6	6	6	4	4	4	4
100	2	2	2	2	2	1	1	1	0
150	0	0	0	0	0	0	0	00	00
200	00	00	00	00	00	00	00	000	000

Note: 18 AWG as indicated above this line could be 20 AWG electrically. 18 AWG is recommended for mechanical strength.

© Cengage Learning 2014

Figure 4-5 Wire selection chart for general-purpose electric circuits: follow manufacturer's recommendations for specific applications.

a cross-linked polyethylene plastic material. Some types of wire insulation are more resistant to abrasion, oil, or temperature than other types of insulation. The thickness of the wire insulation material also varies depending on the type of insulation material, which becomes an important factor with sealed connectors as described later in this chapter. Wire insulation material is chosen by the OEM for the specific application. Replacement wiring should have the same type of insulation material as that utilized by the OEM.

Automotive wire specifications are defined by the **Society of Automotive Engineers (SAE)**. SAE is a worldwide engineering professional organization that develops standards and recommended practices for the automotive industry. You are likely familiar with SAE designations for engine oil viscosity, such as SAE 15W40. SAE standards or recommended practices are designated by a J prefix such as SAE J1128, which is the SAE document that defines low-voltage wire insulation material. The main types of wire insulation used in modern truck electrical systems include TXL, GXL, and SXL. These types are all cross-linked polyethylene, which is the same material used in modern plumbing. Automobiles often use wire with GTP designation, which is a polyvinyl chloride (PVC) material. PVC does not offer as much abrasion or temperature resistance as cross-linked polyethylene, and is not as resistant to fuel or oil.

Circuit Identification. A large bundle of wires can be somewhat intimidating when troubleshooting an electrical problem. To assist the technician, the wire insulation can provide a means of identifying each wire. Most modern OEMs use color-coded wire insulation to identify a particular circuit. For example, an OEM may use pink insulation to signify ignition circuits, while yellow is used for headlamp circuits. Some wire insulation may also have what is referred to as a tracer. A tracer is a second or third color stripe to further identify a particular circuit. Some OEMs may use wire insulation that is all the same color but stamp a unique coded circuit identification number on the wire insulation every few inches. The trend for North American OEMs is to use both color-coded and continuously stamped wires to aid the technician in diagnosis of electrical problems.

Chassis Ground

For electric current to flow, there must be a path from one terminal of the voltage source, such as a battery, to the electrical device, such as a light, and back to the other battery terminal. Thus, it would seem that each electrical device must be supplied with two wires. One wire would originate at the battery-positive terminal and the other wire would end at the battery-negative terminal. Because frame rails are constructed of highly conductive steel, the truck frame or chassis becomes one of the conductors for many electrical devices. In most modern highway trucks, the battery-negative terminal is connected to the truck's frame by a large-diameter cable. Alternatively, this connection to the frame rail may be located near the starter motor. This causes anything made of a conductive material that is bolted to the frame to be connected to the battery-negative terminal through the low-resistance frame.

Tech Tip: Repeated failure of universal joints in the steering linkage and driveshafts may be caused by poor ground connections. Electric current flowing through the point contact surfaces of the needle bearings in universal joints causes arcing resulting in bearing failure.

When the frame is connected to the battery-negative terminal, the truck is said to have a negative **ground** system. The term **earth** instead of *ground* is often used outside of North America. Both *earth* and *ground* refer to the common practice in electric power transmission in which one conductor is connected to the actual surface of the earth (the literal ground) because moist earth is a marginal conductor of electricity.

The truck's steel frame is an excellent conductor of electricity due to its large cross-sectional area. These metal components are like a piece of large-diameter wire connected to the negative battery terminal.

Since the frame is typically connected to the negative battery cable, it is only necessary to supply the positive half of the wiring to a device such as a headlamp. The negative or ground terminal of the headlamp may be connected by a section of wire to the frame or chassis. When the headlamps are switched on, current will flow through the positive wire to the headlamp, through the headlamp filament, through the section of wire connected to the frame or chassis, and ultimately back to the battery-negative terminal through the cable connected to the frame. The use of the vehicle frame or chassis, including cab sheet metal, as the conductor to the negative battery terminal reduces the amount of electrical wiring needed by approximately one-half. Anything made of conductive material on a negatively grounded vehicle that conducts current to the negative battery terminal is said to be grounded.

Typical chassis ground symbols

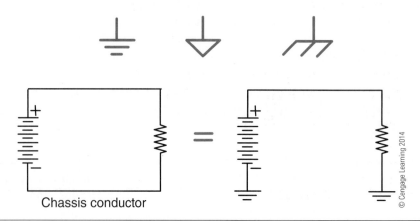

Chassis conductor

Figure 4-6 Chassis or frame ground symbols: a low-resistance circuit is assumed between all ground symbols.

The term *chassis ground* is also often used to describe this negative circuit.

Electrical schematics do not illustrate the vehicle frame or chassis as a conductor. Instead, one of the ground or earth symbols shown in **Figure 4-6** is used to illustrate the connection to chassis ground. Anywhere throughout the circuit diagrams that you observe this symbol, it indicates that a low-resistance electrical connection exists between that ground symbol and any other ground symbol. This simplifies the circuit diagrams considerably because it is only necessary to show a ground symbol and not the complete path back to the negative battery terminal for each electrical device.

Some older trucks may be positive ground instead of negative ground. These vehicles connect the positive battery cable to the frame or chassis. If you are unsure of the ground type of the vehicle you are working on, verify the type of ground of the vehicle before making electrical system repairs or jump-starting the vehicle. This is done by determining which battery terminal is connected to the frame or chassis.

Wiring Harness Routing and Clipping

Wiring harness **routing** refers to the placement of wiring harnesses on a truck. **Clipping** is the attachment of the wiring harness to the vehicle. Trucks are subjected to operating conditions that are often more extreme than passenger cars are subjected to. Vibration, heat, and rotating components such as driveshafts must be taken into consideration when routing and clipping wiring harnesses. Truck cabs are often mounted on air springs, allowing the cab to move up and down several inches relative to the frame. Permitting enough slack in the wiring harness clipping to account for this type of movement must also be taken into consideration.

OEMs determine the original routing and clipping of the wiring harnesses. Wiring harness repairs, major overhauls, and additions to the truck's wiring to install new electrical features may require disturbing the original harness routing and clipping. Therefore, a technician may be required on a regular basis to determine the best routing and clipping for a section of wiring harness.

Wiring Harness Protection. Wiring harnesses installed outside of the cab of the vehicle are almost always covered with some type of protective material to prevent wire damage due to abrasion. Flexible conduit is made of a rugged plastic material designed to withstand abrasion (**Figure 4-7**). Most flexible conduit is split lengthwise to permit installation of wires. The flexible conduit covering the wires permits the harness to be bent for installation in the truck.

Some OEMs may use a special fuzzy cloth tape instead of flexible conduit to protect wiring harnesses from abrasion. This tape is wrapped around the wiring harness in a spiral manner.

Figure 4-7 Flexible conduit protects wire insulation from abrasion.

Some exterior wiring may also have a plastic jacket molded around a group of insulated wires. The jacket acts to protect the insulated wires from damage and typically requires no other protective material, such as flexible conduit. This jacketed cable is often utilized for antilock brake (ABS) wheel speed sensors and trailer lighting cable.

Wiring Harness Attachment. Several types of devices are utilized to secure wiring harnesses to the vehicle. The most common device is the tie strap, which is also known as a strap lock or zip strip. These tie straps are made of a tough plastic material and are available in a variety of sizes. Tie straps are used to secure the wiring harness to rigid mounting points throughout the truck. Some tie straps are designed to attach the wiring harness to holes in the frame rail. Other tie straps may have provision for a mounting screw.

Tie straps must be used with care because they are capable of cinching the wiring very tightly. Cinching a tie strap tightly too close to a wiring connector can cause terminals to be pulled out of the connector or can cause wire insulation to be crushed causing wire conductors to make contact with each other.

Numerous types and sizes of clips are also utilized to secure wiring harnesses. These clips typically have mounting holes so that wiring can be attached to rigid components on the truck. These clips may be made of steel but have rubber padding or plastic coating to prevent cutting into the wiring harness.

Installing a Section of Wiring Harness. In general, the best routing path and clipping techniques for the wiring harness should be the same path used by the OEM. However, changes made to the truck, such as installation of a power take off (PTO) and hydraulic pump, may require repositioning of a harness. Adding new electrical components may also require installation of the associated wiring harness. General routing and clipping guidelines for wiring harnesses are as follows:

- Route the harness to avoid heat sources such as the exhaust pipes and exhaust aftertreatment devices.
- Do not install the harness over sharp edges.
- Attach the harness so that it is supported at least every 24 in. (0.6 m).
- Use extreme care in routing near rotating components such as driveshafts and components that move when the truck is operating, such as suspension and steering.
- Use grommets when the harness passes through holes in sheet metal or frame rails. For wiring harnesses that pass through noncircular openings in frame cross-members, install a section of

split flexible conduit around the circumference of the opening to act as a grommet.
- Provide strain relief at connection points. Strain relief refers to the practice of harness attachment so that a connection point is not under tension.
- Special care must be given to the routing and clipping of positive battery cables between the battery box and the starter motor. This large circuit is not protected by any fuse or other circuit protection device. The positive cable is typically covered with flexible conduit for additional protection against abrasion.

Wiring Harness Repair

Ask several truck technicians the best way to repair damaged wiring and you may get a variety of answers. In general, wiring repair techniques vary considerably. Repairing damaged wiring can be very time consuming. Often, time is of the essence and the truck operator or company just want the truck or trailer repaired rapidly so that it can be used to haul an important load. In such cases, the wiring repair that the technician chooses to perform may be an expedient repair, such as using insulated crimp-type butt connectors like those shown in **Figure 4-8** to splice in a new section of wire. A standard insulated terminal crimping tool is used to compress or crimp these insulated butt connectors (so named because the ends of the wires are butted together). The connectors are available in three standard sizes with color-

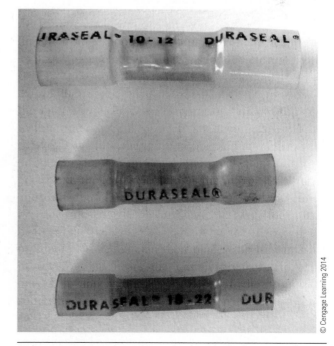

© Cengage Learning 2014

Figure 4-8 Crimp-type butt connectors with heat-shrinkable insulation.

coded insulation in red, blue, and yellow, corresponding to the wire gauge size range of the connectors.

The preferred type of butt connector has a heat-shrinkable insulation sleeve that surrounds the barrel of the crimp. After crimping, a heat source is utilized to melt the insulation sleeve (**Figure 4-9**). The plastic material is designed to shrink, and a sealer material is emitted from the insulation that serves to seal the connection from water and provide some strain relief for the connection. Most experts agree that these heat-shrinkable insulated crimp connectors provide a good quality and expedient wiring repair. Many OEMs now recommend the use of these heat-shrinkable butt splices for wiring repairs.

When time is not as much of a concern, other more durable methods of repairing wiring may be preferred by some shops and technicians. Many experts agree that one of the best methods of joining wires is the use of splice clips to crimp the wires together and then soldering the crimped splice. The connection is then covered with a heat-shrinkable tubing to insulate and seal the connection. If performed properly, this type of repair should last indefinitely. This procedure is shown in **photo sequence 1**.

Tech Tip: Plastic electrical tape is not acceptable for sealing of wiring repair outside the cab for a long-term repair. The tape cannot seal out all water and the adhesive in the tape will dissolve when exposed to fuel, oil, or heat. Also, do not use RTV silicone sealer to seal wiring repairs. The acetic acid released as the sealer cures can cause corrosion. Heat-shrinkable tubing containing sealer is the preferred method of sealing wiring repairs.

Connectors and Terminals

Connectors are utilized to provide a means to connect wiring harnesses to each other and to connect wiring harnesses to switches and other devices such as lights. Several different connectors are used on trucks and trailers. Two of the most common types of connectors are:

- Molded connectors: This type of connector contains terminals that are molded in a rubber or soft plastic material (**Figure 4-10**). The molded

© Cengage Learning 2014

Figure 4-9 Insulation is heated after crimping causing plastic insulation to shrink and seal the repair.

© Cengage Learning 2014

Figure 4-10 Molded connector.

Soldering Two Wires

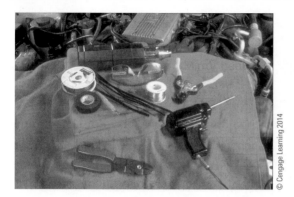

PS1-1 Tools required to solder copper wire: 100-watt soldering iron, 60/40 rosin core solder, crimping tool, splice clip, heat shrink tube, heating gun, and safety glasses.

PS1-2 Disconnect the fuse that powers the circuit being repaired. Note: If the circuit is not protected by a fuse, disconnect the battery ground terminal.

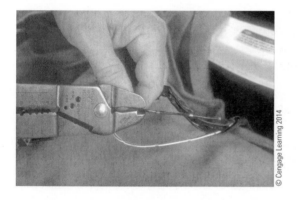

PS1-3 Cut out the damaged section of wire.

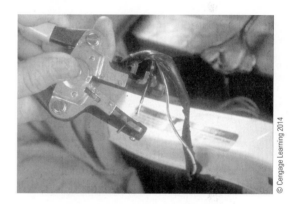

PS1-4 Using the correct-size stripper, remove about 1/2 in. of the insulation from both wires.

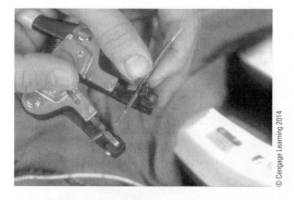

PS1-5 Now remove about 1/2 in. of the insulation from both ends of the replacement wire. The length of the replacement wire should be slightly longer than the length of the wire removed.

PS1-6 Select the proper size splice clip to hold the splice.

PHOTO SEQUENCE 1 (Continued)

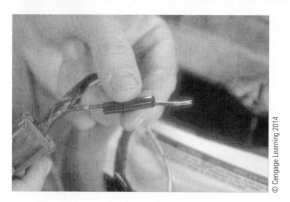

PS1-7 Place the correct size and length of heat shrink tube over the two ends of the wire.

PS1-8 Overlap the two splice ends and center the splice clip around the wires, making sure that the wires extend beyond the splice clip in both directions.

PS1-9 Crimp the splice clip firmly in place.

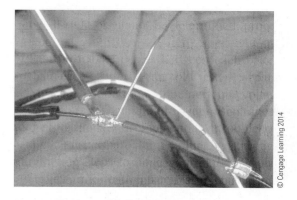

PS1-10 Heat the splice clip with the soldering iron while applying solder to the opening of the clip. Do not apply solder to the iron. The iron should be 180 degrees away from the opening of the clip.

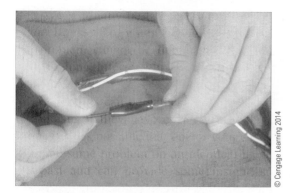

PS1-11 After the solder cools, slide the heat shrink tube over the splice.

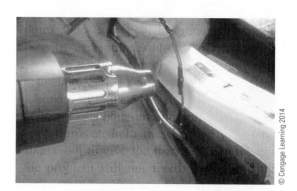

PS1-12 Heat the tube with the hot air gun until it shrinks around the splice. Do not overheat the heat shrink tube.

connector is often designed to be mated to a specific device, such as a trailer sealed lamp assembly.

- Hard-shell connectors: This type of connector is typically made of a plastic material (**Figure 4-11**) and is designed to hold as few as 1 and as many as 80 or more terminals. This type of connector is used to provide interconnection between wiring harnesses and connection to electronic control modules.

Sealed Connectors. Hard-shell connector sets utilized outside the cab are typically sealed to prevent water intrusion. This sealing may be accomplished by individual seals on each wire (**Figure 4-12**) that are designed to be crimped on each terminal and fit tightly into the hard-shell connector body when the terminal is inserted into the connector body. Another means of sealing a connector is with the use of a rubber seal inside the connector body with holes that fit around each wire (**Figure 4-13**). Unused terminal cavities are sealed with a solid plastic plug. Other sealed connectors may have a solid rubber membrane that is pierced when a terminal is inserted in the connector.

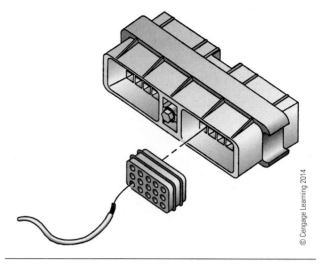

Figure 4-13 Multi-wire connector seal.

Sealed connectors also have some means to seal out water between the mating halves of the connector. A rubber sealing ring around the circumference of the male half of the connector set or inside of the female half is typically used to provide this seal.

Regardless of the connector seal type, the thickness of the wire insulation for replacement wire is important to ensure a good seal. If the insulation material is too thick, it may not be possible to insert the wire into the seal without damaging the seal. If the insulation material is too thin, a gap will exist between the insulation and the seal, which could permit water to enter into the connector.

Terminals. Terminals are the pins (male) and sockets (female) that are crimped to the ends of each wire in a wiring harness. Terminals are made of electrical conductive material such as brass, copper, or steel and may be coated with tin or gold. Terminals are crimped to the exposed copper conductor at the end of each wire so that the terminal becomes electrically conductive. Crimping typically involves the use of a W-shaped die in a pliers-type crimping tool as shown in **Figure 4-14**. Performing a good quality crimp requires practice. The wire insulation is first stripped back just enough so that the bare copper wire strands extend just beyond the terminal's cable core wings (**Figure 4-15**). The terminal's insulation wings must be crimped around the wire insulation to act as a strain relief, so do not strip back too much wire insulation. The crimping tool is then used to crimp the cable core wings to the copper wire strands. The crimping tool must be the correct size for the wire and the terminal so that the correct amount of crimping force is applied. Too much crimping force will break the wire strands; too little crimping force results in a poor connection

Figure 4-11 Hard-shell connector body.

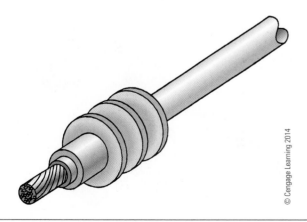

Figure 4-12 Individual wire seals installed before crimping terminals to wire end.

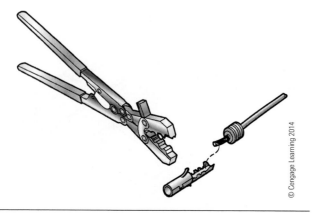

Figure 4-14 Terminal crimping tool.

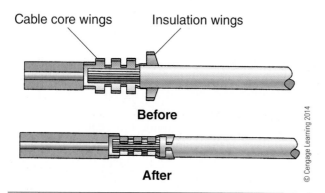

Figure 4-15 Wire end and terminal before and after crimping operation.

and the likelihood that the wire end will pull out of the terminal. After crimping, visually inspect for broken wire strands and gently tug on the terminal to verify the terminal does not pull off the wire end. Once the core crimp is validated, crimp the insulation wings around the wire. Note that some crimping tools are designed to crimp both the core crimp and the wire insulation at the same time. Connectors that use individual wire seals have terminals with very large insulation wings to accommodate the seal.

Figure 4-16 illustrates a Deutsch brand solid barrel terminal and crimping tool. The solid barrel terminals are an open tube at the crimp end, unlike the flat core and insulation wings shown in **Figure 4-15**. A special crimping tool like that shown in **Figure 4-16** is required to crimp these terminals. The crimping force is controlled by a selectable terminal size setting on the tool. The depth of the crimp is adjusted by a knob.

There are many different types of terminal ends. Terminals that are designed to be inserted into hard-shell connectors are either male or female terminals. Typically, female terminals are used in male connectors and vice versa. The male terminal in one connector is designed to fit tightly inside the female

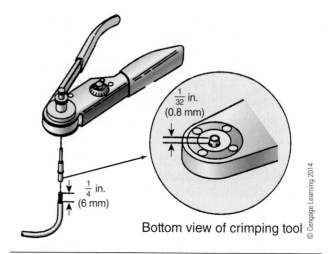

Figure 4-16 Barrel-type terminal crimping tool: crimp position and crimping force must be adjusted prior to use.

terminal of the mating connector so that a low-resistance electrical terminal-to-terminal connection is obtained when connector halves are installed together. Terminals designed to be inserted into sealed connectors may require a wire seal that is crimped to the terminal. The terminal and seal assembly are then inserted into the connector body (**Figure 4-17**).

Many terminals have a tang or barb that is used to lock the terminal into the connector body (**Figure 4-18**). Other terminals have an opening in the terminal designed to permit a locking tab, built into the connector body, to lock the terminal into the connector body. There are also typically secondary locks of some sort to ensure that terminals are retained in the connector body. These secondary locking devices may snap over the rear of the connector and prevent terminals from pushing out of the connector when the connectors are mated. Other types of terminal retention may include

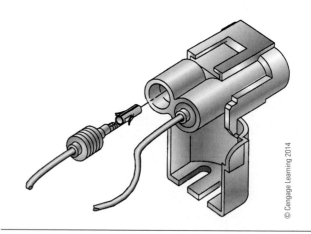

Figure 4-17 Wire with terminal and individual seal is inserted into connector body.

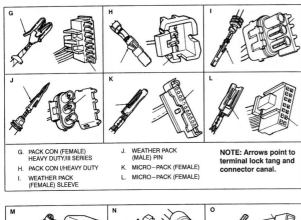

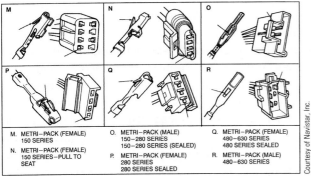

Figure 4-18 Connector and terminal locking devices secure the terminal in the connector body.

plastic wedges that fit inside the connector's mating face to keep the connector locking tabs engaged.

Not all terminals are inserted through the back side or non-mating end of the connector. Terminals known as pull-to-seat terminals require the wire first to be pushed through the connector without a terminal attached (**Figure 4-19**). A terminal is then crimped onto the wire end, and the wire, with attached terminal, is then pulled into the connector as shown in **Figure 4-20**.

Terminals in hard-shell connectors can become corroded through water intrusion or damaged from carrying too much current. Terminals can be removed from most types of connectors for replacement. Some

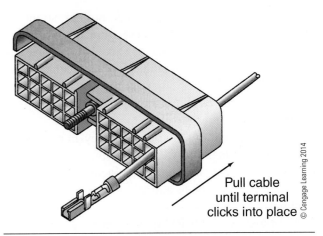

Figure 4-20 Pulling the terminal to seat into place in the connector body after terminal is crimped to wire end.

terminals require special terminal removal tools for extraction from the connector, as shown in **Figure 4-21**. It is important to consult OEM-provided information on the proper method of extracting terminals from a connector to avoid damage to the terminals and the connector.

Some OEMs may offer small sections of insulated wire that are pre-crimped with a specific terminal. These may be referred to as pigtails. The insulation on the non-terminated end of the wire is stripped and crimped to the wiring harness using a conventional butt-splice. These repair components prevent the need to purchase specialized terminal crimping tools for occasional wiring harness repairs.

Terminal and Connector Problems

One problem you may encounter with connectors is that the connector is not fully mated with the mating connector or device. Connectors typically have some type of latch that is only engaged if the connector is

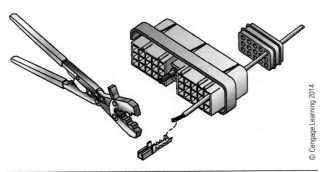

Figure 4-19 Pull-to-seat terminals: terminal is crimped to wire end after inserting wire end into connector body and seal.

Figure 4-21 Special terminal extracting tools.

fully seated. This latch prevents normal strain or movement of the wiring harness from disconnecting the connector from its mate. Make sure when reconnecting a connector that the latching mechanism is engaged. Often, a click sound heard as the connector latch engages serves as an indication that the connector is fully seated.

Connectors can also be damaged by improper removal of terminals or from excessive strain on the wiring such that the terminal locking tab inside the connector is broken and can no longer retain the terminal.

Terminals can fail for several reasons. One of the most common terminal problems is spread or loose female terminal sockets, which result in a poor connection with the male terminal or an intermittent open circuit. This is often caused by previous troubleshooting attempts where a careless technician (or owner) has used a voltmeter lead to probe the terminal. It is important for you as an electrical troubleshooter to use proper diagnostic techniques that do not damage the wiring harness. Never poke a test light or voltmeter lead into a terminal socket. In addition, terminals and connectors are designed only for a limited number of connection and disconnection cycles. Each disconnection causes the fit between the mating terminals to loosen. You should try to minimize the number of times you disconnect and reconnect a connector set during troubleshooting to prevent creating new problems.

Tech Tip: It may not be possible by visual inspection alone to determine if a female terminal socket is spread open. Use a corresponding spare male terminal pin to test for loose female terminal sockets. The terminals should have a snug fit. Some OEMs have special terminal test probes or a "pull-gauge" that can be used to verify the tightness of female terminal sockets.

Bent male terminal pins are another common terminal failure. This typically occurs when the connector halves are not aligned when connected. Although it may be possible to straighten male terminal pins, replacement of the terminal or the device is often necessary. If the device is an engine controller or other electronic component, the cost of a careless connection can be very expensive to repair.

Corrosion is another common terminal failure. Water and road chemicals are a common cause of corrosion, especially at the terminal crimp where the copper strands of the wire are exposed. Another type of corrosion is known as **fretting corrosion**. Fretting corrosion refers to corrosion caused by movement. Terminals are typically plated with tin or gold to resist corrosion. Over time, vibration causes a slight movement between the terminal pin and socket. This movement causes the terminal plating to wear off at the very small contact points where the terminals contact each other. The base metal that makes up the terminal is then exposed to oxygen resulting in the formation of a thin layer of nonconductive material called oxidation between the terminals. This layer of oxidation acts as a series resistance resulting in a reduction in the current flow through the circuit, per Ohm's law. Often times, disconnecting and reconnecting the connector containing terminals with fretting corrosion causes this thin layer of oxidation to be temporarily wiped clean. It is very difficult to see fretting corrosion. You may suspect fretting corrosion if during troubleshooting you disconnect a connector, observe no problems with the terminals or connectors, and everything works again when the connectors are reconnected. Unfortunately, the layer of oxidation will likely form again. The only real solution for fretting corrosion is terminal or component replacement, along with minimizing vibration of the wiring harness by routing and clipping. Some OEMs may also specify the use of a light layer of special dielectric grease on the terminals of connectors that are subject to vibration, which acts as a lubricant to prevent the terminal plating from wearing as a means to control fretting corrosion. Do not substitute common petroleum-based grease for this special dielectric grease. Doing so may result in deformation of the plastic connector components and seals.

Another common terminal failure is terminal push out. This refers to terminals that are not fully seated in the connector. When the connectors are mated, the terminal is pushed out of the connector instead of mating with the terminal in the other half of the connector. The tip of the pushed out terminal may still be making intermittent contact with its mating terminal, resulting in intermittent operation.

Broken copper wire strands at the terminal crimp are another common terminal failure. This often results from the use of an improper crimping tool causing the copper strands of the wire to break at the crimp. This type of failure can result in intermittent electrical problems as the broken ends of the copper wire may make intermittent contact with the terminal. The wire insulation is often still crimped to the terminal, providing what appears to be a good connection. Gentle tugging or wiggling of the wiring near the connector can often reveal this failure.

SWITCHES

Electric switches are devices that interrupt the flow of current in a circuit. Electric switches are found mostly in the cab of the truck and control electrical devices such as lights, horn, and windshield wipers.

The simplest type of switch is referred to as a knife switch. A knife switch is illustrated in **Figure 4-22**. Knife switches are not typically used in modern truck electrical systems, but they can be used to illustrate the action of an enclosed switch.

The moving component of a knife switch is known as a blade. The blade is made of a conductive material such as copper. The two terminals of the knife switch are connected in series with a circuit. A nonconductive material such as plastic covers the end of the blade, which permits the user to move the blade without being shocked. The knife switch is also mounted to a base made of nonconductive material.

The knife switch shown in **Figure 4-22** only has two positions, open and closed. In the open position, electric current in the circuit is interrupted. The open position may also be called the OFF position because a device controlled by the switch would be off with the switch open. In the closed position, electric current is permitted to pass through the switch to the rest of the circuit. Thus, the closed position is also called the ON position.

In a series circuit, it makes no difference where a switch is placed because the current is the same everywhere in a series circuit. An open switch in a series circuit causes the current to be 0A everywhere in that series circuit. However, standard automotive practice is to interrupt the positive side of the circuit that is supplying an electrical device and permanently connect the negative side of the electrical device to chassis ground.

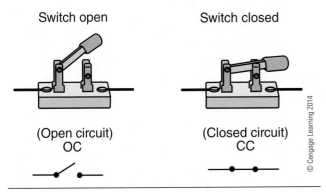

Figure 4-22 Knife switch.

When a switch is closed, the resistance between the terminals of the switch is very low, like a piece of wire. This low resistance permits current to flow through the switch with almost no opposition. In the open state, a switch has an infinitely high resistance between the terminals that stops all current flow in the circuit.

The two pieces of a switch that actually make or break the circuit current flow are known as **contacts** because the two pieces make contact when the switch is closed. One of the contacts is the moving contact, such as the blade in the knife switch; the other contact is the stationary contact.

Switch Types

There are many different types of switches. Switches may simply make or break the flow of electric current in a single circuit, like the knife switch shown previously. Switches that are more complex may control several different circuits at the same time. Switches may be a simple on-off type or may have several positions.

Two major terms, **pole** and **throw**, are commonly used to describe how a particular switch is designed to function.

Poles and Throws. The term *pole* describes the number of input terminals that the switch has. You can also think of the number of poles as being the number of movable contacts inside of the switch because each of the switch's input terminals is connected to a movable contact inside the switch. Because the knife switch shown in **Figure 4-22** only has one input terminal and one movable contact, the simple knife switch is a single pole switch.

The term *throw* refers to the number of positions that a switch can be moved or thrown to and still complete a circuit. The simple knife switch is a single-throw switch because the switch can only be moved to one position and still complete a circuit. The knife switch only completes a circuit in one position (the ON position). The knife switch also controls only one path for current flow.

The terms *pole* and *throw* are used together to describe the basic functionality of a switch. The simple knife switch could be described as a single-pole single-throw switch, which is abbreviated as SPST. The schematic symbol for an SPST switch is shown in **Figure 4-23**.

A switch can have more than one pole. The switch shown in **Figure 4-24** is referred to as a double-pole switch. The dashed line indicates that a mechanical

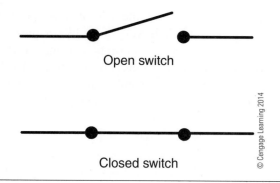

Figure 4-23 Single-pole single-throw (SPST) switch symbol.

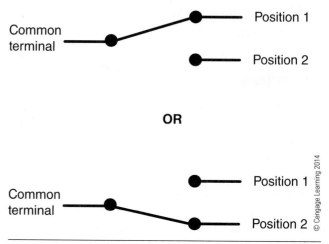

Figure 4-25 Single-pole double-throw (SPDT) switch shown in the two possible positions.

connection exists between the two movable switch contacts but that no electrical connection exists between the two contacts. The mechanical connection is made of an insulating type of material such as plastic. The mechanical connection between the switch contacts causes both switch contacts always to be closed or open at the same time. Both contacts move at the same time, permitting two independent circuits to be controlled by one double-pole switch. Because this switch can also only be moved to one position to cause the switch to complete circuits, the switch is a single-throw switch. In the ON position, the switch is completing circuits. In the OFF position, the switch is not completing any circuits. Therefore, the switch shown in **Figure 4-24** is referred to as a double-pole single-throw switch or DPST. A DPST switch is like placing two SPST switches next to each other and mechanically connecting the two switches so that one lever controls both switches.

A switch can also have more than one throw. A double-throw switch can be moved to two different positions to complete an electrical circuit, as shown in **Figure 4-25**. This switch has only one input terminal and one movable contact, so it is also a single-pole switch. Therefore, this switch is a single-pole double-throw (SPDT) switch. An example of an SPDT switch is a floor-mounted headlamp dimmer switch. A headlamp dimmer switch is used to switch back and forth between low-beam and high-beam headlamps.

Another type of switch is a multiple-pole multiple-throw switch. These switches are often rotary switches. An example of a rotary switch is the ignition (key) switch used on most trucks. An ignition switch is shown in **Figure 4-26**. This key switch has four positions: accessory, off, ignition, and crank. This key switch also has three moving contacts. This type of switch is also known as a ganged switch because each of the moving contacts moves in unison.

Latching or Momentary Contact. Most wall switches found in homes are **latching**-type switches. *Latching* means that the switch will stay in a particular position after being flipped or thrown to that position. It is not necessary to hold a wall switch in the on position to keep a light on. The wall switch latches in either the on or off position to turn the lights on or off.

Other types of switches are **momentary contact**. A momentary contact switch requires the user to depress and hold the switch to close or open the circuit. Momentary contact switches are spring loaded in either the open or the closed state and must be depressed to change states. After the switch is released, it returns to the original state by the spring tension. An example of a momentary contact switch is a car or truck horn switch. A horn is typically only used for brief periods of time. The horn will only sound as long as the horn switch located in the steering wheel is depressed. A spring in the horn switch returns the switch to original or normal state.

Normal State. **Normal state** refers to the state (open or closed) of a momentary contact switch when the switch is not depressed or otherwise disturbed. Think of normal state as the state of the switch if it were being tested with an ohmmeter at the parts counter.

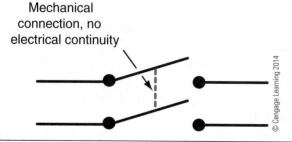

Figure 4-24 Double-pole single-throw (DPST) switch symbol.

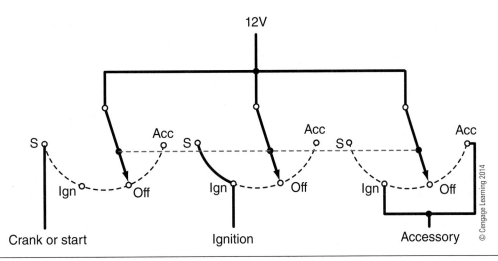

Figure 4-26 Typical ignition switch with three sets of switch contacts.

A switch type where the spring maintains the switch in the closed position is referred to as a normally closed switch. *Normally closed* means the switch is closed (on) unless someone or something intervenes by pressing on the switch in opposition to the spring tension.

A switch in which the spring maintains the switch in the open position is referred to as a normally open switch. *Normally open* means that the switch is open (off) unless someone or something intervenes by pressing on the switch in opposition to the spring tension. A momentary contact horn switch is a normally open switch.

Normally open and *normally closed* are abbreviated NO and NC, respectively. The schematic symbols for momentary contact normally open and normally closed switches are shown in **Figure 4-27**.

All switches shown thus far are designed for the person operating the switch's moving contact to change the switch's positions. Some other switches may use pressure or fluid flow to cause the moving contact of a switch to change states. An example is an engine oil pressure switch. An engine oil pressure switch is threaded into an engine oil gallery where one component of the switch is exposed to engine oil under pressure. An engine oil pressure switch is typically a normally closed switch. The switch is closed when the oil pressure is low.

Note that low engine oil pressure is not considered a normal condition if the engine is running. Do not confuse normally closed or normally open with a normal condition. *Normal state* refers to the state of the switch when tested at the parts counter. There would be no oil pressure applied to the oil pressure switch when it is lying on the parts counter. When the switch is closed, the engine oil pressure warning lamp in the instrument cluster illuminates to warn the operator that the engine oil pressure is low. When the oil pressure is above a designated amount, the spring in the normally closed switch is compressed by the engine oil under pressure, which causes the switch contacts to open. This causes the engine oil pressure warning lamp to turn off.

Switch Problems

The resistance of a closed switch should be as low a value as possible because the switch is typically connected in series with the device being controlled by the switch. Any closed switch resistance results in a voltage drop across the switch and reduces the current flowing in the circuit, just like adding a series resistor to the circuit. The voltage drop across the switch contacts reduces the voltage available for the device controlled by the switch.

The switch contact resistance may be abnormally high because the switch contacts have been exposed to water and have formed a layer of corrosion. For example, if the contacts of a headlamp dimmer switch have corroded on a truck with a 12V electrical system, the contact resistance will be higher than when the switch was new. The contact resistance will cause a voltage drop across the dimmer switch. If 2V is dropped across the closed contacts of the dimmer switch, then Kirchhoff's voltage law indicates that

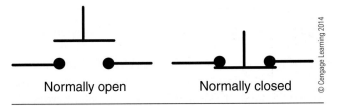

Figure 4-27 Momentary contact switches.

only 10V will be available to the headlamps. This will cause the headlamps to be much dimmer than normal.

Another cause of switch contact resistance is contact burning. The switch contacts make and break (interrupt) the flow of current in a circuit. This making and breaking action causes some arcing across the switch contacts, like an arc welder, when the switch initiates or stops the flow of current in the circuit. The arcing creates heat, which burns the switch contacts. The higher the level of current that is interrupted by the switch, the greater the amount of contact arcing that occurs. Switches are designed to conduct up to a specific amount of current. Exceeding the maximum current specification can cause the switch contacts to burn, resulting in a high level of series resistance.

Switch Troubleshooting. The resistance between the terminals of a closed switch should be very low. Therefore, the voltage drop across a closed switch (switch is in the ON state) should be close to 0V. Conversely, the resistance between the terminals of an open switch should be very high. Therefore, the voltage drop across an open switch (switch is in the OFF state) should be nearly the same as the battery voltage as shown in **Figure 4-28**. This might seem backward, but it is all about Ohm's law and the rules of series circuits. In a series circuit, the voltage divides proportional to the series resistance values. A light bulb, a resistor, or some other electrical component that does work is often referred to as a **load**. Because the load should have a resistance that is much greater than the closed switch resistance, almost all of the voltage should be dropped across the load with the switch closed, while nearly none of the voltage is dropped across the closed, low-resistance switch contacts.

When the switch is open, the resistance between the open switch terminals should be nearly infinite. The resistance of the open switch should be much greater than the load resistance, so virtually all of the voltage should be dropped across the open switch contacts with the switch open.

It is very important to understand why there is nearly 0V dropped across a closed switch and nearly all the total battery voltage dropped across an open switch. To prove this, **Figure 4-29** shows a switch in a 12V circuit that is used to control a 1Ω lamp. The closed switch resistance is given as 0.001Ω (1mΩ). The open switch resistance is given as 1 million ohms (1MΩ). Actual closed and open switch resistances may be much higher or lower than these values, but this should give you an idea of what is happening. If you use these values to calculate the voltage dropped across the switch when the switch is closed and when

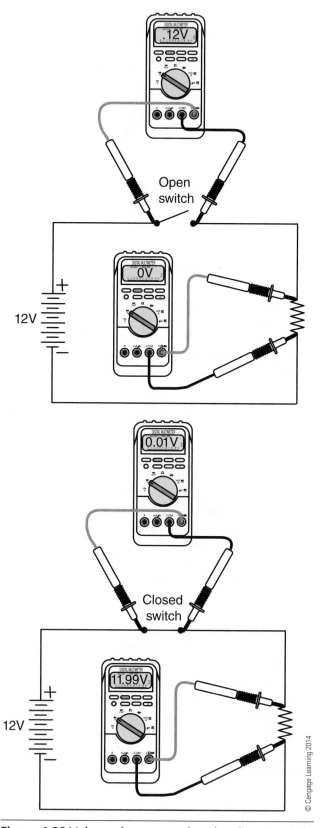

Figure 4-28 Voltage drop across closed and open switch.

the switch is open, you will observe that there is nearly 0V dropped across a closed switch and nearly all the source voltage (12V in this case) dropped across an open switch.

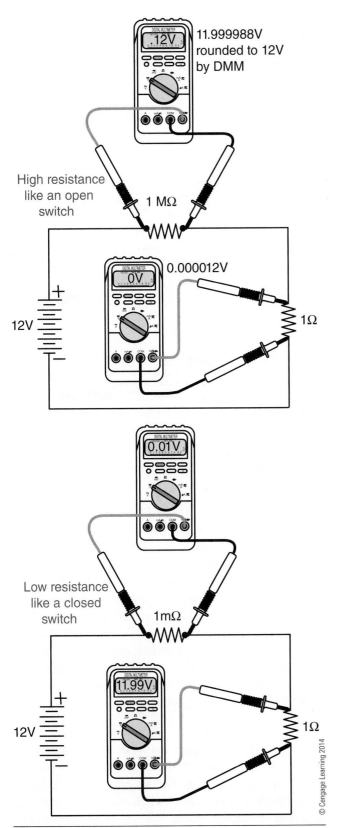

Figure 4-29 Open and closed switch replaced by resistors to illustrate voltage drop across open and closed switch.

Tech Tip: It may seem backward, but you should measure close to 0V across a closed switch (switch is ON) and near battery voltage across an open switch (switch is OFF). Remember that you are measuring the voltage dropped across the switch, not the voltage that is present at the switch output terminal referenced to ground.

Figure 4-30 illustrates one of three different problems in which a light bulb symbolized by the resistor does not illuminate or only illuminates dimly when the switch is closed. Note how the voltmeter is used to measure the voltage at two locations throughout the circuit to troubleshoot the problem. If the voltage across a closed switch indicates near-battery voltage and the lamp does not illuminate, then the closed switch resistance is much greater than the load resistance, causing the majority of the voltage to drop across the closed switch contacts (**Figure 4-30**). This may be due to burned or corroded switch contacts. The high resistance of the defective closed switch reduces or stops all current flow in the circuit.

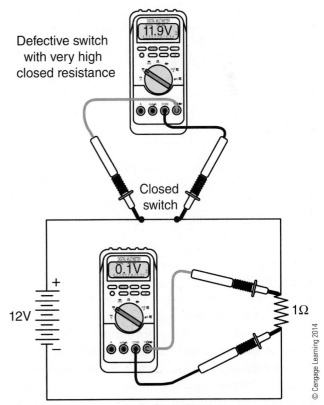

Figure 4-30 Switch with high closed resistance causing low or no voltage at load: note high voltage drop across closed switch.

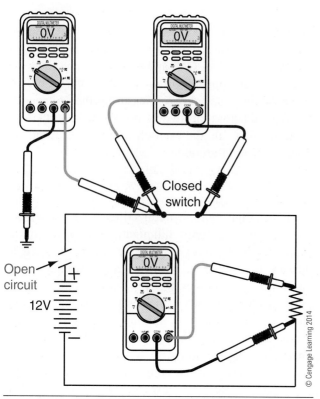

Figure 4-31 Open circuit between switch and battery causing 0V at switch terminals and 0V at load.

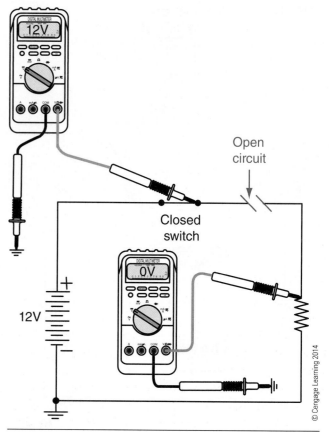

Figure 4-32 Open circuit between switch and load causing voltage at switch terminals but 0V at load.

If nearly 0V is measured across the closed switch terminals and the lamp does not illuminate, determine if 12V referenced to ground is present at the switch supply terminal. If 12V is not available at the switch supply terminal, look for a problem between the supply side of the switch and the voltage source. In the example shown in **Figure 4-31**, the circuit is open between the battery and the switch.

If 12V referenced to ground is present at the switch supply terminal and 0V is dropped across the closed switch, look for a problem between the output side of the switch and the load device ground. In the example shown in **Figure 4-32**, the circuit between the switch and the lamp is open.

Switch Replacement. Most switches provide long life but may wear out over time due to normal use. However, a switch that has not been in service for very long that has failed is cause for further investigation. Instead of just replacing a damaged switch, you should determine the cause of the damage or you may have to repair the vehicle again next week for free. If the switch damage is caused by water leaking onto the switch or circuit current that is higher than the switch's rating, the replacement switch probably will soon fail as well.

Tech Tip: Finding the root cause of a failure minimizes embarrassing come-backs. Component (part) failure due to normal use over a prolonged period of time is expected. However, a component that fails on a newer truck or a component that has been replaced several times should make you suspicious. The failure of a component such as a switch may be due to some other problem elsewhere in the circuit. The component that fails may just be the weakest link in the system.

Switches used as inputs to electronic modules must be rated for low current. It would seem that low current would not be a problem with switches because interrupting high current causes switch contacts to burn due to arcing. However, some amount of contact arcing serves to clean the contact surfaces for common switches. This is the reason that most switches have a specified minimum current rating in addition to the maximum current rating. Using a standard switch in a low-current application (less than 0.1A or 100mA) may result in inconsistent switching. These special low-current switches may have contacts coated with gold or other metal to prevent corrosion. Low-current circuits are sometimes referred to as *dry circuits*.

RELAYS

Relays are electromagnetic switching devices that permit a small amount of current to control a large amount of current. Relays of various sizes and current ratings are used extensively throughout truck electrical systems to permit a switch with a low-current rating or an electronic module to control a high-current circuit. A relay can be considered as a remotely controlled switch.

Relay Operation

Figure 4-33 illustrates the internal components of a typical relay. The relay shown in this example contains a set of normally open switch contacts. A coil of wire is wrapped around a stationary core made of metal. One end of the coil is shown as being grounded in this example. The other end of the coil is connected to a terminal identified as the **control circuit**, which is used to control the relay as explained later in this section. A movable component known as an **armature** is made of metal and is positioned above the stationary core. The power source terminal of the relay in this example is electrically connected to the metal armature. One of the switch contacts is attached to the end of the armature. The other switch contact is stationary and is located below the armature switch contact. The load terminal of the relay in this example is connected to the stationary contact.

A spring is utilized to maintain an air gap between the armature and core to keep the switch contacts held in the open position. When electric current passes through the coil of wire, the stationary metal core inside the coil of wire becomes magnetized, as explained in **Chapter 3**. This causes the metal armature to be magnetically attracted toward the core, in opposition to the tension of the armature spring. The downward movement of the armature toward the magnetized core causes the switch contact mounted to the armature to make an electrical connection with the stationary switch contact. The closed relay switch contacts then provide a path for current to flow from the power source to the load, similar to a closed switch.

Figure 4-34 shows the schematic symbol for a common miniature relay found on trucks. This relay is often referred to as a DIN or ISO relay. This particular relay is about the size of an ice cube and has male blade-type terminals. These relays are plugged into a mating connector, which is often incorporated with a large hard-shell connector called a fuse block or fuse terminal. The footprint or bottom view of the relay (terminal layout) is standardized as are the terminal numbering designations. The footprint design only permits one way for the relay to be plugged into the fuse block. Terminals numbered as 86 and 85 are the coil connections. Typically, terminal 86 is connected to the positive voltage supply and terminal 85 is connected to the negative voltage source (ground), but this polarity is often reversed by some OEMs for no particular reason. The resistance of the wire making up the coil on most miniature relays used in 12V systems is typically 50Ω to 100Ω. The electrical circuit that

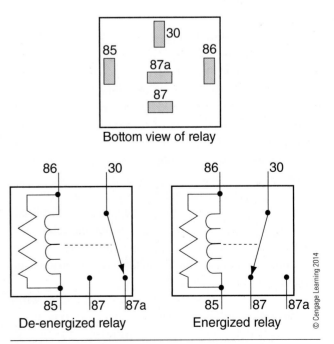

Bottom view of relay

De-energized relay   Energized relay

© Cengage Learning 2014

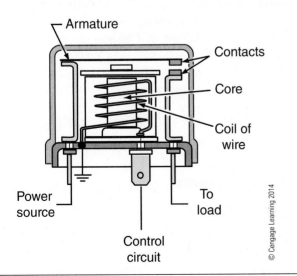

© Cengage Learning 2014

Figure 4-33 Relay components.

Figure 4-34 Common relay terminals and footprint.

provides power and ground to the relay coil is referred to as the control circuit.

Terminal 30 of the miniature ISO (DIN) relay is referred to as the **common** terminal. The common terminal is often connected to a positive voltage supply, but it could also be connected to ground depending on the circuit design. Terminal 87 is referred to as the normally open output of the relay. Normally open indicates that when the relay is sitting on the parts counter, the contacts are open. Therefore, no continuity exists between terminal 30 and terminal 87 until the relay coil is **energized**, which means that sufficient current is flowing through the relay coil to cause the electromagnet to overcome the armature spring tension and change the state of the contacts. This movement of the armature can typically be heard as an audible click and can typically be felt on the relay cover. The armature movement causes the normally open switch contacts to close, providing a low-resistance connection between the common terminal (30) and the normally open relay output terminal (87).

Tech Tip: Memorize the purpose of each terminal number (30, 85, 86, 87, and 87A) on a miniature ISO relay. These ISO relay terminal designations are common throughout the truck industry. Most relays have the terminal number molded into the relay base. This lets you know what value of voltage would be expected in each of the corresponding fuse block terminals when you are troubleshooting a relay circuit.

Like switches, relay contacts use a pole and throw designation to describe the relay configuration. The relay shown in **Figure 4-34** is a single-pole double-throw (SPDT) relay. One set of contacts is normally open; the other set is normally closed. The common terminal (30) acts as a shared switch contact for both the normally open and normally closed contacts, thus the designation as "common." The ISO designation for the normally closed relay terminal is 87A. A low-resistance electrical connection exists between the common terminal 30 and the normally closed terminal 87A when the coil is de-energized. When the relay coil is energized, the electrical connection between the common terminal and the normally closed terminal is opened, like an open switch.

Many other types of relays are available with multiple poles and throws, but the customary type used on trucks is the SPDT type. This SPDT type typically has a normally closed contact, held closed by spring

tension, and a normally open contact that is closed magnetically when the relay coil is energized. When the normally closed contact is open, the normally open contact is closed and vice versa.

Most relays provide a graphic illustration on the cover to show the relay's contact configuration and the definition of the relay's terminals. The schematic symbols vary by relay manufacturer, but some examples of relay cover schematics along with their footprint are shown in **Figure 4-35**. The dashed line indicates that the switch is closed by a magnetic field. The symbol for the relay coil is often shown as a rectangle with a diagonal line, as shown in these examples.

Many other relays have a different numbering system and use a 1, 2, 3, 4, 5 numbering convention. The schematic symbol is typically printed or etched on the relay cover and should indicate the internal connection of each of the relay's terminals.

Relay Coil Suppression

Inductors used on a modern truck, such as a relay coil, typically require some type of suppression device to prevent a high-level negative voltage spike from developing across the coil when current flow through the coil is rapidly interrupted. Not all relays have suppressed coils. A non-suppressed relay may have the same footprint as a suppressed relay, meaning it would plug right in to the fuse block and operate the same way as a suppressed relay. It is very important to use relays with suppressed coils on modern trucks to prevent damage to electronic components on the truck. The schematic for the relay printed on the relay cover typically shows the coil suppression, if present (**Figure 4-36**). Most truck relays have resistor suppression. A resistor in parallel with the relay coil prevents the negative voltage spike from reaching a high level, as explained in **Chapter 3**. Resistor suppression may also be shown as a rectangular box in parallel with the coil instead of the typical zigzag line used to represent a resistor.

Some relays may use an electronic component called a diode as the suppression device instead of a resistor. Diodes are covered in **Chapter 6**.

Relay Problems

Relays can fail for three main reasons:

1. Open circuit (break) or high resistance in the relay coil wiring or connections resulting in little or zero current flow through the coil. This prevents the relay from energizing. This failure

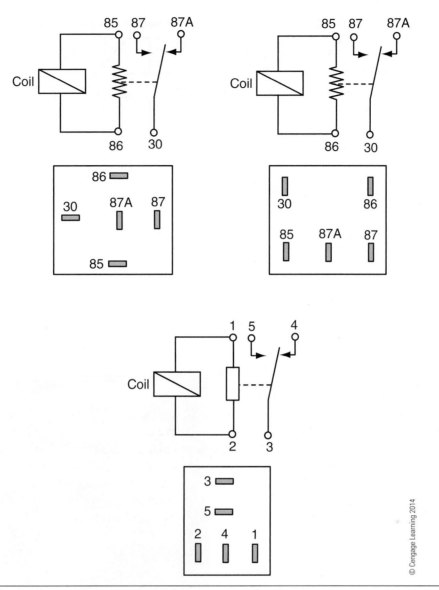

Figure 4-35 Typical relay contact schematics.

causes electrical devices supplied with electricity through the normally open contacts of a relay not to operate.

2. Relay contacts are burned or otherwise damaged, resulting in a high level of series resistance with the load device when the contacts are closed. This can cause low or zero current flow through the closed relay contacts. This failure also causes electrical devices that are supplied current through the normally open contacts of a relay not to operate or to operate erratically.

3. Relay normally open contacts are welded or stuck in the closed (energized) position. This causes the normally open contacts to remain closed even after the relay coil is no longer energized and causes the normally closed contacts to remain open at all times with SPDT relays.

This problem often occurs due to the relay conducting too much current or switching too high a voltage. This failure causes electrical devices that are supplied current through the normally open contacts of a relay to operate when the relay coil is not energized.

Troubleshooting Relays and Relay Control Circuits.
One of the most common methods of troubleshooting an inoperative device controlled by a relay is to listen or feel for the relay to click when the coil is energized or de-energized. The click is caused by movement of the relay armature. If the relay click cannot be heard or felt when the control circuit is powered, the relay coil is probably not being energized. This might be caused by an open circuit in the relay coil control circuit or a defective relay.

© Cengage Learning 2014

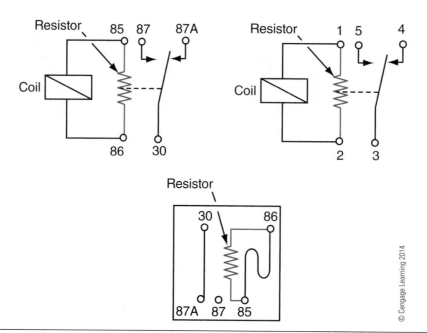

Figure 4-36 Relay with resistor suppression.

Tech Tip: The relay click method is probably one of the most widely utilized methods of determining whether a relay is energizing. Be careful when using this method, though, because more than one relay in the fuse block may "click" at the same time. The click from one relay may also be felt in several other relays, even though they are not being energized.

If the relay clicks when energized, then the relay control circuit and relay coil are probably okay. At this point, troubleshooting an inoperative device controlled by a relay is about the same as troubleshooting a device controlled directly by a switch because the relay is a remotely controlled switch. Most technicians keep spare relays that are known to be good to use for quick troubleshooting.

You should avoid the common practice of replacing a relay with a jumper wire placed between the fuse block terminals corresponding to the relay common and normally open terminals to test the circuit. A mistake in placing the jumper wire could result in damage to an electronic module that may be used to control the relay. Special relay breakout Ts are available, such as that shown in **Figure 4-37**, which permit voltage measurements to be obtained for troubleshooting relays located in a crowded relay panel or fuse block.

When replacing or adding new relays to the truck, it is good practice to mount the relay so that the

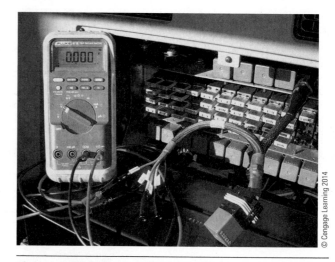

Figure 4-37 Relay breakout T used to test ISO relays and relay circuits.

terminals are pointing downward. This reduces the chances of water making its way into the relay. A small amount of water in a relay can result in contact damage and corroded terminals.

Larger Relays

The miniature relays shown in the previous section are typically used to switch less than 40A of current. Trucks may have circuits in which much more than 40A needs to be controlled, such as the starter motor solenoid, glow plugs, or intake air heater. For these high-current switching applications, a larger relay known as a magnetic switch—or mag switch—is typically used. Magnetic switches have a hollow coil winding that surrounds a metallic plunger (**Figure 4-38**). The spring-loaded plunger is magnetically pulled into the windings when the windings are energized. The movement of the plunger results in the large contact disc (washer) inside the magnetic switch providing electrical contact between the external studs. Magnetic switches control high levels of current and are often included in the starter motor control circuit.

Some magnetic switches are designed for continuous duty, meaning they can be switched on and left on indefinitely. Other magnetic switches can only be used for intermittent duty, such as many starter motor magnetic switches. The duty limitation is based on heat generated by the control coil. Using the wrong type of magnetic switch for the application can cause premature magnetic switch failure.

The magnetic switch in **Figure 4-38** is shown with two control circuit terminals that are internally connected to the coil. One terminal is supplied with a positive source, and the other terminal is supplied with a ground connection. Neither of these terminals is grounded internally. Other similar magnetic switches may only have one external control terminal that is supplied with a positive source. The negative side of the winding is internally connected to the case of the magnetic switch and depends on the grounding of the mounting bracket to provide a ground for the control circuit.

CIRCUIT PROTECTION DEVICES

A **circuit protection device (CPD)** is a device designed to protect an electrical circuit from excessive heat due to higher than normal current flow. The primary purpose of a CPD is to protect the truck's wiring harness. CPDs include **fuses, circuit breakers, fusible links,** and **positive temperature coefficient (PTC) devices**. Each of these devices will be covered in detail.

CPDs are often thought of as safety valves for electricity. A truck air brake system has a pressure relief valve designed to open at about 150 psi (1034 kPa) and release air pressure from the system before damage occurs due to excessive air pressure. This safety valve should not typically open during operation because air brake system pressure should never reach 150 psi unless something has failed in the pressure control system, such as the governor. In a manner similar to a pressure relief valve, CPDs open if current exceeds some value. Like a pressure relief valve in an air brake system, a CPD should only open if a problem occurs.

Current and Heat

Current flow through a resistance, such as a section of wire, causes heat to be produced as indicated by Watt's law in the form shown in **Equation 4-1**.

$$\textbf{Watts} = \textbf{Current}^2 \times \textbf{Resistance}$$
$$\textbf{or}$$
$$\textbf{Watts} = \textbf{Current} \times \textbf{Current} \times \textbf{Resistance}$$

The superscript 2 or exponent in the Watt's law equation indicates that the current is to be multiplied by the current. Thus, increasing the current flow through a piece of wire causes the power dissipated by the wire (heat) to increase at a very rapid or nonlinear rate. For example, increasing the amount of current flowing through a section of wire by a factor of 10 causes the amount of heat that the section of wire must reject to increase by a factor of 100.

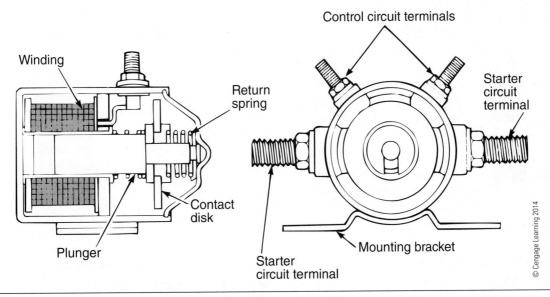

© Cengage Learning 2014

Figure 4-38 Components of a magnetic (mag) switch.

If you have ever looked inside a toaster, you have seen the wire in the toaster glow red-hot. Most insulation material surrounding truck wiring will catch fire if overheated. Causing a wire to conduct too much current can result in the wire glowing red-hot, like the wire in the toaster. This heat can cause the wire insulation material to ignite and start a fire. This leads to an important fact about CPDs.

Important Fact: The primary purpose of a CPD is to keep the wire temperature below the point where it may become overheated.

CPDs are designed to permit current flow in a circuit up to a specific level known as the current rating. All CPDs are normally closed and are designed to open to interrupt the flow of current in the circuit like an open switch.

As indicated, wire has resistance, and current flowing through resistance produces heat per Watt's law. As wire diameter is decreased, the resistance of the wire is increased. Therefore, the smaller the diameter of a wire, the less current the wire can safely conduct without overheating. The OEM typically selects the current rating of a particular CPD based on the smallest-diameter wire (largest AWG number) that is being protected by the CPD. If it becomes necessary to replace a section of wire in a truck's wiring harness, it is important that the replacement wire have the same (or larger) diameter as the original wire engineered by the OEM.

Additional wiring that is added to a truck to power an electrical accessory should always be protected by a CPD. Some types of electrical system failures can cause higher than normal current to flow in a circuit, as will be discussed later in this chapter. The current rating of the CPD must be selected so that the smallest-diameter wire protected by the CPD will not overheat before the CPD opens should current in the circuit become higher than normal. Consult the electrical accessory manufacturer's information for the recommended CPD current rating and wire gauge to be used to supply the accessory.

WARNING *NEVER replace a wire with a wire that has a smaller diameter. The CPD may not adequately protect the smaller-diameter wire from overheating and could lead to a fire. Always protect any new wiring that is added to a truck with a properly sized CPD.*

Current flow through a CPD that exceeds the current rating of the CPD does not instantly generate enough heat to cause the CPD to open. If you have ever held your hand over a fire, you know that your hand will get hotter the longer it is held over the fire. If you pass your hand over the fire very quickly, you might not feel the heat of the fire at all. This is similar to the reason that a combination of both current level and time is necessary to cause a CPD to open. As the current flowing through the CPD increases, the time necessary for the CPD to heat up enough to open decreases. A CPD that is designed to carry 30A may take several minutes to open if the CPD is conducting 40A of current. The same CPD may open in less than a second if the CPD is conducting 100A of current.

Some CPDs are designed to open more slowly or more quickly than other CPDs. The reason for having different times to open characteristics is to permit the CPD to be able to handle high **inrush current** or surge current. Inrush or surge current describes a high level of current flow that occurs when some devices are first switched on. Devices with high inrush currents, like motors, will draw a very high level of current initially which rapidly decreases. Motors draw a high level of current initially because a motor that is not rotating is just a piece of low-resistance wire. A motor that draws 10A while running at normal speed may draw 100A when the motor is first switched on. Once a motor begins to rotate, the current drawn by the motor decreases dramatically. The reason for this will be explained in **Chapter 8**. By the time the motor reaches operating speed, typically within a few seconds, the motor may only be drawing 10A. The CPD that protects a circuit supplying an electric motor must be capable of withstanding the motor inrush current without opening, even though the brief inrush current may be many times greater than the CPD's current rating.

Reasons for Excessive Current Flow in a Circuit

Other than normal surge current, the current is excessive in a circuit for two reasons: higher than normal voltage or lower than normal resistance. Most trucks are either 12V or 24V systems. The system voltage typically does not increase much above 15V or 30V, respectively. If a charging system problem were to occur, the system voltage could increase significantly, causing the current drawn by devices like lights to increase dramatically as well. Therefore, several CPDs on a truck may open because the truck electrical system voltage increased substantially due to a charging system problem. This type of charging system problem in which the voltage is too high will be addressed in **Chapter 7**.

Circuit current will also increase above normal if the total circuit resistance decreases. This is the reason that most CPDs open. Circuit resistance may decrease because someone has added several devices, such as lights or other types of loads, to a circuit. Remember that adding resistance in parallel causes total circuit resistance to decrease. Ohm's law indicates that if resistance decreases while the voltage is held at a constant value, the current flow in a circuit will increase.

Several wall electrical outlets in a house may be wired in parallel with each other. The parallel wall outlets are wired in series with a CPD, such as a circuit breaker located in a fuse box. Switching on several high-current devices—like a toaster, hair dryer, and microwave oven—to outlets that are protected by the same CPD will cause the total circuit resistance to decrease and total current to increase. This current increase may cause the circuit breaker to trip because the combined current of the parallel devices has exceeded the circuit breaker's design amperage. If the circuit breaker did not open when the wiring is carrying too much current, the house wiring might become overheated and cause a fire.

Switching on too many high-current devices at the same time which are all protected by the same CPD causes the CPD to open. In the same way, adding too many parallel loads to a truck electric circuit can cause the current through the CPD to increase such that the CPD opens. The CPD that opens because of excessive current flow is doing its job, protecting the wiring from overheating. A CPD is like a pressure relief or safety valve in a truck air brake system. If a problem with the governor prevents the compressor from cutting out, the safety valve opens at about 150 psi (1034 kPa). You would not remove the safety valve and replace it with a pipe plug if the safety valve kept opening due to a governor problem. Doing so would cause some component in the air brake system to eventually rupture. The same is true of a CPD. Replacing a CPD that keeps opening with a CPD that has too high a current rating for the wiring can cause the wiring to overheat.

WARNING *NEVER replace a CPD, such as a fuse or circuit breaker, with a CPD that has a higher current rating than the OEM's recommendation. Doing so may result in a fire.*

Another way for the circuit resistance to become lower is for a **short to ground** to occur. A short to ground describes current flow in a circuit that has taken a shortcut to chassis ground (battery negative) instead of flowing through the load. This shortcut can occur

when a conductor, such as a wire, contacts the grounded vehicle chassis. Because any metal connected to the chassis is grounded, any wire that is connected to the battery-positive terminal must not contact the chassis. The resistance of copper wire is very low. The resistance from the steel frame rail to the negative battery terminal is also very low. Should the insulation on a wire wear through due to abrasion where the wire crosses the edge of a frame rail, the wire will make direct contact with the frame rail. The total circuit resistance will then be very low because it consists only of wire resistance. This will cause a very large amount of current to flow. The actual amount of current flowing through this short to ground depends on how many strands of wire are contacting the frame, on how much force is being exerted on the wire to frame contact point, on the wire gauge or diameter, and on the length of the wire. Current flow due to a short to ground circuit may be as small as 20 amps or it can exceed several hundred amps. The CPD must interrupt this flow of current or the wiring may overheat very quickly.

A simple series circuit with a lamp bulb controlled by a switch is shown in **Figure 4-39**. A fuse with a 10A rating protects the wiring in this circuit. A fuse is a CPD designed to melt open before the wire temperature gets too hot. The schematic symbol for a fuse is the sideways S symbol shown in **Figure 4-39**. This particular light bulb causes approximately 1.2A of current to flow in this circuit when the switch is closed. This indicates that the light bulb's resistance when illuminated is approximately 10Ω.

A short to ground has been introduced to the simple circuit, as shown in **Figure 4-40**. This short to ground may have occurred because the wire was routed over a sharp metal edge of the cab sheet metal. In this example, the resistance of the short to ground has been approximated to 0.1Ω, even though the actual resistance of a shorted circuit may be much less than this. The short to ground is in parallel with the light bulb.

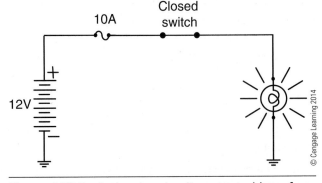

Figure 4-39 Typical series circuit protected by a fuse.

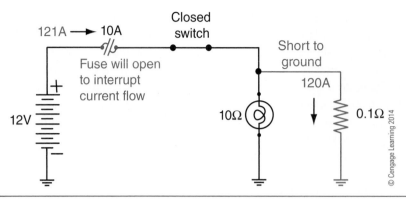

Figure 4-40 Short to ground circuit.

The total circuit resistance with two parallel-connected resistors is found from the product over the sum rule from **Chapter 2**. Solving for total resistance, a value of approximately 0.099Ω is obtained. When the circuit is shorted to ground, the total circuit resistance becomes approximately 0.099Ω. Given a 12V battery, the circuit current with the short to ground present is calculated to be approximately 121A. This 121A flowing through the 10A fuse will cause the 10A fuse to open rapidly before the wire heats up much. Without a fuse present in the circuit, the 120A of current caused by the short to ground would cause the wire to overheat and possibly cause a fire.

In real-life troubleshooting situations, the exact resistance of a short to ground is difficult if not impossible to measure. The values given for the shorted circuit are just an example for illustration purposes.

Most ohmmeters are also not very accurate below about 0.5Ω (500 mΩ). Most electrical heating devices including light bulbs and glow plugs have a very low resistance when the heating device is at ambient temperature. When current flows through the heating device, the resistance increases substantially. This concept of resistance change with temperature change will be discussed in several chapters throughout the text. A high-current device like a diesel fuel filter heater may have a cold resistance of only 0.3Ω. Trying to measure the resistance to ground of the fuel heater circuit to test for a short circuit may make it appear that the fuel heater is shorted to ground when, in fact, the 0.3Ω of resistance may be perfectly normal at ambient temperature.

Tech Tip: Keep the low-resistance accuracy limitation of your ohmmeter in mind when troubleshooting so you do not waste time looking for a problem in a circuit where none exists. Devices such as heating elements and light bulbs have a very low resistance when the device is at room temperature. As the device

heats up, the resistance increases greatly (by a factor of 10 or more). Normal cold resistance of a high-power device like a heating element and a short to ground look very similar when tested using an ohmmeter.

Types of CPDs

The main types of circuit protection devices utilized in electrical systems of modern trucks are fuses, circuit breakers, and fusible links. Each of these devices is designed to be the weak point of a circuit and open before the wiring gets too hot.

Fuses. Fuses are circuit protection devices that contain a piece of small gauge wire or a thin flat metal strap referred to as an element. The fuse element becomes the weakest link of an electrical circuit. This is accomplished by selecting a fuse with a current rating less than the design amperage of the smallest gauge of wire in the circuit that is protected by the fuse. Since current flow creates heat and the fuse element is a small-gauge wire or thin metal strap, the element will heat up more rapidly than the smallest-gauge wire that is protected by the fuse. If an excessive amount of current flows through the fuse for enough time to generate sufficient heat to melt the fuse element, the flow of current in the circuit will be interrupted. Once the element has melted, the fuse is destroyed and must be replaced. A fuse that has opened is often said to have **blown**. A blown fuse is a fuse with a melted element. Examples of blown fuses are shown in **Figure 4-41**.

Fuses have a specific current rating. Fuses for truck electrical systems are available that are designed to open with less than 1A of current flow to those that are designed to handle more than 500A of current flow. Fuses also have a voltage rating. The voltage rating indicates the maximum voltage that the fuse is

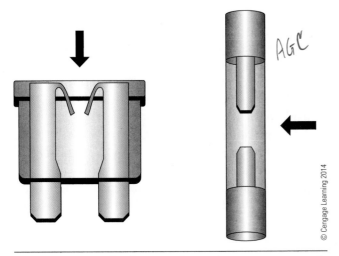

Figure 4-41 Melted open fuse element commonly known as a blown fuse.

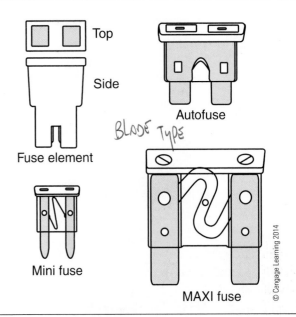

Figure 4-42 Fuses used in truck electrical systems.

for 24V Systems

designed to interrupt. Most fuses designed for truck electrical systems have a voltage rating of 32V.

Fuses are also classified by how long it takes the fuse to open with a given percentage of current overload. Fuses may be classified as fast-acting or time-delay fuses. Any fuse opens based on a combination of current and time. A 20A fuse does not typically open just as soon as current flow through the fuse reaches 20A. Most fuses are designed to maintain a 110 percent overload current indefinitely, so a 20A fuse will probably never blow if 21A of current is flowing through the fuse. At a 150 percent current overload (30A of current flow for a 20A rated fuse), a 20A fuse may blow after 2 seconds for a fuse classified as fast acting or 20 seconds for a fuse classified as time delay. The inrush current of devices controlled by the circuit that the fuse protects determines what type of fuse is necessary.

Most modern truck fuses are blade-type fuses. The blades of the fuse are male terminals. Samples of blade-type fuses are shown in **Figure 4-42**. The male terminals on the blade fuse plug into female terminals mounted in a fuse block or fuse panel. A fuse block or fuse panel is a connector body designed to accommodate several fuses. Trucks may have one main fuse block or several smaller fuse panels located throughout the truck, such as a Freightliner® Power Distribution Module (PDM) shown in **Figure 4-43**. Many fuse blocks are also designed to accommodate relays as well (**Figure 4-44**).

A blade fuse has a colored translucent plastic housing. This allows the fuse element to be visible for inspection. A fuse that is blown may only have a small visible gap in the element with minimal discoloration of the plastic housing if the fuse blew due to a slight overload. However, if a short to ground caused a large

Figure 4-43 Freightliner power distribution module fuse panel.

amount of current to flow through the fuse, the plastic housing may be blackened from the associated heat caused by the short to ground.

The colored fuse housing is color-coded based on the fuse's current rating. The color of the housing for a particular fuse current rating has been standardized in the industry by SAE. Standardized fuse colors throughout the automotive industry increases the likelihood that the correct value of fuse for each circuit is installed in the fuse block.

Figure 4-44 Fuses and relays are often located in the same panel.

Tech Tip: The appearance of a blown fuse can assist you in determining whether the fuse failed due to a moderate level of excessive current, such as too many marker lamps on the same circuit, or whether the fuse failed due to a very high level of excessive current, such as a short to ground. Additionally, the thin fuse element in 5A and smaller fuses may fracture over time due to vibration on severe service trucks, even without any overload ever having occurred. This can cause an intermittent open circuit.

Blade fuses are available in a variety of physical sizes, as shown in **Figure 4-42**. Most modern automobiles and many newer trucks make use of SAE J2077 fuses. The J2077 designation refers to the SAE specification number that defines this fuse. These small fuses are also known by the trade name Mini® or ATM® fuses. These small fuses are classified as fast-acting fuses. This means that the fuse will open relatively rapidly if current flow through the fuse exceeds the design amperage. SAE J2077 fuses are available from 2A to 30A current ratings.

The next larger blade-type fuse is an SAE J1284 fuse. This fuse is also known as an ATC®, ATO®, or Autofuse®. These fuses are somewhat larger than SAE J2077 fuses and were the standard fuse in automobiles and most trucks for many years. These fuses are also classified as fast-acting fuses. SAE J1284 fuses are available from 1A to 40A.

An even larger blade-type fuse is an SAE J1888 fuse. This fuse is also known as a MAXI® fuse. These fuses are classified as time-delay fuses so they are often used to protect circuits connected to devices with high inrush currents, such as electric motors. SAE J1888 fuses are available from 20A to 80A.

Standard blade fuse housing colors for each SAE classification are shown in **Figure 4-45**.

SAE J2077 Mini fuse

Current rating	Color
5	Tan
7.5	Brown
10	Red
15	Blue
20	Yellow
25	White
30	Green

SAE J1284 Autofuse

Current rating	Color
3	Violet
5	Tan
7.5	Brown
10	Red
15	Blue
20	Yellow
25	Natural
30	Green

SAE J1888 MAXI fuse

Current rating	Color
20	Yellow
30	Green
40	Amber
50	Red
60	Blue
70	Brown
80	Natural

Figure 4-45 Standardized blade fuse colors.

Prior to the development of blade fuses, glass cartridge fuses were the standard. A glass cartridge fuse is a small glass tube with a fuse element running the length of the tube. Each end of the glass tube has a metal cap that connects to the fuse element. The fuse current and voltage ratings are stamped into the caps. There are many types of glass cartridge fuses, but the most common types found on older trucks are SFE and AGC types.

AGC fuses all have the same physical dimension regardless of fuse current rating. SFE-type fuses increase in length as the fuse's current rating increases. The fuse block for SFE fuses is designed so that only fuses of the correct length can be placed in the corresponding fuse block cavity. This prevents a fuse with too high a current rating from being installed in a particular fuse block cavity because such a fuse would be too long for that cavity.

Larger bolt-in-type fuses are used to protect large-diameter cables and act as main fuses for power distribution systems. A MIDI® or AMI® fuse is a bolt-in-type fuse with ratings from 30A to 200A. These are color-coded and may have a clear window that permits the fuse element to be seen to determine if the fuse is blown. MIDI fuses are shown in a Freightliner Power Net Distribution Box (PNDB) in **Figure 4-46**.

A very large bolt-in-type fuse known as a MEGA® or AMG® fuse (**Figure 4-47**) is often used to protect the main battery power feed cable that supplies the cab electrical system. MEGA fuses are available with ratings from 40A to 500A.

Circuit Breakers. Circuit breakers are thermal devices that use current flow to heat a thin piece of bimetallic strip, as shown in **Figure 4-48**. If the bimetallic strip is sufficiently heated, the strip will bend or snap, causing a pair of contacts to open and interrupt current flow in a circuit. An open circuit breaker is referred to as being **tripped**. When the circuit breaker cools down, the circuit breaker can be reset, which causes the contacts to close again and restore current flow in the circuit. Because a circuit breaker can be reset, this provides an advantage over a fuse, which must be replaced if blown. Like fuses, circuit breakers have a current rating and voltage rating. Circuit breakers generally require longer to open than a

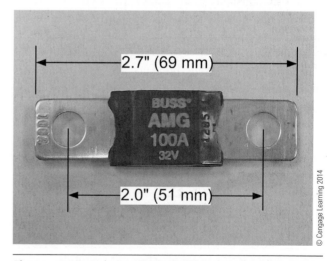

Figure 4-47 Bolt-in AMG MEGA fuses are used to protect large cables and distribution networks.

Figure 4-46 Bolt-in AMI MIDI fuses and ATC fuses in Freightliner power distribution net box.

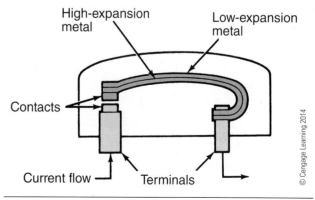

Figure 4-48 Internal components of SAE Type I circuit breaker.

comparable fuse for a given amount of excess current, so they are considered time delay.

The schematic symbol for a circuit breaker (CB) is a sideways C, as shown in **Figure 4-49**. Like a fuse, the current rating of the circuit breaker is often displayed next to the schematic symbol along with identification such as CB3.

Circuit breakers are classified by SAE into three different categories: Type I, Type II, and Type III. The three categories describe the type of circuit breaker reset mechanism.

SAE Type I circuit breakers, such as the example shown in **Figure 4-48**, are automatic reset devices. Type I circuit breakers reset automatically without operator intervention after the bimetallic strip cools down. The reset action causes the contacts to close and restore current flow to the circuit. If the circuit is still drawing too much current, the breaker will open or trip again repeatedly. Type I circuit breakers are commonly used in windshield wiper and headlamp circuits where the automatic reset feature is desirable. However, the constant reset action of Type I circuit breakers can cause the circuit breaker contacts to be damaged from excessive making and breaking of a high-current circuit.

SAE Type II circuit breakers are modified automatic reset devices as illustrated in **Figure 4-50B**. Type II circuit breakers reset automatically after the cause of the excess current in the circuit is no longer present. Type II circuit breakers also reset when the voltage supply for the circuit breaker, such as ignition voltage, is switched off for a few minutes. Type II circuit breakers contain a heating coil, which causes the bimetallic strip to open when the device is conducting too much current. Type II circuit breakers are designed so that current flow through the heating coil continues even after the circuit breaker trips. When the current through the heating coil decreases after a repair is made or power is removed from the circuit, the bimetallic strip cools and automatically closes again like a Type I circuit breaker.

SAE Type III circuit breakers are manual reset devices as illustrated in **Figure 4-50A**. A trip-indicating button or lever pops up when the breaker is tripped. After the bimetallic strip cools sufficiently, the button can be depressed to reset the breaker. Depressing the reset button before the bimetallic strip has cooled enough will cause the button to pop back up when released. If the cause of the excess current is still present after the bimetallic strip has cooled, the circuit breaker will trip again when the bimetallic strip heats up sufficiently.

Like blade fuses, blade-type circuit breakers are common on modern trucks. Most blade-type circuit breakers are designed to fit into the same fuse block cavities as blade fuses. The footprint of the terminals is the same as a blade fuse, but the circuit breaker may be wider, thicker, and taller than the blade fuse. The housing or reset button of most blade-type circuit breakers is color-coded, using the same coding as the respective blade-type fuse. Other circuit breakers have the current rating stamped on the circuit breaker housing and have no color-coding.

It is important when replacing a circuit breaker that the OEM's recommendations for current rating and reset type are observed.

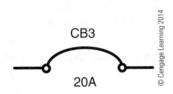

Figure 4-49 Circuit breaker schematic symbol.

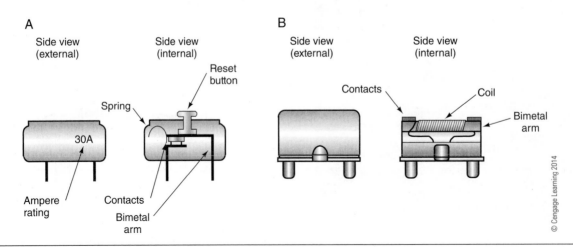

Figure 4-50 SAE Type III circuit breaker (A) and Type II circuit breaker (B).

Fusible Links. A fusible link is a piece of wire with a high-temperature insulation material that is flame resistant (**Figure 4-51**). A fusible link is designed to be similar to the weakest link in a chain. The weakest link in the chain will break before the stronger links in the chain when subjected to excessive force.

The wire gauge of a fusible link is typically selected so that it is smaller diameter than the smallest wire that is directly protected by the fusible link. In the event of a short to ground in a circuit protected by a fusible link, the fusible link heats up more than any wire that is protected by the fusible link. Excessive current flow through the fusible link causes the wire to melt like a large fuse element. The special insulation material used in fusible links does not melt off the fusible link nor does it support a flame, like many types of standard wire insulation. Fusible links are like fuses in that once they have melted, they must be replaced.

Fusible links are somewhat difficult to replace because, unlike a fuse, they do not merely connect to a fuse block like a blade or cartridge fuse. Replacing a fusible link involves cutting the damaged fusible link out of the wiring harness and splicing a new fusible link in its place. This can be a very time-consuming process because fusible links are often located near the starter motor and are difficult to access. However, fusible links are typically limited to protecting high-current circuits such as alternator charging circuits and cab power supply cables. These types of circuits should rarely be subjected to currents high enough to cause the fusible link to burn open. The circuits that fusible links supply are typically branched into other sub-circuits that are protected by smaller CPDs such as fuses and circuit breakers.

Unlike fuses, fusible links do not have a current rating. Instead, fusible links are rated by conductor diameter. The size of the fusible link conductor is stamped on the insulation. Often, fusible links are sold in metric sizes (mm^2) and converted to the closest AWG size.

A fusible link that has burned open must be replaced with another fusible link of the same length and gauge. It is critical that the fusible link be replaced only with fusible link wiring of the correct gauge for the application. The fusible link must be the smallest-diameter circuit to provide adequate wiring harness protection. Standard design practice is for the fusible link to be four AWG sizes smaller (larger number) than the smallest section of wire that is protected by the fusible link.

> **WARNING** *A fusible link that has burned open must only be replaced with the same size of special fusible link wiring. Installing a larger gauge of fusible link is like installing a larger fuse and can result in a fire. NEVER replace a fusible link with standard wire. The insulation material on standard wire can ignite, resulting in a fire.*

The routing of a fusible link must also be placed such that the fusible link is not contacting anything flammable. The excessively high current flow through a fusible link causes the fusible link wire conductor to glow red-hot before it melts to interrupt the circuit. The heat given off by the hot fusible link wire is significant and can cause surrounding material to ignite.

Positive Temperature Coefficient Devices. Positive temperature coefficient (PTC) circuit protection devices have been used for many years in automobiles. PTC refers to the property of most conductors where the resistance of the conductor increases as the temperature of the conductor increases. If the temperature of the conductor versus the resistance of the conductor were plotted on an XY graph, the line would go uphill from left to right as shown in **Figure 4-52**. This uphill direction of the plotted line is referred to as a positive slope in math. The word *coefficient* is also a math term meaning a multiplier. As Watt's law indicates, an increase in current flow causes an increase in wattage or heat dissipated by the conductor. Thus, as current flow increases through a conductor with a positive temperature coefficient, the resistance of the conductor will increase. PTC circuit protection devices make use of this increase in resistance with increased current flow to protect wiring from overheating.

A typical PTC device contains a polymer (plastic) such as polyethylene mixed with carbon or graphite. When the temperature of the PTC device is below the designed "trip point," the device has very little resistance across its terminal like a fuse or circuit breaker. As the device self-heats when current flow through the device increases, the polymer expands

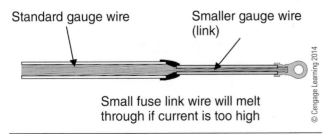

Standard gauge wire   Smaller gauge wire (link)

Small fuse link wire will melt through if current is too high

© Cengage Learning 2014

Figure 4-51 Fusible link.

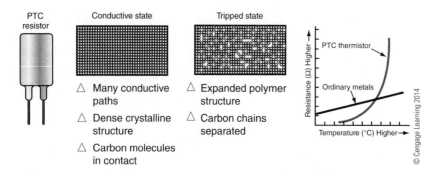

Figure 4-52 PTC device operation: high temperature due to excessive current flow causes the resistance of the device to increase to protect the circuit.

resulting in separation of the carbon chains connecting the two terminals of the device. The resistance of the PTC device increases at a nonlinear rate as the temperature increases, resulting in a reduction of circuit current sufficient to protect the wiring. This is illustrated in **Figure 4-52**. If the condition causing the high-current flow is resolved, the PTC material will cool, resulting in the resistance of the device returning to nearly 0Ω. Thus, the PTC device automatically resets similar to an SAE Type II circuit breaker.

Virtual Fusing. Virtual fusing refers to electronic circuit protection. Virtual fusing will be discussed in detail in **Chapter 13**.

Application of Circuit Protection Devices

The fuse blocks and fuse holders located throughout the vehicle may be referred to as a power distribution center (PDC), power distribution module (PDM), or PNDB. The PDC typically contains CPDs and relays.

A CPD, such as a fuse, may be used to protect several circuits as shown in **Figure 4-53**. Each of the

parallel-connected resistors represents an electrical load such as a light or motor. This type of connection has the advantage of limiting the number of CPDs necessary to protect the wiring. However, if the CPD should open, all electrical devices supplied by the CPD will be inoperable.

In **Figure 4-54**, each load is shown being protected by its own fuse. Thus, an excess current through any one branch will only cause one CPD to open leaving the other branches unaffected.

OEMs typically use a combination of both methods. Devices that are deemed as more important than others or that draw a high amount of current may have their own CPD while less important devices may be grouped together and protected by one CPD.

To protect the large-gauge wiring supplying the fuse block or panel, a high-current CPD such as a fusible link or MEGA fuse is typically used. Typical truck CPD architecture is shown in **Figure 4-55**. Fuse F1 is a 150A fuse, such as a MEGA or MIDI fuse, and protects the large gauge wire, which supplies the fuse block. Typically, this would be a 2 AWG or larger circuit. If the conductors of this 2 AWG wire were

Figure 4-53 One fuse protecting several circuits.

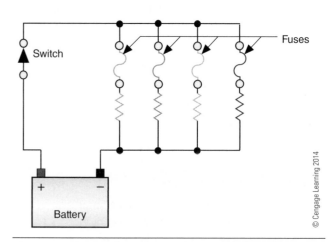

Figure 4-54 One fuse for each parallel branch.

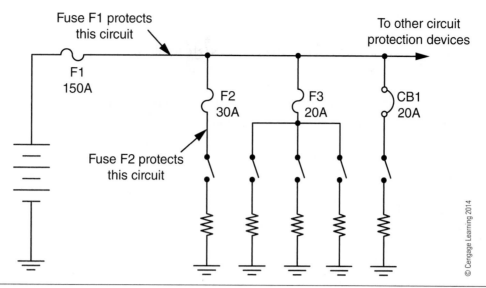

Figure 4-55 Typical truck CPD layout.

shorted to ground, more than 500A or more of current might flow through fuse F1 resulting in a blown fuse F1.

Fuse F2 in **Figure 4-55** is a 30A fuse, which protects the smaller gauge wiring between the fuse and the load, such as a blower motor. Typically, this would be a 10 AWG or larger circuit. If the conductor of the 10 AWG wiring protected by fuse F2 were shorted to ground, 200A or more of current might flow through both F1 and F2. However, only fuse F2 would probably blow. This is because 200A flowing through a 30A fuse should cause it to blow much more quickly than 200A flowing through a 150A fuse.

Using a chain as an analogy for series-connected fuses, imagine a chain that is designed for 4400 lb (2000 kg) has two defective weak links like that shown in **Figure 4-56**. Imagine that the weakest link will fail when subjected to 220 lb (100 kg), while the second weakest link will fail when subjected to 1750 lb (800 kg). If the defective chain where subjected to a 2200-lb (1000-kg) load, only the weakest link would likely fail. Once the weakest link fails, the load falls to the ground; therefore, the second weakest link is no longer overloaded so it does not break.

TROUBLESHOOTING WIRING PROBLEMS

Troubleshooting wiring problems is one of the most common electrical-related tasks required of a truck technician. The most common wiring problems are the result of open circuits or circuits that are shorted to ground.

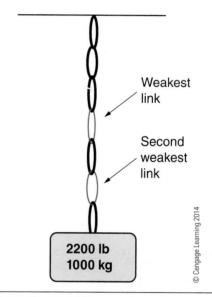

Figure 4-56 Chain with two weak links is similar to two fuses of different values connected in series.

Terminology for Wiring Problems

The term *short circuit* or *short* is often misused. It seems that anytime a wiring problem is suspected, it is blamed on a short circuit. Most often, the cause of an inoperative electrical component is an **open circuit**, not a short circuit. An open circuit is not the same as a short circuit. An open circuit is caused by a severely corroded terminal, a terminal pushed out of a connector, or a broken or cut wire (**Figure 4-57A**). All of these scenarios cause high or infinite circuit resistance, which results in reduced or interrupted current flow in the circuit. A circuit with excessive resistance such as that caused by corroded terminals is similar to an open circuit, but there is still some current flow, which may cause a lamp to be dim or other similar symptom.

There are actually three different types of short circuits. A short to ground typically occurs when the wire insulation has worn through due to abrasion or pinching and the wire conductor is contacting a grounded metal component such as the frame rail (**Figure 4-57B**). This is the most common type of short circuit.

The second type of short circuit is a short to battery positive (**Figure 4-57C**). This can occur due to abrasion between two wires that causes the insulation to wear away on both wires, causing contact between the two conductors. This type of failure can also occur when two wires are crushed together or when wire insulation wears through a sharp edge of a battery-positive connection stud such as that found at the starter motor. This short to battery positive is a rare condition but can cause unexpected operation of electrical components that is difficult to troubleshoot.

The third type of short circuit is a wire-to-wire short, or copper-to-copper short. This condition occurs when the insulation of two or more circuits that are not on continuous battery connection has worn through and the conductors of the circuits are making contact with each other (**Figure 4-57D**). This type of short circuit can be one of the most difficult wiring failures to troubleshoot because closing one switch may cause two seemingly unrelated operations to occur. For example, a wire-to-wire short between the horn circuit and the headlamp circuit would cause the headlamps to illuminate and horn to sound when the horn switch is depressed. Switching the headlamps on would also cause the horn to sound.

An open circuit will not cause a CPD to open while a shorted to ground circuit will probably cause a CPD to open. A wire-to-wire short can cause a range of unexpected operations. A short to battery positive can cause drained batteries and unexpected operation as well. In general, any type of a short circuit is more difficult (time consuming) to diagnose than an open circuit, so it often costs the customer more to have the cause of a short circuit diagnosed and repaired than an open circuit.

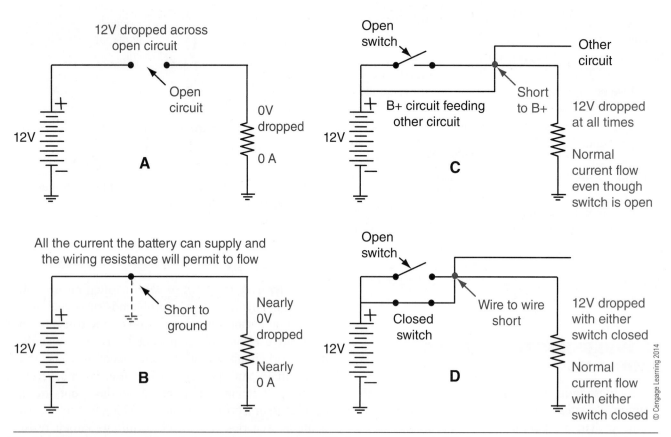

Figure 4-57 Types of wiring defects: (A) Open circuit, (B) short to ground, (C) short to battery positive, (D) wire to wire short.

Tech Tip: A change in the circuit current flow typically indicates a problem. An open circuit or high series resistance decreases the circuit current while a short to ground increases the circuit current, as indicated by Ohm's law.

Finding the Cause of Excessive Current Flow

Where is the best place to start when you have determined that a fuse is blown? What you need is a plan of action to attack the problem. A general diagnostic philosophy for troubleshooting electrical problems is shown in **Figure 4-58**. This type of diagram is called a flowchart. Flowcharts use variously shaped blocks of text with arrows to direct you from one step of a process to another. Flowcharts are very common in electrical troubleshooting guides. A rectangular-shaped flowchart block tells you that some action is to be performed. A diamond-shaped block typically indicates that a yes or no decision must be made. If the answer is yes, then you will go to the block in the direction of the yes arrow. If the answer is no, then you are directed to the block indicated by the no arrow.

To illustrate this process at work, a sample problem will be investigated using the general electrical diagnostic flowchart shown in **Figure 4-58**. In this example, a customer complains that the heated mirrors on his truck do not work. Heated mirrors use a resistive heating element inside the mirrors to minimize fogging and ice buildup on exterior mirrors.

Step 1 shown in **Figure 4-58** is to verify the concern. This step is important because the problem reported by the customer may go through several persons before it ends up on a work order. It is also difficult to get paid for troubleshooting a system that is functioning the way it was designed to function. It may be necessary to find out how a system is supposed to operate, using OEM information in some circumstances. It is possible that the truck operator does not know how the system is supposed to perform on a particular make and model of truck and that no real problem exists.

Performing step 1, the technician verifies that none of the truck's mirrors heat up when the mirror heat switch is closed (on) with the key switch in the ignition position.

BASIC ELECTRICAL / ELECTRONIC DIAGNOSTIC PROCEDURE FLOWCHART

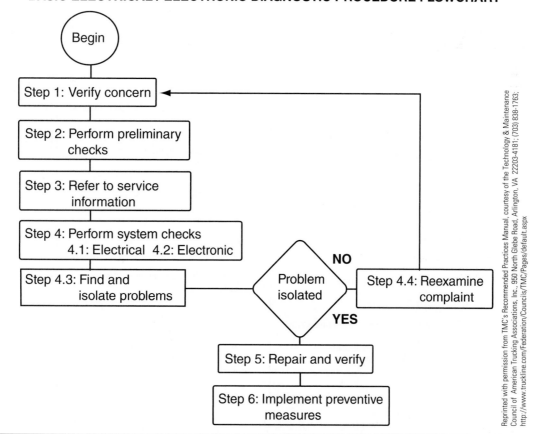

Reprinted with permission from TMC's Recommended Practices Manual, courtesy of the Technology & Maintenance Council of American Trucking Associations, Inc., 950 North Glebe Road, Arlington, VA 22203-4181; (703) 838-1763; http://www.truckline.com/Federation/Councils/TMC/Pages/default.aspx

Figure 4-58 Diagnostic flowchart.

Step 2 of the procedure is to perform preliminary checks. Preliminary checks include a visual inspection. The importance of a visual inspection cannot be overemphasized in electrical and electronic troubleshooting. Quite often, you may find a disconnected connector or other obvious cause of the problem during the visual inspection. Other preliminary checks may include a road test of the vehicle to look or listen for problems, which may also be required to verify the concern in step 1. While operating the vehicle, watch for seemingly nonrelated problems.

A visual inspection of the truck indicates no obvious problem with the mirror heat system. However, the technician is working night shift and notices that the back-up lamps also do not work on the truck when it is brought into the shop. The truck operator may have not complained about this because the truck is used for daytime driving only.

Step 3 of the procedure is to refer to service information. Accurate service information is critical for troubleshooters. Service information includes service

manuals and wiring diagrams, troubleshooting guides, and service bulletins. In some cases, the owner's manual can be a source of information.

Wiring diagrams make use of symbols to represent some of the electrical components covered in this chapter. These symbols may vary between OEMs, so it is a good idea to review the symbols used by the OEM for their wiring diagrams, such as shown in **Figure 4-59**. This information can typically be found in the first few pages of wiring diagrams.

In looking at the wiring diagram for the mirror heat circuit, you observe that a single 20A fuse designated F5 is used to protect both the mirror heat circuit and the back-up lamps. The use of a single CPD to protect more than one circuit is common in trucks, as shown in **Figure 4-60**.

Step 4 of the procedure is to perform systems checks as outlined in the service literature. These systems checks are often shown in the form of a flowchart. For simple circuits such as mirror heat, there may not be a systems check provided in service

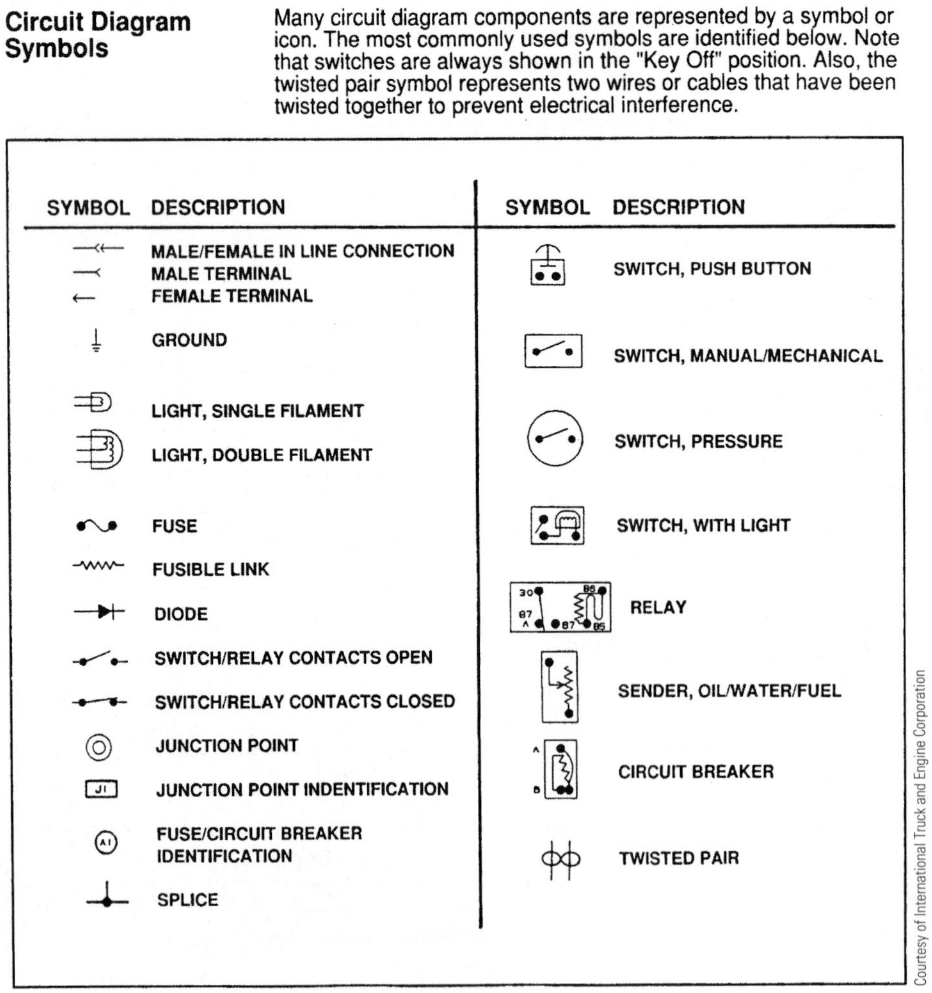

Circuit Diagram Symbols

Many circuit diagram components are represented by a symbol or icon. The most commonly used symbols are identified below. Note that switches are always shown in the "Key Off" position. Also, the twisted pair symbol represents two wires or cables that have been twisted together to prevent electrical interference.

SYMBOL	DESCRIPTION	SYMBOL	DESCRIPTION
	MALE/FEMALE IN LINE CONNECTION		SWITCH, PUSH BUTTON
	MALE TERMINAL		
	FEMALE TERMINAL		
	GROUND		SWITCH, MANUAL/MECHANICAL
	LIGHT, SINGLE FILAMENT		SWITCH, PRESSURE
	LIGHT, DOUBLE FILAMENT		
	FUSE		SWITCH, WITH LIGHT
	FUSIBLE LINK		RELAY
	DIODE		
	SWITCH/RELAY CONTACTS OPEN		SENDER, OIL/WATER/FUEL
	SWITCH/RELAY CONTACTS CLOSED		
	JUNCTION POINT		CIRCUIT BREAKER
	JUNCTION POINT INDENTIFICATION		
	FUSE/CIRCUIT BREAKER IDENTIFICATION		TWISTED PAIR
	SPLICE		

Courtesy of International Truck and Engine Corporation

Figure 4-59 Common wiring diagram schematic symbols.

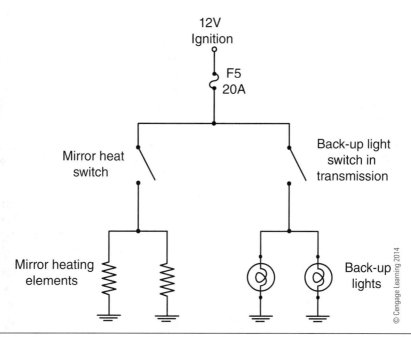

Figure 4-60 Fuse wiring diagram indicating fuse F5 supplies the mirror heat and back-up lamp circuits.

literature because general troubleshooting skills should be sufficient to find the cause of the problem. More complicated electronically controlled systems typically require specific systems checks provided in service literature to locate the cause of the problem.

A substep identified in **Figure 4-58** as step 4.3 is to find and isolate the problems. Troubleshooting typically involves isolating the problem. Again, the wiring diagram and other service information are invaluable in narrowing down the problem.

The mirror heat circuit shown in **Figure 4-60** uses ignition-powered fuse F5 to supply power to the mirror heat switch and the back-up lamps. The technician checks fuse F5 and finds that the fuse is blown.

Like a detective, you must look for the clues left by the problem. One of the clues when the CPD is a fuse is the appearance of the blown fuse. By examining the fuse element, a technician can make a rough determination of whether the fuse opened because of an overloaded circuit or because of a short to ground. If only a small gap exists in the fuse element with minimal discoloration (blackening) of the fuse housing, then the fuse probably opened due to an overload. If instead, the element is not even visible and the housing is blackened, the fuse probably opened due to a short to ground. This clue will assist you in determining what to do next. The technician looks at the fuse and can determine that the fuse is probably blown due to a short to ground because the fuse element is not even visible through the blackened fuse housing.

Using the wiring diagram as a guide, the technician isolates the trouble by switching off each device that is protected by the CPD in question. This will aid in pinpointing the cause of the excess current. In this case, only mirror heat and the back-up lamp circuits are protected by fuse F5, so the technician puts the transmission in neutral to switch off the back-up lamps and switches off the mirror heat switch and the ignition switch. This step will assist the technician in determining which of the circuits is causing the short circuit.

If the CPD is a fuse, temporarily replace the blown fuse with a Type III circuit breaker of *the same or lower current rating*. This prevents wasting fuses during troubleshooting. At this point, the technician switches on each device that is supplied current by the CPD in question. If switching on a particular device causes the circuit breaker to trip, then the portion of the circuit after—or downstream of—the switch controlling the device is suspect. The technician switches the key switch to the ignition position and checks the circuit breaker. In this case, the breaker has not tripped. Next, the technician closes the mirror heat switch and finds that the mirrors now heat up and the circuit breaker does not trip. This indicates that the mirror heat circuit does not appear to be the cause of the problem. The technician then places the transmission in reverse and hears the circuit breaker trip. This makes it appear that the cause of the trouble is somewhere after the transmission back-up lamp switch.

If the technician had just replaced the mirror heat fuse F5 and found that it did not blow again, he might have thought that he had fixed the problem. However, when he backs the truck out of the garage, the shorted to ground back-up lamp circuit causes F5 to blow again. The technician would have given the operator back a truck with the mirror heat still inoperative.

The problem has now been isolated to somewhere past the back-up lamp switch. The technician finds that a wiring harness clip has broken near the transmission, which has permitted the back-up lamp switch wiring harness to fall on the exhaust pipe. The hot exhaust has melted the insulation of the back-up lamp circuit and the bare copper strands are making contact with the metal exhaust pipe, which is grounded.

Because the problem has been isolated, the decision block in the process (the diamond) shown in **Figure 4-58** indicates that step 5 should be performed next. This step is to repair and verify. Often, the repair is the easiest part of the process and may take only a few minutes. After the repair, it is vital to verify that the original complaint has been resolved. You can be assured that your customer will verify that you have resolved the original complaint.

Often during electrical troubleshooting, it is necessary to disconnect wiring harness connectors. Inadvertently leaving a connector disconnected may result in more than just the mirror heat from operating. Verifying that the repairs you have made have resolved the complaint should reduce the chances of such mistakes.

For step 5, the technician repairs the wiring harness and places a new 20A fuse in fuse block cavity F5. The technician then verifies that both the mirror heat and back-up lamps function correctly.

Step 6 in **Figure 4-58**, implement preventive measures, involves both taking steps to prevent the same problem from occurring again and looking the vehicle over for other problems. Having the same problem occur again in a few days does not make for a satisfied customer. In addition, having something else go wrong with the truck a few days after the truck was in your shop, even though it is unrelated to your repair, is not a good reflection upon you as a professional. You may even be accused of causing the next problem that occurs with the truck just because you had worked on the vehicle a few days earlier, even though you had nothing to do with the cause of the new problem.

The technician improves the wiring harness routing and clipping by using a more secure type of harness clip so that this problem should not happen again. The technician also looks the truck over for other potential problems before returning it to the truck operator.

In this example, the problem could be duplicated. The CPD could be made to open (blow). Unfortunately,

in real life, intermittent problems are common. Intermittent shorts to ground of wiring are often caused by wiring that has worn through due to abrasion at some location, and movement while the truck is being driven causes the worn spot in the harness to contact a ground point intermittently. This type of problem may require a great deal of time to diagnose. Once again, the wiring diagrams for the truck can be very useful for this type of troubleshooting along with a thorough visual inspection.

The same problem will be evaluated again, except this time the circuit breaker did not trip when the transmission was shifted into reverse. The technician verifies that the back-up lamps are operational as is the mirror heat. The fuse that was removed indicated that the problem is probably due to a short to ground because the fuse element shows signs of a large amount of current flow. At this point, the technician needs to find the cause of the intermittent short to ground. A visual inspection of all potentially affected wiring for the mirror heat and back-up lights after or downstream of fuse F5 is a good place to start. Special attention should be given to places where the wiring harness is routed over edges of frame rails. During the visual inspection, the technician should leave the applicable circuits energized (switched on) if possible. The idea is to try to duplicate the intermittent short to ground during the inspection. The technician gently pulls and shakes the affected wiring during the inspection in an attempt to re-create the short to ground. An audible device like a buzzer can be temporarily connected in parallel with the circuit breaker to let the technician know if the circuit breaker trips during the inspection, as shown in **Figure 4-61**. If the short to ground is located, the CPD will open and the buzzer will sound (**Figure 4-62**).

Many good digital ohmmeters have a minimum-maximum peak hold setting that can also be useful for locating this type of intermittent problem. This peak hold or glitch detect setting will typically cause an audible alarm in the meter to sound if a new minimum or maximum reading has been detected. To use this setting, disconnect the truck's batteries, remove the fuse in question, and connect the ohmmeter between ground and the load side of the fuse terminal, as shown in **Figure 4-63**. In this case, the mirror heat switch is closed and transmission placed in reverse to close both switches so that the entire circuit is complete. Also, disconnect the loads (mirror heating elements and back-up lamp bulbs) so that the normal resistance to ground through the load is opened. This should cause the ohmmeter to indicate an open circuit until the short

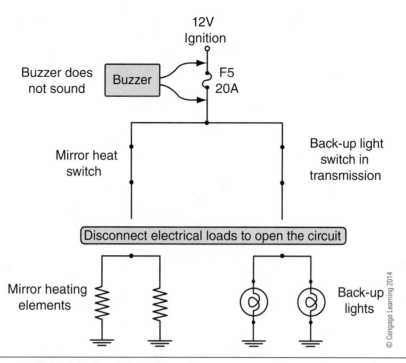

Figure 4-61 Buzzer in parallel with circuit breaker; short to ground not present so buzzer does not sound.

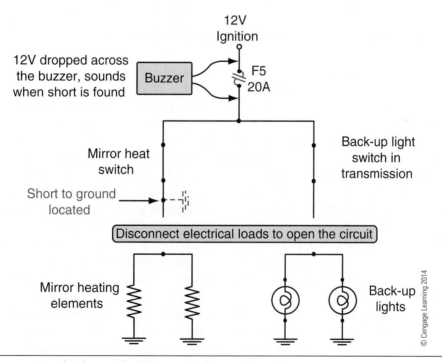

Figure 4-62 Short to ground is located: CPD opens, buzzer sounds.

to ground is found. Set the meter for the fastest peak hold setting, as described in the DMM's operating instructions, and gently move all associated wiring harnesses in an attempt to cause the short to ground. When the short is duplicated, the meter should sound an audible alarm and record a new peak minimum resistance reading, alerting you that you have located the problem.

Tech Tip: The peak hold or glitch detect feature of a DMM is a useful option for a truck technician. Many DMMs designed for automotive service have this feature. Look for this feature when choosing a DMM.

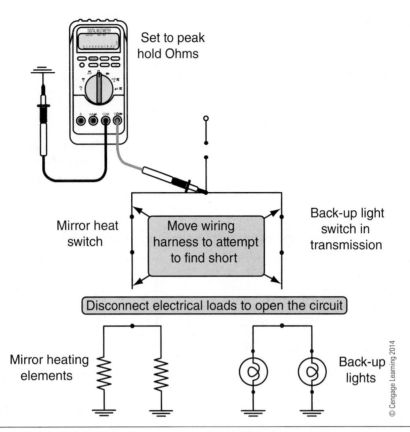

Figure 4-63 Using peak hold with DMM ohmmeter to find intermittent short.

In the best-case scenario, the technician quickly finds the cause of the intermittent short to ground in a location that is easy to repair. In real life, these intermittent shorts can sometimes be very difficult to duplicate. Sometimes the problem must get worse before it can be re-created during troubleshooting.

If the intermittent short to ground cannot be duplicated, try driving the vehicle on rough roads; turn the steering wheel to the stops, and conduct other realistic maneuvers to attempt to cause the problem to occur again. You may also want to talk to the truck operator to determine if he or she can provide any clues. It is usually best to speak to the truck operator yourself instead of going through service writers when the problem cannot be duplicated. Each person that passes along information tends to change the information slightly.

More clues can be gathered by asking your fellow technicians if they have ever run into this particular problem. It is possible that one of your co-workers has seen the same problem on an identical model truck in the past and can provide some insight, such as an obscure location where the wiring harness is prone to rubbing through. This type of "pattern failures" knowledge from experience is often very useful. However, do not skip troubleshooting and go straight

to pattern failures. This type of "diagnosis" does little for your professional development as a troubleshooter. Even if you suspect a pattern failure, go through the troubleshooting steps to hone your skills if you have time. Service information such as technical service bulletins and newsletters, as discussed in later chapters, can also be very useful for troubleshooting intermittent problems. Many pattern failures are identified by the OEMs in updated service information to aid technicians in diagnosis of known problems.

Case-based reasoning is another common term for pattern failure diagnostics. Some OEMs are applying case-based reasoning to their web-based diagnostic procedures. Once a particular component or system is known to be a common source of failure, the diagnostic procedures for that problem are modified so that testing for that known issue becomes one of the first items checked.

In some cases, the intermittent short to ground cannot be duplicated. In these cases, you may have little choice but to give the truck back to the operator with the problem still not fixed. Make sure that the replacement CPD is the correct type and current rating. Never install a CPD with a rating higher than that recommended by the OEM. Instruct the truck operator to watch for the problem to determine if he or she can provide any insight. You may want to advise the truck operator to take some spare fuses of the correct amperage. Also, inform the truck operator not to replace the fuse with a fuse that has a higher current rating should the fuse blow again.

Finding the Cause of High Resistance in a Circuit

High resistance in circuits results in decreased current flow. In the case of an open circuit, all current flow in the circuit is interrupted. Finding the source of high resistance in a circuit involves the use of a DMM voltmeter to measure the voltage present referenced to a good chassis ground at various points throughout the energized circuit. Many technicians still use a tool called a test light for this type of troubleshooting. A test light consists of a 12V light bulb in a clear screwdriver-type handle (**Figure 4-64**). A wire with an alligator clip provides a means to connect the test light to ground or a positive voltage source. This wire is connected to one terminal of the light bulb. The other light bulb terminal is connected to a sharpened metal tip, like a punch. Connecting the alligator clip to ground and touching anything that is at a potential of 12V will cause the light bulb to illuminate (**Figure 4-65**). Connecting the alligator clip to a positive voltage source and touching the metal tip to a grounded component will also cause the light bulb to illuminate.

Test lights were the main electrical diagnostic tool used by truck mechanics in the past. However, the use of test lights has become somewhat controversial in recent years. Many shops have banned the use of test lights entirely. Electronic control modules on trucks can be damaged through the relatively high current flowing through a test light bulb. This is especially true when the alligator clip is connected to a positive voltage source and the test light is used to determine if a circuit is grounded.

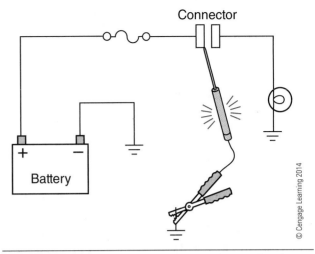

Figure 4-65 Test light indicating presence of voltage.

In addition to potentially damaging electronic components, the sharp tip of the test light is sometimes misused to puncture wire insulation to determine if a circuit is powered or grounded. This practice can cause serious wiring harness damage and should be avoided. Terminals can also be damaged when the test light tip is forced into a female terminal socket causing the socket to spread open. All troubleshooting examples in this book make use of a DMM instead of a test light because the use of test lights is discouraged by many OEMs.

To illustrate the effects of high resistance in a circuit, a sample lighting circuit shown in **Figure 4-66** will be evaluated. Closing the switch causes current to flow through the light bulb. Any series resistance that is added to the circuit will cause the current through the light bulb to decrease. This added resistance could be anywhere in the circuit. To find the location of this resistance, a DMM voltmeter with the negative lead connected to a good chassis ground is used to measure the voltage that is present throughout the circuit. The expected voltage measurements with the switch open and with the switch closed are shown in **Figure 4-66**.

A source of high series resistance is introduced in **Figure 4-67**. This series resistance causes the light bulb not to illuminate or to illuminate dimly with the switch closed, depending on the amount of resistance. Two resistors that are connected in series form a voltage divider circuit. The source voltage will divide proportionally to the resistance values of the two series resistors. A DMM voltmeter is utilized to find the source of the resistance by measuring the voltage referenced to ground throughout the circuit. This technique requires that the circuit not be opened anywhere to take measurements, such as disconnecting connectors. Opening the circuit to measure the voltage

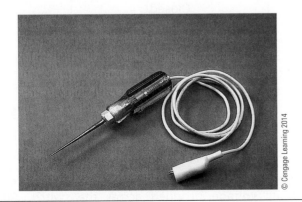

Figure 4-64 Test light—most manufacturers discourage the use of a test light for troubleshooting.

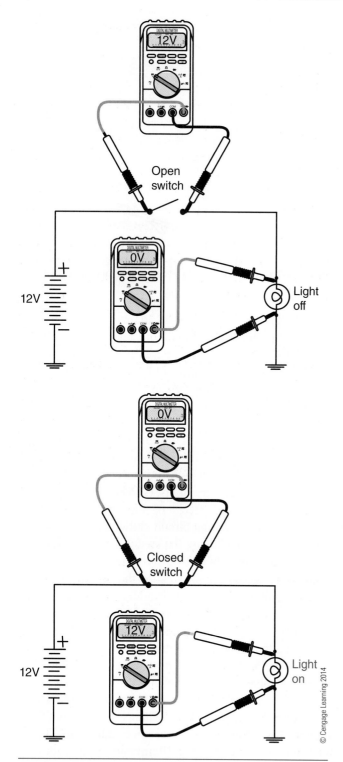

Figure 4-66 Circuit with no defects, open switch and closed switch.

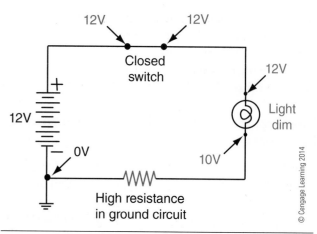

Figure 4-67 High resistance in light bulb ground circuit causing light to be dim. All voltages shown are referenced to a good chassis ground.

light bulb. This indicates that the source of the high resistance in this circuit is in the ground circuit.

In the example shown in **Figure 4-68**, the switch contacts have burned or corroded and are acting as an open circuit. With the switch closed, the light bulb does not illuminate. Using a DMM to measure voltage referenced to a good chassis ground causes the DMM to indicate 12V on the battery side of the switch and 0V at the load side of the switch. Placing the DMM test leads across the switch would cause a reading of 12V, indicating that the rest of the circuit is intact and the cause of the open circuit is a defective switch.

A defective switch could also add a series resistance if the contacts are burned or corroded yet still permits some current to flow through the rest of the circuit (**Figure 4-69**). Using a DMM to measure

places the DMM in series with the circuit and can lead to misdiagnosis, as explained in later chapters. The DMM is moved throughout the circuit and indicates the voltages referenced to a good chassis ground, as shown in **Figure 4-67**. Note that there is 2V difference in the voltage measured to ground on either side of the

12V dropped across defective switch
with very high closed resistance

Figure 4-68 Open circuit caused by defective switch with all voltage dropped across the switch. All voltages shown are referenced to a good chassis ground.

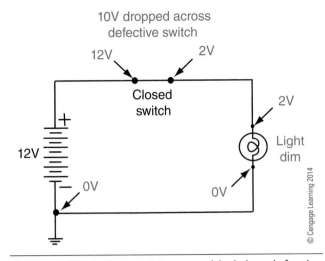

Figure 4-69 Series resistance added by defective switch causes a series voltage divider circuit. This reduces the voltage dropped across the light bulb causing the bulb to illuminate dimly.

voltages referenced to a good chassis ground indicates that 10V has been dropped across the defective switch contacts, leaving only 2V to be dropped across the light bulb, as predicted by Kirchhoff's voltage law.

EXTRA FOR EXPERTS

This section covers additional information related to electrical components.

Relay Coil Hold-In and Pull-In Voltage

One thing to be aware of when troubleshooting relays is the difference between coil hold-in voltage and the coil pull-in voltage. The voltage across the relay coil that is necessary to cause enough current to flow to create magnetic force strong enough to overcome the armature spring is typically much less than the relay coil rating. For example, if the voltage applied to a relay coil were supplied with an adjustable voltage source that starts at 0V and is slowly increased, the relay with a 12V coil would probably overcome the spring tension and pull the armature in when the coil supply voltage reaches approximately 6V. The voltage that causes the relay normally open contacts to close is referred to as the pull-in voltage.

Once the armature is pulled in, the magnetic field must be reduced substantially to cause the relay armature to return to the normal de-energized position. On most common 12V relays, the voltage supplied to the coil can be decreased to approximately 3V before the relay spring overcomes the magnetic force and the relay contacts snap open. The voltage across the relay coil when the mechanical spring force causes the contacts to open again is referred to as the hold-in voltage. The difference between relay hold-in voltage and relay pull-in voltage is something to keep in mind while troubleshooting.

Summary

- Truck wiring is divided into a series of sections called harnesses. Harnesses are mated using connectors. Connectors contain terminals that provide the electric contact between sections of wire.

- Wire diameter for North American–built trucks is typically defined by the AWG number. The smaller the AWG number, the larger the wire diameter.

- Wire has resistance. The resistance of wire is directly proportional to the wire length and inversely proportional to the wire diameter.

- The frame rails, sheet metal, and other metal components of a typical truck are called chassis ground. The negative battery terminal is connected to the chassis ground. The chassis ground acts as a conductor for many truck electric circuits.

- Wiring harness routing refers to the placement of a wiring harness on a truck. Clipping is the attachment method used to attach the harness to the chassis.

- Electric switches are devices used to interrupt the flow of current in a circuit. There are many different types of switches. The functionality of a particular switch can be defined by the terms *pole* and *throw*.

- Normal state of a switch describes momentary contacts that are open or closed when the switch is not depressed or otherwise disturbed. A normally closed switch has continuity until the switch is depressed. A normally open switch has continuity only when the switch is depressed.

- A closed switch should have near 0V dropped across it; an open switch should have near-battery voltage dropped across it.

- Relays are electromagnetic switching devices. A small amount of current flow through a coil forms an electromagnet. The electromagnet causes the relay contacts to open or close, in much the same way as a switch. A relay is like a remotely controlled switch.

- Relay control circuit describes the electrical circuit that powers the relay coil. When a positive voltage source and ground are provided by the relay control circuit, the relay coil is energized.

- Circuit protection devices are designed to protect the truck's wiring harness. Circuit protection devices include fuses, circuit breakers, and fusible links.

- Inrush or surge current describes the high level of current that is drawn by some devices for a brief period of time when first switched on, such as motors and light bulbs. Surge current may be many times greater than the normal current drawn by a device after a brief period of time.

- Adding additional loads such as lights to a circuit causes the current through the circuit protection device to increase. Never replace a circuit protection device with one that has a higher rating than the OEM's recommendations.

- Fuses are single-use devices that have an element designed to open when current flow through the fuse exceeds the fuse rating. A combination of time and current causes the fuse element to open.

- Circuit breakers use a bimetallic strip to open a set of normally closed contacts. The three main classes of circuit breakers are SAE Type I, Type II, and Type III.

- Fusible links are made of wire with a special insulation material designed not to burn. The fusible link is designed to be the smallest wire gauge in the circuit that it is protecting.

Suggested Internet Searches

Try the following web sites for more information:
http://www.amp.com
http://www.belden.com
http://www.delphi.com
http://www.colehersee.com
http://www.eaton.com
http://www.te.com
http://www.bosch.com
http://www.littelfuse.com
http://www.bussman.com

Review Questions

1. Which is a true statement regarding terminals and connectors?

   A. A connector is always made of a conductive material and a terminal is made of a nonconductive material.

   B. Connectors are always sealed and terminals are unsealed.

   C. Connectors are always male; terminals are always female.

   D. Connectors retain (hold) terminals.

2. Which is the largest-diameter wire of the choices below?

   A. 20 AWG

   B. 4-0 AWG

   C. 4 AWG

   D. 0 AWG

3. Which of the following is a true statement?

   A. A 10-ft (3 m) section of 10 AWG wire has more resistance than a 20-ft (6 m) section of 10 AWG wire.

   B. A 10-ft (3 m) section of 10 AWG wire has less resistance than a 10-ft (3 m) section of 20 AWG wire.

   C. Wire resistance is the same regardless of wire diameter.

   D. Wire insulation always has a low resistance.

4. Technician A says that the engine block is designed to be isolated from chassis ground on most modern trucks. Technician B says that the cab is properly grounded through the steering gearbox and steering shaft. Who is correct?

   A. A only

   B. B only

   C. Both A and B

   D. Neither A nor B

5. The insulation of a wire in a truck chassis wiring harness has been damaged, which has eventually caused an open circuit because of corrosion. The corroded copper wire has turned green and is blackened for several feet beyond the damage due to oxidation. Which of these is the best way to repair this section of wiring?

   A. Remove the corrosion with sandpaper and solder the wire together, covering with electrical tape to seal.

   B. Use a crimp-type butt splice to crimp the ends of the corroded wire together and cover with RTV.

   C. Cut out the damaged section of wire and replace with new wire, using splice clips, soldering, and heat shrink tubing containing sealer.

   D. Twist the wires together and cover with electrical tape.

6. A single-pole single-throw (SPST) switch is being tested at the parts counter with an ohmmeter. Which of the following statements is true?

   A. With the switch open, the resistance across the switch terminals should be near 0Ω.

   B. With the switch closed, the resistance across the switch terminals should be infinitely high (OL).

   C. With the switch open, the resistance across the switch terminals should be infinitely high (OL).

   D. An ohmmeter should never be used to test a switch unless the switch is installed in the circuit and voltage is present.

7. An oil pressure switch is typically a normally closed type of switch that changes states at approximately 7 psi (48 kPa). Such a switch is being tested at the parts counter with an ohmmeter. Oil pressure is "normally" about 20 psi (138 kPa) at idle for this engine. What would be the expected ohmmeter reading across the switch terminals when a new switch is tested at the parts counter?

   A. Near 0Ω

   B. Infinitely high resistance (OL)

   C. At least 10Ω

   D. An ohmmeter should never be used to test a switch unless the switch is installed in the circuit and voltage is present.

8. A 10A Type III circuit breaker in the truck fuse block that protects the trailer marker lamp circuit keeps tripping after the trailer marker lamps are switched on. It takes about 1 minute for the breaker to trip. The circuit breaker does not trip if the trailer is not connected to the truck. Technician A says that a hard short to ground in the trailer marker lamp wiring may be causing the circuit breaker to keep tripping. Technician B says that there may be too many trailer marker lamps, which causes perhaps 15A of current to flow through the 10A circuit breaker and causes it to trip. Who is correct?

   A. A only

   B. B only

   C. Both A and B

   D. Neither A nor B

9. The trailer stop lamps do not illuminate when the brake pedal is depressed, but the truck stop lamps are operational. The circuit diagram for this truck indicates that the stop lamp switch controls the truck stop lamps directly and supplies the high side to the trailer stop lamp relay control circuit. The relay that controls the trailer stop lamps does not click when the brake pedal is depressed. Which of these would probably *not* be the cause of the inoperative trailer stop lamps?

   A. Defective trailer stop lamp relay

   B. Open trailer stop lamp relay control circuit ground

   C. Defective stop lamp switch

   D. Open circuit between the stop lamp switch and the trailer stop lamp relay coil high side

10. Which of these is an improper use of a test light?

    A. Piercing the wire insulation to verify that voltage is present.

    B. Using the test light to test for voltage at a sensor used in an electronic diesel engine control system.

    C. Jamming the sharpened tip of the test light into the back of a sealed connector to test for the presence of voltage.

    D. All of the above are improper uses of a test light on a modern truck.

11. The primary purpose of a suppression resistor in a relay is which of the following?

    A. Increase current flow through the coil.

    B. Decrease the magnetic field.

    C. Reduce the level of the negative voltage spike when the relay is switched off.

    D. Prevent relay contact bouncing.

12. When should a fuse with a current rating greater than the OEM's recommendations be installed in a fuse block?

    A. When a fuse blows intermittently and the cause cannot be found.

    B. When the truck's owner signs a waiver form.

    C. When additional parallel-connected loads have been added.

    D. A larger than recommended fuse should never be installed.

13. Which truck electrical device would probably be controlled using a normally open momentary contact switch?

    A. Headlamps

    B. Turn signals

    C. Horn

    D. Windshield wipers

14. Pressure switches in the primary and secondary air systems are used to cause a warning lamp to illuminate and an alarm to sound on a truck with air brakes when pressure drops below 60 psi (414 kPa). The switches are wired in parallel so that either one will provide a ground for the warning devices when pressure drops below 60 psi to act as a warning of low air pressure. What type of switch are these pressure switches?

    A. Normally open

    B. Normally closed

    C. Normally latched

    D. Normally unlatched

15. A truck electrical circuit is protected by a polymer PTC device. Technician A says that polymer PTC devices contain a bimetallic spring that snaps open when current flow through the device increases. Technician B says that resetting a polymer PTC device requires a special reset tool. Who is correct?

    A. A only

    B. B only

    C. Both A and B

    D. Neither A nor B

CHAPTER 5

Batteries

Learning Objectives

After studying this chapter, you should be able to:

- Work safely around batteries.
- Explain the fundamentals of how a lead acid battery produces a voltage.
- Describe series-aiding and series-opposing voltage source connections.
- Explain the process of battery discharge and recharge.
- Describe how battery internal resistance affects battery terminal voltage.
- Define typical battery ratings such as CCA and reserve capacity.
- Safely recharge a battery using a battery charger.
- Test conventional and low-maintenance batteries.
- Jump-start a truck using the correct procedures.
- Service a truck battery box.
- Describe recombinant batteries.
- Discuss lithium-ion batteries utilized in hybrid electric vehicles.

Key Terms

BCI group number

carbon pile

cell

charging

cold cranking amps (CCA)

conductance

discharging

electrolyte

hydrometer

internal resistance

no-load voltage

plates

recombinant

reserve capacity

self-discharge

series-aiding

series-opposing

specific gravity

state of charge

surface charge

INTRODUCTION

Batteries store and transform energy. Batteries store energy in the form of chemical energy. The stored chemical energy is transformed into electrical energy that causes a difference in electrical potential (voltage) across the battery's terminals. This voltage causes a direct current to flow when a complete electrical circuit is connected across the battery's terminals. A battery can only supply direct current (DC). Trucks and automobiles typically use a type of battery known as a lead acid battery. The acid is sulfuric acid (H_2SO_4). The lead and the acid together cause chemical reactions to occur, which produce a voltage.

LEAD ACID BATTERY SAFETY

The lead acid battery is potentially one of the most hazardous components of a truck. Most batteries used in trucks contain sulfuric acid.

WARNING *Sulfuric acid is a hazardous substance that can cause blindness and serious chemical burns to the skin. Always wear approved eye protection and protective clothing when handling, charging, servicing, or testing batteries.*

Battery cases can fracture if the battery is dropped, causing sulfuric acid to pour out of the battery. Keep baking soda near stored batteries to neutralize the sulfuric acid, and later flush with water.

Batteries produce an explosive gas when charging or discharging. If the gas should explode, the battery case and internal components can turn into shrapnel and spray sulfuric acid over a large area with tremendous force. Always test or charge batteries in a well-ventilated area in order to dissipate the explosive gas formed within the battery. Never create sparks, open flames, or smoke around batteries, even batteries that are just being stored.

A wrench or other metal object that makes contact with the battery-positive terminal and negative terminal or chassis ground can result in a hazardous situation. The object can become heated to a dangerous level within seconds and potentially cause the battery to explode or start a fire.

Unwanted batteries must be disposed of properly in accordance with local ordinances. The lead and sulfuric acid that are used in truck batteries are hazardous to personal safety as well as the environment.

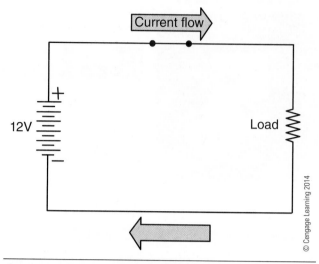

Figure 5-1 Battery discharging.

LEAD ACID BATTERY FUNDAMENTALS

When a lead acid battery is supplying current to an electrical circuit, the battery is said to be **discharging**. **Figure 5-1** shows a battery supplying current to a load. The direction of conventional current flow is from the battery-positive terminal, through the load, and back to the battery-negative terminal, as shown by the arrows in **Figure 5-1**. As the battery discharges, the limited supply of stored chemical energy in the battery is being depleted.

The chemical energy in truck batteries can be replenished by causing a direct current to flow through the battery in a direction opposite from the battery discharge current, as shown in **Figure 5-2**. When the battery is receiving such a current, the battery is said to be **charging**. This charging current is supplied by an engine-driven generator that is commonly known as

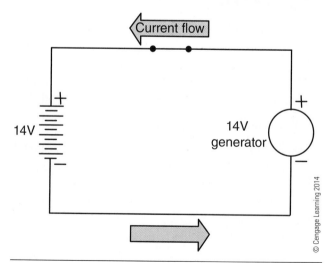

Figure 5-2 Battery recharging.

an alternator. The charging current supplied by the alternator causes the battery to reverse the process and convert electrical energy back into chemical energy. The current supplied by the alternator powers the vehicle's loads, such as lights, and recharges the battery. The discharging and charging process is repeated each time the truck's engine is started.

Batteries serve four main purposes in a truck's electrical system:

1. Batteries supply electric current when the truck's engine is not running, to power ignition-off accessories such as hazard flashers.
2. Batteries supply the current for the starter motor to crank the engine.
3. Batteries supply current to the truck electrical system when the engine is running and the combined current demand of devices such as the wiper motor, blower motor, and lights exceed the truck's alternator output capability. The batteries also permit the truck to be driven for a brief period of time should the charging system fail.
4. Batteries stabilize the vehicle's electrical system. The truck's batteries smooth out voltage spikes caused by switching, such as the negative voltage spikes associated with coils.

Batteries are also used to supply power during overnight parking for cab electrical accessories and HVAC systems, such as Freightliner's ParkSmart™ system.

Most modern truck batteries are categorized as low water loss batteries. These low water loss batteries are further classified as maintenance-free or low-maintenance batteries. However, older technology batteries known as conventional batteries may still be used in some equipment. The difference between a low water loss battery and a conventional battery is the type of materials utilized in the battery construction. Conventional batteries also have filler caps to permit the addition of water as a maintenance item. The maintenance-free varieties of low water loss batteries are sealed and have no such provision for adding water. Maintenance-free batteries are designed never to need water over the life of the battery. Non-sealed low-maintenance varieties of low water loss batteries have provisions for adding water, but the design and battery construction significantly reduce the amount of water that the battery requires, compared to conventional batteries.

The chemical reaction that generates voltage in both conventional and low water loss batteries is nearly the same. However, there are some differences in how a

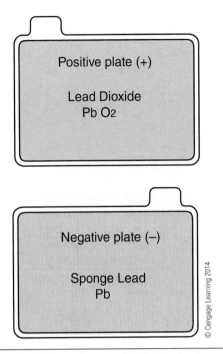

Figure 5-3 Positive plate and negative plate.

low water loss battery and a conventional battery are installed, serviced, and tested. This chapter will concentrate mostly on the maintenance-free variety of low water loss batteries, the most common type of battery used in modern trucks.

Internal Battery Components

The internal structure of a battery is composed of lead. The lead is what makes a battery such a heavy component and is necessary for the chemical process that produces electricity. The sulfuric acid mixed with water acts as an **electrolyte**. An electrolyte is a substance that mixes with water to form a solution or mixture that conducts electricity. Pure water alone is not a good conductor of electricity.

The fundamental voltage-producing component in a battery is a pair of **plates**. The plates are constructed of two different types of lead (**Figure 5-3**). One type of plate is a positive plate and is constructed of a compound made of a combination of lead and oxygen called lead dioxide (PbO_2). The other type of plate is a negative plate and is made of sponge lead (Pb). The lead material is pasted to a holding structure known as a grid, which is the foundation on which the plates are constructed. The grid for both positive and negative plates in low water loss batteries is usually made of lead calcium and looks like a small chain link fence or expanded metal section (**Figure 5-4**). The grid holds the soft lead material in place. Together, the grid and the pasted-on lead material make up a positive or negative plate.

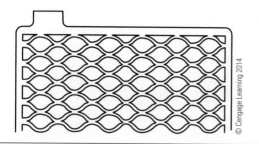

Figure 5-4 Grid.

When a positive and a negative plate are placed close to each other and are submersed in a sulfuric acid and water electrolyte solution, a chemical reaction takes place that causes a difference in potential (voltage) between the two plates to be produced. Each of these combined pairs of plates is referred to as a **cell**. This chemical reaction causes each pair of plates in a lead acid battery to initially produce about 2.1V per cell. Common truck batteries are known as flooded cell or wet cell batteries. This refers to the liquid electrolyte that floods the lead plates.

Because a difference in potential (voltage) exists between a positive and a negative plate, contact between the closely spaced plates due to vibration or other movement would result in a short circuit. To prevent the closely spaced plates from contacting and shorting out, an electrically insulating nonconductive separator is inserted between each positive and negative plate. The separator is a thin sheet of porous material such as treated paper or fiberglass. The porosity of the separator permits the liquid electrolyte

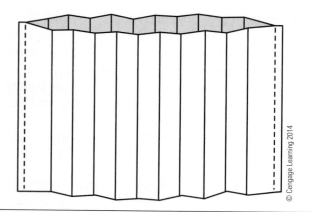

Figure 5-5 Envelope separator.

to pass between the plates, yet keeps the plates from making electrical contact with each other. Envelope-type separators are common, like that shown in **Figure 5-5**. The plates are placed in the envelope separator in the same way that a letter is placed in a paper envelope for mailing.

Combining Plates to Make a Cell

Ten or more negative plates and the same number of positive plates are connected in parallel to form a cell or element, as shown in **Figure 5-6**. The plates are stacked together with a separator between each alternating positive and negative plate, like a slice of Swiss cheese between alternating slices of turkey and roast beef in a thick sandwich. The stack of plates and separators makes up a cell. The plate straps shown in

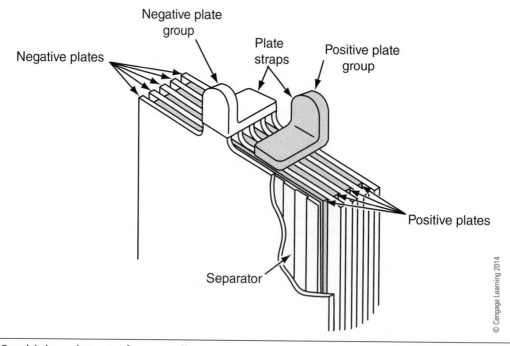

Figure 5-6 Combining plates to form a cell.

Figure 5-6 provide the parallel connection among the plates in a cell or element.

Parallel-Connected Voltage Sources

In a parallel resistor circuit, the voltage measured across each resistor that is connected in parallel is always the same. This leads to an important fact regarding voltage sources such as batteries:

> **Important Fact:** Voltage sources that are placed in parallel with each other will assume a common voltage.

Because voltage sources that are connected in parallel assume a common voltage, only voltage sources of nearly the same value are placed in parallel with each other. This parallel connection is made by connecting all positive terminals together and all negative terminals together. The voltage produced by a single positive and negative plate when the cell is fully charged is 2.1V. Thus, the voltage produced by each cell with several pairs of parallel-connected plates is also 2.1V, regardless of how many plates are contained in each cell. Placing several sets of plates in parallel with each other by connecting all of the positive plates together and all of the negative plates together provides a larger lead surface area for the electrolyte to act upon, compared to a single set of plates. This parallel connection of the individual plates increases the current that the cell can supply, but the voltage per cell remains at 2.1V.

WARNING *Only voltage sources of nearly the same value should be connected in parallel, and the connection must be such that all positive terminals are connected together and all negative terminals are connected together. Otherwise, the two voltage sources try to obey the law of parallel circuits and become the same voltage. This equalization of the two voltage sources can result in a very high level of current flow between the two voltage sources. If a 12V battery and a 6V battery are connected in parallel, the batteries can explode as they both attempt to become a common voltage.*

Series-Connected Voltage Sources

A flashlight that uses two 1.5V D cell batteries has a light bulb designed for 3V. If you have replaced the batteries in a two-cell flashlight, you know that the

batteries must be installed so that the positive end of one of the batteries is inserted into the flashlight so that it contacts the bulb terminal. The second battery must be inserted into the flashlight in the same direction as the first battery, or the lamp will not illuminate. The second battery is inserted so that its positive terminal contacts the first battery's negative terminal. Thus, the positive terminal of one battery and the negative terminal of the other battery are contacting each other, as shown in **Figure 5-7**. The voltmeter in this example indicates that there is 3V measured between the positive terminal of the upper battery and the negative terminal of the lower battery.

The two 1.5V batteries connected in this series arrangement produce a total of 3V because of another important fact related to electricity:

> **Important Fact:** Voltage sources that are connected in series—so that the positive terminal of one voltage source is connected to the negative terminal of another voltage source—produce a total voltage across the combined voltage sources that are equal to the sum (addition) of the individual voltage sources.

The two batteries in the flashlight example are connected in series, like resistors in a series circuit. When the positive terminal of one battery is connected to the negative terminal of the next battery, the voltage measured across the terminals of the combined batteries increases. When voltage sources are connected in series so that the positive terminal of one source is connected to the negative terminal of the next voltage source, the connection is referred to as **series-aiding**.

Unlike parallel-connected voltage sources, series-connected voltage sources do not have to be the

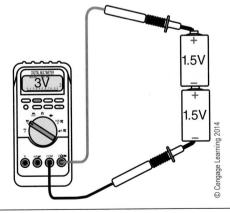

Figure 5-7 Flashlight batteries connected in series-aiding configuration to produce 3V.

same value. A 6V flashlight battery and a 1.5V flashlight battery can be connected in a series-aiding configuration, resulting in a total of 7.5V across the two batteries.

Voltage sources can also be connected in series so that the positive terminal of one voltage source is connected to the positive terminal of another voltage source. This leads to another important fact about batteries:

Important Fact: Two voltage sources connected in series—so that the *positive* terminal of one voltage source is connected to the *positive* terminal of another voltage source—produce a total voltage across the negative terminals of the combined voltage sources equal to the largest voltage source minus the smallest voltage source. The same is true if the two negative terminals of the voltage sources are connected in series and the voltage is measured across the two positive terminals.

This type of connection, where like terminals are connected together in series, is referred to as a **series-opposing** connection. If one of the batteries in a two-battery flashlight is inserted in the flashlight upside down, as shown in **Figure 5-8**, the bulb will not illuminate. The two batteries cancel out each other's voltage, resulting in zero net voltage and zero current through the flashlight bulb.

Series-opposing connections are not the norm in a truck electrical system, but you should be aware of the consequences because an inadvertent series-opposing battery connection could be made when trying to install the four or more batteries used in many trucks. Accidentally connecting batteries in a series-opposing connection is not uncommon. The idea of series-opposing voltage sources will also

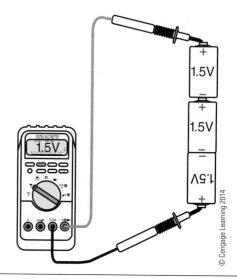

Figure 5-9 Two series-opposing connected batteries cancel each other out, producing a net voltage of 1.5V for three series-connected batteries.

assist in understanding battery recharging in later sections of this chapter. The concepts of series-aiding and series-opposing voltages will also come up again in the study of alternators and motors.

More than two voltage sources can be connected in series. A three-battery D cell flashlight has a bulb designed for 4.5V because three 1.5V batteries connected in a series-aiding arrangement produce about 4.5V.

If one battery is installed upside down in a three-battery flashlight, will the bulb illuminate? The answer is yes, but not very brightly. Installing one battery upside down will cause two of the batteries to cancel each other out, thus leaving a net voltage of only about 1.5V (**Figure 5-9**). This 1.5V produces only enough current to cause the flashlight bulb designed for 4.5V to illuminate dimly.

D cell flashlight batteries have only one cell that produces about 1.5V, as compared to 2.1V for a lead acid battery. If six 1.5V batteries are connected in a series-aiding arrangement, then the voltage across the positive terminal of the first battery and the negative terminal of the last battery will be 9V. This is how a 9V battery, like that used in a smoke detector, is manufactured. A 9V battery like the one shown in **Figure 5-10** contains six series-aiding connected 1.5V cells.

Combining Battery Cells or Elements in Series

A 12V truck battery contains six cells or elements, with a voltage of about 2.1V per cell, that are connected in a series-aiding arrangement by one of the methods shown in **Figure 5-11**. This series-aiding connection among cells provides a voltage of about

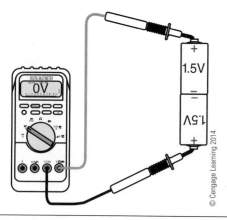

Figure 5-8 Flashlight batteries connected in series-opposing configuration to produce 0V.

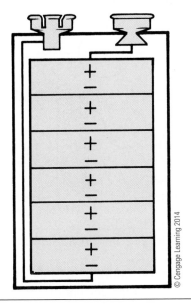

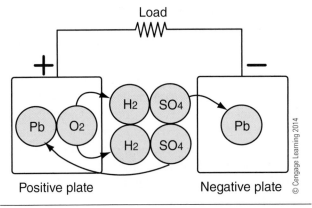

Figure 5-12 Battery discharge chemical reaction.

Figure 5-10 Nine-volt battery with six cells connected in series-aiding configuration.

12.6V across the battery's external terminals (2.1V × 6 cells = 12.6V). The parallel-connected positive plates of one cell are connected in series to the parallel-connected negative plates of the next cell and so on, until all cells are connected in series.

CHEMICAL REACTION IN BATTERIES

Lead acid batteries produce voltage and are recharged by a chemical reaction. A basic understanding of this chemical reaction should assist you in visualizing the operation of a lead acid battery.

Chemical Reaction during Battery Discharge

The initial voltage produced by each pair of plates in a fully charged lead acid battery is about 2.1V.

Several pairs of plates are connected in parallel to form a cell. The six series-aiding connected cells combine to produce about 12.6V. However, the voltage produced by the plates decreases as the battery discharges. An electrolyte is a solution made with water that conducts electric current. The sulfuric acid electrolyte starts out as sulfuric acid diluted with water. Sulfuric acid is a compound made of hydrogen (H) and sulfate (SO_4). The chemical reaction during the discharge phase causes the sulfate (SO_4) in the electrolyte to combine with the lead (Pb) in the negative plate to transform the negative plate into a compound called lead sulfate ($PbSO_4$). This is illustrated in **Figure 5-12**.

A reaction during battery discharge also takes place in the positive plate composed of lead dioxide (PbO_2). The sulfate (SO_4) in the electrolyte combines with the lead (Pb) in the positive plate to transform the positive plate into lead sulfate ($PbSO_4$). Additionally, the oxygen (O) in the positive plate combines with the hydrogen (H) in the electrolyte to transform the sulfuric acid in the electrolyte into water (H_2O).

Initially when the battery is fully charged, the two plates are dissimilar materials (lead and lead dioxide) and the electrolyte contains sulfuric acid. As the

Intercell connections

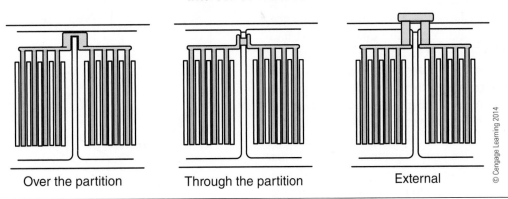

Over the partition   Through the partition   External

Figure 5-11 Cell interconnection.

battery is discharged, the positive and negative plates are both transforming into the same lead sulfate material and the sulfuric acid is being transformed into water. As this occurs, the chemical reaction decreases, causing each pair of plates to produce less voltage. The chemical energy stored in the battery is being depleted during the discharge phase. Eventually, the plates are composed of nearly the same material (lead sulfate) and the electrolyte has become mostly water, causing the plates to stop producing any measurable voltage because all the chemical energy has been transformed into electrical energy.

Tech Tip: Oxygen is referred to as a diatomic element. In the absence of other elements, a single atom of oxygen (O) always combines with another atom of oxygen to form O_2. There are six other diatomic elements which exhibit this same behavior: bromine (Br), iodine (I), nitrogen (N), chlorine (Cl), hydrogen (H), and fluorine (F). Together, these spell BrINClHOF (Brinclhof), which can be used as a mnemonic aid.

Chemical Reaction during Battery Charging

The chemical reaction that occurs during discharge can be reversed by applying a voltage across the battery terminals from an external source that is greater than the voltage being produced by chemical reaction in the battery. This voltage being applied to the battery causes a current to flow through the battery in a direction opposite from the battery discharge current. This current is known as a charging or recharging current and it flows through each series-connected cell. This charging current also obeys the laws of parallel circuits and splits up to flow through each pair of parallel-connected plates via the electrically conductive electrolyte. This charging current flow also causes the lead sulfate in both the positive and negative plates to separate back into lead (Pb) and sulfate (SO_4) through the magic of chemistry, as shown in **Figure 5-13**. The sulfate in the plates gradually returns to the electrolyte and the lead remains on the plates. The reaction also causes the water in the electrolyte to separate into hydrogen and oxygen. The oxygen in the electrolyte combines with the lead in the positive plate to gradually transform the positive plate back into lead dioxide. The transfer of oxygen from the electrolyte to the positive plate and the addition of the sulfate to the electrolyte from both plates

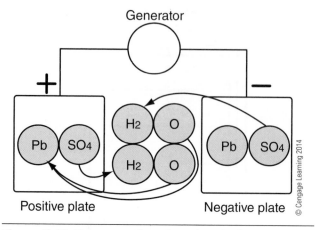

Figure 5-13 Battery recharging chemical reaction.

cause the amount of sulfuric acid in the electrolyte solution to increase and the amount of water in the electrolyte solution to decrease. When all of the sulfate in the plates has been returned to the electrolyte, the charging process is complete, thus restoring the chemical energy in the battery to its maximum level.

EXTERNAL BATTERY COMPONENTS

The cells are installed in the battery case; this exterior container is usually made of polypropylene and holds the battery cells and electrolyte. The case has separate partitions for each cell, as shown in **Figure 5-14**. The partitions keep the electrolyte for each cell in a separate container. The bottom of the case has support braces for the cell assembly to sit atop. This provides an open space beneath the cells for

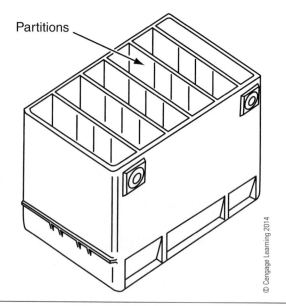

Figure 5-14 Case partitions separate cells.

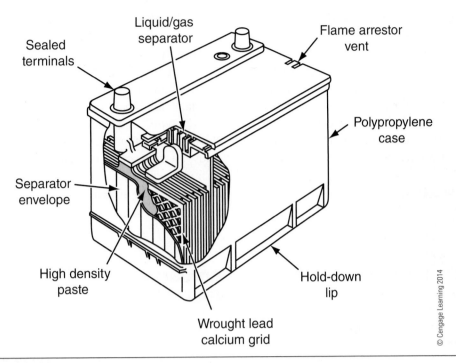

Figure 5-15 Assembled battery components.

sediment to settle. The sediment consists of pieces of battery plate material that flake off over the life of the battery. If the sediment level rises so that it contacts the plates, it can short the plates together, relegating the battery to scrap.

The case of maintenance-free batteries is sealed, with the exception of small vents that are necessary to permit any hydrogen or oxygen gas that develops during charging and discharging to escape to the atmosphere. A cut-away drawing for a complete maintenance-free battery is shown in **Figure 5-15**. Some new generation batteries are completely sealed with no vents, as described in a later section of this chapter.

Battery Terminals

The external terminals of the battery pass through the case and provide a means to connect the vehicle electrical system to the battery. Two major types of battery terminals are used on trucks, as shown in **Figure 5-16**. The side battery terminals shown in **Figure 5-16** are seldom used in medium- or heavy-duty trucks but may be found in some older light-duty trucks with diesel engines.

Some truck batteries have tapered lead posts like those found on many older automobile batteries. The positive post is slightly larger in diameter than the negative post, to minimize the chance of misconnecting the batteries. These tapered lead posts are still found on some older trucks and trucks outside the

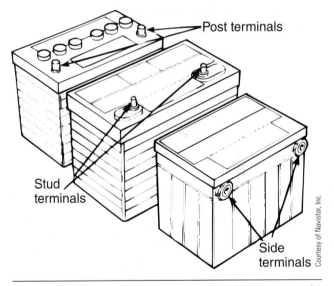

Figure 5-16 Types of battery terminals: most trucks use stud-type terminals.

North American market but are not commonly used in modern North American trucks.

The most common type of battery terminal used by the North American heavy-duty truck industry is the stainless steel 3/8" NC threaded male stud terminal. Both the negative and positive terminals are equipped with this same 3/8" NC stud. Threaded stud terminals permit multiple battery cables with common ring terminals to provide an easy means of connecting batteries. The stud has sufficient height to permit multiple ring terminals to be stacked together for interconnection of batteries.

Instead of common ring terminals, trucks may have sealed terminal connections on the battery cables like those shown in **Figure 5-17**. These sealed terminals are similar in appearance to the cable ends used with automotive side terminal batteries. These terminals are stackable to permit interconnection with other batteries. The sealed terminals reduce the development of corrosion on the battery terminals by sealing out electrolytic gas and water.

Terminal Marking

The positive and negative terminals of the battery are identified by a + and − marking of some sort. This marking may be molded into the battery case or printed on a label. Some batteries may also have a color identification to distinguish the positive from the

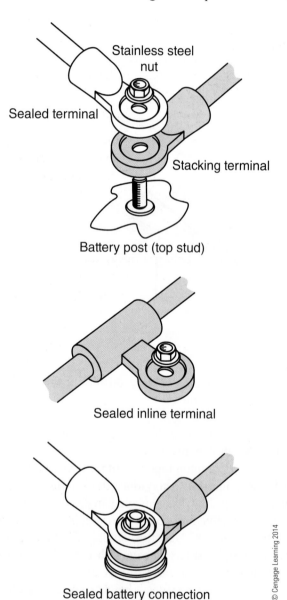

Stainless steel nut

Sealed terminal

Stacking terminal

Battery post (top stud)

Sealed inline terminal

Sealed battery connection

© Cengage Learning 2014

Figure 5-17 Sealed battery cable terminals.

negative cable. In the color identification method used in automotive batteries, positive is always red and negative is always black. This convention is the opposite of that used in accounting, in which negative numbers are shown in red ink and positive numbers are in black ink. In house wiring, the black wire is the "hot" wire, and the white wire is the neutral or "ground" conductor. Do not confuse accounting or house wiring color-coding with automotive battery terminal color-coding. Connecting batteries to the truck electrical system with reversed polarity for even a second can cause significant damage to the electrical system.

CONNECTING MULTIPLE BATTERIES

While most automobiles only have one battery, trucks may require four or more 12V batteries connected in parallel, series, or series-parallel.

Parallel-Connected Batteries

Batteries that are connected in parallel provide more available current than a single battery while still supplying the same terminal voltage as a single battery. The increased available current is necessary for cranking a diesel engine. Most medium-duty trucks require at least two parallel-connected batteries to meet the high current demands of the starter motor. Many heavy trucks with larger displacement diesel engines, especially those that operate in colder climates, require four or more parallel-connected batteries. The parallel connection still provides 12.6V but increases the amount of current that can be supplied by a single battery of the same rating. This is the same concept as using several parallel-connected plates in each cell to produce additional current per cell.

Series-Connected Batteries

Some trucks and construction equipment may have 24V systems. Most military trucks are also 24V systems. It is common for these systems to use two series-connected 12V batteries to provide 24V. The pair of series-connected 12V batteries may also be connected in parallel with another pair of series-connected 12V batteries on vehicles with large displacement engines. This series-parallel connection provides 24V.

Some students have trouble understanding how the positive terminal of one battery can be connected to the negative terminal of another battery without causing a short circuit. After all, if a wire connected the positive post of a 12V truck battery to the negative post of the *same* battery, a large amount of current would flow, resulting in a tremendous amount of heat, flash, and possibly an exploding battery. The reason that the positive terminal of one battery can be connected to the negative terminal of another battery is that there is no path for current to flow from the positive terminal of the top series-connected battery to the negative terminal of the bottom series-connected battery, as shown by the DMM used as an ammeter in **Figure 5-18**. Until a path is provided between these unused terminals, no current flows through the connecting cables installed between batteries. This is why a battery cable with almost zero resistance can be connected between the positive terminal of one battery to the negative terminal of the next battery and nothing happens.

However, once a path for current flow is provided between the unused terminals of the batteries as shown in **Figure 5-18**, conventional current will flow from the positive terminal of the topmost battery to the negative terminal of the bottommost battery, as shown by the DMM ammeter indicating 8A of current flow in **Figure 5-19**. If this path for current flow existed when the last battery connection was made, the load current would pass through the last series battery connection that was made, resulting in a large amount of arcing depending upon the load value. This is the reason that all vehicle electrical devices must be switched off when connecting or disconnecting batteries, regardless

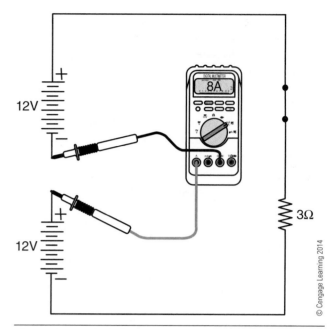

Figure 5-19 Closed switch provides a path for current flow; current flows between series-connected batteries. Always switch off all electrical components before connecting or disconnecting battery terminals.

of the type of battery configuration. Any device that is left in the on position will provide a path for current flow. The result will be a large arc when current flow is initiated (when connecting batteries) or interrupted (when disconnecting batteries).

Tech Tip: Some arcing is normal when connecting battery cables because of electronic modules performing an initialization routine and internal capacitors charging when power is first applied. Do not assume that there is a key-off current draw just because you observe an arc when the last battery connection is made.

BATTERY INTERNAL RESISTANCE

The greater the amount of current that a battery supplies to a circuit, the lower the battery terminal voltage becomes. If you have ever left your headlamps on while cranking the engine on your car, you may have noticed that the headlamps got dimmer while the starter motor was operating. The headlamps became dimmer because the battery terminal voltage lowered as the battery supplied a large amount of current to the starter motor. The reason that the battery terminal voltage cannot be maintained during engine cranking is because of the battery's **internal resistance**.

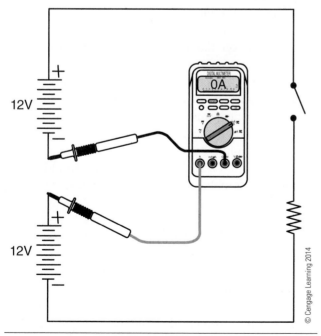

Figure 5-18 No path for current flow due to open switch, zero current flow between series-connected batteries indicated by ammeter.

To illustrate the effects of battery internal resistance, a simple series circuit is shown in **Figure 5-20**. In this circuit, the current is the same everywhere in the circuit and the available 12.6V from the battery is dropped proportionally to the resistor value as indicated by Ohm's law. Because resistor R1 has a value of 1Ω and resistor R2 has a value of 2Ω, the voltage dropped across R1 will be 4.2V and the voltage dropped across R2 will be 8.4V. The polarity symbols on R1 and R2 indicate which end of the resistor is closest to the battery-positive terminal and which end is closest to the battery-negative terminal. The arrow shows the direction of conventional current flow.

In **Figure 5-21**, resistor R1 is redrawn so that it is contained within the battery case instead of located in the external circuit. R1 is shown inside the battery case to represent the internal resistance of the battery. R1 is still in series with the battery and R2. Ohm's law does not change depending on how resistors in series are grouped, so the expected voltage drop across resistor R2 should still be 8.4V because the same 4.2A of current flows in the circuits shown in **Figure 5-20** and **Figure 5-21**. The polarity symbols on the resistors are the same in both figures, but R1 is now drawn upside down in **Figure 5-21** because it has been moved to the left side of the drawing so that it is contained within the battery case.

The battery symbol shown contained inside the battery case in **Figure 5-21** represents a perfect or

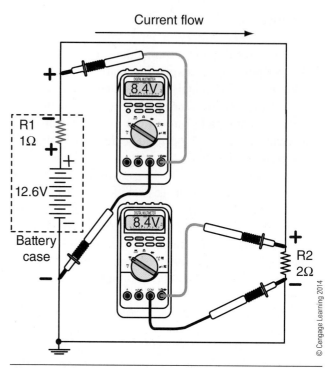

Figure 5-21 Redrawn 1Ω resistor R1 inside battery case indicating high battery internal resistance.

ideal voltage source that does not have any internal resistance. Such an ideal voltage source does not really exist, but the concept of an ideal voltage source connected in series with the internal battery resistance permits the voltage measured at a real battery's terminals to be explained. The current that flows in the circuit shown in **Figure 5-21** is 4.2A. As the current flows through the internal battery resistance, a voltage of 4.2V is dropped on the internal resistance. This voltage drop on the internal resistance leaves only 8.4V across the battery terminals.

If the 2Ω resistor R2 were replaced with a 0.75Ω (3/4Ω) resistor, as shown in **Figure 5-22**, the total circuit resistance would decrease from 3Ω to 1.75Ω (1Ω internal resistance + 0.75Ω load). The current in the circuit would then increase to 7.2A as predicted by Ohm's law (12.6V/1.75Ω = 7.2A). This would cause the voltage dropped across R1 to increase to 7.2V, leaving only 5.4V across the battery terminals. The 5.4V would of course cause 7.2A of current to flow through the 0.75Ω R2 because the current is the same everywhere in a series circuit.

As the current supplied by the battery increased due to the reduction of resistor R2 to 0.75Ω, the voltage measured across the battery's terminals decreased due to voltage drop across the battery's internal resistance represented by R1. This example demonstrates why the battery's internal resistance causes the battery's terminal voltage to decrease as the

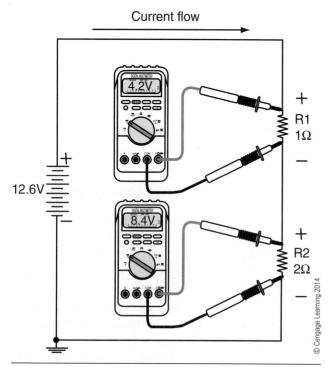

Figure 5-20 Simple series circuit with two resistors to illustrate battery internal resistance.

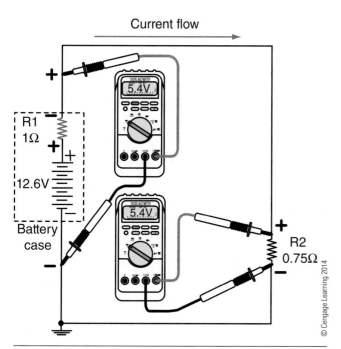

Figure 5-22 Load resistor R2 decreased to 0.75Ω causing increased current flow and decreased battery terminal voltage.

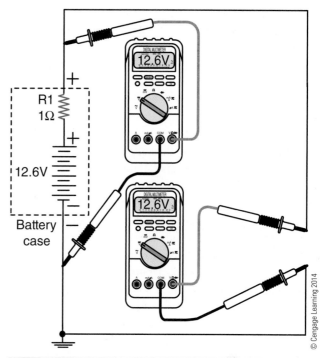

Figure 5-23 Load resistor R2 removed; no-load voltage is the same as battery terminal voltage because virtually 0A of current is flowing through the circuit.

battery supplies additional current. The cause of the decrease in battery terminal voltage as the battery supplies additional current is just Ohm's law and the rules of series circuits applied to the battery's internal resistance.

If R2 is completely removed from the circuit and is replaced with a voltmeter, the voltmeter would indicate about 12.6V, as shown in **Figure 5-23**. A voltmeter connected across the battery terminals would also indicate about 12.6V. The battery terminal voltage increases because the internal resistance of the DMM voltmeter that replaced R2 is 10MΩ or more. Nearly zero current flows in the circuit with R2 removed and the voltmeter inserted in series in its place. Since nearly zero current is flowing in the circuit, the voltage dropped across the battery's internal resistance represented by R1 is nearly 0V, effectively leaving the full 12.6V across the battery's terminals. This measurement across a battery's terminals with no load connected to the battery is referred to as **no-load voltage**. No-load voltage is also known as open-circuit voltage (OCV) because it is a measurement of voltage where no current is being supplied by the voltage source (except the very small, insignificant amount that flows through the voltmeter's internal resistance).

Actual battery internal resistance is usually much smaller than the 1Ω shown in the previous examples. The internal resistance of a battery may be in the region of thousandths of an ohm (milliohms). The battery's actual internal resistance is influenced by many factors. The internal resistance of a battery is always there, but it is usually such a small value of resistance that it is ignored, like the resistance of wire. However, when the load is very large (small value of resistance), such as a starter motor, the internal resistance of the battery becomes a major factor in the battery terminal voltage.

Causes of Battery Internal Resistance

The internal resistance of a battery is not a constant value. Part of a battery's internal resistance is due to the resistance of the plate connection straps, the cell connection straps, and the terminals. All of these resistances remain constant. However, other factors that cause battery internal resistance are the resistance of the electrolyte and a layer of sulfate that builds up on the plates. These components of the internal resistance vary with the battery **state of charge**, temperature, and battery age. As the battery discharges, the electrolyte becomes water and the plates become lead sulfate. Contrary to popular belief, pure water is not a very good conductor of electricity. The layer of sulfate that builds up on the lead plates is also not a good conductor of electricity, resulting in an increase in the battery's internal resistance as the battery becomes discharged.

The battery temperature is also a large factor in the battery's internal resistance. The resistance of the electrolyte is affected by temperature. As temperature decreases, the resistance of the electrolyte increases, resulting in a substantial increase in battery internal resistance.

The internal resistance also increases as a battery ages. A layer of sulfate builds up on the plates over time that is not removed through charging. Pieces of the active lead material also flake off, reducing the plate surface area.

A battery's internal resistance cannot be measured using an ohmmeter because an ohmmeter should never be used in a circuit that has a voltage source present. Internal resistance is also usually such a small value that most ohmmeters could not accurately measure such a small resistance anyway. It is also not possible to separate the battery's internal resistance from the ideal battery voltage source except on paper, as shown in the previous examples.

Because battery internal resistance is usually such a small value, why worry about it? The reason is that even a small resistance has an impact when current levels are very high. A truck diesel engine starter motor may draw in excess of 1000A of current continuously (3000A initially) while the starter motor is cranking a cold diesel engine. By comparison, a single truck headlamp only draws about 5A. The difference in the current draw of the starter motor and the headlamps indicates that the resistance of the starter motor is much less than the resistance of the headlamps. The terminal voltage across two fully charged new 12V batteries connected in parallel in subzero temperatures can easily briefly drop to less than 8V when the starter motor is first energized. The reason for the drop in battery terminal voltage during engine cranking is the battery's internal resistance and the low value of resistance of the load, in this case the starter motor. The very large amount of current drawn by the starter motor causes a large voltage drop on the battery's internal resistance, resulting in a significant decrease in battery terminal voltage.

If battery internal resistance were not an extremely small value, the high amount of current drawn during engine cranking would drop the battery terminal voltage so low that the starter motor would not operate. A battery that has developed a high level of internal resistance may appear to function correctly when relatively small loads such as headlamps are switched on but will "fall on its face" as the terminal voltage drops to a couple of volts when the starter motor is energized.

BATTERIES AND TEMPERATURE

Temperature extremes present some real challenges to lead acid truck batteries.

Low Temperatures and Batteries

By far, the most demanding task for a truck battery to perform is to crank a diesel engine in cold weather. Engine oil at $0°F$ $(-18°C)$ is thick or viscous like pancake syrup that has been placed in a freezer. The starter motor must be able to cause the crankshaft to rotate through this thick oil at 150rpm or greater in order to be able to start a cold diesel engine. The engine oil pump is also trying to pump this thick oil, so it also consumes a considerable amount of the available torque supplied by the starter motor.

The electrical energy provided by the battery is converted to mechanical energy by the rotating starter motor. Cold climate conditions require that the batteries provide for a greatly increased demand during cold period starts.

A diesel engine is a compression ignition engine. The heat generated by compressing air is used to ignite the diesel fuel that is injected into the combustion chamber. In cold weather, the engine must be cranked for a longer period of time in order to start (produce ignition) than in warm weather because the cold metal engine components along with the cold intake air may reduce the temperature of the compressed air such that it cannot ignite the thickened diesel fuel. This means that the batteries must supply starter motor current for a longer period in cold weather than in warm weather as each cylinder compression warms the combustion chamber and adjoining metal surfaces. Engine cranking times of 20 seconds or more at $0°F$ $(-18°C)$ are not uncommon in diesel engines, although preheating devices like glow plugs and intake air heaters (grid heaters) can reduce this time considerably. However, these preheating devices are electric powered and require power from the battery to heat the air or combustion chamber. This also diverts a portion of the battery's stored energy when starting an engine in cold temperatures.

To make matters even worse at cold temperatures, the electrochemical reaction within a battery that transforms the chemical energy into electrical energy decreases with temperature. A cold battery cannot supply as much current as a warm battery. The internal resistance of a cold battery also increases as temperature decreases, which further reduces energy available to the starter motor. The effect of cold weather on batteries is summarized in **Figure 5-24**.

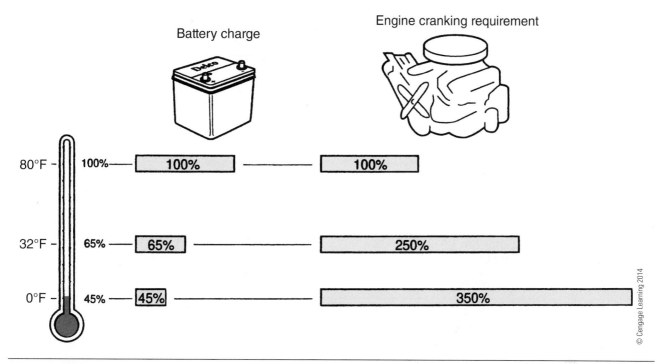

Figure 5-24 Engine cranking requirements versus battery current capability in cold weather.

One additional item of concern in cold temperatures is electrolyte freezing. The electrolyte in a fully charged battery has a freezing point of –75°F (–59°C). As a battery discharges, the electrolyte becomes a higher concentration of water. This reduces the freezing point of the electrolyte substantially. Electrolyte in a nearly discharged battery can freeze at 20°F (–7°C). If the electrolyte freezes, the associated expansion of the frozen electrolyte could split the plastic battery case or otherwise damage the plate structure.

WARNING *A battery that is frozen must never be recharged or jump-started because it could explode. Frozen batteries must be thawed at room temperature before recharging.*

Trucks used in cold climates may use battery heaters or blankets to keep the batteries warm when the truck is parked (**Figure 5-25**). These devices are plugged into an AC power line and keep a battery warm enough to provide sufficient cranking voltage to start a cold engine. Other types of AC-powered heaters such as oil pan heaters and block (coolant) heaters assist in the starting of an engine in very cold climates. Many fleets have posts with electrical outlets situated in front of the trucks so that the various truck heaters can be plugged in overnight.

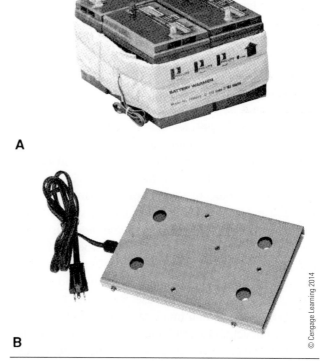

Figure 5-25 Battery heaters used in extreme cold temperatures.

High Temperatures and Batteries

Hot weather also causes battery problems. In high ambient temperatures, a battery loses its charge at a

Rate of Self-discharge
at Various Temperatures

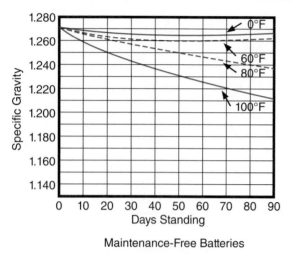

Maintenance-Free Batteries

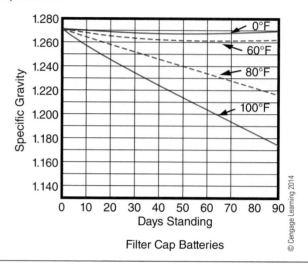

Filter Cap Batteries

Figure 5-26 Battery self-discharge versus temperature.

more rapid rate than in cool temperatures through a phenomenon known as **self-discharge**. Self-discharge refers to the slow loss of charge that occurs when a battery is not being recharged, such as a battery in storage or a truck that is not operated for an extended period of time. Low water loss batteries have a lower level of self-discharge than conventional batteries, as shown in **Figure 5-26**, but all batteries self-discharge with time. If a battery is left in the discharged state for too long, the plates form hard sulfate crystals that coat the plate surface and render the battery useless. Such hard sulfation cannot be removed by normal recharging.

You may have heard or will someday hear that a battery cannot be placed directly on concrete because the concrete will "pull the charge out of the battery." In the early days of automobiles, battery cases were made of wood covered with tar and other materials. These batteries had a very high self-discharge rate, regardless of where they were placed. However, setting such a battery directly on concrete would cause the battery to self-discharge at an even more rapid rate due to electrolyte leakage through the wooden battery case. This increase in self-discharge when a battery is placed on concrete is no longer true of modern batteries due to the improved case materials. However, the legend lives on and many experienced technicians still will not set a modern battery directly on concrete.

Hot weather also causes the temperature of the electrolyte to increase, resulting in additional gassing of the electrolyte during normal in-vehicle charging than in cold weather. The water evaporates from the electrolyte, leaving the plates exposed to the atmosphere. Battery plates will quickly form hard sulfate crystals when not covered with electrolyte. Loss of

water in a maintenance-free battery typically requires battery replacement. High battery temperature also deforms the plates and causes accelerated development of corrosion on positive plates, which can result in internal short circuits.

BATTERY RATINGS

Batteries have several ratings. Most of these ratings are standardized by the Battery Council International (BCI) and Society of Automotive Engineers (SAE).

Voltage Ratings

Truck batteries are mostly rated at 12V, like automobile batteries. The actual no-load terminal voltage of a fully charged 12V battery is about 12.6V. Some older truck batteries have three cells and are rated at 6V with a no-load terminal voltage of about 6.3V. Batteries with 12 cells and a rating of 24V are used in some equipment. However, most 24V trucks use series-connected 12V batteries.

Cold Cranking Amperes

The **cold cranking amperes (CCA)** rating of a battery is the most important battery rating for trucks operated in cold climates. If you have ever replaced the battery in your car, the parts salesperson may have tried to convince you to upgrade to a battery with a higher CCA rating than that with which the car was originally equipped.

The CCA rating of a battery is defined as how much current a new, fully charged battery can provide continuously for 30 seconds at a temperature of 0°F (−18°C) and still maintain a terminal voltage greater than 1.2V per cell (7.2V terminal voltage for a typical 12V truck battery). Truck battery CCA ratings range from 550 CCA for mild climates to 1000 CCA or greater for cold climates. A fully charged, new 1000 CCA battery at a temperature of 0°F (−18°C) will maintain at least 7.2V across the terminals when a 1000A load is placed across the terminals for 30 seconds. The CCA rating of a battery generally decreases with battery age as the battery's internal resistance increases and pieces of the active plate material flake off the plates.

When batteries are placed in parallel, this causes the total CCA ratings to be added together. Three 700 CCA batteries connected in parallel have a combined rating of 2100 CCA.

One of the potential downsides of a higher CCA battery in the same size case is that the increase in plate count may require the plates to be thinner than the plates of a lower CCA battery. This may reduce the life of the battery in some applications. Two 900 CCA batteries in parallel have 1800 CCA. Three 600 CCA batteries in parallel also have 1800 CCA. If battery box space permits, using additional lower-rated CCA batteries may provide better life than that offered by fewer higher-rated CCA batteries. The additional battery will also provide a decreased overall internal battery resistance because placing resistors in parallel decreases total resistance. The reduction in total internal battery resistance results in a higher terminal voltage under a load such as cranking an engine.

The CCA rating is often misunderstood. In heavy truck applications, high CCA batteries and standard CCA batteries both provide about the same amount of cranking current. The difference is that the higher CCA battery will provide a sustained level of current output for a longer period of time during cranking than standard CCA batteries. In very cold weather, standard CCA batteries may not be able to crank a hard-starting engine long enough to get the engine started. Higher CCA batteries will provide cranking current for a longer period of time in cold weather than lower CCA batteries. However, this extended cranking time provided by higher CCA batteries can cause starter motor damage. A starter motor should never be operated for longer than 30 seconds, with at least 2 minutes between cranking attempts to permit the starter motor to cool. Otherwise, the starter motor can be damaged through excessive heating. Installing higher CCA batteries leads to the risk of starter motor damage due to extended cranking times unless the motor is equipped with a thermal overcrank protection, as described in **Chapter 8**.

Installing high CCA batteries on some medium-duty trucks may also result in higher incidence of cranking system component failure such as ring gears because starter motor torque is increased by the increase in cranking current in some applications. Consult OEM information for recommended battery CCA requirements.

Cranking Amperes

Cranking amperes (CA) are similar to CCA except the current is measured with the battery at a temperature of 32°F (0°C). This rating is not typically used with truck batteries, but you should be aware of the difference between CA and CCA. The CA rating is typically about 1.25 times higher than the CCA rating.

Reserve Capacity

One task of the vehicle's battery system is to provide the electrical energy to operate the vehicle for a period of time should the charging system fail. The **reserve capacity** rating indicates how many minutes that new, fully charged batteries are capable of supplying a continuous load of 25A and still maintain a terminal voltage greater than 1.75V per cell (10.5V for a 12V battery) at a temperature of 80°F (27°C). As the battery discharges, the terminal voltage drops as cells begin to produce less than 2.1V per cell, while the plate materials become the same lead sulfate composition and the electrolyte becomes water. A battery that is supplying 25A and only has 10.5V across its terminals is nearly discharged and will not be able to keep an electronically controlled diesel engine operational for much longer. In some applications, the engine electronic control module (ECM) may force engine shutdown prior to the battery terminal voltage ever reaching the 10.5V reserve capacity rating.

Older mechanical diesel engines could be operated without an electrical system. It was not uncommon in the past for trucks that developed charging system problems on the road to be jump-started and kept running continuously until they made it back home or to a repair shop. However, modern diesel engines with electronically controlled fuel systems will not operate without an adequate electrical power supply.

Increasing the number of parallel-connected batteries increases the total reserve capacity. Like the CCA rating, placing batteries in parallel causes the reserve capacity ratings of the individual batteries to be combined together. The reserve capacity rating gives an idea of how long a truck with a charging system problem can be driven with a minimal amount of electrical accessories, such as lights, being operated. Electronic engine controls, including injectors, can draw 25A on some engines. In rainy nighttime operation, the headlamps, windshield wipers, and blower motor draw much more current than the 25A specified in the reserve capacity rating. On a modern truck, a charging system failure means the truck will not be operating very long on the battery's reserve capacity.

BCI Group Number

Automotive batteries are typically assigned a **BCI group number**. The Battery Council International (BCI), in conjunction with the Society of Automotive Engineers (SAE), establishes industry standards for batteries. The BCI group number describes the physical dimensions of the battery, the terminal layout arrangement (i.e., positive on left or right side), and the terminal type (tapered post, threaded post, etc.). The BCI group number is useful information to have when replacing batteries. A sample of a BCI group data for truck batteries is shown in **Figure 5-27**.

BATTERY RECHARGING

To replenish the battery's chemical energy, it is necessary to use an external voltage source to cause a recharging current to flow through the battery. The direction of this charging current is the opposite of battery discharge current. An engine-driven alternator is the voltage source used to recharge the battery on the truck during normal driving. However, an engine-driven alternator is not much good if the engine will not crank due to discharged batteries. The engine-driven alternator also may not be able to fully recharge a battery that has a low state of charge, such as a truck that had to be jump-started due to dead batteries. Batteries with low state of charge may take a day or more of continuous recharging by the alternator to reach an adequate state of charge. A battery charger is typically used to recharge batteries that have discharged to a low state of charge, as it is capable of providing a higher rate of recharge than a typical truck alternator.

Battery chargers convert power line voltage such as 120V AC into a direct current to recharge 6V, 12V, or even 24V batteries. Numerous models of battery chargers are available with a variety of features. Battery chargers are discussed in detail later in this section.

Modern battery chargers are capable of supplying a voltage that is much greater (i.e., 18V) than the voltage

BCI GROUP NUMBERS, DIMENSIONAL SPECIFICATIONS AND RATINGS

BCI Group Number	Maximum Overall Dimensions						Post/ Terminal	Assembly Figure No.	Performance Ranges	
	Millimeters			Inches					Cold Cranking Performance Amps. @ 0° F (-18° C)	Reserve Capacity Minutes @ 80° F (27° C)
	L	W	H	L	W	H				
HEAVY-DUTY COMMERCIAL BATTERIES 12-VOLT (6 CELLS)										
4D†	527	222	250	20 3/4	8 3/4	9 13/16	A, B, A2	8	490-590	225-325
6D	527	254	260	20 3/4	10	10 1/4	A, L	8	750	310
8D†	527	283	250	20 3/4	11 1/8	9 13/16	A, A2, B,B1	8	850-1250	235-465
28	261	173	240	10 1/4	6 13/16	9 7/16	T	18	400-535	80-135
29H	334	171	232	13 1/8	6 11/16	9 1/8	A	10	525-840	145
30H	343	173	235	13 1/2	6 13/16	9 1/4	A	10	380-685	120-150
31	330	173	240	13	6 13/16	9 7/16	A, T, M, M1	18	455-950	100-200
HEAVY-DUTY COMMERCIAL BATTERIES 6-VOLT (3 CELLS)										
3	298	181	238	11 3/4	7 1/8	9 3/8	A	2	525-660	210-230
4	334	181	238	13 1/8	7 1/8	9 3/8	A	2	550-975	240-420
5D	349	181	238	13 3/4	7 1/8	9 3/8	A	2	720-820	310-380
7D	413	181	238	16 1/4	7 1/8	9 3/8	A	2	680-875	370-426
SPECIAL TRACTOR BATTERIES 6-VOLT (3 CELLS)										
3EH	491	111	249	19 5/16	4 3/8	9 13/16	A	5	740-850	220-340
4EH	491	127	249	19 5/16	5	9 13/16	A	5	850	340-420
SPECIAL TRACTOR BATTERIES 12-VOLT (6 CELLS)										
3EE	491	111	225	19 5/16	4 3/8	8 7/8	A	9	260-360	85-105
3ET	491	111	249	19 5/16	4 3/8	9 13/16	A	9	355-425	130-135
4DLT	508	208	202	20	8 3/16	7 15/16	A	16L	650-820	200-290
12T	179	177	202	7 1/16	6 15/16	7 15/16	A, L	10	460	160
16TF	421	181	283	16 9/16	7 1/8	11 1/8	A	10F	600	240
17TF	433	177	202	17 1/16	6 15/16	7 15/16	A	11L	510	145
GENERAL-UTILITY BATTERIES 12-VOLT (6 CELLS)										
U1	197	132	186	7 3/4	5 3/16	7 5/16	X	10	120-295	23-40
U1R	197	132	186	7 3/4	5 3/16	7 5/16	X	11	220-235	25-37
U2	160	132	181	6 5/16	5 3/16	7 1/8	A3, X	10	120	17

Charts shown are reprinted with permission from Battery Council International

Figure 5-27 Sample BCI group number pages.

provided by the alternator (14.2V). Therefore, when recharging batteries that are installed in a vehicle, it is recommended that the batteries be isolated from the vehicle's electrical system (i.e., disconnect the vehicle's positive and negative battery cables before recharging). This reduces the potential for damage to electronic components.

Details of Battery Recharging

To recharge a battery, a charging current must flow through the battery in a direction opposite of that when a battery is being discharged (**Figure 5-28**). To cause a charging current flow through a battery, a voltage source must be connected across the battery's terminals. This voltage source must supply a voltage greater than the voltage being produced by the battery alone through its chemical reaction. The voltage supplied by the battery charger and the voltage being produced by the battery's chemical reaction become series-opposing voltages.

As a charging current passes through the battery, a chemical reaction occurs and causes the positive and negative plates to transform back into lead oxide and sponge lead, respectively, while the electrolyte becomes a higher concentration of sulfuric acid. This charging process takes time.

A battery also takes a period of time to become discharged. The time that it takes for a fully charged battery to discharge depends on how much current the battery is supplying. A fully charged battery that is only supplying 1A of current to a single marker lamp bulb will take about 10 times longer to discharge than a battery that is supplying 10A of current

to the low-beam headlamps. The amount of current that is supplied by a battery for a period of time is expressed in amp-hours.

To calculate how long it will take a battery to recharge, you would need to know many things, such as the battery capacity, the battery state of charge, the characteristics of the battery charger, charging current, and so on. In the time it takes to do these calculations, you could already have the battery connected to a battery charger and be working on something else. In general, it takes a combination of time and current to recharge a battery.

Additional information about battery recharging can be found in the Extra for Experts section at the end of this chapter.

Overcharging

As a battery is being recharged, some of the electrolyte is converted to hydrogen and oxygen gas. With conventional batteries, the amount of gassing that occurs during recharging is much greater than with low water loss batteries, due to the composition of the plate grids. This is the reason that conventional batteries have a means to replace the water lost by gassing during recharging. The amount of gassing that occurs with low water loss batteries is much less than with conventional batteries. However, if too high a charging current is used to charge the battery, the water in the electrolyte will be lost even in low water loss batteries. If a battery is being overcharged, a sulfur or rotten egg smell is usually evident, and violent gassing inside the battery may be heard. If electrolyte is lost in a low water loss battery that has no provisions for replacing water, the battery must be replaced.

WARNING Never *drill holes in a sealed battery case to add water or electrolyte. This can defeat the safety features of the battery that prevent battery explosion.*

When recharging a battery, the temperature of the battery increases. If the temperature gets too high from an excessive charge rate, the battery plates can warp or short together, also resulting in the need to replace the battery. In severe cases of overcharging, the battery can explode, causing sulfuric acid to spray over a large area.

It is important to monitor a battery that is being recharged to prevent overcharging. It may be necessary to reduce the charging voltage or current limit to reduce battery temperature.

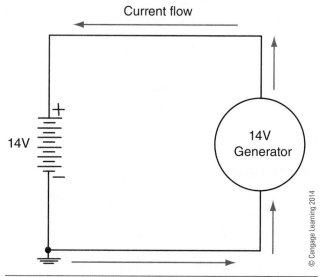

Figure 5-28 Direction of current flow passing through a battery to recharge the battery.

Charging a Very Discharged Battery

The internal resistance of a battery that has a very low state of charge (<10 percent) is very high. This high internal resistance of a discharged battery makes it difficult to recharge the battery. Ohm's law says that as the total series circuit resistance goes up, the total circuit current goes down. The battery shown in **Figure 5-29** has a very low state of charge and is only capable of producing 10V instead of the 12.6V produced by a fully charged battery. The battery's internal resistance is shown as 1kΩ, which is a very large amount of internal resistance. The 14V generator in parallel with the battery is only capable of sourcing about 0.004A (4mA) of charging current through the battery due to the battery's internal resistance.

The internal resistance of a severely discharged battery may gradually decrease after an extended period of charging and permit additional current to flow through the battery as the battery is recharged, but this process can be very slow (many hours). To reduce the time necessary to recharge a battery with a very low state of charge, set the battery charger to the highest voltage output available for the voltage rating of the battery being charged. Some 12V battery chargers can output 18V to initiate recharge in a much-discharged battery. The higher voltage will increase charging current flow through the battery.

Depending upon the battery charger, you will need to monitor closely the current accepted by a battery with a very low state of charge. When the internal resistance starts to lower, the current supplied by the battery charger will begin to increase rapidly. The voltage output of the charger should then be decreased to prevent electrolyte loss or overheating the battery.

If the battery has been discharged for too long a period, the plates form a layer of crystalline sulfate that cannot be removed by charging. When this occurs, the battery internal resistance will never decrease, so the battery cannot be recharged and must be replaced.

Battery Chargers

Many modern battery chargers are a constant-voltage taper-rate type. This type of battery charger maintains the output voltage to a selectable level. As a battery recharges with a constant-voltage-type charger, the charging current decreases as explained in the Extra for Experts section.

Battery chargers typically have a timer mechanism that stops the charging of the battery after a selectable amount of time. The timer also usually has an override setting that leaves the battery charger on continuously. It is important not to overcharge a battery, as discussed in the previous section on overcharging batteries, because the water in the electrolyte solution is evaporated and in most batteries cannot be replaced.

One of the more difficult things for many technicians in training, and even experienced technicians, to learn is the importance of reading instructions. Not all battery chargers operate in the same way. Some battery chargers may be a constant current type, a constant voltage type, or even a fully automatic type. It is important to read the OEM's instructions so that maximum performance can be obtained from the equipment. In the past, technicians seemed to take pride in not having to read instructions to figure out how to operate a piece of equipment. Times have changed, things are much more complex than in the past, and sensitive electronic components can be damaged or stressed by careless actions so that they fail two miles away from your shop. Read and follow the instructions. If you cannot find the instructions, look online. Most companies provide downloadable copies of operating instructions for their products at no charge.

Battery Recharging Safety

Safety when working around batteries cannot be overemphasized. Batteries are some of the most hazardous items with which a technician is required to work. When a battery is recharging, explosive hydrogen gas is given off, even with modern low water loss batteries. An exploding battery causes sulfuric acid to spray over a large area, which can blind persons in the

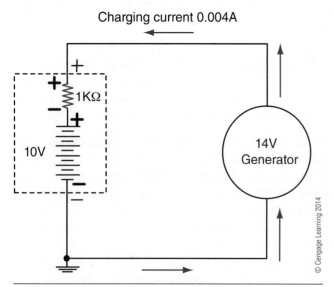

Charging current 0.004A

Figure 5-29 Battery with high internal resistance not accepting much charging current.

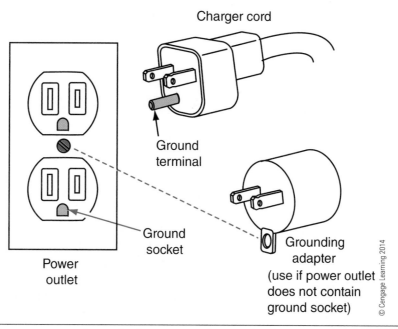

Charger cord

Ground
terminal

Ground
socket

Power
outlet

Grounding
adapter
(use if power outlet
does not contain
ground socket)

© Cengage Learning 2014

Figure 5-30 Battery charger ground terminal prevents electric shock.

vicinity and cause serious chemical burns. Always wear safety goggles, not just ordinary eyeglasses, when working around batteries.

Batteries can contain an enormous amount of potential energy. Electric shock is not a significant concern for 12V batteries. However, burns from jewelry or tools that make contact between the battery posts or between the positive terminal and ground are a real concern. Most technicians, especially those who work around electricity, do not wear jewelry at work, including rings. Gold is one of the best conductors of electricity. A gold chain that drops out of your shirt while you are bent over a battery can turn to molten gold in a flash, leaving a scar around your neck for life where the chain had been. Military personnel should also be aware of dog tag chains when working around batteries for this same reason.

Make sure that the battery charger AC cable grounding-prong is present (round male terminal on the power plug shown in **Figure 5-30**) and that the battery charger is plugged into a grounded outlet. This grounding path can protect you from electrocution. Do not use a battery charger that is not properly grounded.

Battery Charger Connections

The vehicle battery cables, including cables for electronic controllers that connect at the battery, should be disconnected from the batteries when using a battery charger. This is necessary because the battery charger's output voltage during recharging may be much higher than the typical vehicle alternator charging voltage. The increased charging voltage may damage electronic components on the vehicle. Additionally, the voltage supplied by battery chargers is not a pure DC level. Some AC "ripple" exists with most battery chargers. This AC voltage component may cause problems with the truck's electrical system components, including voltage regulators for the charging system.

Battery charger clamps should never be directly connected to stud-type battery terminals. There is not enough surface area on the stud for battery charger clamp contact. Special charging posts that thread onto the battery terminal studs should be utilized. These provide a larger surface area for the battery charger clamps to attach.

The order of connecting and disconnecting the battery charger is important. Causing sparks around a battery that is venting an explosive gas is not a good idea. To prevent causing sparks near the battery, follow this procedure:

1. Always switch off and unplug the charger before making or interrupting any connections at the batteries.
2. Connect the battery charger cables to the battery terminal adapters.
3. Plug the battery charger into a wall outlet.
4. Switch on the battery charger to the appropriate settings.

Reverse this order when disconnecting the battery charger.

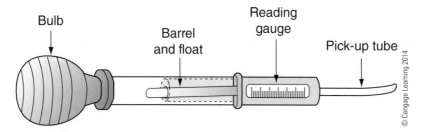

Figure 5-31 Hydrometer used to measure battery electrolyte specific gravity.

BATTERY TESTING

Battery testing is a common task for truck technicians. The tools and techniques for battery testing are described in this section.

Specific Gravity

In the past, when truck batteries were all conventional batteries with filler caps, the most important tool for battery testing was a **hydrometer**. A hydrometer is similar in appearance to a turkey baster, as shown in **Figure 5-31**. The hydrometer has a glass tube with an internal float marked with graduations. A squeeze bulb at the top of the tube permits the extraction of a small amount of battery electrolyte (**Figure 5-32**) to measure the **specific gravity** of the electrolyte by the height that the float is suspended in the electrolyte (**Figure 5-33**). *Specific gravity* is a unitless chemistry term and refers to the weight of the volume of any liquid compared to an equal volume of pure water. The weight of the liquid is divided by the weight of an equal volume of pure water to determine the specific gravity. Therefore, if the liquid is heavier than water, the specific gravity will be greater than 1. If the liquid is lighter than water, the specific gravity will be less than 1. Pure water has a specific gravity of 1 because any number divided by itself is always equal to 1.

Pure sulfuric acid has a specific gravity of 1.835. This means that 1 gallon of sulfuric acid is 1.835 times heavier than 1 gallon of pure water. Electrolyte is a solution of water and sulfuric acid, typically mixed with 36 percent sulfuric acid and 64 percent water by weight. This causes the specific gravity of the electrolyte to be in the range of 1.265 to 1.280 at 80°F. All battery electrolyte specific gravity readings are referenced to 80°F. If the electrolyte temperature is not

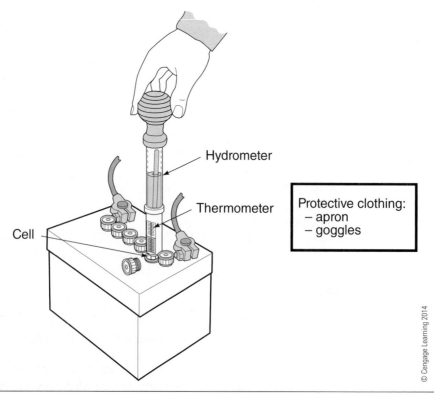

Figure 5-32 Drawing electrolyte from a conventional battery.

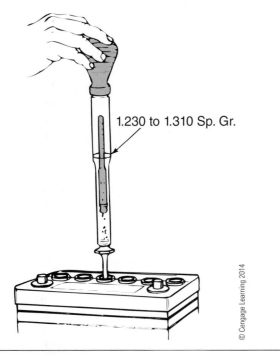

Figure 5-33 Reading a hydrometer.

80°F (27°C), the specific gravity reading must be adjusted to the 80°F value, using the chart shown in **Figure 5-34**.

As a battery discharges, the electrolyte becomes water. Thus, the specific gravity of the electrolyte is an indicator of the state of charge for each cell. A fully charged cell may have a specific gravity of 1.265 corrected to 80°F. As discharge occurs, the specific gravity of the electrolyte decreases, as shown in **Figure 5-35**.

Another tool used to measure specific gravity is an optical refractometer like that shown in **Figure 5-36**. This tool is also used to test engine coolant freeze protection. Newer models may also include a scale for measuring the concentration of urea in diesel exhaust fluid (DEF), as will be explained in **Chapter 14**.

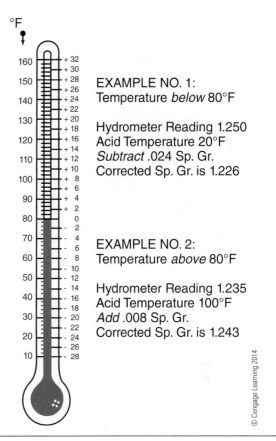

EXAMPLE NO. 1:
Temperature *below* 80°F

Hydrometer Reading 1.250
Acid Temperature 20°F
Subtract .024 Sp. Gr.
Corrected Sp. Gr. is 1.226

EXAMPLE NO. 2:
Temperature *above* 80°F

Hydrometer Reading 1.235
Acid Temperature 100°F
Add .008 Sp. Gr.
Corrected Sp. Gr. is 1.243

Figure 5-34 Specific gravity temperature correction chart.

A drop of electrolyte is placed on a glass window and the specific gravity is shown when viewed through the eyepiece. **Figure 5-37** indicates a specific gravity of 1.275 for this example.

Built-In Hydrometer for Low Water Loss Batteries

The specific gravity of the electrolyte in a cell indicates how much voltage the cell is capable of producing. As the specific gravity of the electrolyte

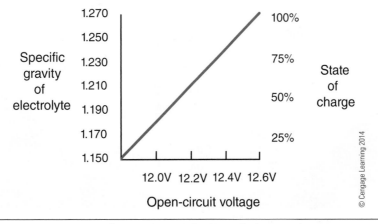

Figure 5-35 Temperature-corrected specific gravity and state of charge chart.

Figure 5-36 Optical refractometer.

decreases due to the electrolyte becoming a lower concentration of sulfuric acid, the voltage that the cell can produce also decreases. Since most low water loss batteries do not have cell filler caps, there is no way to measure the specific gravity of each cell because the electrolyte cannot be extracted from the battery. Many low water loss batteries have a built-in hydrometer. This built-in hydrometer uses a colored ball (typically green) in a glass rod as shown in **Figure 5-38** to provide a rough indication of the state of charge. A built-in hydrometer only indicates if the state of charge of one cell is above about 65 percent state of charge. The built-in hydrometer will also indicate if the electrolyte is low in the one cell containing the hydrometer. If everything is fine in the cell with the built-in hydrometer but problems exist in other cells, the built-in hydrometer can give a false indication of the battery state.

When a battery is new and fully charged, the specific gravity readings of all cells typically have the same value. As the battery ages, one or more cells may begin to become less effective at the chemical reactions involved in charging and discharging. The lead material also begins to flake off the plates, rendering the cell less effective. Because the specific gravity of each cell in low water loss batteries cannot be measured, the terminal voltage at no-load and loaded conditions is used to determine the battery's health and state of charge. Specific gravity and the voltage that a cell produces are related. As the specific gravity of a cell decreases, the voltage that the cell produces will also decrease.

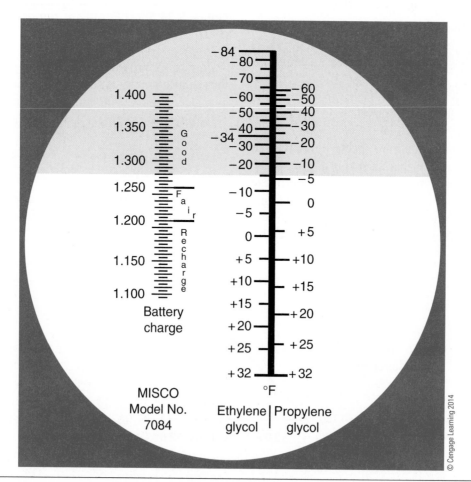

Figure 5-37 Specific gravity measurement with optical refractometer.

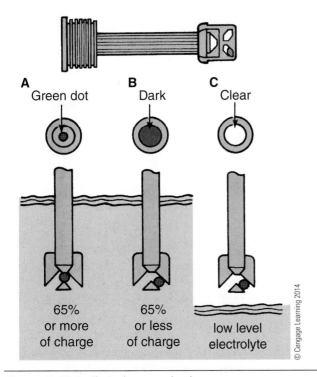

A Green dot   **B** Dark   **C** Clear

65% or more of charge   65% or less of charge   low level electrolyte

Figure 5-38 Built-in battery hydrometer.

Terminal Voltage as a State of Charge Indicator

Because most modern batteries cannot be opened to test the specific gravity, the no-load terminal voltage of a battery can be utilized to provide an indication of battery state of charge. This assumes that all cells are producing roughly the same voltage and that the battery has not been recently charged.

The chart shown in **Figure 5-39** shows that only a 200mV (0.2V) decrease in no-load terminal voltage represents about a 25 percent decrease in the battery's state of charge. Thus, a small measurement error in battery terminal voltage can cause a large error in determining the battery state of charge.

Stabilized Open Circuit Voltage	State of Charge
12.6 volts or more	Fully charged
12.4 volts	75% charged
12.2 volts	50% charged
12.0 volts	25% charged
11.7 volts or less	Discharged

Figure 5-39 Open-circuit battery terminal voltage versus state of charge.

Surface Charge

A common term used when talking about batteries and charging systems is **surface charge**. Surface charge refers to hydrogen gas bubbles that form on the plates during battery charging. These bubbles are rapidly dissolved when the battery is discharged. The bubbles also dissipate over a long period of time (hours). The surface charge causes each cell to produce a higher than normal voltage until the hydrogen bubbles dissipate. For example, most charging systems attempt to maintain about 14V across the battery terminals during charging. Immediately after a battery has been charged by the alternator, the battery terminal voltage may exceed 13.5V for a brief period of time due to surface charge. Remember that the terminal voltage of a fully charged 12V battery is about 12.6V.

The battery no-load terminal voltage is used to evaluate the battery state of charge and condition. Surface charge can cause a nearly discharged battery to appear to be fully charged. Surface charge is removed by placing a prescribed load on the battery for a period of time after it has been charged or by letting the battery sit for several hours after it has been recharged before beginning any testing. Surface charge must be removed to accurately test any battery.

Battery Test Equipment

Conventional battery test equipment consists of a voltmeter, an ammeter, and a large low-resistance, high-power variable resistor called a **carbon pile**. A carbon pile is a series of thin carbon plates stacked together. Carbon is a poor conductor of electricity due to its four valence electrons. Each carbon plate has a resistance like resistors that are wired in series. The total resistance of the carbon pile is varied by squeezing the carbon plates together. The tighter the plates are squeezed together, the lower the total resistance of the carbon pile. The total resistance of the carbon pile with the plates tightly clamped together is much less than 1Ω. The carbon pile resistor permits a very low resistance yet high wattage load to be applied to the battery for testing.

The three tools—voltmeter, ammeter, and carbon pile—are often combined to make up a charging/starting system tester as shown in **Figure 5-40**. This tool is often referred to as a VAT for volt-amp tester. A fan is usually contained in the tester to cool the carbon pile resistor during operation. Large cable clamps, like those used for jumper cables, provide the connection to the battery for the voltmeter and provide the connection to the carbon pile. The starting/charging system's tester inductive ammeter clamp is connected

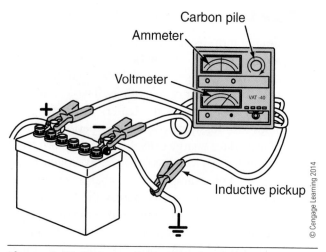

Figure 5-40 Carbon pile battery load tester.

around one of the load tester cables. The ammeter permits monitoring of the current, supplied by the battery, which is flowing through the carbon pile. For charging system testing, the inductive ammeter clamp is instead placed around one of the battery cables to measure charging current, as shown in **Figure 5-40**. Charging system testing will be discussed in **Chapter 7**.

Another important tool for testing stud-terminal-type batteries is a set of adaptors that thread onto the battery studs, like those shown in **Figure 5-41**. These adapters provide a large surface area for the carbon pile alligator clamps to attach. These adapters should also be used when connecting a battery charger to a stud-terminal-type battery.

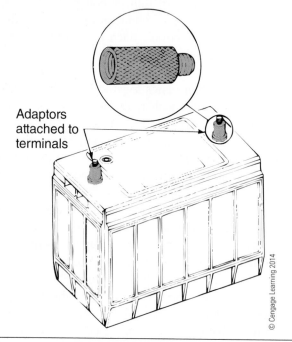

Figure 5-41 Battery charging and testing stud adapters.

Maintenance-Free Battery Testing

Prior to testing, surface charge must be removed if the batteries have been recently charged by a battery charger or by the truck's alternator. Surface charge is removed by loading the battery, using a carbon pile resistor or other means. Remove surface charge by loading a battery to one-half of the battery's CCA rating for 15 seconds. Permit the battery to stabilize for several minutes before testing.

After surface charge is removed, if applicable, maintenance-free battery testing can be divided into four steps:

1. Visually inspect battery for damage such as leaking electrolyte and obvious terminal damage. Replace any battery that has damage. Do not attempt to repair a leaking battery case.
2. Examine built-in hydrometer (if applicable) for an approximation of battery electrolyte level. A clear hydrometer indicates that the cell with the hydrometer is low on electrolyte. Such a battery must be replaced. If electrolyte level appears to be all right or cannot be determined, proceed to step 3.
3. Disconnect the cables between each battery so that each battery can be tested separately. Measure the no-load open-circuit voltage of each battery to determine state of charge. If the open-circuit voltage is less than 12.4V, recharge the battery using a battery charger and retest. If the open-circuit voltage is greater than 12.4V after any surface charge has been removed, proceed to step 4.
4. Estimate the electrolyte temperature by measuring the battery case temperature. Test each battery separately under load by applying a load equal to one-half the CCA rating for 15 seconds. Observe the battery terminal voltage after 15 seconds with the load still applied and compare it to the chart shown in **Figure 5-42**. If the battery terminal voltage is above the value shown in **Figure 5-42**, the battery appears to be acceptable and can be kept in service. If the battery terminal voltage is less than the value shown in **Figure 5-42**, recharge the battery and retest. If the battery fails the test a second time, replace the battery.

Battery Conductance Testing

Another type of small hand-held battery tester utilizes a measurement of a battery's **conductance** as a means of determining the health of the battery without

Battery Temperature (F)	Minimum Test Voltage
70°	9.6 volts
60°	9.5 volts
50°	9.4 volts
40°	9.3 volts
30°	9.1 volts
20°	8.9 volts
10°	8.7 volts
0°	8.5 volts

© Cengage Learning 2014

Figure 5-42 Temperature versus acceptable load test voltage chart.

the need for a large carbon pile. Conductance is the opposite of resistance and has units of siemens (S) or mhos. Note that mho is the reverse spelling of ohm. Conductance testers apply a low-level AC signal to the battery to calculate the conductance. These testers operate on the principle that the higher the conductance, the lower the battery's internal resistance. As discussed earlier in this chapter, the internal resistance increases as a battery ages.

A Midtronics® conductance tester is shown in **Figure 5-43**. Unlike the carbon pile tester, a conductance tester does not place a heavy electrical load on the battery to test. The only electrical connection to the batteries is by alligator clamps attached directly to the battery posts. A few pieces of information must first be entered into the tester such as the battery temperature, the CCA rating, and the number of batteries. It is important to enter this information correctly because the test results may not be correct if, for example, the battery temperature is entered incorrectly. Recall that battery temperature is another factor that affects battery internal resistance. After entering the information, the test of the batteries is complete within a few seconds and results are displayed like those shown in **Figure 5-44**.

Many fleets use battery conductance testing to monitor the state of battery health on each truck to predict that battery replacement is required before a failure occurs.

JUMP-STARTING

Jump-starting is sometimes necessary when batteries do not have sufficient state of charge to crank an engine fast enough or long enough to start. However, successfully jump-starting a truck with a diesel engine in cold weather is not always possible due to the tremendous energy requirements of the starter motor. This is especially true if the batteries are severely discharged.

The first rule of jump-starting is to wear safety goggles. Incorrect connections when jump-starting can result in battery explosion. Specific steps for jump-starting are as follows:

1. Make sure the electrolyte in the discharged batteries is not frozen. Do not attempt to jump-start a frozen battery because it can explode or damage the vehicle electrical system that is providing the boost. Because most battery cases are sealed, it is not possible to look inside the battery to observe if electrolyte is frozen. If batteries have a very low state of charge and the temperature is below 20°F, then the battery electrolyte may be frozen. Do not take any chances; remove suspect frozen batteries and place them inside the shop to warm up before jump-starting. The warm batteries have a much better chance of starting the vehicle anyway.

2. Make sure both trucks have batteries with the same voltage rating. Do not try to jump-start a battery with one that has a different voltage rating (6V, 12V, or 24V). Also, never use an arc welder to jump-start a truck. This is just asking for trouble and may result in extensive damage to the truck's electronic modules.

3. Make sure that the two trucks are not touching. Otherwise, current could flow through the point where the vehicles are making contact.

4. Turn off all electrical accessories on both trucks.

5. Connect jumper cables shown in **Figure 5-45** as follows (assuming 12V negative ground systems on both vehicles):

   a. Attach positive jumper cable to the positive terminal of the charged booster battery (in the vehicle that runs). Keep the other end of this positive cable in your other hand to prevent contacting the chassis ground. Some trucks may have a jump-start stud near the battery box. If so, use the jump-start stud instead of connecting directly to the batteries.

   b. Attach the other end of the positive jumper cable to the positive terminal of the discharged battery or to the jump-start stud, if applicable.

   c. Attach negative cable to the negative terminal of the charged battery (the vehicle that runs) while keeping the other end of this negative cable in your other hand.

The **Internal Batteries Status Indicator**, which appears in the screen's top left corner, lets you know the status and charge level of the analyzer's 6 1.5V batteries. The X shown in the figure shows that the EXP is powered by the battery you're testing to conserve the internal batteries.

Press the two **Soft Keys** linked to the bottom of the screen to perform the functions displayed above them. The functions change depending on the menu or test process. So it may be helpful to think of the words appearing above them as part of the keys. Some of the more common soft-key functions are SELECT, BACK, and END.

When you first connect the EXP to a battery it functions as a voltmeter. The voltage reading appears above the left soft key until you move to other menus or functions.

In some cases, you can use the **Alphanumeric Keypad** to enter numerical test parameters instead of scrolling to them with the **ARROW** keys.

You'll also use the Alphanumeric Keys to create and edit customer coupons. The keypad includes characters for punctuation. To add a space, press the **RIGHT** and **LEFT ARROW** keys simultaneously.

The **Title Bar** shows you the name of the current menu, test tool, utility, or function.

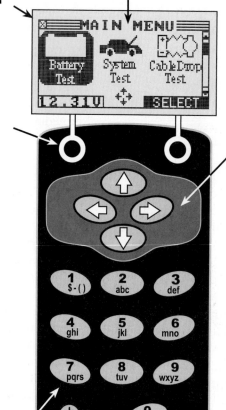

Press the **POWER** button to turn the EXP on and off. The EXP also turns on automatically when you connect its test leads to a battery.

Whichever way you turn on the EXP, it always highlights the icon and setting you last used for your convenience.

The **Selection Area** below the **Title Bar** contains items you select or into which you enter information. The area also displays instructions and warnings.

The **Directional Arrows** on the display show you which **Arrow Keys** to press to move to other icons or screens. The Up and Down Directional Arrows, for example, let you know to press the **UP** and **DOWN ARROW** keys to display the screens that are above and below the current screen.

The Left and Right Directional Arrows let you know to use the **LEFT** or **RIGHT ARROW** keys to highlight an icon for selection.

Another navigational aid is the **Scroll Bar** along the right side of the screen. The position of its scroll box tells you which menu screen you're viewing.

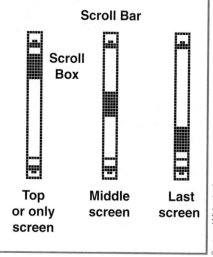

Figure 5-43 Midtronics EXP battery tester.

d. Attach the final negative cable to the frame or some other heavy ground location of the vehicle that is to be jump-started. It is important that this connection be made away from the battery to prevent the associated arc from causing a battery explosion.

If the engine starts, remove the cables in the reverse order as installed.

It may be necessary to run the engine in the vehicle providing the booster batteries for a considerable time to recharge the batteries. The current required by the starter motor on the disabled vehicle is usually much

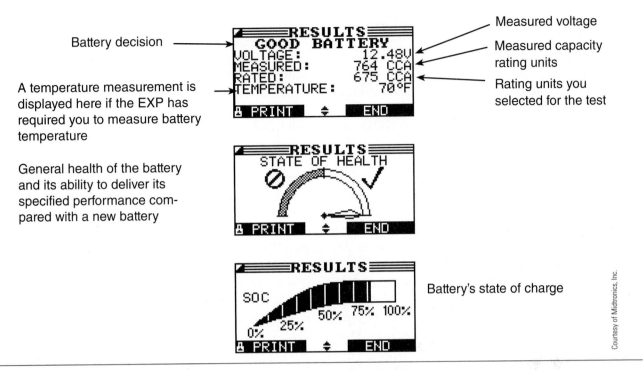

Battery decision

Measured voltage

Measured capacity rating units

Rating units you selected for the test

A temperature measurement is displayed here if the EXP has required you to measure battery temperature

General health of the battery and its ability to deliver its specified performance compared with a new battery

Battery's state of charge

Courtesy of Midtronics, Inc.

Figure 5-44 Midtronics test results screens.

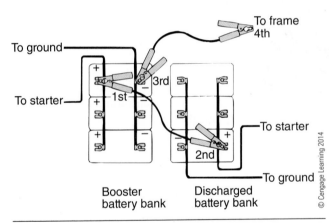

Figure 5-45 Cable attachment order for jump-starting a 12V truck with another 12V truck.

© Cengage Learning 2014

greater than jumper cable connections alone can provide, so the discharged batteries will have to be partially recharged before attempting to crank the engine.

Many road service trucks have a generator set designed for jump-starting trucks. This equipment works much better than jump-starting using booster batteries because of the generator set's higher current and voltage capabilities. Be sure to follow all OEM instructions to avoid damage to the equipment and to the vehicle being jump-started.

BATTERY SYSTEM SERVICE

Battery system service consists of visual inspection and cleaning of the battery box, the battery cables and terminals, the battery hold-downs, and the batteries. Battery maintenance includes adding water to conventional or low-maintenance batteries.

Installing a battery in an automobile is usually not too difficult. Most cars only have one battery and one set of battery cables. However, installing multiple truck batteries, wired in a variety of ways, requires attention because an incorrect connection can result in battery explosion or serious arcing.

Battery Connections

Installing a maze of battery cables to four powerful truck batteries can be intimidating at first. To reinstall the batteries correctly, it is always a good idea to mark cables before removal and make a sketch of how the batteries were originally installed, or use a digital camera to take a picture before disconnecting any cables.

Some trucks may have smaller connections directly at the battery-positive and ground terminals for the engine and automatic transmission electronic controllers. These connections must usually be disconnected first, before removing the main battery cables, and reconnected last, after other cables have been reconnected. Otherwise, the small-gauge wires for these electronic modules may be the only ground and power wires connected to the vehicle's electrical system. Attempting to operate a high-power device like the starter motor with only these small-gauge power or ground connections may cause damage to the applicable module or the power or ground wiring for the module.

Before disconnecting any battery cables, make sure everything on the truck is switched off. This prevents arcing when the connection at the batteries is opened. When disconnecting battery cables on a truck with a negative ground system (most modern trucks are negative ground), remove the battery negative cables first. The negative cables on a negative ground system are removed first to prevent the tool, such as a wrench, from providing a short circuit between the positive battery terminal and chassis ground. By disconnecting the negative battery terminals first, the chassis will no longer be grounded. Thus, any contact between the tool and the chassis will not result in a short circuit across the battery terminals. After the negative cables are removed, the positive cables can then be removed.

Before reconnecting batteries, make sure everything in the vehicle is turned off. Otherwise, connecting the battery cable will complete all electrical circuits that are switched on, causing a large current surge and associated arc.

When reconnecting the batteries, install the positive cables first. The negative cables are connected last. Once again, this reduces the chances that the metal wrench will provide a short circuit to chassis ground. Torque 3/8" stud terminals to 10ft-lb (13.5Nm) or as recommended by battery manufacturer. Tightening stud terminals too tightly can cause the stud to break loose inside the lead mounting post, requiring battery replacement.

Battery Hold-Downs

Batteries are heavy due to the lead content. Although lead is heavy, it is also soft and easily damaged. The lead components in a battery are subject to breakage if the battery is permitted to bounce around in the battery box. Battery hold-downs secure the battery to the battery box. The battery hold-downs also keep the battery-positive post from touching any grounded metal component such as a frame rail. Such contact could result in a fire or a battery explosion.

Most truck battery boxes use J bolts or some other type of long threaded stud to secure a nonconductive bar across one of the edges of the battery. The base plate type of hold-downs common on automobiles is not typically used in trucks because most trucks use more than one battery.

Water Replenishment

Conventional and non-sealed low-maintenance batteries have a cell filler cap atop each cell. The filler caps on conventional batteries are threaded into the case or snap into place as part of a three- or six-gang cap. Conventional batteries are vented through small vents in the filler caps. Non-sealed low-maintenance batteries may require removal of a label to access the filler caps (**Figure 5-46**). Sealed maintenance-free batteries have no filler caps and must be replaced when the electrolyte level is low.

The amount of water used by a battery varies with charging voltage and battery temperature. High charging voltages and high ambient temperatures increase the amount of water that the battery consumes.

To add water to batteries, first put on safety goggles. The sulfuric acid in the battery can blind you. A special self-leveling bottle or syringe bulb type of filler is used to add distilled water without spilling. Only distilled water should be added to a battery. Minerals in tap water can cause a significant decrease in battery life. Add water only up to the battery's defined full level or to a level no greater than 1/2" (13 mm) above the cell separators. Overfilling cells will leave no room for expansion and will cause the electrolyte to vent as the battery is charged.

Never add electrolyte or sulfuric acid to batteries that have already been in service. Battery operation mostly causes the water in the electrolyte to be lost, not the sulfuric acid. Adding additional sulfuric acid results in too strong of a concentration when the battery is recharged, and this will damage the battery.

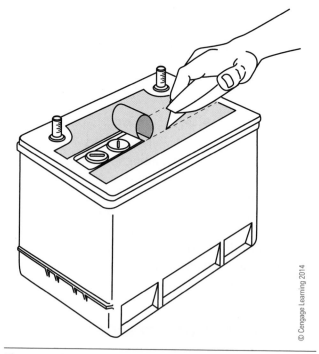

© Cengage Learning 2014

Figure 5-46 Removing label to access filler caps on low-maintenance batteries.

Some new conventional batteries may be shipped without any electrolyte. Prior to use, electrolyte must be added. The electrolyte is typically purchased premixed in a separate container with a plastic tube used to fill the battery. If electrolyte is not premixed, always pour the acid slowly into the water to mix; never pour the water into the acid. Pouring water into acid may result in splattering of the acid. The use of safety goggles when working with batteries is necessary. Even the slightest amount of sulfuric acid in your eyes is painful, to say the least.

Maintaining a Battery System

Battery service includes checking and adding water (if applicable), cleaning terminals of corrosion and oxidation, cleaning the battery top cover, and cleaning the battery box. As with any other battery-related task, safety goggles must be worn. Corrosion-resistant gloves and long-sleeved shirts are also recommended.

Battery terminals form a layer of corrosion due to sulfuric acid fumes. This corrosion causes resistance and dissolves cable connectors. Excessive corrosion can also attack the copper battery cable strands, requiring battery cable replacement. If excessive corrosion is found, suspect a battery that is being overcharged.

The first step in battery service is a visual inspection. Look for excessive corrosion, broken hold-downs, corroded or rusted battery boxes, or any other sign of trouble (**Figure 5-47**). This includes the battery

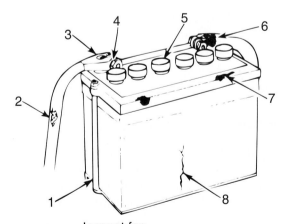

Inspect for:
1 Loose holddowns
2 Defective cables
3 Damaged terminal posts
4 Loose connections
5 Clogged vents
6 Corrosion
7 Dirt or moisture
8 Cracked case

© Cengage Learning 2014

Figure 5-47 Battery maintenance inspection items.

box mounting. Look for loose battery box mounting bolts, broken welds, or cracks in the heavy battery box.

If it will be necessary to remove the batteries to service them, start by disconnecting the battery ground cables (assuming a negative ground system). This includes any connections to electronic engine controllers, which are typically connected directly to the battery terminals. Next, the battery-positive cables can be removed. Make sure to mark cables before disassembly and note how the batteries are installed. Make a sketch so that you can get the cables hooked back up the right way. Pay close attention to the positive and negative markings on the batteries. Truck batteries typically do not have the polarity marked as well as automotive batteries do and are easy to install backward.

After removing battery cables, remove the battery hold-downs. It may be necessary to use a penetrating oil to loosen rusted hardware. To avoid the risk of battery explosion, never use a torch to heat this hardware. The batteries can then be lifted out of the battery box. Use care in lifting because truck batteries are very heavy.

Remove rust and corrosion from the battery box, cables, terminals, and mounting hardware with a wire brush. Make sure that you have eye protection and chemical-resistant gloves for this work. The corrosive particles cause severe pain if they get in your eyes or on your skin. A mixture of baking soda and water can be utilized to neutralize any sulfuric acid that is deposited on the cables and battery terminals. Do not let this solution enter the battery cells.

Rinse the batteries and battery box and permit to dry. Paint the battery box if necessary to prevent rust from forming or spreading. Reinstall the battery cables in the opposite order in which they were removed. Make sure to install the ground cables last with a negative ground system. This is important because any contact between a wrench and chassis ground while the positive cable is being tightened would result in the wrench providing a high-current path to ground.

RECOMBINANT LEAD ACID BATTERIES

Recombinant lead acid batteries refer to a type of completely sealed battery that never requires the addition of water (**Figure 5-48**). The term *recombinant* refers the oxygen produced by the positive plate recombining with hydrogen produced by the negative plate to form water. Recombinant batteries are also referred to as valve-regulated lead-acid (VRLA) batteries.

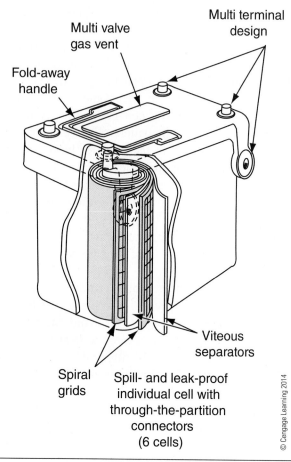

Multi terminal design

Multi valve gas vent

Fold-away handle

Viteous separators

Spiral grids

Spill- and leak-proof individual cell with through-the-partition connectors (6 cells)

© Cengage Learning 2014

Figure 5-48 Recombinant battery.

Although there is no vent to the atmosphere, a safety valve is designed to open should the internal pressure of the battery become too high. Recombinant batteries experience virtually no corrosion deposit formation at the terminals because there is no out-gassing during recharging. Other advantages of recombinant batteries over flooded cell batteries include longer service life, but they have a much higher initial cost. Recombinant batteries are much more resistant to vibration than flooded cell batteries and can provide much higher terminal voltages during engine cranking.

Gelled Cell Batteries

Gelled or gel cell batteries are recombinant batteries that use a thick gelatin-like electrolyte. This has an advantage over flooded-cell batteries in that because there is no vent and no liquid electrolyte, the battery can be used in applications where the vehicle may not always be level such as off-road work trucks. Gel cell batteries can also be discharged to a very low state of charge frequently without significant damage to the battery as compared to standard flooded-cell lead acid batteries. This is referred to as a deep-cycle. Some trucks with sleeper cabs may have isolated deep-cycle

batteries used to power accessories when the engine is not running.

Special automatic battery chargers are required to recharge gel cell batteries to prevent overpressurizing the battery resulting in opening of the regulation valve. Charging voltage must be maintained between 13.8V and 14.1V. Gel cell batteries are also typically heavier than comparable flooded cell batteries.

Absorbed Glass Mat Batteries

Absorbed glass mat (AGM) batteries are recombinant batteries in which the electrolyte is absorbed into thin fiberglass mats. The plates and mats may be flat like those in conventional batteries but at least one AGM manufacturer winds the plates and mats into a spiral giving the battery case a unique shape.

AGM batteries are typically lighter than comparable flooded cell batteries and also have very low self-discharge rates. AGM batteries are also more tolerant to overvoltage than gel cells. Like gel cell batteries, AGM batteries do not have liquid electrolyte so they are excellent for off-road usage. AGM batteries also perform well in extreme cold conditions and have a very low internal resistance. The result is exceptional engine cranking performance.

HYBRID ELECTRIC VEHICLE (HEV) BATTERIES

The Eaton hybrid electric vehicle system discussed in **Chapter 15** utilizes lithium-ion battery technology to produce 340V. Lithium-ion batteries are used as the power source for many consumer electronic devices such as laptop computers and cell phones as well as power tools.

Lithium-Ion Battery Safety

Although there is no maintenance required for HEV lithium-ion battery packs other than replacement of the entire sealed battery pack, you should still be aware of some of the potential hazards. Lithium is the lightest of all metals, but it is flammable and reacts in the presence of water to form hydrogen and lithium hydroxide. The high voltage of the lithium-ion battery pack can produce a fatal electric shock so all OEM precautions must be followed. You should not attempt to work on the high-voltage components of an HEV until you have had proper training.

Since there is no maintenance required, you should never attempt to open the sealed battery pack. In the event of a battery fire, do not attempt to put out the fire yourself without specialized training and equipment to include a self-contained breathing apparatus and a class D fire extinguisher. The gasses produced by a lithium-ion battery fire are hazardous. Notify your local fire department of the emergency and evacuate the area. As indicated, lithium reacts when exposed to water so make sure the fire department knows that a lithium-ion battery is present so that they can take the proper precautions.

In the event of accidental puncture of the battery pack resulting in spillage of the electrolyte, you should not attempt to clean up yourself but, again, notify your local fire department of the situation. If any electrolyte does contact your skin or eyes, rinse with water for 15 minutes and seek immediate medical assistance.

A lithium-ion battery pack removed from a truck is considered hazardous material. If it becomes necessary to ship a lithium-ion battery pack, appropriate regulations for shipment of hazardous materials must be followed. Dispose of lithium-ion battery packs in accordance with local regulations.

High-voltage HEV wiring is covered with an orange-colored insulation. Never cut or splice into high-voltage wiring. Never attempt to repair this high-voltage wiring.

A high-voltage service disconnect switch is present on the battery pack. Use this service switch when performing any vehicle service that requires disconnection from the high-voltage source.

Lithium-Ion Battery Chemistry

The two "plates" in a lithium-ion battery utilized in HEVs are lithium manganese oxide ($LiMn_2O_4$) and carbon (C). The electrolyte is a lithium salt in an organic solvent (not water). Each lithium-ion cell produces about 3.6V. Forty-eight lithium-ion cells are connected in series to produce approximately 170V per battery pack. Two battery packs are connected in series to produce 340V.

Lithium-Ion Battery Maintenance

There is no maintenance or service of lithium-ion battery packs. No external recharging of the battery packs should be attempted except as specified by the manufacturer. If replacement of the battery packs is required, consult the OEM's service information for the specific steps.

EXTRA FOR EXPERTS

This section provides additional details about batteries that you may find interesting after gaining some experience working with batteries.

Details on Battery Discharging

The term *amp-hours* is really an indication of the number of coulombs being supplied by the battery. A coulomb is the charge associated with a large number of electrons (6.25×10^{18} electrons). The definition of 1 ampere, as explained in earlier chapters, is 1 coulomb passing a stationary point in 1 second. Thus, current (amps) has units of coulombs per second. The elements of time (hours and seconds) "cancel out" when amps are multiplied by hours, leaving only coulombs. However, the amp-hours unit is more descriptive and easier to work with than coulombs and is commonly used to describe a battery that is charging or discharging.

For example, 1A of current flowing from the battery for 12 hours is 12 amp-hours. 12A of current flowing from the battery for 1 hour or even 1440A of current flowing from the battery for 30 seconds are both also 12 amp-hours. Twelve amp-hours is also the same as 43,200 coulombs, given that by definition a single coulomb is 1 amp-second (3600 seconds in 1 hour, 12A × 3600 seconds = 43,200 coulombs). Therefore, 12A of current flowing from a battery for 1 hour causes 43,200 coulombs to "leave" the battery.

To replace the 12 amp-hours or 43,200 coulombs, 12 amp-hours of recharging is necessary. This is not taking losses into account, such as heat loss, that occur each time energy is converted from one form to another, but it is a close approximation. Replacing the 12 amp-hours can be accomplished by recharging the battery so that 1A of charging current flows through the battery for 12 hours or recharging so that 12A of charging current flows through the battery for 1 hour. Regardless of the time and the current, 43,200 coulombs must be "shoved" back into the battery to recharge the battery.

Most modern battery chargers have a constant-voltage taper-rate design. As the name implies, the voltage across the battery terminals is held to some constant value while the charging current (rate) tapers off as the battery recharges.

The current accepted by a battery changes during the recharging cycle as does the voltage across the battery's terminals (even with a constant-voltage charger). The current accepted by the battery during charging changes for two reasons. When a battery is discharged to a 25 percent state of charge, the battery's internal resistance is much greater than when the battery is at a 90 percent state of charge. The battery's internal resistance would seem to cause the charging current to be less with a battery that has a 25 percent state of charge and a high internal resistance than a battery that has a 90 percent state of charge and a low internal resistance. However, the opposite is true.

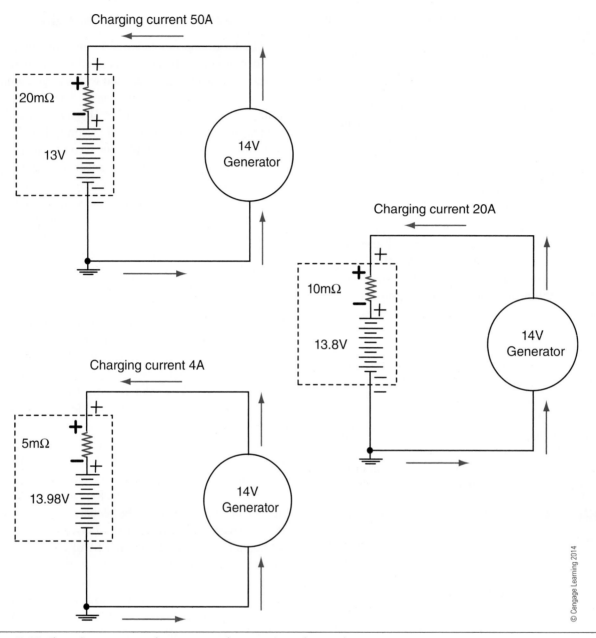

Figure 5-49 Charging current decreases as battery is recharged.

A battery that has a 25 percent state of charge will have a much higher charging current than a battery that has a 90 percent state of charge when the same voltage is applied across the terminals. The reason for the apparent contradiction of Ohm's law is that the voltage produced by the battery (the ideal voltage source portion that does not include internal resistance) acts as a series-opposing voltage to the battery charger. This series-opposing voltage of the battery often has a greater effect on the level of charging current accepted by the battery than the battery's internal resistance (except batteries that are completely discharged).

In the earlier discussion on series-opposing voltages, a three-cell flashlight has a bulb designed

for 4.5V. If one of the flashlight batteries is installed upside down, the series-opposing voltages cause the net current flowing through the flashlight bulb to decrease, resulting in a very dim bulb. Series-opposing voltages cancel each other out; two of the 1.5V flashlight batteries are canceling each other out, leaving just 1.5V to cause the current to flow through the flashlight bulb.

In a similar manner, a progressively larger portion of the voltage supplied by the battery charger is canceled out by the battery's own increasing "ideal" voltage as the battery recharges to a higher level (**Figure 5-49**). The "ideal" voltage sources shown inside the battery cases include surface charge so the battery terminal voltage is greater than 12.6V.

The increasing internal "ideal" voltage inside the battery is a series-opposing voltage to the voltage supplied by the generator. This causes the net charging current flowing through the battery to decrease as the battery is recharged. The current "accepted" by the battery decreases as the battery is recharged for this reason, even though the battery's internal resistance is decreasing as the battery is recharged.

A battery with a 25 percent state of charge only has a terminal voltage of about 12V, while a 100 percent state of charge battery has a terminal voltage of about 12.6V. When the voltage supplied by the battery charger remains at a constant level, the increasing internal "ideal" battery voltage as the battery recharges causes the charging current to decrease or taper off. This is where a constant-voltage taper-rate (current) battery charger gets its name.

Stages of Battery Recharging

Battery recharging can be broken into three distinct stages of operation:

1. Bulk stage
2. Absorption stage
3. Float stage

Recharging a battery with a 25 percent state of charge with a constant-voltage taper-rate battery charger will be described.

The first stage of battery recharge is the bulk stage. In the bulk stage, the battery accepts the greatest amount of charging current because the internal "ideal" battery voltage is low and cannot offer much opposition to the charging voltage supplied by the battery charger. All battery chargers have a limit as to how much current they can supply. To keep from exceeding this current limit, the battery charger reduces its output voltage to a value less than the selected charging voltage (even with a constant-voltage battery charger). In the bulk stage, the charging current is a constant value; that is, the maximum the battery charger can supply for the selected charger settings. The charging voltage is increasing during the bulk stage to offset the increasing internal "ideal" battery voltage as the battery recharges, as shown in **Figure 5-50**.

The second stage of battery recharging is the absorption stage. In the absorption stage, the battery charging voltage reaches a constant value of perhaps 15V. Charging voltage is typically one of the battery charger's settings that can be selected by using a knob. The charging current begins to decrease in the absorption stage, even though the battery's internal resistance is lowering. The decrease in charging current

is due to the series-opposing effect of the increasing internal "ideal" battery voltage as the battery recharges (**Figure 5-51**). The battery must be monitored

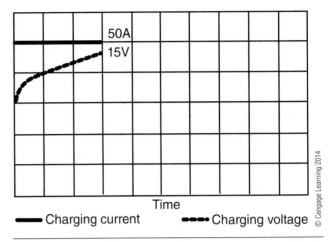

Figure 5-50 Bulk stage; charger at current limit, voltage increasing.

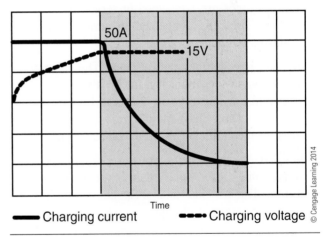

Figure 5-51 Absorption stage; constant voltage, charging current decreases.

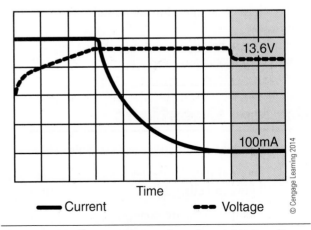

Figure 5-52 Float stage; reduced constant voltage, reduced constant charging current.

in this stage so it does not needlessly vent electrolyte. When the battery is nearing a full charge, the charging voltage must be reduced to complete the charge cycle.

The third and final stage of battery recharge is the float stage. In the float stage, the battery charger voltage is reduced to about 13.6V. This causes the charging current to further reduce to perhaps 100mA (1/10th of an amp), as shown in **Figure 5-52**. This stage provides a deep, internal charge of the battery plates, not just near the surface of the plates. The low current prevents the venting of electrolyte because of the reduction in charging voltage. A battery can be left in this float state for an extended period of time without damaging the battery.

Summary

- Batteries store and transform energy.

- Batteries are some of the most hazardous components found in trucks.

- Lead acid batteries contain two different types of lead along with an electrolyte made of sulfuric acid and water.

- The two types of lead are formed into plates. A group of plates is connected in parallel to form a cell. The cells are connected in series. A 12V battery has six cells connected in series. Each cell produces a voltage of about 2.1V.

- Batteries have an internal resistance. This internal resistance varies depending on many factors, including temperature and battery state of charge. Battery internal resistance causes the battery terminal voltage to decrease as the current supplied by the battery increases.

- No-load or open-circuit voltage refers to the terminal voltage of a battery that is not supplying any current. This no-load voltage can be used to indicate the battery state of charge.

- There are several battery ratings. CCA indicates how much current a fully charged battery can supply continuously for 30 seconds and still maintain a terminal voltage of 7.2V. Reserve capacity is another rating and indicates how long (in minutes) a battery can supply a 25A load and still maintain a terminal voltage of 10.5V.

- Batteries are recharged by causing a current to flow through the battery. A combination of time and current flow is necessary to recharge a battery.

- Specific gravity refers to the weight of electrolyte compared to an equal volume of water. The measurement of specific gravity using a hydrometer can be used to determine battery state of charge in conventional batteries.

- Surface charge refers to gas bubbles on the plates that can make a battery appear to be at a higher state of charge than it actually is. Surface charge must be removed prior to testing a battery.

- A carbon pile tester is used to load test a battery.

Suggested Internet Searches

Try these web sites for additional information on batteries:

http://www.batterycouncil.org
http://www.acdelco.com

http://www.exide.com
http://www.jci.com
http://www.batteryuniversity.com
http://www.optimabatteries.com

Review Questions

1. A 12V lead acid battery has six cells. These six cells are connected together in what type of arrangement?

   A. Internal parallel connection between cells.

   B. Positive cells are connected in series alternately with negative cells.

   C. Three positive cells are connected to the positive terminal; three negative cells are connected to the negative terminal.

   D. Internal series-aiding connection between cells.

2. A lead acid battery is completely discharged. What can be said about the plate material and the electrolyte?

   A. Positive and negative plates have transformed into nearly the same lead material and the electrolyte has become mostly water.

   B. Positive and negative plates have transformed into sulfur and the electrolyte has become pure sulfuric acid.

   C. The electrolyte has become pure sulfuric acid and the plates have transformed into different compounds of lead.

   D. The plates have transformed into different compounds of lead and the electrolyte has become a mixture of water and sulfuric acid.

3. A sealed maintenance-free battery has lost electrolyte as shown by the built-in hydrometer. What should be done with the battery?

   A. Drill holes in the top of the battery and refill with water.

   B. Recharge battery until the electrolyte level returns to normal.

   C. Take the battery out of service and dispose of properly.

   D. Throw the battery in a dumpster.

4. Batteries are being replaced in a truck with a 12V negative ground electrical system that has two 12V batteries. Technician A says that batteries should be connected in parallel by connecting the positive terminal of one battery to the negative terminal of the other battery. Technician B says that the last battery cable connection should be the positive cable connection at the starter motor to reduce arcing at the batteries. Who is correct?

   A. A only

   B. B only

   C. Both A and B

   D. Neither A nor B

5. Which battery would probably have the highest internal resistance, assuming all batteries are at the same temperature?

   A. A maintenance-free battery where a DMM indicates a no-load voltage of 12.6V across the battery terminals.

   B. A conventional battery with all cells having a specific gravity of 1.250 or higher.

   C. A conventional battery with all cells having a specific gravity of about 1.230.

   D. A low-maintenance battery where a DMM indicates a no-load voltage of 12.2V across the battery terminals.

6. Which is *not* a true statement?

   A. CCA rating indicates a battery's cold cranking performance.

   B. The reserve capacity rating indicates how long a battery will provide a specific terminal voltage while supplying a specific current should the charging system fail.

   C. Placing batteries in parallel causes both the total capacity and total CCA to increase.

   D. The CCA rating and the CA rating are typically the same value for most truck batteries.

7. The CCA rating of each of four identical batteries connected in parallel is 650 CCA. What is the total CCA rating for the four batteries?

   A. 650 CCA

   B. 162.5 CCA

   C. 1300 CCA

   D. 2600 CCA

8. A truck with nearly dead batteries has been jump-started. Technician A says that driving the truck for a couple of hours without shutting off the engine will be sufficient to fully recharge the batteries. Technician B says that the batteries should be recharged using a battery charger and tested. Who is correct?

A. A only

C. Both A and B

B. B only

D. Neither A nor B

9. Surface charge is being discussed. Technician A says that surface charge should be removed before load testing batteries. Technician B says that a battery with a no-load terminal voltage of 13V likely has surface charge present. Who is correct?

A. A only

C. Both A and B

B. B only

D. Neither A nor B

10. How should a carbon-pile-type load tester (VAT) be used?

A. Connect the load tester's alligator clamps directly to the battery's 3/8" stud-type terminals.

B. Load to three times the battery's CCA rating for 15 seconds.

C. Disconnect the interconnection cables between batteries and load test each battery separately at one-half the battery's CCA rating for 15 seconds to test.

D. Recharge the battery to make sure sufficient surface charge is present before testing because surface charge is necessary for an accurate load test of the battery.

11. Which of the following is potentially unsafe when working with batteries?

A. Using a torch to heat battery hold-downs that are rusted.

B. Recharging a battery with frozen electrolyte.

C. Adding pure sulfuric acid to a battery to restore specific gravity.

D. All of the above are unsafe practices.

12. The BCI group number of a battery may indicate which of the following battery attributes?

A. Physical dimensions of the battery

B. Cold cranking amp performance range

C. Type of terminals

D. All of the above

13. Which tool would be used to load test a battery?

A. Hydrometer

C. Distilled water

B. Jumper cables

D. Carbon pile

14. A particular truck with a 24V electrical system has four 12V batteries. The four batteries are connected in what type of arrangement?

A. Series

C. Series-parallel

B. Parallel

D. Series-opposing

15. Which battery cable should be disconnected first on a truck with a 12V negative ground system that has multiple batteries?

A. All of the negative cables.

B. All of the positive cables.

C. One of the positive cables and then one of the negative cables, alternating with each battery.

D. It does not matter which cable is disconnected first.

6 Basic Electronics

Learning Objectives

After studying this chapter, you should be able to:

- List the various types of resistors.
- Determine whether a diode is forward biased or reverse biased in a circuit.
- Test a diode using a DMM.
- Explain the function of a zener diode.
- Describe the operation of a bipolar transistor.
- Explain the difference between a transistor being used as an amplifier and a transistor being used as a relay or switch.
- Describe the operation of a field effect transistor (FET).
- Discuss the concept of a high side driver and a low side driver.

Key Terms

anode

base

cathode

collector

covalent bond

diode

doping

drain

electronics

emitter

field effect transistor (FET)

forward biased

gate

high side driver

JFET

junction

light-emitting diode (LED)

low side driver

MOSFET

N-type

negative temperature coefficient (NTC)

P-type

potentiometer

reverse biased

rheostat

saturation point

semiconductor

silicon

sinking

source

sourcing

transistors

zener diode

zener voltage

INTRODUCTION

This chapter is an introduction to **electronics**. In previous chapters, electrons were controlled using switches, relays, and other mechanical devices that have moving components that are used to direct the flow of electrons (current). Electronics refers to the control of electrons using electricity, without any visible moving components.

In the 1960s, electronics started appearing on automobiles. The electronic voltage regulator was one of the first electronic components to be used in automobiles and was soon followed by electronic ignition. Both of these advancements provided substantial performance improvement and increased reliability over the electromechanical systems that they replaced. Since then, electronics has become a major part of the modern automobile and truck. Every major modern truck system now has some tie to electronics, from antilock brake systems (ABS) to electronic diesel engine controls.

RESISTORS

Resistors are one of the most basic electrical components and are an essential component of most electronic circuits. Resistors are used for three main purposes in an electrical or electronic system:

- Resistors limit the current flow through a circuit.
- Resistors divide voltage.
- Resistors give off heat.

There are two main types of resistors—fixed and variable.

Fixed-Value Resistors

Fixed-value resistors are resistors that have a fixed or set value of resistance. These resistors are typically made of carbon or a metal oxide material. An outer jacket of insulating material surrounds the fixed resistors, as shown in **Figure 6-1**. Notice that the wattage rating of the resistor increases with its physical size. Resistors transform electrical energy into heat. Power is measured in units of watts. Resistors have a wattage rating that indicates how much heat the resistor can safely dissipate. A physically larger resistor has more surface area than a smaller resistor, permitting more heat to be dissipated; thus, the higher wattage rating.

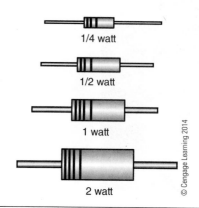

Figure 6-1 Fixed-value resistors: physical size of resistor increases as wattage rating of resistor increases.

Resistor Color Code

To identify a resistor's value, a color code is often used. This color code consists of a series of colored bands around the girth of the resistor, as shown in **Figure 6-2**. The color code permits visual identification of a resistor's approximate value.

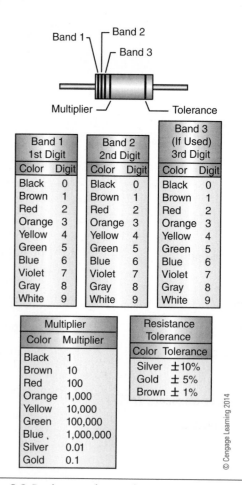

Band 1 1st Digit		Band 2 2nd Digit		Band 3 (If Used) 3rd Digit	
Color	Digit	Color	Digit	Color	Digit
Black	0	Black	0	Black	0
Brown	1	Brown	1	Brown	1
Red	2	Red	2	Red	2
Orange	3	Orange	3	Orange	3
Yellow	4	Yellow	4	Yellow	4
Green	5	Green	5	Green	5
Blue	6	Blue	6	Blue	6
Violet	7	Violet	7	Violet	7
Gray	8	Gray	8	Gray	8
White	9	White	9	White	9

Multiplier		Resistance Tolerance	
Color	Multiplier	Color	Tolerance
Black	1	Silver	±10%
Brown	10	Gold	± 5%
Red	100	Brown	± 1%
Orange	1,000		
Yellow	10,000		
Green	100,000		
Blue	1,000,000		
Silver	0.01		
Gold	0.1		

Figure 6-2 Resistor color code.

The resistor color code also indicates the tolerance value of the resistor. Tolerance describes the range of acceptable values for the resistor. For example, a 1kΩ resistor with a ±5 percent tolerance can be anywhere between 950Ω and 1050Ω.

Other Fixed Resistors

Fixed resistors may also be made of wire, as in the case of a wire-wound resistor. Wire has resistance, as discussed in previous chapters. Wire-wound resistors are typically used in low-resistance, high-wattage applications, such as HVAC fan motor resistor blocks (**Figure 6-3**). A wire-wound resistor is simply a wire that is wound into the shape of a coil. These resistors can get very hot during operation and are designed to have air flowing over them to dissipate heat.

The wire-wound resistors shown in **Figure 6-3** are arranged in a stepped resistor network. Stepped resistors are fixed resistors that are placed in series with one another. This stepped resistor configuration is often used to control the speed of a motor, such as an HVAC fan motor. Stepped resistor networks permit only a limited number of discrete (distinct) values of resistance. Electric motors used for fan motors rotate faster as the voltage applied to the motor is increased. The fan speed control switch determines the amount of voltage applied to the motor by stepping through the series resistors. This series resistance controls the speed of the motor, as shown in **Figure 6-3**. A portion of the voltage is dropped

across the resistor, and the remainder of the voltage is supplied to the motor.

In the low-speed position shown in **Figure 6-3**, 3Ω of resistance is placed in series with the motor, causing the motor to operate at a low speed. In the medium-speed position, the series resistance is reduced to 2Ω, which increases the voltage supplied to the motor, resulting in an increased motor speed. The medium-high position further reduces the series resistance to 1Ω, resulting in an additional increase in motor speed. In the high-speed position, no series resistance is provided by the resistor network, permitting full system voltage to be supplied to the motor, which results in the motor operating at its highest speed.

Variable Resistors

The resistance value of a variable resistor can be modified after it is installed in a circuit. A variable resistor is similar to a stepped resistor network except that any value of resistance within a specific range can be obtained, instead of the discrete or distinct levels of resistance provided by a stepped resistor network. For example, the stepped resistor network shown in **Figure 6-3** provides four distinct values of resistance corresponding to the four stopping points.

The rotating switch contact of a variable resistor is known as a wiper. The wiper makes contact with a resistive track, which is often a wire-wound resistor. The track is typically circular shaped to correspond to the arc formed by the moving wiper. The location of the wiper on the resistive track dictates the resistance value of the variable resistor (**Figure 6-4**). Moving the wiper causes the resistance value to change. The volume knob of a radio may be connected to a variable resistor wiper. Rotating the knob causes the variable resistor's resistance value to change. This resistance change results in an increase or decrease in the volume.

A variable resistor typically has three terminals, as shown in **Figure 6-4**. One common schematic symbol for a type of variable resistor is also shown in **Figure 6-4**. The two terminals labeled A and C are connected to either end of the resistive track. This resistive track is a fixed resistor and has a set value. For example, the resistance between terminal A and terminal C may be fixed at 10kΩ for a particular variable resistor. This resistance will not change as the wiper is rotated. The wiper is connected to terminal B. With the wiper travel beginning at a position where it is contacting terminal A, moving the wiper

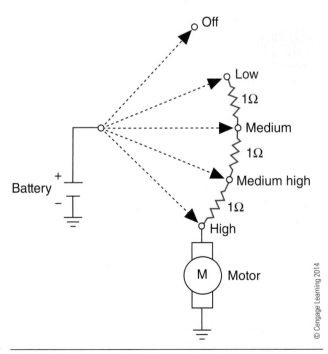

Figure 6-3 Stepped resistor network.

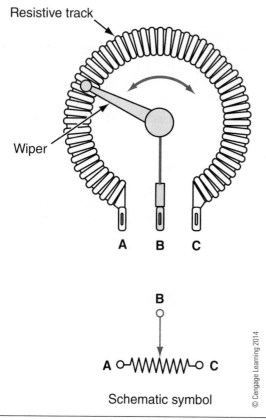

Figure 6-4 Infinitely variable resistor.

through the range of travel causes the resistance between terminal B and terminal A to increase continually from 0Ω to 10kΩ (**Figure 6-5**). This same motion causes the resistance between terminal B and terminal C to decrease from 10kΩ to 0Ω as the wiper makes contact with terminal C at the other end of the wiper's travel.

When the wiper terminal (B) and only one of the other terminals (A or C) of a variable resistor are being used, the variable resistance device is known as a **rheostat**. A rheostat can be used to control the current in a lighting circuit such as instrument panel illumination, as shown in **Figure 6-6**. Adding series resistance to the instrument panel lamp circuit causes the intensity of the lamps to decrease. The series resistance of the rheostat drops a portion of the available voltage, leaving a reduced voltage available for the instrument panel lamps. The reduced voltage available at the instrument panel lamps causes the current flow through the lamps to decrease, thus decreasing illumination level. This is similar to the stepped resistor network, except there is an infinite number of resistance level "steps."

When all three terminals of a variable resistor are used, the device is known as a **potentiometer** or *pot*. A potentiometer is used to divide voltage. A potentiometer is wired so that a specific value of voltage,

such as +5V and ground, is connected across terminal A and terminal C of the potentiometer as shown in **Figure 6-7**. The voltage measured between terminal B and ground will change continuously from 5V to 0V as the potentiometer is rotated through its full travel. The voltage varies because the potentiometer is acting as a voltage divider. A voltage is dropped on the portion of the resistive track between the 5V terminal and the wiper terminal. The remainder of the voltage is dropped on the portion of the resistive track between the wiper terminal and the ground terminal. For example, if the wiper of a 10kΩ potentiometer is set so that 2kΩ of resistance exists between the wiper terminal and ground (terminals B and C), the voltage measured between the wiper and ground would be 1V when 5V is applied across terminals A and C of the potentiometer (**Figure 6-7**). The voltage measured between terminals B and C of the potentiometer is variable and changes as the potentiometer is rotated. Potentiometers are often used as accelerator position sensors with electronically controlled diesel engines. The potentiometer converts the accelerator position into a voltage value that is used by the engine electronic control module (ECM) to determine the driver's request for power from the engine. This is covered in more detail in later chapters.

SEMICONDUCTORS

A **semiconductor** is a material that is neither a good nor a poor conductor of electric current. The most common semiconductor materials used in electronic components are made from **silicon**. Silicon (Si) is an element found in sand and stone. Note that *silicon* is not the same as *silicone*, which is a polymer used in sealants and lubricants. Semiconductors typically have a **negative temperature coefficient (NTC)** meaning that as the temperature of the material increases, the resistance of the material decreases. This is the opposite of most conductors like copper which have a positive temperature coefficient (PTC). Pure silicon has four valence electrons like carbon. However, pure silicon at low temperatures is such a poor conductor of electric current that is essentially an electrical insulator. The reason for this is that pure silicon forms a crystal structure as shown in **Figure 6-8**. In this crystal structure, the four valence electrons of each silicon atom mesh with the valence electrons of adjacent silicon atoms to fill the valence band of each atom with eight electrons, as shown by the silicon atom at the center of **Figure 6-8**. This sharing of electrons between atoms is called a **covalent bond**. Recall that an insulator is a material with five or more valence electrons.

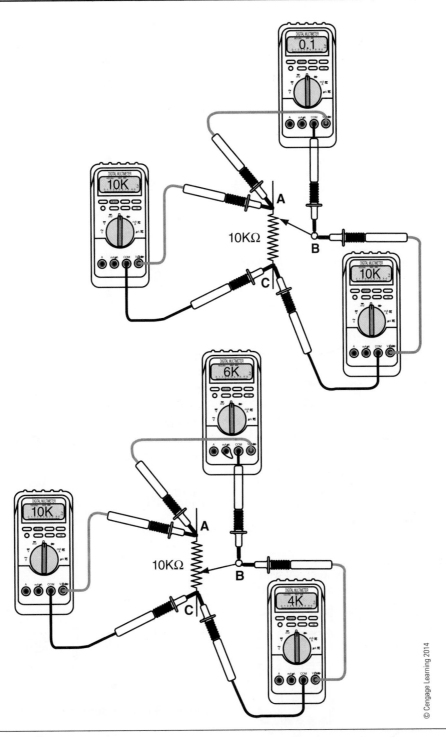

Figure 6-5 Variable resistor operation.

The covalent bonding of silicon atoms results in virtually no free electrons, so at low temperatures a pure silicon crystal acts similar to an insulator with eight valence electrons.

However, adding impurities to the silicon results in changes to the properties of the material at the molecular level, causing the silicon material to become a useful semiconductor. The addition of impurities to silicon is called **doping**.

Two different types of impurities are added to silicon to form two different mixtures of silicon. One mixture has an excessive amount of electrons and is called **N-type** material for negative material. Arsenic (As), which has five valence electrons, is an impurity that can be mixed with pure silicon to form N-type material. **Figure 6-9** illustrates an arsenic atom in the center of the silicon crystal. With limited exception, the rule of the universe is that eight electrons are the

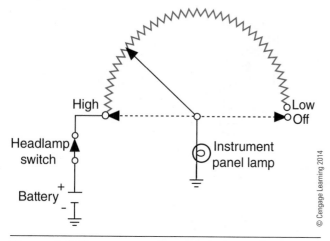

Figure 6-6 Rheostat used to control panel illumination level.

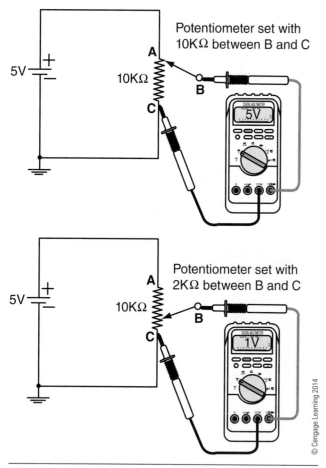

Figure 6-7 Potentiometer.

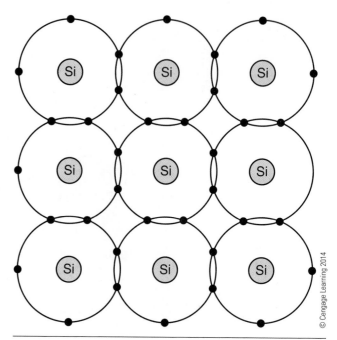

Figure 6-8 Pure silicon (Si) crystal covalent bonds act as an electrical insulator at low temperatures.

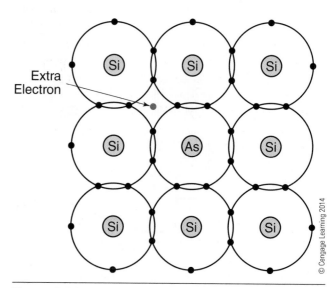

Figure 6-9 N-Type semiconductor doped with arsenic (As).

maximum number of electrons that can be present in the valence band. The extra electron shown cannot fit into any of the full valence bands. Another mixture of silicon has an excessive amount of holes and is called **P-type** material for positive material. Boron (B), which has three valence electrons, is an impurity mixed with pure silicon to form P-type material.

Figure 6-10 shows a boron atom in the center of the silicon crystal. The hole shown is a shared valence band with only seven electrons, so there is room for one more electron.

When the P-type and N-type materials are fused together so that a **junction** or boundary between the P-type material and the N-type material exists, some amazing things happen at the electron level when a voltage is applied to the junction. The conductivity of the semiconductor material changes as a voltage is

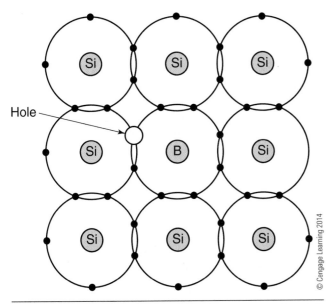

Figure 6-10 P-Type semiconductor doped with boron (B).

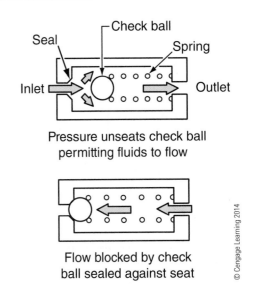

Figure 6-11 One-way hydraulic check valve.

applied to either permit an electric current to flow or to be interrupted. Thus, electricity controls electricity without any visible moving components.

The physics that occur at the electron level of a PN junction and within P-type and N-type material is beyond the scope of this book. A detailed understanding of semiconductor theory requires a study of quantum mechanics, which almost no one understands anyway. However, it is not necessary to have more than a basic understanding of electronics to become a good troubleshooter of truck electronic system components.

DIODES

Diodes are two-terminal semiconductor devices that act as the electrical equivalent of a hydraulic system one-way check valve. A one-way check valve permits hydraulic flow in one direction and inhibits flow in the opposite direction (**Figure 6-11**). Like a check valve, diodes permit electric current to flow in

one direction and block current flow in the opposite direction.

Unlike hydraulic check valves that contain a ball, spring, and seat, diodes contain no visible moving components. Many types of diodes are formed from doped silicon and contain a PN junction.

The schematic symbol for a common diode is shown in **Figure 6-12A**. The arrow indicates the direction that the diode will permit conventional (positive to negative) current flow. **Figure 6-12B** shows a plot of the current flow through a diode versus the voltage dropped across the diode.

If a diode is inserted in a circuit so that the diode permits current flow, the diode is said to be **forward biased** (**Figure 6-13**). If the diode is inserted in the circuit so that the diode blocks current flow, the diode is said to be **reverse biased** (**Figure 6-14**). When a diode is forward biased, the positive side of the voltage source is connected to the P-type material terminal of the diode—known as the **anode**—and the negative side of the voltage source is connected to the N-type material terminal of the diode, known as the **cathode**.

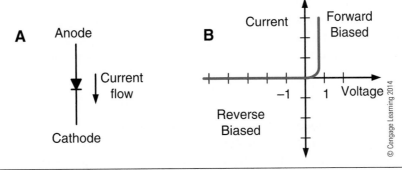

Figure 6-12 Diode voltage versus current curve and schematic symbol.

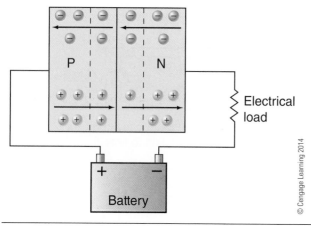

Figure 6-13 Forward-biased diode permits current flow through the load.

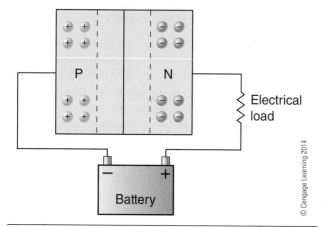

Figure 6-14 Battery polarity reversed, reverse-biased diode blocks current flow through the load.

Most diodes have a stripe printed around the girth of the cathode end to signify which end of the diode is the cathode. The stripe also corresponds to the line end of the diode symbol (**Figure 6-15**).

Tech Tip: The line placed at the bottom of the triangle of the diode symbol looks like a minus sign. This line end of the diode corresponds to the cathode or negative terminal of the diode. This may be helpful in remembering which end of a diode symbol is positive and which end of a diode symbol is negative for a forward-biased diode.

To assist in understanding how a diode functions, a hydraulic check valve will be evaluated in greater detail. A hydraulic check valve consists of a ball, a spring, and a seat (**Figure 6-16**). At very low hydraulic

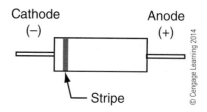

Figure 6-15 Stripe indicates cathode end of diode.

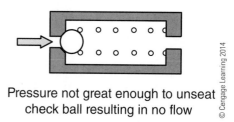

Pressure not great enough to unseat check ball resulting in no flow

Figure 6-16 Check valve with zero flow in normal flow direction due to supply pressure not great enough to overcome spring force to unseat check ball.

pressures, the spring tension keeping the ball against the seat may be greater than the force of the hydraulic fluid under pressure that is trying to open the check valve. Until enough hydraulic pressure builds up to cause the spring to compress enough to move the ball off the seat, no fluid will flow through the check valve in the direction of normal fluid flow permitted by the check valve. The hydraulic pressure necessary to unseat the check ball is typically very low.

When the hydraulic pressure is sufficient, the check ball will lift off its seat and begin to permit hydraulic fluid to flow. Even with the check ball fully lifted off the seat, the check valve will cause some restriction in the hydraulic circuit in the direction of normal flow, resulting in a small pressure drop across the check valve.

A diode behaves in a manner similar to a check valve. Most silicon diodes will not permit current flow in the forward-biased direction until approximately 0.7V is dropped across the diode. This 0.7V can be thought of as the diode "turn-on" voltage. This 0.7V continues to be dropped across the diode anytime the diode is forward biased. Note that some diodes may have turn-on voltages as low as 0.4V, but the concept is still the same. The circuit in **Figure 6-17** has a resistor, a 12V battery, and a forward-biased diode. The diode has 0.7V dropped across it, leaving the remaining 11.3V to be dropped across the 1kΩ resistor. If the resistance value were changed to 500Ω, the resistor would still have 11.3V dropped across it and the diode would still have about 0.7V dropped across it. The diode is not a fixed-value resistor, so it does not

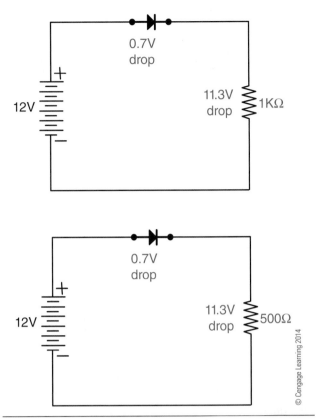

Figure 6-17 Forward-biased diode has 0.7V drop leaving 11.3V drop across load resistor.

seem to "obey" Ohm's law. Increasing current flow in the circuit by increasing the voltage supply or reducing the resistor value still causes about the same amount of voltage to be dropped across the diode. The plot shown in **Figure 6-12B** illustrates this behavior. Virtually no current flows through the diode until the voltage dropped across the diode exceeds the approximate 0.7V "turn-on" voltage.

Tech Tip: One way to think of a forward-biased diode is that it is like a self-controlled variable resistor that will automatically become whatever resistance value is necessary to maintain a voltage drop across the diode of approximately 0.7V.

Diodes have maximum current ratings that must be observed. Current flow through a diode causes heat. Too much heat is detrimental to any electronic component. A series resistor must always be used with a diode to limit current and to provide some other device on which to drop the remainder of the available voltage. The forward-biased diode should only have a

voltage drop of about 0.7V and the rest of the voltage must be dropped across some other component, per Kirchhoff's voltage law. Connecting a forward-biased diode across a 12V battery without a series resistor present in the circuit forces the diode to obey the laws of parallel circuits and have 12V dropped across it. This will typically destroy the diode instantly due to the large amount of current that will flow through the diode. This may occur so fast that you have no indication of a problem.

Tech Tip: Electronic components are *very* unforgiving of mistakes. Using improper troubleshooting techniques can result in permanent damage to costly electronic components. Follow the OEM's troubleshooting recommendations to avoid damaging electronic components.

In the reverse-biased direction, the diode acts like a hydraulic check valve that is blocking the flow of hydraulic fluid. The voltage measured across a reverse-biased diode would be the same as the voltage measured across an open switch placed in the same location of the circuit as the reverse-biased diode (**Figure 6-18**). All the voltage is dropped across an open switch because the resistance of the open switch is extremely high (near infinity).

Like an open switch, diodes have a limit to the amount of voltage that they can block in the reverse-biased direction. This rating is called the inverse voltage rating. This rating specifies how high a reverse voltage that the reverse-biased diode can block before it breaks down and permits current to flow. If this inverse voltage rating is exceeded, the common diode will be destroyed.

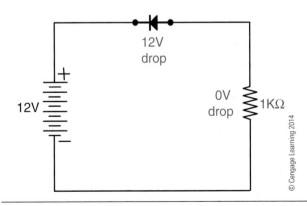

Figure 6-18 Reverse-biased diode has all of the available voltage drop; zero current flow through resistor resulting in 0V dropped across resistor.

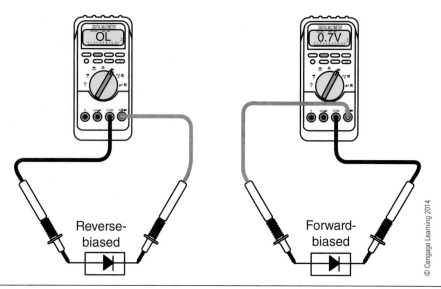

Figure 6-19 Diode testing using DMM.

Testing Diodes

Digital multimeters (DMMs) or analog ohmmeters are used to test diodes. Most DMMs have a special setting for diode testing that is typically identified by the diode symbol. To test a diode, the positive lead of the DMM is connected to the anode (positive) terminal of the diode and the negative lead of the DMM is connected to the cathode (negative) end of the diode (**Figure 6-19**). The DMM will cause current to flow through the forward-biased diode. The DMM display will typically indicate the forward-biased voltage drop of the diode (between 0.4V and 0.7V for most diodes). A diode that is shorted will indicate much less than 0.4V, while a diode that is an open circuit will indicate OL on the DMM display or something similar to indicate an open circuit. The DMM leads are then reversed to test the diode in the reverse-biased direction. The DMM should indicate OL or something similar to indicate an open circuit in the reverse-biased direction because the diode is blocking the flow of current. A DMM that displays a low voltage in the reverse-biased direction indicates that the diode is shorted, meaning the diode is acting as a conductor in both directions.

The special diode test setting of the DMM must be used and not the standard ohms settings. Most DMMs do not supply enough voltage in the standard ohms settings to test the diode in the forward-biased direction, thus the special diode test setting.

Diode Applications

Diodes are used for many different purposes in truck electrical systems. Common diodes are often called rectifier diodes. This is because one of the important tasks of a diode is to rectify alternating current into direct current. The alternator produces an alternating current (AC). The truck electrical system is direct current (DC). Rectifier diodes are used to convert the alternating current to direct current. This will be discussed in detail in **Chapter 7**.

Diodes are also used as suppression devices for coils such as those found in air conditioning (A/C) compressor clutches, solenoids, and relays. The voltage produced by a collapsing magnetic field that surrounds an inductor at the time when current flow through an inductor is rapidly interrupted results in the production of a reverse polarity voltage, as described by Lenz's law. If a diode is placed in parallel with the coil so that the diode is reverse biased by the normal solenoid supply voltage, the diode will be forward biased by the negative voltage spike associated with a collapsing magnetic field surrounding the coil. The result is that the negative voltage spike is held to a level of about −0.7V, as illustrated in **Figure 6-20**. Many devices, such as air solenoids and A/C clutch coils, have this built-in diode suppression. It is important that the polarity of the voltage supplied to such devices with diode voltage suppression be observed, such as when bench testing a device. If a 12V battery were incorrectly connected with reverse polarity to a device that utilizes diode suppression, such as an A/C clutch, the suppression diode would be forward biased with virtually no series resistance in the circuit to limit current flow through the diode. The diode would act similar to a short circuit until it burns open from the excessive amount of current. This may occur so rapidly that you do not even realize that the suppression diode has been destroyed. The component, such as an A/C clutch coil, would still function without a

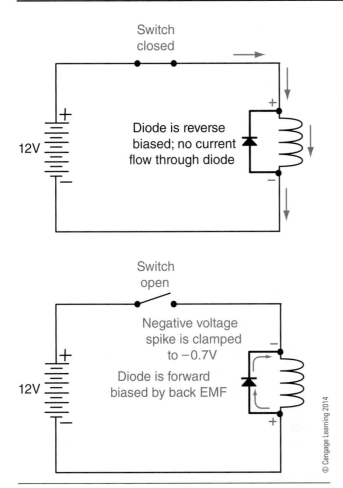

Figure 6-20 Diode across coil shown as reverse biased by the battery voltage and forward biased by back EMF (CEMF).

functional suppression diode, but may cause intermittent electrical problems or damaged electronic components.

Light-Emitting Diodes

A **light-emitting diode (LED)** is a type of diode that emits visible light energy when forward biased. LEDs are used for numerous lighting applications on modern trucks. The schematic symbol for an LED shown in **Figure 6-21** is similar to a common diode except that arrows pointing away from the diode indicate the light that is being emitted.

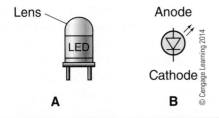

Figure 6-21 Schematic symbol for LED.

An LED emits a specific color of light such as shades of red, green, yellow, blue, and white. The color of the light emitted by the LED is due to the type of material that is mixed with the silicon resulting in a range of wavelengths or frequency of the visible light.

Most LEDs require from 1.5V to 2.2V of forward-biased voltage drop to illuminate. Like an incandescent light bulb, an LED will illuminate more brightly as the current through the forward-biased LED is increased. Like a common diode, LEDs also have a maximum current rating that must not be exceeded. A series current-limiting resistor is always used with an LED.

Connecting an LED without a series current-limiting resistor across a 12V battery will destroy the LED. Most DMMs have provisions for testing diodes that also work for testing LEDs. The DMM is set to the diode test mode and the meter leads are connected to the LED. In the forward-biased direction, the DMM will supply enough voltage to switch the LED on, causing the LED to illuminate. The voltage dropped across the LED is displayed in the DMM digital display. If the test leads are swapped so that the LED is reverse biased, the LED should not illuminate.

LEDs have become very popular as exterior lighting for trucks and trailers. Other uses for LEDs include instrument panel cluster warning lamps such as ABS malfunction warning. The long life of LEDs makes them ideal as instrument panel cluster warning lamps.

Zener Diodes

Zener diodes are special diodes that behave like common rectifier diodes in the forward-biased direction. However, zener diodes behave much differently from common diodes in the reverse-biased direction. A zener diode has a rating referred to as the **zener voltage**, designated V_Z. Reverse-biased voltage levels across the zener diode that are less than the zener voltage rating cause the zener diode to block the flow of electric current in the same manner as a common diode. However, increasing the voltage dropped across a reverse-biased zener diode up to the diode's zener voltage rating causes the zener diode to switch on and conduct current in the reverse-biased direction. The voltage dropped across the reverse-biased zener diode will then be maintained at the zener voltage rating, even if the source voltage increases. Zener diodes are often used in voltage regulator circuits for this reason.

The schematic symbol for a zener diode is shown in **Figure 6-22A**. The symbol also typically indicates the zener voltage. **Figure 6-22B** shows a plot of the current flow through a 6V zener diode versus the voltage dropped across the zener diode.

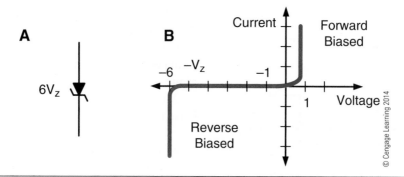

Figure 6-22 Zener diode voltage versus current curve and schematic symbol.

To assist in describing the function of a zener diode, a simple hydraulic pressure relief valve will be examined. A pressure relief valve is used to limit or regulate the pressure in a hydraulic circuit. A pressure relief valve blocks the flow of hydraulic fluid until the system pressure exceeds the pressure relief valve opening pressure. This opening pressure is a function of the relief valve spring tension. The outlet port of the pressure relief valve typically provides a path back to the sump. The valve shown in **Figure 6-23** is designed

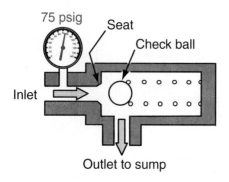

Pressure greater than spring force unseats check ball, permitting fluid to flow to sump limiting pressure to 75 psi

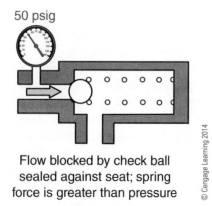

Flow blocked by check ball sealed against seat; spring force is greater than pressure

Figure 6-23 Pressure relief valve.

to open when the pressure applied to the hydraulic fluid reaches 75 psi (517 kPa). Below 75 psi, no hydraulic fluid flows through the valve. When the pressure applied to the hydraulic fluid approaches 75 psi, the hydraulic pressure begins to overcome the spring force holding the valve closed. The open valve will then maintain approximately 75 psi of pressure in this hydraulic system. Pressure relief valves may be found in engine lubrication systems to limit the maximum engine oil pressure.

In the forward-biased direction, the zener diode behaves like a check valve similar to a common diode. Approximately 0.7V is dropped across the zener diode in the forward-biased direction. In the reverse-biased direction, a zener diode will not permit any current to flow until the voltage dropped across the diode reaches the zener voltage, similar to a pressure relief valve. Thus, a hydraulic circuit with a check valve facing one direction and a pressure relief valve facing the other direction plumbed in parallel would be similar to a zener diode, as shown in **Figure 6-24**. Hydraulic fluid can easily flow from right to left in this example because of the standard check valve. However, fluid cannot flow from left to right because the check valve blocks flow. The pressure relief valve also blocks flow from left to right, until pressure applied to the hydraulic fluid reaches the opening pressure of the pressure relief valve. When pressure applied to the hydraulic fluid at the pressure relief valve's inlet port reaches the valve's opening pressure, the pressure relief valve will open to permit fluid to flow back to the sump. The pressure relief valve will remain open to maintain the hydraulic system pressure at approximately 75 psi with this particular pressure relief valve.

In a manner similar to the check valve, the zener diode shown in **Figure 6-25** would permit conventional current to flow from the right side to the left side of the diode if the polarity of the voltage sources shown were reversed. However, the zener diode in this example will block conventional current flow from left

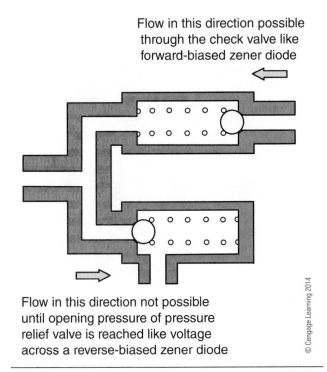

Flow in this direction possible through the check valve like forward-biased zener diode

Flow in this direction not possible until opening pressure of pressure relief valve is reached like voltage across a reverse-biased zener diode

© Cengage Learning 2014

Figure 6-24 Hydraulic check valve similar to a forward-biased zener diode; pressure relief valve similar to a reverse-biased zener diode.

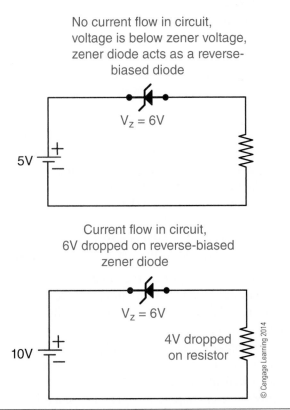

No current flow in circuit, voltage is below zener voltage, zener diode acts as a reverse-biased diode

$V_z = 6V$

5V

Current flow in circuit, 6V dropped on reverse-biased zener diode

$V_z = 6V$

10V

4V dropped on resistor

© Cengage Learning 2014

Figure 6-25 Reverse-biased 6V zener diode blocking current (top) and permitting current to flow (bottom).

to right if the voltage dropped across the reverse-biased zener diode is less than the zener voltage (6V in the example shown in **Figure 6-25**). When the voltage dropped across the reverse-biased zener diode rises to the zener voltage rating, the zener diode will then conduct current in the reverse-biased direction and maintain the zener voltage across its terminals, provided a series resistor is present to drop the remainder of the voltage (**Figure 6-25**). Increasing the voltage supplied by the voltage source would cause the voltage dropped across the zener diode to continue to remain at the zener voltage value. The remainder of the available voltage would be dropped across the series resistor.

....

Tech Tip: Think of a reverse-biased zener diode supplied with a voltage above the zener voltage as a self-controlled variable resistor that will automatically become whatever value of resistance is necessary to maintain the zener voltage across its terminals.

....

TRANSISTORS

Transistors are electronic devices that amplify a small electric signal into a larger electric signal. Prior to the invention of transistors in 1947, vacuum tubes were used as amplifiers. Before the widespread use of transistors in the 1960s, switching on a device such as a radio or television meant waiting for the vacuum tube heaters to warm up for a minute or so before the device would function. Vacuum tubes were also not very reliable and were very large, requiring the television or radio also to be large. Because these tubes were made of glass and very easily broken, they were seldom used in cars or trucks.

The computers and all other electronic devices of today would not be possible without the miniaturization made possible by transistors. Transistors are an important part of most truck electronic control modules. Even though most truck technicians will typically not diagnose a problem to the electronic component level, a basic understanding of transistors will assist you in understanding what is occurring in the inside of an electronic module, which will make you a better troubleshooter.

Bipolar Transistors

A bipolar transistor is a three-terminal device composed of P-type material and N-type material, like that used in a diode. There are two major types of

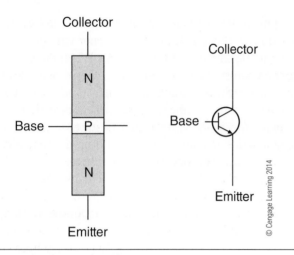

Figure 6-26 NPN bipolar transistor.

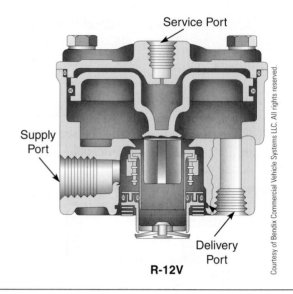

R-12V

Figure 6-27 Air brake system relay valve is similar to a transistor.

bipolar transistors: NPN and PNP. Because the NPN type is the most common type utilized in automotive applications, it will be used in all examples in this chapter. An NPN transistor is like a sandwich in which a slice of P-type material is the meat placed between two slices of N-type material bread. Each of the three terminals of the transistor is connected to a corresponding layer of material (**Figure 6-26**). The P-type material slice of an NPN transistor is called the **base**. One of the N-type slices is referred to as the **collector**; the other N-type slice is called the **emitter**. The emitter, base, and collector are abbreviated E, B, and C, respectively. The schematic symbol for an NPN transistor is shown in **Figure 6-26**. A PNP transistor symbol is very similar except the arrow is pointing inward instead of outward. An easy way to remember which symbol is an NPN transistor is to remember <u>N</u>ot <u>P</u>ointing i<u>N</u> with regard to the arrow.

Transistors are used for two main purposes:

- To amplify low-level signals into higher-level signals.
- To act as an electronic switch similar to a relay.

Bipolar Transistors as Amplifiers

A transistor that is being used as an amplifier operates in ways that are similar to the action of an air brake system relay valve. A relay valve like that shown in **Figure 6-27** is commonly used in air brake systems on trucks with tandem rear axles for the primary (rear) brakes and for the trailer brakes. When the foot valve is depressed in a typical air brake system, pressurized air proportional to the force applied to the foot valve is typically directed to the front (secondary) brake chambers. The front brakes are close to the foot valve and the volume of the two front chambers are typically relatively small. However, the rear (primary) brakes

and the trailer brakes are much further away from the foot valve and the volume of air needed to apply these brakes is much greater than that needed for the front brakes. If the air supplied to the rear and trailer brake chambers were directly supplied by the foot valve, there would be a considerable time lag between when the brake foot valve was depressed and when the brakes actually applied. Instead of routing the primary application pressure through the foot valve, the foot valve may only supply a control pressure for the rear brakes. The control pressure from the foot valve is directed to a relay valve located near the rear brakes. The relay valve has three ports in addition to an exhaust port:

1. Supply port, which is a large-diameter air line supplied directly from the primary air tank.
2. Service port, which is typically a smaller-diameter air line. The service port is connected to the foot valve outlet and acts as a control pressure.
3. Delivery port, which is also a large-diameter air line connected to the rear brake chambers.

The relay valve amplifies the small volume of pressurized air delivered from the foot valve to the service port of the relay valve into a proportionally high volume of air that is directed to the rear brake chambers to apply the brakes. The pressure of the air measured at the service port is about the same as the pressure of the air measured at the delivery port of the relay valve (less the valve crack pressure). The use of a relay valve speeds up the primary brake application time. Trailers also use a relay valve in conjunction with a trailer-mounted air supply tank. A relay valve is especially important in trailers due to the distance of 50 ft (15 m) or more the air must travel from the foot valve to the trailer service brake chambers.

An NPN bipolar transistor used as an amplifier acts in ways that are very similar to an air brake relay valve. The three terminals of the transistor and corresponding ports of the relay valve are as follows:

1. The collector is like the air supply port of the relay valve plumbed to the primary air tank. The collector is connected to the positive terminal of the supply voltage.
2. The base is like the service port of the relay valve plumbed to the foot valve. The base is the small control signal that will be amplified.
3. The emitter is like the air delivery port of the relay valve plumbed to the service chambers. The emitter is connected to the load device, such as a solenoid, and the load is connected to the negative terminal of the supply voltage.

The NPN transistor collector is made of N-type material, the base is made of P-type material, and the emitter is also made of N-type material. The junction where the base and the collector meet is like a PN junction found in a common diode, as is the junction where the base and the emitter meet. When the base is supplied with a positive voltage greater than about 0.7V, conventional current flows from the base to emitter like a forward-biased diode because the base is supplied with a positive voltage and the emitter is connected to ground through the load. No current flows from the base to the collector because the base to collector PN junction (diode) is not forward-biased.

The current that flows from the base to emitter acts as a control current, similar to the way the foot valve supplies a control pressure to a relay valve in an air brake system. The base to emitter current determines how much current the transistor will permit to flow from the collector to emitter and ultimately through the load. The transistor amplifies the current that flows from the base to the emitter 100 times or more. This amplified current flows from the collector to the emitter.

A sample transistor circuit is shown in **Figure 6-28**. A fixed resistor in the circuit supplying the base of the transistor limits the amount of current flowing from the base to the emitter. The base current limiting resistor is 100kΩ and the load resistor is 500Ω in this example. Together, these resistors limit the base current to 112μA with a 12V battery acting as the voltage source.

The physics of semiconductor theory cause this particular transistor to permit a current to flow from the collector to the emitter that is 100 times greater than the current flowing from the base to the emitter. This collector-emitter current flows through the load to ground. The amount of current flowing through the load is the sum of both the base-emitter current and the collector-emitter current, per Kirchhoff's current law. However, in this example 99 percent of the load current is the collector-emitter current.

The transistor shown in **Figure 6-28** illustrates that a small amount of current in the base to emitter circuit is amplified by a factor of 100 in this particular transistor and that all of this current flows through the load device. Increasing the base-emitter current by reducing the base resistor value would cause the collector-emitter current to increase proportionally as well, but only to a point.

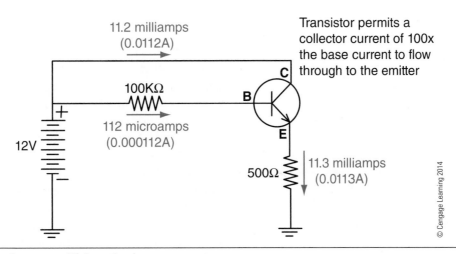

Figure 6-28 Transistor amplifying the base current.

Bipolar Transistors as a Switch or Relay

Increasing the base-emitter current causes an increase in the current that flows from the collector to the emitter. At some point as the base to emitter current is increased, the transistor will not cause any additional current to flow from the collector to the emitter. This point is called the **saturation point**.

When a transistor reaches the saturation point, it is no longer acting as an amplifier but more as a relay. A relatively small amount of current in the base to emitter circuit causes a much larger amount of current to flow through the load. This is similar to the small current flow in a relay coil that causes the relay contacts to close and conduct a large amount of current. The base to emitter current for the upper transistor shown in **Figure 6-29** is zero amps resulting in zero current flow from the collector to emitter. This is like the normally open contacts of a relay when there is zero current flow through the relay coil. The base to emitter current for the lower transistor shown in **Figure 6-29** is above the saturation point causing the maximum possible collector to emitter current. This is like an energized relay coil causing the normally open contacts to close.

To permit a very small current to control a much higher current, two bipolar transistors can be configured as shown in **Figure 6-30** to form what is known as a Darlington pair. If both T1 and T2 in **Figure 6-30** amplify their respective base currents by 100, then a current of 100μA that enters the base of T1 will be amplified by 100 to become a 10mA base current for T2. Transistor T2 amplifies this 10mA base current by 100 to become a 1A current flowing through the collector to emitter of T2. Therefore, the

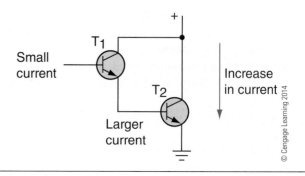

Figure 6-30 Darlington pair permits a very small current to control a very large current.

Darlington pair has amplified the initial base current 10,000 times (100×100).

This use of a transistor as a relay or switch is the basis for gasoline engine electronic ignition introduced in automobiles in the early 1970s. The transistor replaced the ignition contact points as the means to make and break the current flow through the primary winding of the ignition coil. Because there are no moving components or physical contacts, the transistor can provide very fast and precise switching for the life of the vehicle. This also led to the use of microprocessors to determine when the optimal time for spark ignition should occur to reduce exhaust emissions, decrease fuel consumption, and increase performance. A very small signal from the microprocessor is used to ultimately control a high-current transistor.

Transistors used as relays or switches are very common on modern trucks. ABS modulators, automatic transmission control valves, and diesel engine fuel injectors are all controlled by transistors.

Additional information on transistors is covered in the Extra for Experts section at the end of this chapter.

Field Effect Transistors

Field effect transistors (FETs) are transistors that are controlled by voltage. Bipolar transistors, studied earlier, are controlled by current. The term *field* refers to an electric field, like that stored in a capacitor. FETs are widely utilized in automotive electronics. Both bipolar and FETs are semiconductor devices with no internal moving components and typically have three terminals. One important advantage of FETs over bipolar transistors is that because they are voltage controlled and not current controlled, the power consumption of the device is much less than a bipolar transistor. This is not so important when only a few transistors are necessary, but power consumption

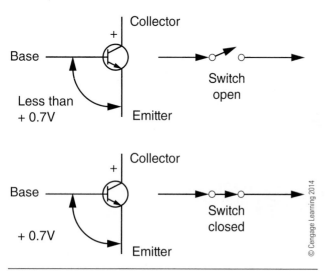

Figure 6-29 NPN transistor used as a switch.

becomes very important when a very large number of transistors are contained within one electronic control module. The three terminals of a typical FET correspond to the three terminals of a bipolar transistor as shown.

Bipolar	FET
Collector	Drain
Base	Gate
Emitter	Source

Recall in a bipolar transistor, the current flowing from the base through the emitter controlled the amount of current that would flow from the collector to the emitter. In an FET, the voltage difference between the **gate** and the source controls the conductivity or "resistance" between the **drain** and the **source**. Although this "resistance" cannot be measured with a DMM, it is okay to think of an FET as a voltage-controlled resistor for the purposes of understanding the concept. However, virtually zero current flows between the gate and the source, unlike the base current in a bipolar transistor.

Junction Field Effect Transistors. There are many different types of FETs. The simplest FETs are known as junction FETs or **JFETs**. There are two main types of JFETs—P-channel and N-channel. The P and N refer to the polarity of the voltage used to control the FET. The schematic symbol for an N-channel JFET is shown in **Figure 6-31**. A P-channel JFET schematic symbol is drawn with the arrow pointing in the opposite direction.

As mentioned, FETs are controlled by the difference in the voltage between the gate and the source. For JFETs, the "resistance" between the drain and source decreases as the difference in voltage between the gate and source decreases. Therefore, when the

source and the gate are both at the same voltage, such as 0V, the "resistance" between the drain and source is at its minimum value. This is like a normally closed relay contact with no current flowing through the relay control coil. Thus, to turn a JFET "on," the same level of voltage is applied to both the gate and the source.

To turn a JFET "off," a voltage is applied to the gate. For P-channel JFETs, a positive voltage difference between the gate and the source is used to control the JFET. The higher the voltage at the gate of the JFET compared to the voltage at the source of the JFET, the higher the "resistance" between the JFET drain and source. This is also similar to a normally closed relay contact, which is opened when current flows through the relay control coil. This concept is illustrated in **Figure 6-32** for a P-channel JFET. The higher the value of the voltage between the gate and source, the more the conductive channel between the drain and source is pinched or closed off until it becomes completely blocked to act as an open circuit. Increasing the voltage difference between the gate and the source terminals is like pinching a garden hose, which results in an increase of the restriction of the hose. Eventually, all water flow can be stopped if the hose is sufficiently pinched. For N-channel JFETs, a negative voltage is used to control the FET in the same manner.

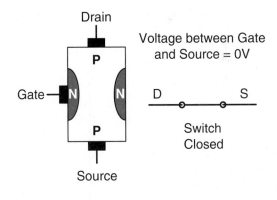

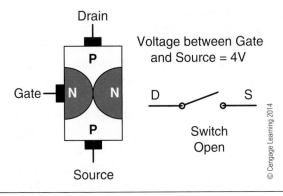

Figure 6-32 P-channel JFET is similar to a normally closed relay.

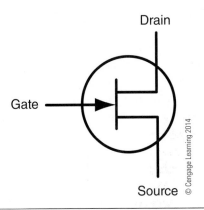

Figure 6-31 N-channel JFET schematic symbol.

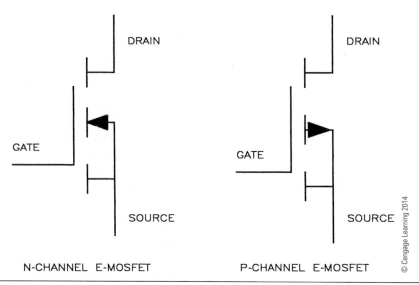

Figure 6-33 Enhancement-mode MOSFET symbols.

MOSFET. One of the most common types of FET found in a truck electronic control module is called a **MOSFET**. This stands for metal oxide semiconductor FET. The symbols for an N-channel and P-channel enhancement-mode MOSFET are shown in **Figure 6-33**. An enhancement-mode FET is similar to a normally open relay, which is the opposite of JFETs. The conductive path between the drain and source is an open circuit when the gate and the source are at the same voltage level. As the voltage between the gate and source increases, the "resistance" between the drain and source decreases. MOSFETs can be designed so that the "resistance" between the drain and source is very low when the device is switched "on" and very high when the device is switched "off." The switching action also occurs very rapidly (microseconds) after the voltage changes at the gate. This makes MOSFETs ideal for switching high currents with precise timing for devices like electronically controlled diesel engine fuel injectors.

Drivers

Large high-power MOSFETs are commonly utilized as drivers for high-current devices. The term *driver* is often used to describe a transistor used as a switch. Transistors utilized as electronic switches may either provide a path to ground (negative) or provide a positive voltage.

When the transistor is providing a path to ground, the device may be described as a **low side driver**. *Low side* indicates negative or ground. When a device provides a path to ground, it is also said to be **sinking** current. Therefore, a low side driver sinks current. A load device such as a solenoid that is controlled by a low side driver would typically have one terminal

connected to a positive voltage source and the other terminal connected to the low side driver as shown in **Figure 6-34**. Conventional current is flowing from the load into the low side driver.

It is common to show a high or low side driver as a simple two-terminal box as shown in **Figure 6-34** instead of a detailed symbol such as that shown in **Figure 6-33**. Obviously, the driver device must be equipped with a gate or base control circuit at a minimum. However, a driver is often a "chip," which contains many other components in addition to a transistor and may have several terminals to indicate current flow through the device and other status information. This detailed information is important for engineers, but is not necessary for basic troubleshooting or understanding of the circuit.

When the transistor is supplying a positive voltage, the device may be described as a **high side driver**. *High side* indicates a positive voltage. When the device provides a positive voltage, it is said to be **sourcing** current. Therefore, a high side driver sources current.

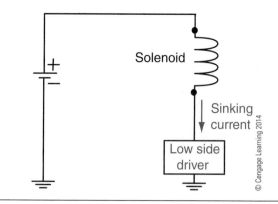

Figure 6-34 Low side driver sinking current to ground.

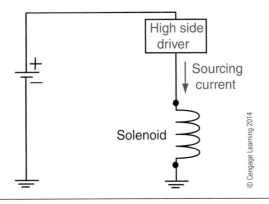

Figure 6-35 High side driver sourcing current to the load.

A load device such as a solenoid that is controlled by a high side driver would typically have one terminal connected to ground and the other terminal connected to the high side driver, as shown in **Figure 6-35**. Conventional current is flowing out of the high side driver to the load.

Most OEMs simply illustrate an electronic module as a box within a circuit diagram with no indication markings as to what is inside the electronic module. The content of the electronic module is considered proprietary. However, looking at the circuit diagram and observing how an electrical device is being controlled and how the device is connected typically makes it possible to determine whether the electronic module is sinking or sourcing current. Knowing whether the controlling device inside of a module is a high side driver or a low side driver can explain why an OEM's troubleshooting information is directing you to perform specific troubleshooting steps.

··

Tech Tip: A device controlled by a low side driver located inside an electronic module will typically be connected to a positive voltage at one terminal and the electronic module at the other terminal. A device controlled by a high side driver located inside an electronic module will typically be connected to ground at one terminal and the electronic module at the other terminal.

··

Both high and low side drivers are utilized to control truck electrical system devices. These driver transistors, like most any other electrical component, are designed to conduct a limited amount of current. Smaller drivers, such as those designed to control relay coils, may be rated for 1A or less while larger drivers may sink or source 20A or more of current directly.

Causing these driver devices to carry more current than the maximum current rating may destroy the driver. This is one reason why proper troubleshooting techniques are very important. Inadvertent or indiscriminant shorts to ground or shorts to battery-positive voltage when troubleshooting an electronic module can cause permanent damage to the module.

EXTRA FOR EXPERTS

This section provides additional transistor theory.

More on Bipolar Transistor Amplifiers

Transistors are used to amplify small signals, such as an audio signal. An audio signal has a changing voltage amplitude and polarity. The low-level audio signal reproduced within a CD player is amplified many times over by transistors. The amplified signal is then further directed to the speaker system.

Besides audio systems, transistor amplifiers are utilized in truck electronic control modules to amplify low-level signals supplied by sensors. For example, an amplifier is used to transform the small signal generated by an exhaust temperature sensor to a useable signal to drive the exhaust temperature gauge known as a pyrometer within the instrument panel cluster.

The transistor illustrated in **Figure 6-36** amplifies the base-emitter current by a factor of 100. The transistor does not produce current; it just regulates the amount of current that flows from the collector to the emitter, similar to the action of a variable resistor. The reduction in the opposition to current flow between the collector and the emitter is actually what causes a transistor to amplify. In the example shown in **Figure 6-36**, the opposition to current flow between the collector and the emitter causes a voltage drop of 6.4V across the collector and emitter. This opposition to current flow between the collector and emitter could be thought of as a resistance, although you could not use an ohmmeter to measure its value. The voltage drop of 6.4V indicates that the transistor is acting as an amplifier, not as a switch or relay.

When there is zero base-emitter current present, there is also zero collector-emitter current. This indicates that the opposition to current flow between the collector and the emitter is very high when there is zero base-emitter current. No current flows from the collector (N-type material) to the base (P-type material) when there is zero base-emitter current because this PN junction between the collector and base acts as a reverse-biased diode. This reverse-biased collector-base

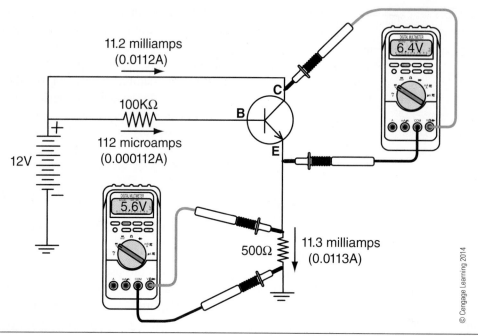

Figure 6-36 Voltage dropped across transistor collector to emitter indicates transistor is acting as an amplifier.

PN junction prevents any current from flowing from collector to emitter similar to a reverse-biased diode.

When current is flowing from the base to the emitter, the properties of the P-type base material changes. This change permits some current to flow from the collector and through the base to the emitter. As base to emitter current flow increases, the collector to emitter current flow also increases. It is as though the base-emitter current controls a variable resistor between the collector and emitter, similar to the way

the control pressure in the air brake relay valve controls a variable pressure source between the primary air supply tank and the primary air brake chambers.

More on Bipolar Transistors as a Switch or Relay

If the base resistance in **Figure 6-36** were reduced from 100kΩ to 47.5kΩ, the base to emitter current would also increase per Ohm's law. This increase in

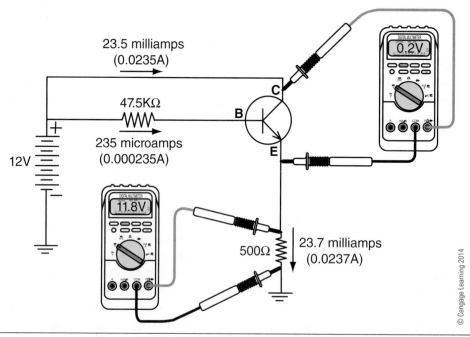

Figure 6-37 Transistor saturated; increasing base current will not result in any additional collector-emitter current; transistor is acting as a switch in the ON position.

base to emitter current would result in a decrease in the opposition to current flow or "resistance" between the collector and emitter, as well as a decrease of the voltage dropped between the collector and emitter as the transistor amplifies the base current as shown in **Figure 6-37**.

If the base resistor value were further reduced, at some point the base current would rise to such a level that the opposition to current flow between the collector and emitter would drop to its minimum possible value. This causes the voltage dropped across the collector and emitter to be about 0.2V (200mV). Any additional increase in base current after this point would not result in any additional current flow through the collector. This is called the saturation point of the

transistor. The saturation point is the point at which the transistor will not provide any additional current amplification as base to emitter current is increased because the opposition to current flow between the collector and emitter is at its minimum value. At the saturation point, the transistor is acting as a closed switch between the collector and emitter. This is analogous to the current through a relay coil. Once sufficient current flows through a relay coil to cause the normally open contacts to close, increasing the current flow through the relay coil will not cause any additional current to flow through the relay's normally open contacts.

Summary

- Resistors are used for three main purposes: to limit current flow, to divide voltage, and to give off heat. Resistors may be fixed value or variable value.

- A rheostat is a two-terminal variable resistor. A rheostat is often used to control the amount of current flowing in a circuit, such as in the case of an instrument panel dimmer control. A wiper moves over a resistive track, causing a change in the amount of resistance between the two terminals.

- A potentiometer is a three-terminal variable resistor. A potentiometer is typically used to divide voltage.

- A diode is a two-terminal electronic device that acts like a one-way check valve for current flow. When current is flowing through the diode, the diode is said to be forward biased. When the diode is blocking the flow of current, the diode is said to be reverse biased.

- Diodes and many other electronic devices are made of silicon. Silicon mixed with impurities can form N-type material and P-type semiconductor material. Placing these two types of silicon next to each other causes a junction to form. This PN junction is the basis of many electronic components.

- The positive end of a diode is known as the anode; the negative end of a diode is known as the cathode. A stripe around the girth of the diode signifies the cathode end of the diode.

- Diodes can be tested outside of a circuit using a DMM or an analog ohmmeter. The diode should have very high resistance when reverse biased by the ohmmeter, and very low resistance when forward biased by the ohmmeter.

- A light-emitting diode (LED) is a special diode that gives off visible light when forward biased. An LED always requires some means of limiting current flow through the LED, unlike a standard light bulb.

- A zener diode is a special diode that acts like a standard rectifier diode in the forward-biased direction. In the reverse-biased direction, the zener diode will permit current flow after the zener voltage is reached.

- A transistor is a device that amplifies a small signal into a large signal. A transistor can be used as an amplifier in devices such as a radio. A transistor can also be used like a relay.

- Bipolar transistors are three-terminal devices and contain two PN-type material junctions. The three terminals are identified as the base, collector, and emitter. Bipolar transistors are current-controlled devices. Bipolar transistors can be NPN or PNP type.

- Field effect transistors (FETs) are three-terminal devices. The three terminals are identified as the source, gate, and drain. FETs are voltage-controlled devices.

- Transistors that are used as switches may be called drivers. Drivers that provide a path to ground for a device when switched on are known as low side drivers. A low side driver is said to sink current. Drivers that supply a positive voltage to a device when switched on are called high side drivers. A high side driver is said to source current.

Suggested Internet Searches

Try these web searches for more information on electronics:

http://www.ti.com

http://www.siemans.com

Review Questions

1. An electric heater is plugged into a 120V AC wall socket. The electric cord between the heater and the wall socket is not hot to the touch. However, the heating element in the electric heater can be seen to glow red-hot. Which is the most accurate statement?

   A. The heating element is hot because the resistance of the heating element is much less than the copper wire in the cord.

   B. The voltage dropped on the electric cord is much higher than the voltage dropped on the heating element.

   C. The electric cord does not feel hot because of the plastic insulation covering the wires.

   D. The heating element has a higher resistance than the electric cord.

2. An accelerator (throttle) position sensor used on an electronically controlled diesel engine is typically what type of device?

   A. Fixed resistor

   B. Rheostat

   C. Potentiometer

   D. Stepped resistor

3. A diode and a 1k ohm resistor are placed in series with a 12V battery. The diode is forward biased. Which of the answers below is closest to the amount of voltage that is dropped across the diode?

   A. 0.7V

   B. 11.3V

   C. 12V

   D. 2V

4. A common diode is being tested with a DMM. Technician A says that using the ohms scale is the best way to test a diode using a DMM. Technician B says that common diodes cannot be tested using any type of DMM. Who is correct?

   A. A only

   B. B only

   C. Both A and B

   D. Neither A nor B

5. An LED is similar to a light bulb except for which difference?

   A. An LED requires a series resistor to limit current and a light bulb is typically used in a circuit without a series resistor.

   B. An LED will only illuminate if the voltage polarity is correct; current flow in either direction will cause a light bulb to illuminate.

   C. An LED gives off very little heat; a light bulb may become very hot.

   D. All of the above are valid differences between LEDs and light bulbs.

6. A zener diode with a zener voltage rating of 6V is placed in series with a 1k ohm resistor and a 12V battery. The diode is placed in the circuit so that it is forward biased and a DMM is used to measure the voltage across the zener diode. The zener diode is then removed and installed in the opposite direction so that it is reverse biased and a DMM is used to measure the voltage drop across the zener diode. What are the two voltage readings?

   A. 12V with zener diode forward biased and 6V with zener diode reverse biased

   B. 11.3V with zener diode forward biased and 6.7V with the zener diode reverse biased

   C. 0.7V with the zener diode forward biased and 12V with the zener diode reverse biased

   D. 0.7V with the zener diode forward biased and 6V with the zener diode reverse biased

7. An NPN transistor is being used as a low side driver for a relay coil (transistor is sinking current). One side of the relay coil is connected to the low side driver. The other side of the relay coil would probably be connected to which of the following?

   A. Ignition or battery positive voltage

   B. The load device

   C. Battery negative or ground

   D. Either A or C

8. Which device in a modern truck electrical system would probably not contain any transistors?

   A. Electronic engine control module

   B. ABS control module

   C. Pyrometer module

   D. Steering gearbox

9. Which device of those listed below would be the one most likely to contain a zener diode?

   A. Headlamp

   B. Suppression resistor across relay coil

   C. Voltage regulator

   D. Potentiometer

10. The stripe on a diode indicates what?

    A. Wiper

    B. Anode end

    C. Tolerance

    D. Cathode end

11. An LED is used for which of the following purposes on a modern truck?

    A. Voltage regulation

    B. Exterior lighting

    C. Reverse biasing

    D. Forward biasing

12. A transistor acting as a relay or switch that provides a path to ground is known as what type of device?

    A. Highway driver

    B. High side driver

    C. Voltage divider

    D. Low side driver

13. Resistors have a wattage rating. Which of the following is true?

   A. The wattage rating indicates the resistance in watts.

   B. The wattage rating indicates how close to the indicated resistance value that the resistor will be.

   C. The wattage rating indicates if the resistor is AC or DC.

   D. None of the above.

14. A diode is being tested using a DMM with a diode test feature. The DMM indicates 0V when the diode is tested in both directions. What does this indicate?

   A. The diode is open.

   B. The diode is forward biased.

   C. The diode is reverse biased.

   D. The diode is shorted.

15. Technician A says that FETs are controlled by the base current and bipolar transistors are controlled by the voltage difference between the gate and source. Technician B says that in an NPN transistor circuit, the base current plus the collector current equals the emitter current. Who is correct?

   A. A only

   B. B only

   C. Both A and B

   D. Neither A nor B

CHAPTER

7 Charging Systems

Learning Objectives

After studying this chapter, you should be able to:

- Explain how an alternator produces a regulated DC voltage.
- Describe how electrical contact is made through the rotating rotor windings.
- Trace the path of current flow from the stator windings through a rectifier bridge.
- Explain how a voltage regulator controls the strength of the rotor's magnetic field.
- Define the purpose of the diode trio or field diode assembly.
- Perform a preliminary inspection of a charging system.
- Disassemble and test the internal components of a typical alternator.
- Test a truck charging system.
- Perform a parasitic current draw test.

Key Terms

alternator	full-field	ripple
brush	hertz	rotor
delta	load dump	slip ring
diode trio	parasitic loads	stator
duty cycle	phase	voltage regulator
excitation	pulse width modulation (PWM)	windings
field diode	rectifier	wye (Y)
frequency	residual magnetism	

INTRODUCTION

The truck's charging system supplies the energy to recharge the truck's batteries. The charging system also acts as the primary voltage source, when the engine is running, to supply the truck's electrical system.

The main component in the modern truck's charging system is the alternating current (AC) generator, which is commonly referred to as an **alternator**. An alternator produces an alternating current, as the name implies. The alternating current is converted into a direct current (DC) by diodes. The voltage that appears at the output terminal of an alternator is a direct current.

A modern truck charging system consists of the following components:

- Alternator with internal rectifier assembly
- Voltage regulator, which is often contained within the alternator
- Batteries
- Cables

Some of the concepts introduced in this chapter will also provide an understanding of the principles of sensors, which measure engine speed, as well as some features of hybrid electric trucks.

ALTERNATOR FUNDAMENTALS

An alternator operates on the principles of electromagnetic induction. The alternator converts rotating mechanical energy provided by the engine into electrical energy.

Creating Voltage with Magnetism

Moving a conductor through a magnetic field so that magnetic lines of force are cut by the conductor causes a voltage to be induced in the conductor, as explained in **Chapter 3**. The conductor must cut through or interrupt the magnetic lines of force to cause a voltage to be induced in the conductor. To review the concept of electromagnetic induction, note that a horseshoe-shaped magnet like that shown in **Figure 7-1** has a magnetic field indicated by the magnetic lines of force. Moving the conductor through

the magnetic field so that magnetic lines of force are "cut" causes a voltage to be induced in the conductor. Conversely, a voltage would also be induced in the conductor if the conductor were held stationary and the horseshoe magnet were moved so that the magnetic lines of force cut through the conductor. The relative motion between the conductor and the magnetic lines of force causes a voltage to be induced in the conductor.

This concept of cutting magnetic lines of force by relative movement between the conductor and the magnetic field can be illustrated by wind. If on a calm day you were to stick your hand out of the window of your car while traveling 30 mph (48 km/h), you would experience a 30-mph wind on your hand. The car is moving relative to the stationary air and your hand is cutting through the air like a conductor cutting through magnetic lines of force. If a storm comes along with 30-mph winds and you were to park your car facing into the wind and stick your hand out the window again, the 30-mph wind on your hand would feel no different than it did when the car was moving. The wind would now be passing through your hand like magnetic lines of force that are in motion and cutting through a stationary conductor.

The polarity of the voltage induced in the conductor depends on the direction that the conductor is moved through the magnetic field (or vice versa) and the direction of the arrows on the magnetic lines of force. With the arrows on the magnetic lines of force pointing in the downward direction shown in **Figure 7-1** and the conductor moved through the magnetic field from right to left, the polarity of the voltage induced in the conductor is as shown in **Figure 7-1**. If the conductor is moved through the same magnetic field but in

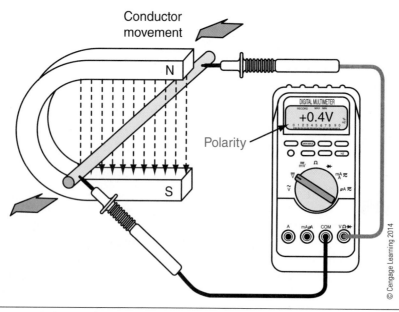

Figure 7-1 Cutting magnetic lines of force with a conductor to induce a voltage: conductor is moving from right to left through the magnetic field.

the opposite direction, the polarity of the voltage induced in the conductor also changes, as shown in **Figure 7-2**. Flipping the horseshoe magnet over so that the direction of the arrows on the magnetic lines of force would be pointing up would reverse the polarity of the voltage induced in both cases. The direction of movement of the conductor with respect to the direction of the arrows on the magnetic lines of force determines the polarity of the voltage induced in the conductor. To summarize, the polarity of the voltage induced in the conductor reverses when the relative motion between the conductor and the magnetic field reverses direction.

If a resistor were connected between the two ends of the conductor and the conductor moved through a magnetic field so that magnetic lines of force were being cut, the voltage induced in the conductor would cause an electric current to flow through the resistor.

Inducing More Voltage

An important fact related to electromagnetic induction is as follows:

....

Important Facts: The greater the number of magnetic lines of force that are cut by a conductor per second, the greater the voltage that is induced in the conductor.

....

Increasing the number of magnetic lines of force that are cut by a conductor per second can be accomplished by two different means. One way is to increase the speed that the conductor moves through the magnetic field (or vice versa). This is like the wind that you feel on your hand outside your car window becoming stronger when you drive faster.

The other way to increase the number of magnetic lines of force that are cut per second by a conductor is to increase the strength of the magnetic field. The strength of a magnetic field is measured by the number of invisible magnetic lines of force within a given area. Therefore, increasing the strength of the magnetic field increases the number of magnetic lines of force that the conductor is cutting through per second. To illustrate this, air is thicker (more dense) at sea level than at high altitudes. Although it is not recommended, if you were to stick your hand out of the window of a race car traveling 200 mph (322 km/h) at sea level, the wind felt on your hand would be much stronger than it would be if you stuck your hand out the window of an aircraft flying 200 mph relative to the ground at 30,000 ft (9145 m). This is because the air is thinner as altitude increases and the "wind" felt on your hand would be much less at 30,000 ft than at sea level, even though you would be traveling at the same relative speed to the stationary air in both instances. This is similar to the concept of increased magnetic lines of force within the same area inducing a higher voltage in a conductor.

Alternating Current

If you have ever ridden in a car that needed the shock absorbers (suspension dampers) replaced, you have experienced simple harmonic motion. The springs of the car and the mass (weight) of the car combine to form a spring-mass system, which oscillates up and down at a prescribed rate.

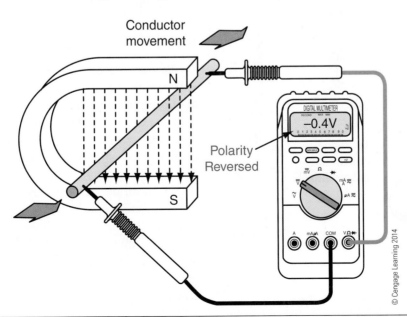

© Cengage Learning 2014

Figure 7-2 Conductor has been moved from left to right through the magnetic field causing a reversal of the polarity of induced voltage.

If a heavy lead ball were attached from a coil spring that was hung from the ceiling, gravity would cause the ball to extend the coil spring from its natural or free length, as shown in **Figure 7-3**. Springs and lead balls may seem out of place in a study of electricity, but such a spring-mass system can be used to describe alternating current (AC).

Because the ball is initially just suspended from the spring, there is no motion. To start the motion, it is necessary either to lift the ball upward or to pull the ball downward and release. In **Figure 7-3**, the ball is pulled downward 6 in. (15 cm) and then released. The spring tension due to the extended spring causes the ball to travel upward, in opposition to gravity. At some point, the ball will reach a height where it is not traveling downward or upward, which is the maximum height the ball will reach. The ball will then begin to travel downward because of gravity. The ball will pass through the original free-hanging or rest position as the motion of the ball causes the spring to extend. The ball will continue downward until it reaches a point that is slightly less than 6 in. below the initial rest point. The ball will then stop traveling downward and reverse directions to travel upward again because of the spring tension.

If a felt-tipped marker were attached to the ball and a roll of paper were passed near the ball while it is in motion at a uniform rate of 1 in. (25 mm) per second, the motion of the ball with respect to time would be recorded on the paper by the felt-tipped marker (**Figure 7-4**). The pattern that would be drawn on the moving paper roll by the felt-tipped marker is a sinusoidal (sine) waveform.

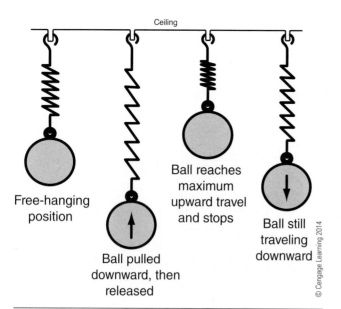

Figure 7-3 Lead ball pulled downward from rest position to stretch spring more than gravity alone and released.

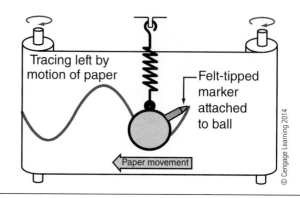

Figure 7-4 Motion of ball drawn on paper.

If the paper were examined in detail after the experiment, it would indicate the height or amplitude of the ball at any given time. If the free-hanging position of the ball is considered as the zero amplitude point, values above the free-hanging point are considered positive values of amplitude; values below the free-hanging point are considered negative values of amplitude. If the paper were being unrolled in front of the moving ball at a uniform rate of 1 in. per second, vertical lines drawn on the paper 1 in. apart would each represent 1 second in time. Horizontal lines also spaced 1 in. apart could be used to indicate the amplitude of the ball. Together, the horizontal and vertical gridlines indicate the amplitude of the ball at each moment of time (**Figure 7-5**).

This experiment introduces the idea of a waveform in which amplitude is referenced with respect to time. Time is typically shown as the horizontal or X-axis, and amplitude is typically shown as the vertical or Y-axis. Amplitude values above the X-axis zero line are positive values, and amplitude values below the X-axis zero line are negative values.

Voltage Waveforms

An oscilloscope is a tool used to display a changing voltage with respect to time. An oscilloscope is like the roll of moving paper used in the spring-mass experiment in that both display or record the amplitude of a changing waveform with respect to time. An oscilloscope has a video display screen that is divided into a grid (**Figure 7-6**). The voltage amplitude of an alternating voltage source is displayed on the vertical axis (Y-axis) of the grid and time is displayed on the horizontal axis (X-axis) of the grid. This permits the amplitude of the voltage source at any given time to be observed. The oscilloscope can be adjusted to change the values of each division of both the horizontal and vertical axes. This permits making adjustments to the oscilloscope to measure a variety of voltage waveforms. The oscilloscope display can be frozen so that the voltage waveform can be analyzed.

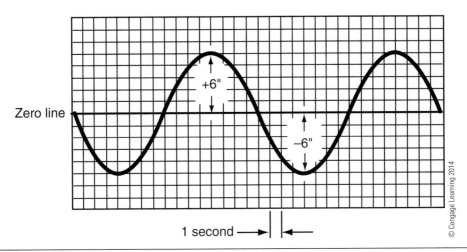

Figure 7-5 Gridlines drawn on paper indicate units of time (sec) for horizontal axis and height (in.) on vertical axis.

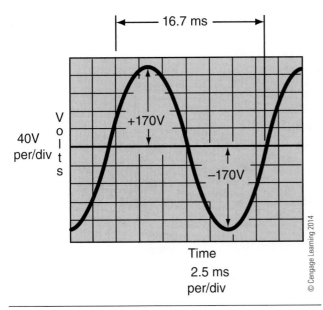

Figure 7-6 Oscilloscope provides a means to display the amplitude of a changing voltage at a particular time.

Automobile technicians have used oscilloscopes to troubleshoot gasoline engine ignition systems for decades. Portable hand-held oscilloscopes have become an essential tool for automobile technicians to diagnose electrical problems. In recent years, diesel technicians have also begun using portable oscilloscopes to troubleshoot the various electronic control systems on modern trucks as trucks have become increasingly more complex, perhaps even more so than automobiles.

If an oscilloscope were used to measure the voltage waveform between the two terminals of a typical North American 120V AC wall socket, the waveform would look like that shown in **Figure 7-6**. The waveform is a sine waveform, like that obtained from the spring-mass experiment. Notice that one-half of the waveform is above 0V, and the other half of the waveform is below 0V. The highest or peak amplitude of the waveform is about 170V, while the lowest amplitude is about −170V. See the Extra for Experts section at the end of this chapter for an explanation as to why the peak value is 170V and not 120V.

An oscilloscope can be thought of as a voltmeter that graphically displays the difference in voltage between the oscilloscope leads, similar to a DMM displaying the difference in voltage between its two test leads in the DC circuits studied in earlier chapters. A DMM voltmeter can be used to measure the voltage dropped across a resistor in a DC circuit. If the DMM voltmeter is connected so that the positive test lead is connected to a point in the circuit where the voltage level is more positive (closer to the battery positive terminal) than the point in the circuit where the negative test lead is connected, the voltmeter will indicate a positive value for the difference in voltage levels between the two voltmeter test leads. This is illustrated in **Figure 7-7A**. If the DC voltage source is then disconnected and reinstalled in the opposite direction so that the polarity of the voltage is reversed without changing the way that the DMM voltmeter is connected, the voltmeter would indicate a negative value for the voltage dropped across the resistor, as shown in **Figure 7-7B**. The current flow through the resistor would also reverse or alter directions because of the change in the polarity of the voltage source.

The voltage measured across the two terminals of a typical North American wall socket has a polarity that is continually changing directions. The shape of the changing voltage as displayed on an oscilloscope is a sine waveform that is positive one-half of the cycle and negative the other half of the cycle. Any current that would flow through a circuit supplied by this continually

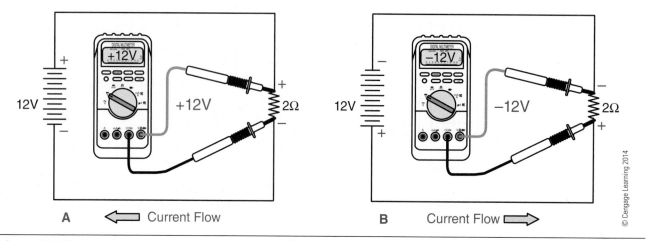

© Cengage Learning 2014

Figure 7-7 Battery connection reversed causing direction of current flow to reverse and DMM to indicate a negative voltage.

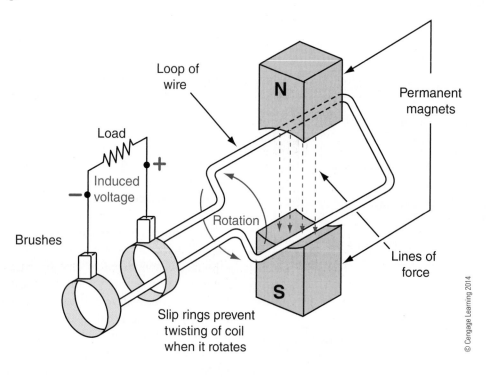

© Cengage Learning 2014

Figure 7-8 Simple alternator.

changing voltage source would flow in one direction when the voltage source is positive and reverse directions when the voltage source is negative. The polarity of the voltage is continually changing, which causes any resulting current flow caused by the voltage to alternate the direction of current flow accordingly. This is where the term *alternating current* or *AC* originates.

The time for one complete cycle of the voltage waveform measured at the wall socket is shown as about 16.7 ms (0.0167 s). This equates to 1/60th of a second. Thus, this waveform would repeat itself 60 times a second. The number of times that a waveform repeats per second is referred to as the **frequency**. Frequency is measured in units of **hertz**. Hertz (Hz) indicates the

number of complete cycles per second that a waveform occurs. The waveform shown in **Figure 7-6** has a frequency of 60 Hz. You are probably familiar with the term *hertz* from FM radio. The numbers displayed on an FM radio tuner indicate megahertz (MHz).

Producing an AC Voltage

Producing a steady DC voltage through electromagnetic induction would be very difficult to achieve. Instead, an AC voltage can be produced easily with rotary motion, such as that provided by a diesel engine.

A simple alternator is shown in **Figure 7-8**. This alternator consists of two permanent magnets and a loop of wire with two **slip rings** attached to each end of the loop. Slip rings are conductive rotary electrical joints. The slip rings permit the loop to rotate yet still make electrical contact with the load resistor through the brushes. An electrical **brush** is a conductor that provides an electrical connection between a stationary and moving component. Brushes used in alternators provide an electrical connection to the rotating slip rings.

In the cross section of the loop of wire in the magnetic field of the simple alternator shown in **Figure 7-9**, position 1 is the starting position. As the loop is rotated counterclockwise from position 1, no magnetic lines of force are being cut by the conductor because the initial movement of the rotating loop is parallel to the magnetic lines of force. The circular path followed by the loop as it rotates from position 1

to position 2 causes the loop to begin cutting through magnetic lines of force. This cutting of magnetic lines of force causes a voltage to be induced in the loop. Because only a few magnetic lines of force are initially being cut by the loop, the level of the voltage induced into the loop is low. However, the circular path followed by the loop causes the number of magnetic lines of force cut by the loop to increase continually as the loop continues to rotate toward position 3. The cross section of the loop shows that the induced voltage is causing a current to flow in the direction indicated by the dart symbols shown in the cross section of the conductor. As the loop rotates to position 3, the circular motion of the loop is now mostly perpendicular (90 degrees) to the magnetic lines of force, causing the maximum number of magnetic lines of force to be cut by the loop.

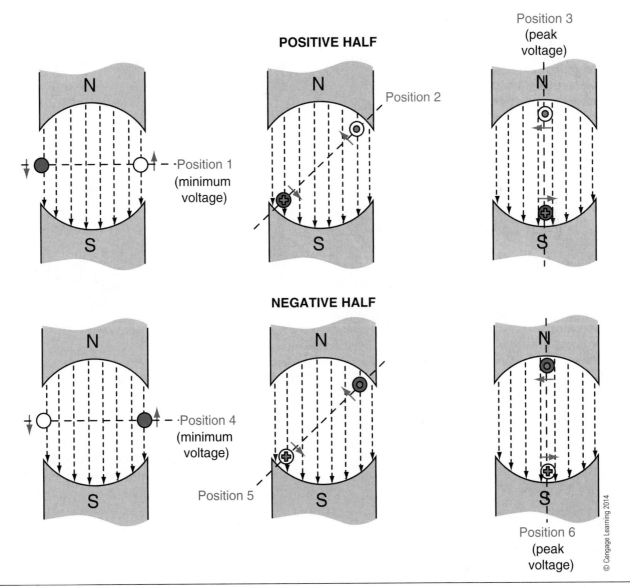

Figure 7-9 Single loop rotating in magnetic field at six different positions.

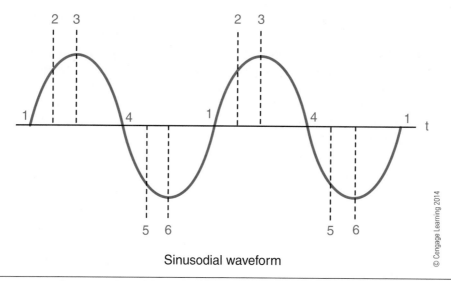

Sinusodial waveform

© Cengage Learning 2014

Figure 7-10 Voltage waveform produced by a single loop rotating in a magnetic field.

This results in the highest or peak voltage being generated. As the loop rotates from position 3 to position 4, the voltage induced in the loop begins decreasing as fewer magnetic lines of force are cut by the circular path of the loop. At position 4, the loop is once again not cutting any magnetic lines of force because the movement of the loop is parallel to magnetic lines of force. The loop rotates from position 4 to position 5, causing the voltage induced in the loop to have an increasing reversed polarity as magnetic lines of force are being cut in a different direction. At position 6, the maximum number of magnetic lines of force is once again being cut by the loop, but in the opposite direction, causing the peak level of reverse polarity voltage to be induced in the loop. The reverse polarity voltage induced in the loop decreases as the loop rotates from position 6 back to position 1 because a decreasing number of magnetic lines of force are being cut by the circular path of the loop. At position 1, the level of the reversed polarity voltage induced in the loop drops to zero to begin the cycle again.

Each complete revolution of the loop in the magnetic field produces a single sine voltage waveform. **Figure 7-10** illustrates two complete revolutions of the loop through the magnetic field. The numbers shown in **Figure 7-10** indicate the amplitude and polarity of the voltage produced at each of the six positions of the loop shown in **Figure 7-9**.

The simple alternator in the previous example illustrates how a sine voltage waveform is generated, but this type of alternator is not very practical for several reasons. For example, the large air gap of the magnetic field would require very large and powerful magnets. As indicated, it really does not matter if the conductor is

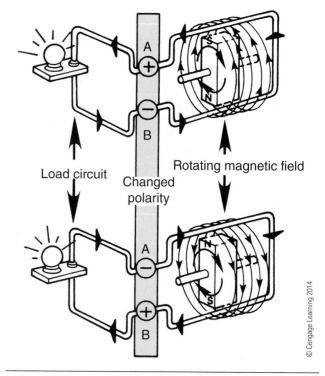

Figure 7-11 Magnet rotating inside of conductive loop induces an AC voltage in the loop.

moved through the magnetic field or the magnetic field is moved through a stationary conductor. Either way will cause voltage to be induced in the conductor. Therefore, alternators use a stationary conductor and rotate a magnet inside the stationary conductor loop, as shown in **Figure 7-11**. This rotary motion also causes a sine voltage waveform to be induced in the conductor. The sine waveform is positive half of the time and negative the other half of the time, like the voltage waveform that would be measured at a wall socket.

Reluctance, introduced in **Chapter 3**, describes the opposition to the path of lines of magnetic force. Iron has a much lower reluctance than air. To improve further the performance of the stationary conductor alternator, placing the conductor loop inside an iron frame as shown in **Figure 7-12** causes the magnetic lines of force to be concentrated in the low-reluctance iron frame. The conductor loop is placed between the frame and the magnet. Therefore, as the poles of the magnet pass the conductor loop, the iron frame causes more magnetic lines of force to cut through the conductor, resulting in an increase of the voltage that is induced in the conductor. The conductor loop and the iron frame form a **stator**. The stator in an alternator remains stationary, as indicated by its name.

To improve the simple alternator still further, replacing the permanent bar magnet with an electromagnet will permit the amplitude of the voltage that is induced in the stator winding to be controlled. The electromagnet consists of a coil of wire wrapped around a piece of steel. Because the electromagnet will be required to rotate, slip rings and brushes supply current to the rotating electromagnet. The electromagnet is mounted to a shaft located at its center. The rotating electromagnet assembly in an alternator is called a **rotor**. This is easy to remember because the rotor is the component in the alternator that rotates. The rotor is designed so that the air gap is minimal between the tip of the electromagnet and the conductor in the stator. **Figure 7-13** shows this

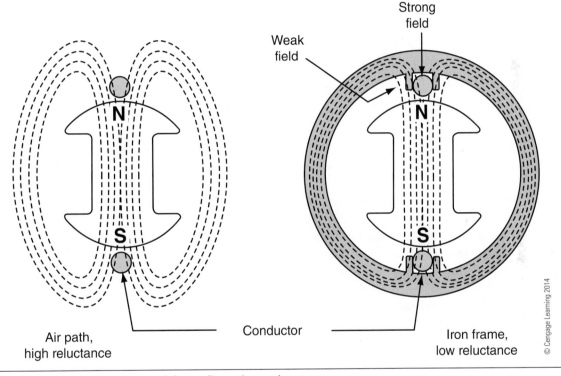

Figure 7-12 Magnetic field lines of force flow through stator.

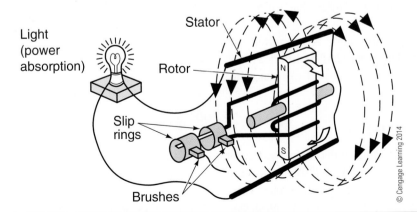

Figure 7-13 Electromagnet rotating inside of conductive loop induces an AC voltage in the loop.

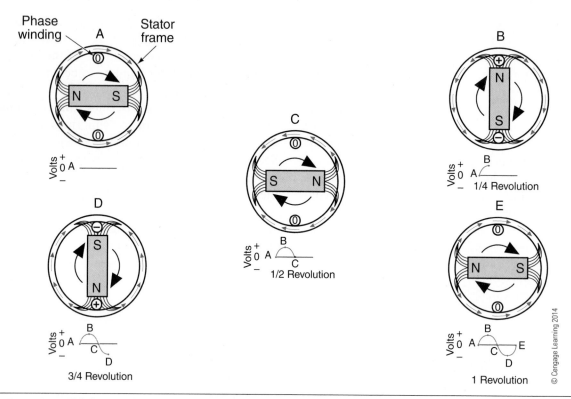

Figure 7-14 Sinusoidal AC voltage waveform produced by rotating electromagnet inside of a conductive loop.

simple alternator, without the frame. Unlike the first example of an alternator, this design is much more practical.

Using an electromagnet for the rotor permits the output voltage of the alternator to be controlled. By increasing or decreasing the current through the electromagnet, the magnetic field strength can be controlled. Remember that the amount of voltage induced in the conductor depends on the number of magnetic lines of force cut per second by the conductor.

An AC voltage is induced in the conductor as the rotor rotates inside the stator, as shown in **Figure 7-14**. This is similar to the sequence shown in **Figure 7-9** except that the magnetic field is rotating inside the loop, instead of the loop rotating inside the magnetic field. Both methods produce a voltage that follows a sine waveform.

The rotor shaft in the alternator is supported on either end by bearings. The rotor shaft is fitted with a pulley, and an engine drive belt is used to rotate the alternator pulley.

Three-Phase Voltage

The waveform of the voltage induced in the stator winding by the simplified alternator as displayed on an oscilloscope corresponds to the rotation of the rotor. One revolution of the rotor in the simplified alternator produces one complete cycle of a sine wave voltage waveform.

Because there are 360 degrees in a circle, a single cycle of the waveform can be divided into 360 degrees as well. If the starting point of the waveform is described as being 0 degrees, the waveform can be divided into four equal time segments corresponding to 90 degrees, 180 degrees, 270 degrees, and 360 degrees (same as 0 degrees), as shown in **Figure 7-15**. These correspond to rotor positions shown in **Figure 7-14**, views A–E.

The simple alternator shown in **Figure 7-13** is useful for illustrating fundamental operating principles. However, this alternator is still not very efficient.

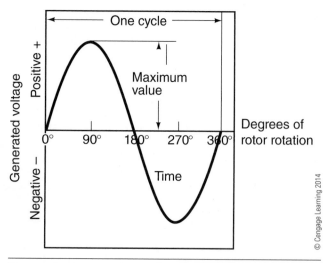

Figure 7-15 Rotation of 360 degrees produces a complete cycle of a sine waveform.

This simple alternator only has one conductor loop, which only produces maximum positive or negative voltage twice per rotation when the tip of the electromagnet passes directly in front of the conductor at 90 degrees and at 270 degrees. The term used to describe a voltage trace in relationship to time with other voltage traces is called **phase**. Because this simple alternator only produces one voltage trace, it is described as being a single-phase alternator. The wall socket found in homes is also a single-phase AC voltage.

Most of the rotor's motion is not producing very much induced voltage with the simple single-phase alternator. If two other loops were added to the stator at equally spaced intervals, three separate sine voltage waveforms could be produced as the rotor is rotated inside the stator. Each loop would be separated by 120 degrees because the 360 degrees of a complete circle equally divided by 3 is 120 degrees. These three loops identified as A, B, and C are shown installed in the stator in **Figure 7-16**.

Each revolution of the rotor now produces three different sine voltage waveforms, as shown in **Figure 7-17**.

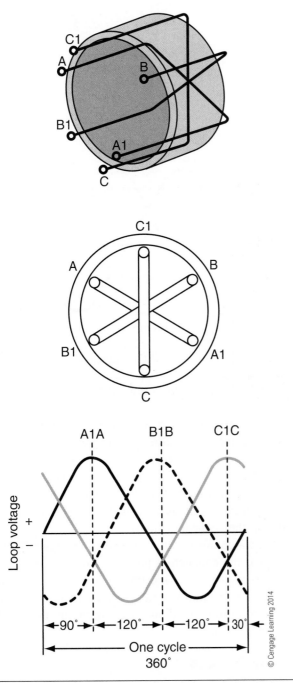

Figure 7-16 Three conductive loops connected together and spaced around the stator.

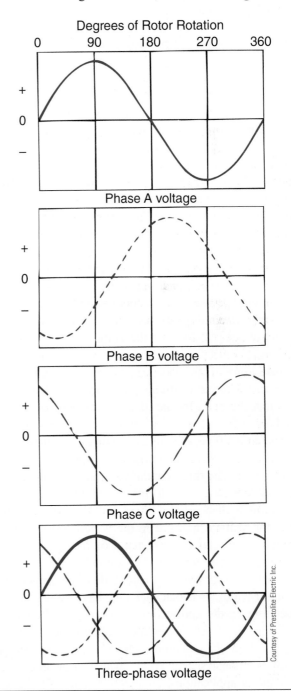

Figure 7-17 Three-phase sine waveform voltage trace.

The three waveforms spaced 120 degrees apart are known as a three-phase voltage. Three-phase voltage provided by the electric power companies is used in industrial applications for large electric motors and machinery but is typically not used in homes. Hybrid electric trucks use a three-phase motor/generator, as will be discussed in **Chapter 15**.

Up to this point, the conductor loops have been shown as a single loop of wire. In reality, coils of wire called **windings** are used to construct a phase instead of a single conductive loop in order to cause more magnetic lines of force to cut the conductor. This makes the alternator more efficient. The iron frame portion of the stator is also not just a piece of solid iron but instead is constructed of thin sections of iron laminations. The laminations are bonded together to minimize heat losses in the stator known as eddy currents. Electric motors and transformers also use this lamination technique for this same reason.

The three windings are joined together in two different types of configurations in modern truck alternators. The three windings may be connected so that a triangle shape is formed; this is known as a **delta-**connected stator and is shown in **Figure 7-18**. The dots show the connection point for the phases. *Delta* is the name of the Greek alphabet character that looks like a triangle. The term *delta* is also used to describe the triangle shape formed by a river as it nears an ocean, such as the Mississippi River delta.

The voltage waveform generated measured across winding A, winding B, and winding C is a three-phase voltage. Notice that the voltage induced in the windings of the phases are not referenced to ground because none of the windings are directly connected to ground. The induced voltage is measured from one corner of the delta to one of the other two corners of the delta.

The other type of winding connection method is known as a **wye** or Y-connected stator as shown in **Figure 7-19**. The three windings all have one common point of connection that forms a Y shape. The voltage waveform is also a three-phase voltage. As with the delta, the voltage induced in the windings is not referenced to ground because none of the windings or phases is directly connected to ground.

The reason for two different methods of connecting the stator windings is that each type has its advantages, depending on the application or use of the alternator. For example, a Y-connected stator can produce a higher output voltage than a comparable delta-connected stator at low rotational speeds (rpm), which is beneficial in some diesel applications. High-current rated alternators are typically delta-connected stators.

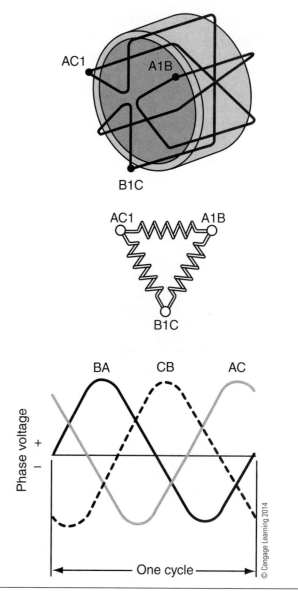

Figure 7-18 Delta-connected stator.

A Y-connected stator is shown in **Figure 7-20A**. Note that the conductors are formed into windings around the entire circumference of the iron frame. The three phases as well as the Y center junction connection—referred to as a neutral junction—can be seen in this illustration, as well as the stack of laminations that make up the stator iron frame.

A delta-connected stator is shown in **Figure 7-20B**. Note that the connection points for the corners of the delta are shown in this illustration. The delta-connected stator has no neutral junction because there is no place where all three windings are common, unlike the Y-connected stator.

Rotor Details

To maximize the voltage-producing ability of the alternator, the rotor electromagnet is not a single set of

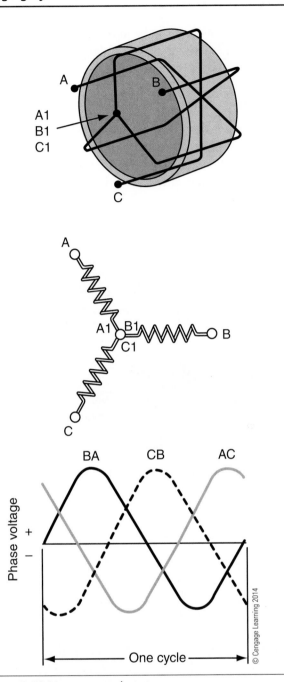

Figure 7-19 Y-connected stator.

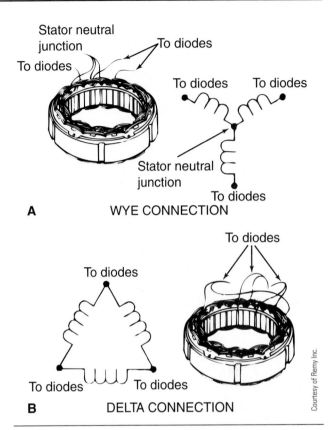

Figure 7-20 Stator winding arrangements.

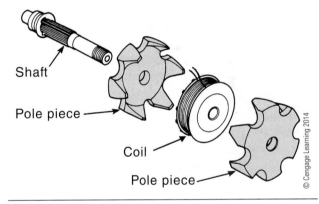

Figure 7-21 Rotor components consist of coil and pole pieces installed on rotor shaft.

north-south poles. Instead, the rotor may have six or more sets of magnetic poles (12 poles total). The rotor is constructed of two iron pole pieces that are pressed onto a shaft, like the six-pole set rotor shown in **Figure 7-21**. Between the two pole pieces is the rotor field coil. The interlocking rotor pole pieces do not make contact with each other. When current flows through the coil, one of the pole pieces becomes the north pole of the electromagnet; the other pole piece becomes the south pole of the electromagnet, as shown in **Figure 7-22**. Note in this illustration the slip rings and brushes that provide the electrical connection with the rotor field coil.

The magnetic field surrounding a simple rotor with eight (total) poles is shown in **Figure 7-23**. Notice that the magnetic lines of force are shown leaving each north pole tip and traveling to the two adjacent south pole tips. The magnetic lines of force are shown pointing in opposite directions between adjacent poles. As each of eight concentrated areas of magnetic lines of force cut through a stationary conductor winding in the stator, a positive half of a sine waveform will be induced in the winding, followed by a negative half of a sine waveform as the next concentrated area of magnetic lines of force cuts through the winding.

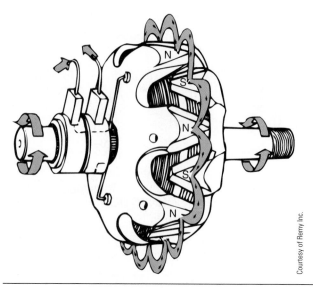

Figure 7-22 Alternating north and south poles of rotor pieces produce a rotating magnetic field.

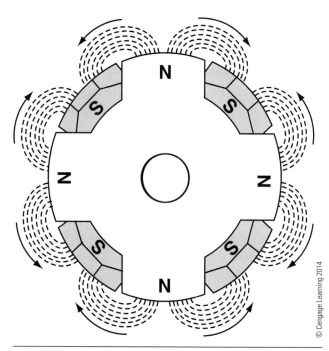

Figure 7-23 Magnetic fields between adjacent rotor pole pieces surround the rotor.

The voltage induced in the stator windings reverses polarity due to the difference in direction of the magnetic lines of force between each of the eight poles that make up the rotor. A complete sine waveform (positive half and negative half) is induced in the conductor winding four times per rotor revolution for the rotor shown in **Figure 7-23**.

Each revolution of the rotor shown in **Figure 7-23** causes portions of four sine waveforms to be induced in each of the three phases. In the same way, a rotor consisting of seven north and seven south poles causes seven sine waveforms to be induced in each of the three phases per rotor revolution. Many truck alternators contain rotors that have seven sets of north-south poles.

Stator Details

A typical delta-connected stator containing windings for all three phases is shown in **Figure 7-24**. The three terminals shown are the AC voltage output of the stator. The stator shown in **Figure 7-25** is missing

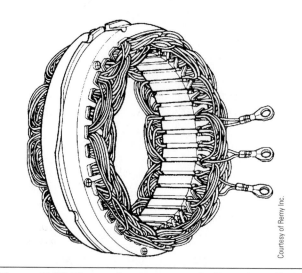

Figure 7-24 Three-phase stator windings installed in laminated iron frame.

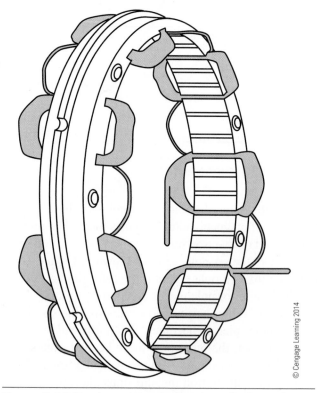

Figure 7-25 Single stator winding installed in a frame with seven coils to illustrate spacing.

the other two phases to illustrate how the windings are distributed around the stator frame. Note that there are seven sets of loops that are formed out of one long piece of wire. All seven loops are wired in series and are equally spaced around the stator frame. The number of sets of loops in each phase is the same as the number of rotor pole sets, so this stator is designed to be used with a seven-pole set (14-pole) rotor. Therefore, each of the seven sets of loops is aligned with one of the rotor's concentrated magnetic field areas simultaneously. The voltage produced in each of the sets of loops is series aiding. This means that the voltage induced in each of the seven sets of loops is added together to form the total voltage induced in each phase of the stator.

The air gap between the rotor poles and the stator is designed to be as small as possible to maximize efficiency. Remember that the strength of a magnetic field decreases at a nonlinear rate as the distance increases. The rotor is supported on both ends of its shaft by bearings. The bearings are pressed into the front housing and rear housing or end covers of the alternator, as shown in **Figure 7-26**.

Converting AC to DC

The alternating current produced by the alternator cannot be used by the truck's electrical system. The alternating current must be converted to a direct current. This is performed using diodes. Diodes are often called rectifiers for their ability to rectify or straighten out AC. Standard rectifier diodes only permit current to flow in one direction. When the diode is permitting current to flow, it is said to be forward biased. When the diode is blocking current flow, it is said to be reverse biased as explained in **Chapter 6**.

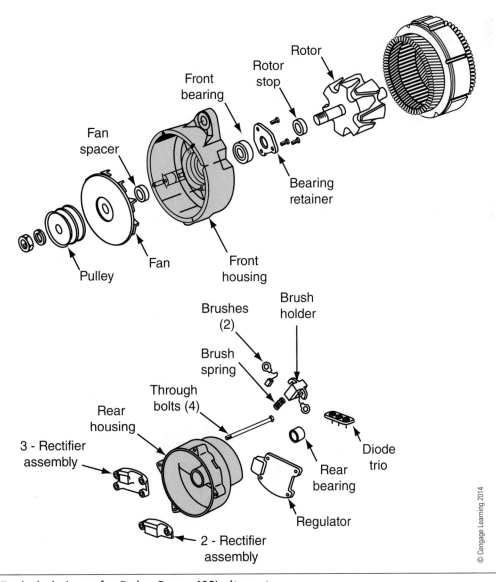

Figure 7-26 Exploded view of a Delco Remy 40SI alternator.

In **Figure 7-27A**, a single-phase AC voltage is induced in the coil shown on the left side of the figure. The induced voltage in the top figure has a polarity such that the top of the coil is positive and the bottom of the coil is negative. The output of the coil is connected to a configuration of four diodes known as a bridge **rectifier** or rectifier bridge. When the top of the coil is positive, the colored diodes are forward biased while the white diodes are reverse biased. Conventional current flows from the top of the coil through diode 2, through the resistor, and through diode 3 to the bottom of the coil. No current flows through diodes 1 and 4 because they are reverse biased.

When the voltage induced in the coil reverses polarity (**Figure 7-27B**), the bottom of the coil is positive while the top of the coil becomes negative. Conventional current flows from the bottom of the coil through diode 4, through the load resistor in the same direction as above, and through diode 1 to the top of the coil. The voltage waveform that appears across the load resistor is not really DC, but it now only has positive polarity, as shown in **Figure 7-28B**. The top of the resistor in **Figure 7-27** is positive for both waveforms, while the bottom of the resistor is negative for both waveforms. This type of waveform in which the negative cycle is flipped over to the positive side is known as a full-wave rectified waveform.

The Y- or delta-stator windings produce a three-phase AC voltage, as shown in **Figure 7-29**. Therefore, six diodes will be necessary to rectify the negative portions of the three AC waveforms produced by the stator into full-wave positive waveforms, as shown in **Figure 7-30**. Notice that three of the diodes have been connected to chassis ground and three are connected to the battery positive output terminal of the alternator. The six diodes rectify the three-phase alternator output into a full-wave positive waveform like that shown in **Figure 7-31**.

The rotor may have seven sets of north and south magnetic poles. Each revolution of a 7-pole set rotor causes seven sine waveforms to be produced in each of the three phases of the stator windings. This indicates that portions of 21 sine waveforms are produced with each complete rotor revolution with a 7-pole set rotor. The full-wave rectified waveform measured at the output of the rectifier assembly produced by this large of a number of waveforms causes a slightly pulsating DC voltage to be produced at the alternator output (**Figure 7-32**). The small humps in the waveform that remain are referred to as **ripple**. The amplitude variation of the ripple voltage shown in **Figure 7-32** is exaggerated to illustrate the concept. The ripple is further reduced by the use of a capacitor connected between the alternator output terminal and ground. A capacitor stores energy, as explained in **Chapter 3**. As the voltage starts to drop slightly between each

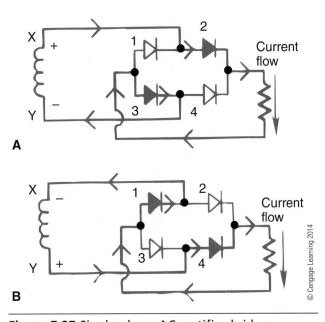

Figure 7-27 Single-phase AC rectifier bridge.

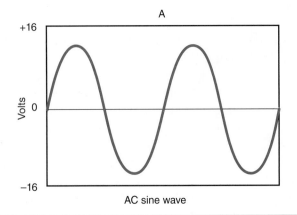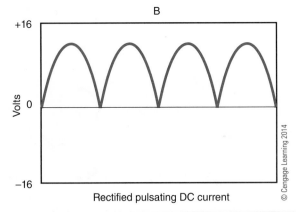

Figure 7-28 Full-wave rectification reverses the polarity of the negative portion of a sine waveform.

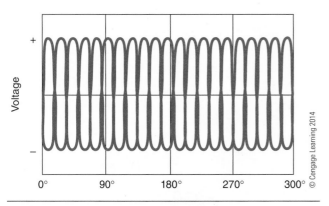

Figure 7-29 Three-phase stator voltage output waveform prior to rectification.

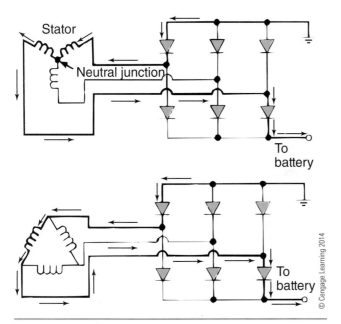

Figure 7-30 Three-phase rectifier requires six diodes.

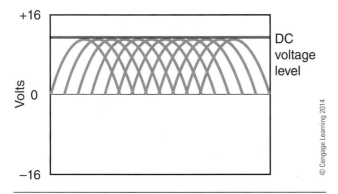

Figure 7-31 Rectified three-phase voltage.

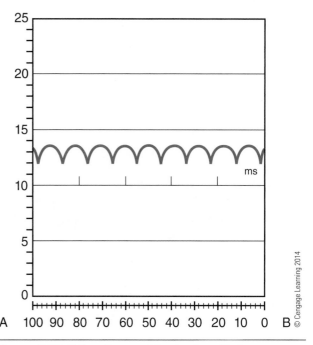

Figure 7-32 DC voltage with ripple: ripple exaggerated for illustration purposes.

The diodes used for rectification of an AC voltage are typically contained within the alternator housing (internally rectified). However, some alternators may use a separate rectifier assembly that is mounted near the alternator (externally rectified). The diodes used for rectification must conduct high levels of current so they must dissipate a large amount of heat. All six diodes are usually formed into a complete assembly that is serviced as one component (**Figure 7-33**). This assembly has an integral heat sink and three studs to provide connection with the three phases of the Y- or delta-stator windings. Three of the diodes, called the negative diodes, are oriented so that they provide a path from chassis ground to the stator windings. Three other diodes provide a path from the stator windings to the alternator positive output terminal and are called the positive diodes. The positive diodes use the insulated heat sink to dissipate heat, and the negative diodes use the grounded heat sink to dissipate heat (**Figure 7-34**). The insulated heat sink is insulated from the grounded alternator housing with plastic or fiber spacers and washers. The three AC output terminals of the stator shown back in **Figure 7-24** are attached to the three terminals of the rectifier shown in **Figure 7-34**.

One common cause of rectifier diode damage is reverse battery polarity. If the batteries are hooked up backward by mistake for even a brief instant, the rectifier diodes in the alternator will all be forward biased with almost no resistance in the charging

pulsation of the rectified waveform, the capacitor discharges to maintain a steady output voltage. The capacitor immediately recharges at the next positive peak pulsation. The large number of waveforms being rectified along with the capacitor causes the alternator output voltage to be a direct current.

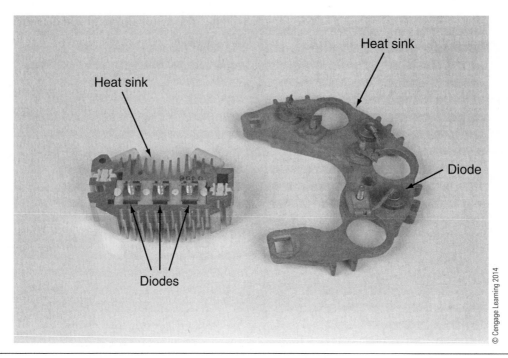

Figure 7-33 Typical rectifier assemblies installed in heat sink.

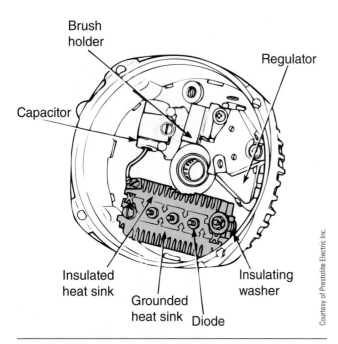

Figure 7-34 Rectifier heat sinks installed in alternator; stator terminals are attached to rectifier studs.

circuit path to limit the current. The huge arc at the battery terminal that occurs when the final connection is made if batteries are connected with reverse polarity makes it obvious that something is wrong. Reverse polarity can also occur from incorrect connections when jump-starting. Reverse battery polarity can cause significant electrical system damage by destroying diodes in electronic modules, suppression

diodes across inductors, and other devices. Excessive current may cause diodes to become an open circuit or they may be damaged such that they conduct current in both directions. This is referred to as a shorted diode because it is acting as a piece of wire. **Figure 7-35** illustrates typical alternator output voltage waveforms observed with an oscilloscope when one or more rectifier diodes are open or shorted. Such damage will result in poor charging system performance, as will be discussed later in this chapter.

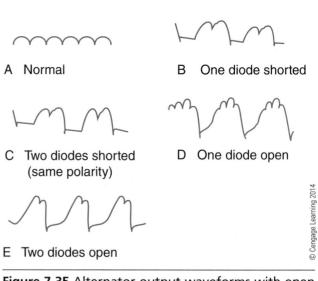

Figure 7-35 Alternator output waveforms with open or shorted rectifier diodes.

Tech Tip: Take the time to make sure that you are making battery connections correctly or you may have to explain to a customer why they should have pay to replace expensive electrical system components that were fine before you replaced the batteries or jump-started the truck. Good luck with that story.

Voltage Regulation Fundamentals

Too high of a charging voltage is detrimental to battery life because electrolyte is lost from gassing. Too low of a charging voltage will cause a battery to not recharge fully, which will also decrease battery life. Electrical system components such as light bulbs and electronic components are also sensitive to too high or too low of a system voltage. Alternator output voltage must be held within a narrow region. Most modern truck alternators are designed to maintain an output voltage of about 14.2V with a 12V electrical system.

The output voltage of the alternator can be controlled in two ways: by regulating the alternator speed or by regulating the strength of the rotor magnetic field. Regulating alternator speed is not practical because the engine is used to rotate the alternator rotor and the engine speed changes continually as the truck is driven. Instead, alternator voltage output is regulated by continually increasing or decreasing the rotor's magnetic field based on the alternator output voltage. A **voltage regulator** is used to maintain a specific alternator output voltage by controlling the amount of current flowing through the rotor field coil. Controlling this current flow controls the strength of the rotor's magnetic field. If the alternator output voltage is too low, the voltage regulator increases rotor current to cause the rotor's magnetic field to increase. If the alternator output voltage is too high, the voltage regulator decreases the rotor current to cause the rotor's magnetic field to decrease.

Prior to the development of automotive electronics (1960s), voltage regulators were electromechanical devices containing vibrating contact points that were mounted outside of the alternator housing in a separate enclosure. These devices controlled the current flowing through the rotor field coil by using contact points to continually make and break the flow of current through the field. Electromechanical voltage regulators were made obsolete by electronic voltage regulators many years ago. However, an understanding of a simple mechanical voltage regulator is useful for understanding the concept of an electronic voltage regulator.

The rotor field coil shown in **Figure 7-36** is supplied with a ground on one side through one of the slip rings and a brush. The other side of the rotor field coil is connected to an open switch through the other rotor slip ring and a brush. The switch is connected to battery positive, so closing the switch causes current to flow through the rotor field coil and create a magnetic field surrounding the coil. HD alternator rotor field coils on higher-amperage alternators typically have a wire resistance of 1.5Ω. Therefore, in a 12V system Ohm's law indicates that approximately 8A of current will flow through the 1.5Ω rotor field coil with the switch closed. With the switch open, zero current will flow through the coil.

This simple voltage regulator consisting of a switch in series with the rotor field coil shown in **Figure 7-36** will require someone to watch the output voltage of the alternator with a voltmeter while the switch is opened and closed to regulate the alternator output voltage. When the alternator output voltage is too high, the switch must be opened to interrupt current flow through the field. When the alternator output voltage drops to a value that is too low, the switch must be closed to cause current to flow through the field. This crude method of controlling the output voltage will cause the alternator output voltage to pulsate between approximately 12V with the switch open and perhaps 16V or more with the switch closed. If the person controlling the switch did not pay attention and left the switch closed, the output voltage of the alternator could go to a very high voltage (50V or more). This occurs because the rotor field coil is supplied by the

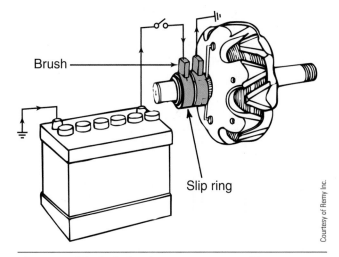

Figure 7-36 Switch shown as controlling current through rotor field coil through slip rings and brushes.

Brush

Slip ring

Courtesy of Remy Inc.

system voltage. An increase in system voltage causes the current through the rotor to increase. An increase in rotor current causes an increase in the rotor's magnetic field strength (more magnetic lines of force in a given area), which causes the alternator to produce a higher voltage and so forth.

If the person controlling the switch fell asleep and left the switch open too long, then the battery alone would be supplying the energy to operate the vehicle's electrical system and would not be recharged by the alternator.

A common method of controlling instrument panel illumination lamp dimming on modern cars and trucks is to use an electronic module to cycle the lamps off and on at a rapid rate, such as 100 times per second. This is so fast that the human eye cannot detect that the instrument panel lamps are pulsating. When the panel illumination is turned down to a low illumination level, the lights may be switched ON for one-third of the time and switched OFF two-thirds of the time, as illustrated in **Figure 7-37A**. A DC voltmeter connected across the lamp would indicate the average voltage.

Therefore, the meter would indicate that the average voltage supplied to the lamps is 4V, which is one-third of 12V.

Increasing the brightness of the panel illumination lighting by rotating the light switch knob causes the electronic control circuit to increase the amount of time that the lights are illuminated to perhaps two-thirds of the time, as shown in **Figure 7-37B**. This causes the average voltage to increase to 8V and the level of illumination to increase accordingly.

This type of control in which the voltage supply is cycled off and on at a rapid rate is called **pulse width modulation (PWM)**. *Pulse width* refers to the width of the ON portion of the pulse compared to the OFF portion of the pulse as displayed on an oscilloscope, as illustrated in **Figure 7-38**. The *width* refers to the time that the pulse is in the ON state because width on an oscilloscope is the horizontal or time axis. The amplitude of the ON pulse is the full supply voltage (i.e., 12V) and the OFF portion is 0V.

The ON time of a PWM signal is referred to as the **duty cycle**. Duty cycle is expressed in a percentage (%)

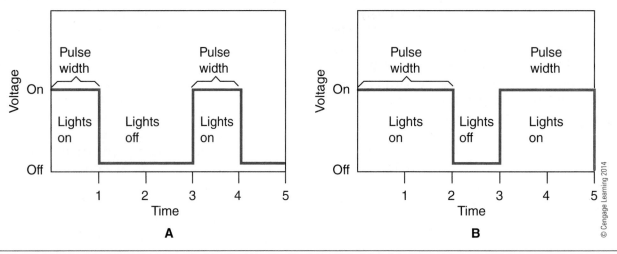

Figure 7-37 (A) Lights on for one time period and off for two time periods resulting in dim lamp output. (B) Lights on for two time periods and off for one time period resulting in brighter lamp output.

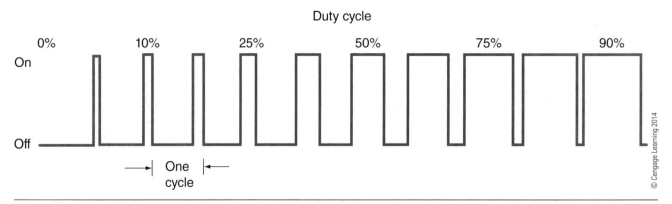

Figure 7-38 Pulse width modulation (PWM). Duty cycle is the percentage of on-time per cycle.

of a complete cycle. A 100 percent duty cycle is a continuous on signal while a 10 percent duty cycle is on 10 percent of the time and off 90 percent of the time.

PWM is utilized for an ever-increasing number of modern truck features such as daytime running lights (reduced headlamp illumination) and blower motor speed control. PWM lends itself well to electronic control. A transistor can be switched off and on at a high rate electronically to produce PWM. This eliminates the need for high-power resistive devices like rheostats, which would be very difficult to control with an electronic circuit.

Modern truck voltage regulators use PWM to control the rotor field current. A transistor is switched off and on at a rapid rate to control field current in response to alternator output voltage. The transistor is used as a switch or relay, not as an amplifier, meaning it is all the way on or all the way off. Most modern truck voltage regulators are contained within the alternator and are referred to as internal or integral regulators.

The voltage supply for the rotor field coil or **excitation** voltage is typically obtained directly from the stator output by way of a three-diode assembly commonly known as a **diode trio** or a **field diode** assembly shown in **Figure 7-39**. The diode trio consists of three rectifier diodes within a single package mounted inside the alternator. The anode (positive end) of each of the three diodes is connected to each of the three phases of the stator windings. The cathodes (negative ends) of the diodes in the diode trio are all connected together. The output of the diode trio is a three-phase rectified voltage.

Because the alternator is producing its own field current, such alternators are referred to as self-exciting. However, the stator does not produce any voltage unless the rotor is magnetized and is rotating inside the stator. Because the diode trio supplies the excitation current for the rotor, the rotor is not magnetized unless the stator is producing voltage. This seems to be a "which came first, the chicken or the egg?" type of scenario. If the rotor is not magnetized and the

alternator pulley is rotated, the stator will not generate any voltage. If the stator does not generate any voltage, then there will be zero voltage available to cause current to flow through the rotor field coil to generate the magnetic field necessary to produce voltage in the stator. Therefore, the **residual magnetism** of the rotor pole pieces is responsible for generating enough voltage in the stator windings after starting the engine to produce the initial voltage necessary to power the field coil until voltage is produced in the stator. Residual magnetism is like magnetic memory of the steel used to make the rotor pole pieces. If you have ever rubbed a magnet on a piece of steel, you know that the piece of steel becomes magnetized for a period of time. This is residual magnetism. The residual magnetism of the rotor can be lost under some circumstances, which causes the alternator not to produce any voltage. This loss of rotor residual magnetism will be addressed later in this chapter.

Electronic Voltage Regulators

The electronic voltage regulator controls the current in the rotor field through PWM to maintain an alternator output voltage of approximately 14V. Each time a device with a high current load—such as the headlamps—is switched on, the output voltage at the alternator drops momentarily as a portion of the alternator output current is diverted to the load. The voltage regulator circuit responds to the drop in the alternator output voltage by increasing the duty cycle of the rotor field current, thus regulating the alternator output voltage. The details of the operation of the electronic components in the regulator, including a zener diode and transistors, are covered in the Extra for Experts section at the end of this chapter.

Electronic voltage regulators are typically contained within the alternator housing and can be replaced as a unit, if defective. Some alternators have provisions for bypassing the voltage regulator to determine if the cause of low alternator output is due to the regulator or some other problem. This test is said to **full-field** the alternator and is explained in more detail

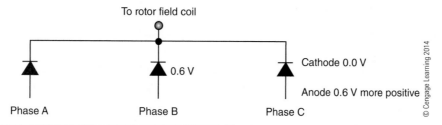

Figure 7-39 Diode trio or field diodes.

later in this chapter. Full-fielding is not recommended on many modern electrical systems with the alternator installed on the vehicle due to the resulting high voltage or the manner in which the charging system is monitored by the voltage regulator.

> **CAUTION** *Always consult the OEMs recommended practices for specific instructions on the proper diagnostic procedures.*

Truck voltage regulators may have a means to adjust alternator output voltage for charging system optimization. Medium-duty trucks utilized for local deliveries may be started several times per day. The engine may not be operated long enough to keep the batteries recharged between deliveries. Increasing alternator output voltage slightly may enable the batteries to last longer in these frequently started vehicles. At the other extreme, an over-the-road truck may only be started once per day and only operated in warm climates. The alternator output voltage can be decreased slightly on these trucks to prevent overcharging the batteries and to increase battery life. Details on adjustable regulators are also addressed in the Extra for Experts section.

Alternator Ratings

Most trucks in North America have 12V electrical systems and use a 12V alternator. However, some trucks may have a 24V electrical system and require a 24V alternator. The operation of the 24V alternator is the same as the 12V alternator, except that internal alternator components are designed for the higher voltage. The voltage regulator in the 24V alternator is designed to regulate output voltage to about 28V.

Besides the voltage rating, one of the most important ratings of an alternator is the current rating. The alternator current rating specifies the approximate amount of current the alternator is capable of producing. The maximum current supplied by the alternator varies with alternator speed (**Figure 7-40**). The pulley diameter on the alternator and the driving pulley on the engine affect the alternator speed. The pulley diameter is sized to match the charging system performance requirements of the truck. Some alternators are utilized in applications that require a high current output at idle, such as school buses, and may use an alternator designed to provide a high level of current at lower engine speed. Any rotational speed information supplied by the OEM regarding alternators refers to the alternator speed, not the engine speed. Typical alternator pulley-to-crankshaft design ratios cause the

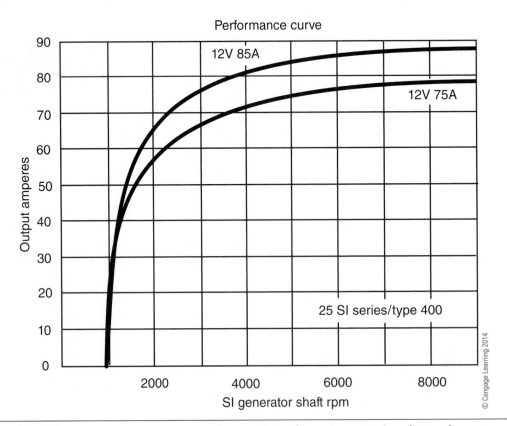

Figure 7-40 Typical alternator output current referenced to alternator rotational speed—not engine speed.

alternator to rotate about three times faster than the engine. Decreasing the alternator pulley diameter can cause an alternator to produce more current at lower engine speeds by increasing alternator speed. However, decreasing alternator pulley diameter may also cause the alternator to rotate at too high of a speed at maximum engine speed, possibly damaging the alternator. Decreasing the alternator pulley diameter also reduces the amount of drive belt to pulley contact area, which may cause the belt to slip under load.

ALTERNATOR TERMINALS AND CIRCUITS

Alternators may have several different terminals while some alternators are referred to as single-wire alternators and only have a positive output terminal and ground.

Positive Output Terminal

On most alternators, the alternator positive output terminal is a 5/16 in. (8 mm) or larger stud-type connection. A plastic sleeve keeps the terminal from contacting the grounded alternator case. This output terminal may be indicated with BAT or B+ markings. Many truck alternators are single-wire systems. These alternators only require a positive output terminal (and a ground terminal).

The alternator B+ output cable is sized to match the alternator current rating and typically uses a fusible link or large fuse as a circuit protection device. The alternator B+ output cable is often connected to the cranking motor B+ stud, which acts as a junction point. The cables between the cranking motor B+ stud and the battery are usually the largest-diameter cables on the vehicle and are very low resistance. Therefore, connecting the alternator output cable to the cranking motor B+ stud is about the same electrically as connecting the alternator output directly to the battery-positive terminal.

Ground Terminal

Most truck alternators have a chassis ground stud and use a cable to provide a good connection to the starter motor negative terminal or engine block. Although the case of the alternator is typically grounded by its mounting, just depending on the alternator mounting points to provide a good ground is not desirable due to the high current involved. The alternator ground terminal may be identified by a ground symbol or B– marking. The alternator ground cable is typically connected to the cranking motor ground stud or frame rail, which both have a very low resistance path back to the battery negative terminal.

Relay Terminal

Other alternator outputs may include a relay or AC terminal. The relay terminal is connected to one of the stator outputs. The voltage present at the relay terminal, measured with respect to ground with the engine running, is a pulsating positive voltage. The relay terminal is sometimes used to energize a special relay coil to power some accessory device when the engine is running or is used as a tachometer input signal. The relay terminal may be indicated by an R or an AC marking.

Indicator Lamp Terminal

An indicator lamp terminal is available on some truck alternators. A charge indicator lamp in the instrument cluster is supplied with ignition voltage and is wired to this indicator lamp terminal (**Figure 7-41**). A resistor inside the alternator identified as R5 is connected between the alternator indicator lamp terminal and the grounded alternator housing. The indicator lamp terminal is connected to the output of the diode trio.

If the alternator is functional, the voltage on both sides of the charge indicator lamp filament will be about 14V so zero current will flow through the charge indicator lamp. If the alternator is not functional, such as in the case of a broken drive belt, there will be zero voltage at the output of the diode trio. With zero voltage present at the output of the diode trio, the alternator indicator lamp terminal will be grounded through resistor R5. This will cause the indicator lamp to illuminate to warn the truck operator of a charging system problem. This indicator lamp will also illuminate as a lamp check if the key switch is in the ignition position and the engine is not running. This indicator lamp terminal is typically identified as the "I", "L", or 1 terminal.

Remote Voltage Sense Terminal

The output voltage measured between an alternator's B+ terminal and ground may be 14V. The internal voltage regulator will attempt to maintain this 14V. However, with an alternator output voltage of 14V, 500mV of charging circuit voltage drops would result in only 13.5V measured across the battery terminals. This may not be sufficient voltage to maintain an adequate battery state of charge.

Many alternators now utilize a remote voltage sense terminal to compensate for changing circuit

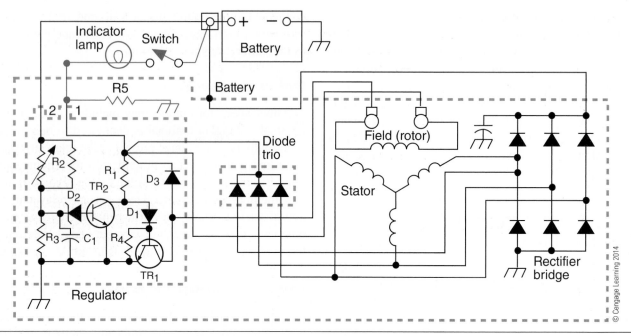

Figure 7-41 Indicator lamp terminal circuit.

voltage drops. The remote voltage sense terminal provides an indication of the voltage at the batteries or other location in the charging circuit selected by the OEM. The internal voltage regulator uses the voltage present at the remote voltage sense terminal as a reference voltage, instead of the alternator output voltage. Therefore, the alternator output voltage may be 14.5V measured at the alternator's B+ terminal while 14V is present across the battery terminals.

On some 2010 and later Freightliner models, the alternator's remote voltage sense terminal is connected to the power net distribution box (PNDB) located on the outside dash panel. The remote sense circuit is protected by a fuse in the PNDB.

Many replacement alternators have remote voltage sense capability. If this option is desired for a vehicle that was not originally equipped with remote voltage sense, it is important that the circuit be protected by a fuse. A 5A fuse is typically sufficient for the 16 AWG remote sensing circuit because only a small amount of current is flowing through this circuit. The fuse must be placed at the connection point in the circuit where the voltage is to be measured, such as the battery positive terminal.

On Leece-Neville alternators, the remote sense terminal is identified as the "S" terminal. Delco Remy alternators may indicate "Remote Sense" or the remote sense terminal is designated as the 2 terminal, as indicated in **Figure 7-41**. See the Extra for Experts section for more information on the operation of the voltage regulator as related to remote voltage sense.

Voltage Regulator Terminals

Alternators that use external voltage regulators may have a terminal identified as "FIELD". This terminal is connected to an external voltage regulator per the OEM's instructions.

Some alternators that are otherwise self-excited may have an "I", "IG" or "IGN" internal voltage regulator turn-on terminal. This terminal is typically provided with ignition voltage. In some cases, this terminal acts both as an indicator lamp terminal and an internal voltage regulator turn-on terminal.

BRUSHLESS ALTERNATORS

Brushes make contact with the rotor slip rings. In time, the brushes wear out, as do the slip rings. With the exception of a loose drive belt, most alternator problems are caused by worn brushes. Many over-the-road trucks travel more than 150,000 miles (241,402 km) per year and would require frequent brush replacement. Brushless alternators have been developed that, as the name implies, do not have brushes or slip rings. These brushless alternators still use a rotor to generate a rotating magnetic field like brush-type alternators. However, the field coil is stationary and designed so that it does not rotate with the rotor. Instead, the stationary field coil sits inside a hollowed-out rotor assembly, as shown in **Figure 7-42**. The field coil is directly connected to the voltage regulator.

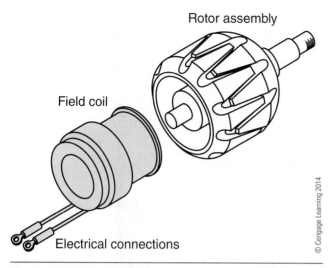

Figure 7-42 Brushless alternator rotor and field coil.

The rotor in a brushless alternator is hollowed out and fits over the stationary field coil. The north and south pole pieces in the rotor are constructed of low-reluctance steel like a brush-type rotor. However, a high-reluctance material such as aluminum may be cast between the pole pieces to hold the rotor together, yet still keep the north and south pole pieces of the rotor magnetically isolated. Current flow through the stationary field coil causes a magnetic field to surround the field coil. The magnetic lines of force produced by the field coil are directed through the low-reluctance rotor to the stator windings. This causes a voltage to be induced in the stator windings in the same manner as a brush-type rotor.

A cutaway view of a Delco Remy® heavy-duty brushless alternator is shown in **Figure 7-43**.

Brushless-type alternators are physically larger than comparably sized brush-type alternators and have a higher initial cost. However, this cost may be offset by the reduced maintenance and downtime associated with a brushless alternator on a high-mileage truck.

DUAL-VOLTAGE ELECTRICAL SYSTEMS

Some trucks use 24V cranking motors while the remainder of the truck's electrical system is a standard 12V system. In the past, a complicated series-parallel switch system was used to provide the 24V necessary for the cranking motor. This switch placed one (or more) battery in series with another battery (or more) to crank the engine. The batteries were placed back in parallel with each other after cranking so that all batteries could recharge. This switch system is no longer utilized in modern trucks. An example of this series-parallel system is shown in **Figure 7-44**.

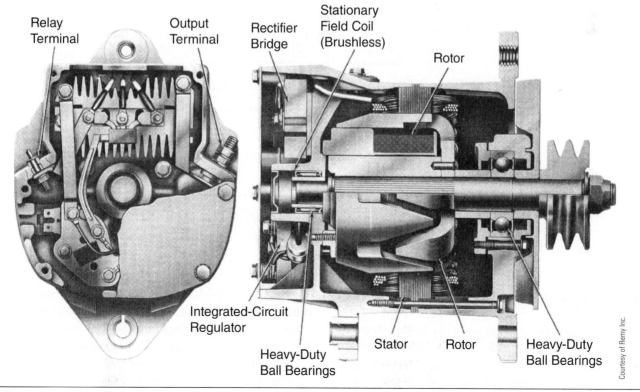

Figure 7-43 Cutaway view of a Delco Remy heavy-duty brushless alternator.

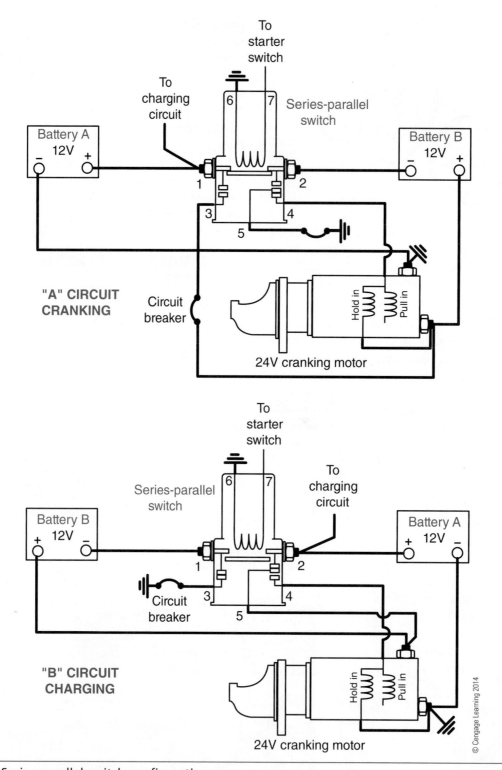

Figure 7-44 Series-parallel switch configurations.

Transformer Rectifier Alternators

The complicated series-parallel switch system can be replaced by using a dual-voltage alternator. The main output terminal of the alternator supplies 14V, like a typical alternator. Another terminal provides an output voltage that is double the main terminal voltage.

This 28V charging voltage is obtained through a transformer that is contained within the alternator. A transformer is used to step up or step down one level of AC voltage into another AC voltage, as discussed in **Chapter 3**.

The dual-voltage alternator consists of a standard three-phase 12V alternator with an attached three-phase transformer. The three-phase transformer is connected in parallel with the three phases of the stator windings, as shown in **Figure 7-45**. This causes the same voltage that is induced in each phase of the stator windings to be applied to the primary windings of the transformer. The transformer is designed as a 1:2 step-up transformer. This causes the voltage across each phase of the secondary windings to be double that of the stator windings. A six-diode rectifier bridge like that in the 12V alternator is used to rectify the transformer output into a DC voltage that is double that of the 12V alternator output.

Figure 7-45 illustrates two batteries connected in a series-aiding configuration identified as C battery and S battery. The voltage across either battery's terminals

will be near 14V with the engine running. However, the voltage measured from the positive terminal of the C battery to chassis ground will be approximately 28V. The only load device that is connected to this higher voltage is the cranking motor. The rest of the truck electrical system is supplied by the S battery and the 12V alternator output, just like a standard 12V system vehicle.

Modern Dual-Voltage Systems

Series-parallel switches and dual-voltage alternators have been mostly replaced on modern trucks by electronic battery equalizers and DC-to-DC converters. A battery equalizer permits a single-output 24V alternator to recharge two series-connected 12V batteries.

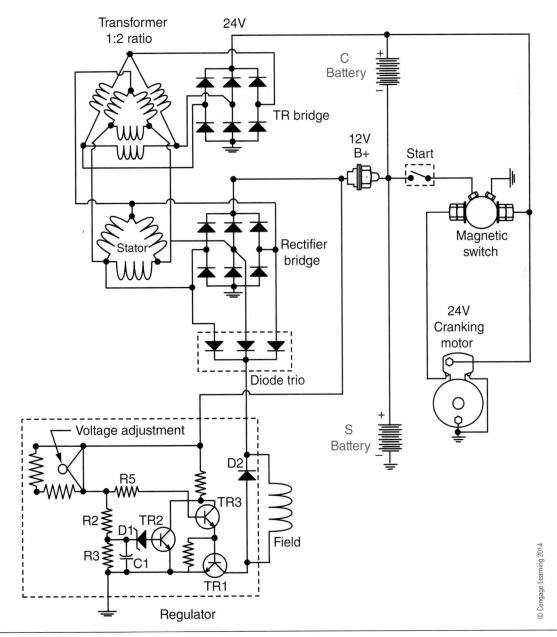

Figure 7-45 Transformer rectifier dual-voltage alternator schematic.

The input terminal of the battery equalizer is supplied with 24V from the alternator output terminal. The output terminal of the battery equalizer has a voltage that is equal to one-half of its input terminal voltage. The output of the battery equalizer is connected to the positive terminal of the vehicle's 12V electrical system battery, similar to the 12V output terminal of the TR alternator shown in **Figure 7-45** being connected to the "S" battery. The purpose of the battery equalizer is to provide exactly one-half of the 24V alternator's output voltage to both of the series-connected 12V batteries to adequately recharge the batteries.

A DC-to-DC converter is used to increase 12V up to 24V on trucks with a 12V charging system. The input terminal of the DC-to-DC converter is supplied with 12V from the alternator output terminal. The output terminal of the DC-to-DC converter has a voltage that is double its input terminal voltage. The 24V output of the DC-to-DC converter can be used to supply 24V devices and accessories. Most trucks and trailers outside of North America have 24V electrical systems. DC-to-DC converters are commonly used to provide 24V trailer lighting capability on trucks with 12V electrical systems. See the Internet links at the end of this chapter for more information on battery equalizers and DC-to-DC converters.

CHARGING SYSTEM PROBLEMS

Charging system problems may result in either an overcharged or an undercharged battery.

An overcharged battery is caused by a charging system that is maintaining the charging voltage at too high of a level. Common complaints from too high of a charging voltage include:

- Voltmeter reads too high
- Sulfur smell
- Wet batteries and high water usage (conventional batteries)
- Lamps that are too bright and burn out frequently and turn signals that flash too rapidly
- Decreased battery life

An undercharged battery may be caused by a charging system that is maintaining the charging voltage at too low of a level, along with other possible causes. Common complaints from too low of a charging voltage include:

- Voltmeter reads too low
- Slow cranking speeds or no crank
- Lamps that are too dim and turn signals that flash too slowly
- Decreased battery life

Preliminary Checks

Prior to any charging system testing, the batteries must be individually tested and recharged, if necessary. The batteries must be at least at a 75 percent state of charge before the charging system is tested. The temporary installation of test batteries with at least a 75 percent state of charge may be necessary if batteries are discharged or otherwise fail testing due to low electrolyte or other causes. Inspect and clean all battery terminals before testing if necessary. Batteries should also be near room temperature for accurate charging system testing.

Tech Tip: Testing a charging system using batteries that are not sufficiently charged may lead you to an incorrect diagnosis. Recharging batteries can take a considerable amount of time. Having sufficiently charged batteries that you can temporarily install for testing can save you and your customer valuable time.

A good visual inspection of charging system components is the next step. Through a visual inspection, you can often quickly spot problems. It is embarrassing to spend a great deal of time troubleshooting a problem only to have a co-worker walk past and notice something obviously wrong, such as a missing alternator drive belt. A visual inspection of the charging system includes the alternator mounting hardware and drive belt. Often, charging system problems are caused by a loose alternator drive belt. Adjust the drive belt to the proper tension before testing. V-belt tension is typically adjusted by some type of adjustment bolt that causes the alternator or some other component to pivot on one end of the mounting point. Proper belt tension is important— too loose and the belt will slip, causing belt damage and low charging system performance; too tight and bearings in the alternator and other components that use the alternator drive belt can be damaged. Special belt-tension gauges should be used to adjust drive belt tension.

Many trucks now have spring-loaded automatic belt tensioners and ribbed (flat) serpentine drive belts like that shown in **Figure 7-46**. These require no adjustment, but the tensioner may lose spring tension over time and require periodic replacement. Belt alignment is important for serpentine belts; misalignment can result in component bearing failure. Consult OEM information for belt alignment specifications.

The visual inspection also includes looking for loose, corroded, or broken cable terminals and disconnected connectors throughout the charging system.

© Cengage Learning 2014

Figure 7-46 Delco Remy 28SI pad-mount alternator with serpentine drive belt and automatic tensioner.

CAUTION *Each OEM has procedures for testing the charging system that should always be followed. The procedure that follows is only an example and cannot be used for all vehicles.*

Another item to check before testing the charging system is the charging circuit resistance. The charging circuit includes the positive and ground cables between the alternator and the batteries. The cranking motor positive stud is sometimes used as the main positive junction point for the entire vehicle electrical system. The alternator positive output cable is typically connected to the cranking motor positive stud, which is in turn connected to the battery-positive terminals through the large-diameter positive battery cables. The resistance of the positive and negative (ground) circuits between the alternator and the battery must be very low because a large amount of current flows through this circuit. This resistance is only a few milliohms (thousandths of an ohm) so it cannot be accurately measured using a standard ohmmeter.

Instead of direct measurement of resistance in low-resistance circuits, a technique is used in which the voltage that is dropped on the low-resistance cables is measured with a known amount of current flowing through the cable. Recall that *voltage drop* means the difference in voltage between one point and another point, similar to the use of a differential pressure gauge to find the difference in pressure between two points of a hydraulic circuit. The greater the voltage dropped on the charging system cables with a known level of current flowing through the cables, the higher the resistance of the cables. High resistance in the charging

system circuit limits the amount of current the alternator can supply to the batteries and the rest of the vehicle electrical system.

The voltage dropped on the charging system positive and negative cables is obtained by using a carbon pile to cause the alternator-rated current to flow through the alternator positive and negative circuits with the engine not running. The carbon pile is connected between the alternator positive output terminal and the alternator ground terminal, as shown in **Figure 7-47**. A voltmeter is also connected between the alternator positive output terminal and negative output terminal. Another voltmeter is connected across the battery terminals. A clamp-on ammeter is clamped around the alternator positive output cable to monitor current in the circuit. The carbon pile is then loaded to a value equal to the alternator current rating (about 120A on most standard trucks). This causes current to flow from the batteries to the carbon pile through the alternator output cables with the engine not running. The current is flowing through the alternator output cables in the opposite direction of when the alternator is supplying the current to the electrical system, but direction of current flow is not important in this case. With rated current flowing, the difference in the voltage measured across the batteries and the voltage measured across the alternator output terminal and alternator ground terminal will provide the total amount of voltage dropped on the alternator output cables and battery cables. Subtract the battery terminal voltage from the voltage measured at the alternator to find this voltage drop value. The voltage drop on the combined positive and negative charging circuit should be less than 0.5V.

If the voltage drop on the total alternator output circuit is more than 0.5V with rated alternator current flowing through the charging circuit, then the cause of the excessive voltage drop must be found. Connect the positive lead of a voltmeter to the battery-positive terminal and the negative lead of the same voltmeter to the alternator positive output terminal, as shown in **Figure 7-48**. A voltmeter test lead extension may be necessary to reach both points. Load the carbon pile again to cause the rated alternator current to flow through the charging circuit and read the voltage dropped on the positive circuit between the alternator and the batteries. The voltage drop on the positive charging circuit should be less than 0.25V with rated current flowing.

Move the voltmeter so that the positive lead is connected to the alternator negative output terminal (ground) and the negative lead is connected to the battery-negative terminal and load the carbon pile to

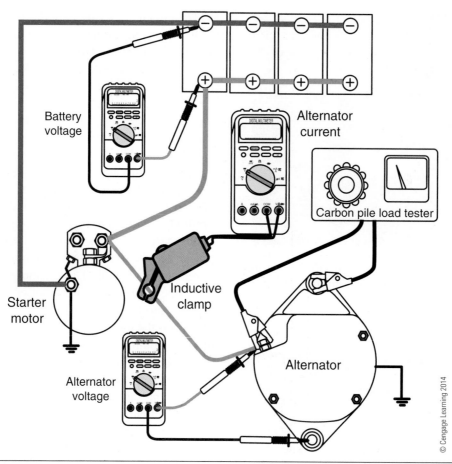

Figure 7-47 Testing alternator charging circuit voltage drop using carbon pile load tester.

the rated alternator current again. The voltmeter will display the voltage dropped on the negative alternator circuit. The voltage drop on the negative charging circuit should be less than 0.25V with rated alternator current flowing through the charging circuit.

Correct the cause of the excessive resistance by cleaning and tightening connections or replacing cables as needed, and retest for voltage drop.

Testing the Charging System

After the visual inspection, the alternator charging circuit voltage drop test, and verification of the battery state of charge, charging system testing can begin. Again, consult the specific OEM information for the recommended test procedures.

Charging system testing typically consists of measuring the alternator output voltage and current under loaded and unloaded conditions and comparing the results to the OEM's specifications. If the alternator performance does not meet the specifications, the cause of the problem is then determined.

The first test is to determine the unloaded alternator output voltage and current measurements with

adequately charged batteries in a room temperature shop environment as follows:

1. Connect a voltmeter across the alternator positive and negative terminals and install a clamp-on ammeter around the alternator positive output cable so that the current arrow on the inductive probe is pointing away from the alternator. Remember that the alternator is supplying the current for the truck's electrical system so the direction of conventional current flow is from the alternator positive output, through the vehicle electrical system including the batteries, and back to the alternator ground or negative cable. Place the probe so that it is at least 12 in. (30 cm) from the alternator if possible so that the strong magnetic field of the rotor does not influence the inductive ammeter reading.

2. Start the engine with all electrical loads off and run the engine at about 1500 rpm for 2 minutes to allow the batteries to recover from cranking.

3. Observe the voltage and current meters with all electrical accessories on the truck switched off and with the engine running. What you

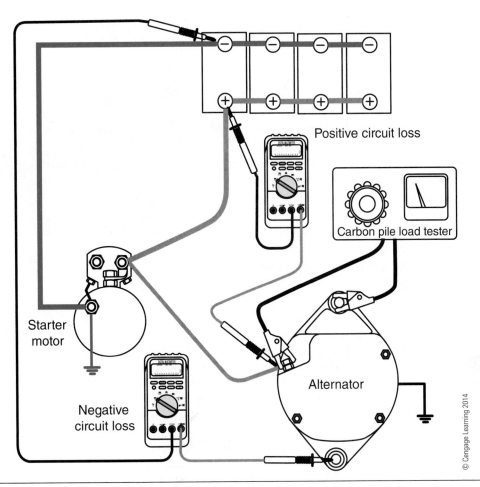

Positive circuit loss

Carbon pile load tester

Starter motor

Alternator

Negative circuit loss

© Cengage Learning 2014

Figure 7-48 Determining individual alternator cable voltage drops.

should observe on a 12V system after 2 minutes is 13.5V to 14.5V measured across the alternator output and ground terminals with an output current that decreases to a value much less than 40A.

If unloaded alternator output current is greater than 40A after 2 minutes with the engine running at 1500 rpm, make sure all electrical accessories are switched off. If all accessories are switched off, it is possible that a high-current-draw device like the glow plugs or intake air heater is being energized by a faulty control circuit. Use a clamp-on ammeter to determine where the excessive unloaded alternator output current is flowing and correct the problem before proceeding. If the clamp-on ammeter indicates that the majority of the 40A or more of un-loaded alternator output current is flowing into the bat-teries as a charging current after 2 minutes of engine running at 1500 rpm, and the cranking time necessary to start the engine was not excessively long, then the bat-teries might not be at a sufficient state of charge for testing or may be defective.

Using the results of the unloaded alternator output voltage test, go to the test indicated in **Figure 7-49**.

Alternator Output Voltage	Perform
13.5V to 14.5V	Test 1
Less than 13.5V	Test 2
Greater than 14.5V	Test 3

© Cengage Learning 2014

Figure 7-49 Alternator output voltage test determi-nation.

Test 1: Acceptable Unloaded Charging Voltage

If the charging voltage at the alternator under light electrical load is 13.5V to 14.5V, the next step is to determine the maximum alternator output current by loading the system using a carbon pile.

1. Attach a clamp-on ammeter around the alter-nator positive output cable.
2. Connect a carbon pile load tester across the battery terminals.

3. Start the engine and increase engine speed to about 1500 rpm. Switch off all accessories and lights.

4. Permit the engine to run for about 2 minutes and monitor the alternator output current with all vehicle loads off.

5. Adjust the carbon pile until the ammeter indicates the highest value. This is the maximum alternator output current. Quickly return the carbon pile to the unloaded position to prevent damage to the carbon pile.

If the maximum alternator output current is within 85 percent of its rating, and the alternator regulates its output voltage to 14V ± 0.5V under normal operating conditions, then the charging system appears to be in good working order. If the complaint is related to slow cranking or dead batteries, then there may be excessive key-off current draws referred to as **parasitic loads** that are causing the batteries to discharge when the truck is not being operated. Testing for these parasitic loads is addressed in a later section.

If the alternator voltage and current are determined to be within specification but the truck has a history of charging system problems, it is possible that the alternator is not sized correctly for the application. Some trucks are utilized as work platforms and may spend a substantial amount of time with the engine idling and a large number of electric loads energized, such as work lights. These loads may exceed the alternator current rating, especially at idle speed. An alternator with an increased current rating may be required to resolve a charging system complaint on these trucks.

If the maximum alternator output current is less than 85 percent of its rating, then the next step is to perform a full-field test of the alternator, but only if specified by the OEM. Full-fielding means to bypass the voltage regulator and cause the rotor field coil to be supplied with full charging system voltage (100 percent duty cycle). This test is used to determine if the low current output is caused by the voltage regulator or the other internal components of the alternator.

The full-fielding technique varies from model to model, and the OEM information must be consulted for the specific method. As indicated, full-fielding is not recommended by many OEMs for modern electrical systems and may not even be possible on some alternators with internal voltage regulators. On applicable alternators with internal voltage regulators, a small piece of wire or other tool that is grounded is inserted into a specific full-field access hole as shown in **Figure 7-50**. This action grounds the negative side of the rotor field winding, thus bypassing the

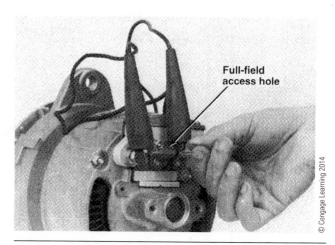

Figure 7-50 Full-fielding the rotor—only perform full-field test if specified by OEM.

voltage regulator. If full-fielding causes the alternator current output to be at least 85 percent of the rated alternator current output, then the voltage regulator system is the likely cause of the low charging current. If full-fielding the alternator does not increase output current to within 85 percent of current rating, then an internal alternator problem such as worn brushes, a damaged stator, or a faulty rectifier or diode trio is indicated.

WARNING *Only perform a full-field test on-vehicle if specified by the OEM. Use extreme caution when full-fielding an alternator. The output voltage can rise to a very high level in a brief time and cause damage to the electrical system. It also may be very difficult to access the full-fielding access hole on some alternators. Be careful when working around rotating components to avoid injury.*

Test 2: Low Alternator Output Voltage

If the unloaded alternator output voltage is less than 13.5V and alternator output current nears 0A, it is possible that the rotor has lost its residual magnetism. This is common after disassembly of the alternator or after a long period of not operating the truck. The rotor must have enough residual magnetism to initiate the charging process. Consult the specific OEM information for the method to restore residual magnetism. In general, a jumper wire is briefly connected between the alternator positive output terminal and another alternator terminal, such as the R or relay terminal. This procedure may be referred to as "flashing the field," due to the sparks that may be seen.

Some alternators may have adjustable voltage regulators and the low alternator output voltage may be caused by a misadjusted voltage regulator. Consult the OEM's information for instructions on adjusting the voltage regulator.

If low alternator output voltage is not caused by a loss of residual magnetism or a misadjusted voltage regulator, the next step is to full-field the rotor field winding, as previously described. If full-fielding causes the alternator output voltage to increase substantially, then the problem is probably due to a faulty voltage regulator. If full-fielding does not cause the alternator output voltage to increase, then an internal alternator problem is indicated. Such a problem may be caused by something as simple as worn brushes. Brushes and even the voltage regulator can be replaced on some alternators without disassembly of the alternator. Otherwise, disassemble and test the alternator components such as the diode trio and rectifier diodes per the OEM's instructions or replace the alternator. Most shops now typically replace the entire alternator assembly with a new or remanufactured alternator instead of disassembling, testing, and repairing an alternator.

Test 3: High Charging Voltage

If the charging voltage voltage measured at the battery terminals is above 14.5V at room temperature with a 12V system, the charging voltage is too high. This indicates a possible problem with the voltage regulation system or the remote voltage sensing circuit, if applicable. The voltage regulator requires an accurate measurement of system voltage to maintain the alternator output voltage at the optimal value. If the voltage regulator does not sense voltage correctly due to excessive resistance in the sensing circuit, then the regulator will regulate at too high of a voltage. This problem may also be caused by an internal short to ground of the field coil or other failure that causes the voltage regulator to be bypassed. Alternators with adjustable regulators may also be adjusted incorrectly, causing this overcharge problem.

Specific OEM information should be consulted for troubleshooting this high-charging-voltage problem. In general, checks of the voltage-sensing circuit are performed. Alternators with external regulators may have several troubleshooting steps, while alternators with internal voltage regulators typically require alternator disassembly for further testing.

Practices to Be Avoided

In the past, it was common for truck technicians to test alternator performance by disconnecting the battery's positive cable with the engine running. If an alternator problem existed, the voltmeter would indicate a very low voltage and the engine might have also stalled. If voltage remained at a normal level with the battery cable disconnected, then it was assumed that the alternator performance was acceptable. This type of testing should never be performed on a modern truck.

Disconnecting the battery cable with the engine running causes a **load dump** to occur when the charging current flowing into the battery is interrupted. Load dump describes a very high voltage being produced by the alternator when a loss of connection to the battery occurs. The alternator output voltage can reach levels of 35V or more for a brief period of time due to the interruption of the charging current flowing through the battery. The high voltage is produced by the rapid collapse of the magnetic field surrounding the rotor windings as the electrical load rapidly decreases when the battery is disconnected. This high voltage caused by load dump can easily damage electronic components in a modern truck.

Another practice of the past was to temporarily install charged batteries from one truck into another truck with discharged batteries. After using the charged batteries to start the truck, the charged batteries were then disconnected with the engine running and the truck was then driven without any batteries present in the electrical system. This practice will cause a load dump when the charged batteries are disconnected, but also causes the alternator to operate in a condition referred to as batteryless operation. One of the purposes of the batteries in a truck electrical system is to maintain a smooth, stable charging system voltage by acting as a large capacitor. Without batteries in parallel with the alternator output, the voltage regulator causes the alternator output voltage to overshoot and undershoot the desired 14V output voltage anytime the alternator load changes. These voltage excursions can damage electronic components.

CAUTION *Never disconnect any battery or alternator cable with the engine running. The rapid change in alternator output current can result in a very high alternator output voltage, which can destroy modern truck electronics. Never operate an engine without batteries connected in parallel with the alternator output. This batteryless operation will cause the alternator output voltage to become very unstable, which could result in damage to electronic components.*

KEY-OFF PARASITIC LOADS

A key-off parasitic load is current that is drawn from the batteries when the key switch is in the off position. All modern trucks have a certain amount of parasitic load due to electronic modules. Most electronic modules draw a small amount of current (microamps or milliamps) to maintain memory such as for the clock or preset stations in a radio. Over time, these parasitic loads will use enough of the battery's stored energy that the engine cannot be cranked fast enough to start. With fully charged batteries, the truck should be able to crank after not being operated for a month or more if the parasitic current is not too high.

A problem such as a faulty luggage compartment light switch may keep a light on and cause a high level of key-off parasitic current. This may cause the truck not to crank after sitting over the weekend. Testing the charging system on such a truck after recharging batteries will probably indicate that the charging system is operating correctly.

Testing for key-off parasitic loads on a modern truck requires some specific techniques to avoid a misdiagnosis. In the days before electronic modules, the technician would simply disconnect the negative battery cable with the key in the off position and watch for an arc when the negative battery cable was reconnected, indicating that something was remaining on with the key in the off position. Another method of the past to determine if parasitic loads were present was to place a test light in series between the open ground cable terminal and the battery ground terminal. If the test light illuminated, then there were parasitic loads present.

Neither of these methods for testing for the presence of parasitic loads is valid for modern trucks. When battery power is first supplied to the truck's various electronic modules, the modules will often draw several amps of current as capacitors in the module are charged for a second or so. Devices like electronic instrument panel clusters will often perform a hard reset of the gauge motors when resupplied with power and can cause a couple of amps of current draw for a few seconds until the initialization completes. A few amps of current will flow when the batteries are first reconnected on a modern truck, causing a visible arc as the connection is made, even if there is nothing wrong with the truck. This current should drop to a few milliamps after a minute or so as the electronic modules go into a low power consumption or "sleep" mode. Placing a test light in series with the disconnected battery cables as a means of testing for parasitic current draw will reduce the voltage available to electronic modules, causing them not to go into this sleep mode and to continue to draw current at a high rate due to the presence of the test light. This will keep the test lamp illuminated, leading the technician to

believe that a large parasitic draw is present, when in fact none exists.

The correct method of measuring parasitic current draw on modern trucks is shown in Photo Sequence 2. The clamp-on ammeter is used first to measure parasitic load current to make sure that the current level is not so high that the series-connected ammeter will be damaged. The accuracy of a clamp-on ammeter when measuring low current levels is not sufficient alone for making accurate parasitic load current measurements. A series-connected ammeter must be used to perform an accurate parasitic load current measurement. Most DMMs have an ammeter that is capable of measuring 10A directly. Check your specific ammeter before measuring parasitic load current.

If the final parasitic load current is less than approximately 50mA (0.050A), there probably is not a sufficient parasitic load current to cause a problem. It is also possible that the cause of the parasitic current draw is intermittent, such as a truck operator leaving a light switched on in the sleeper or a luggage compartment switch that fails intermittently.

In most trucks with three parallel-connected batteries, a parasitic load current of 50mA should permit the engine to start after 30 days of not operating. However, trucks that are not used very often may require installation of a battery disconnect switch to avoid discharged batteries. It is important to follow the OEM's recommendations for installation of a battery disconnect switch on a modern truck with an electronically controlled diesel engine, especially 2010 and newer trucks which use diesel exhaust fluid (DEF).

Another cause of parasitic load current is conductive minerals, such as road salt, on the top of the battery between the terminals. This parasitic current will not show up on the ammeter during the parasitic load test. Internal battery short circuits caused by sediment or plate shorting are another form of parasitic current. Because the batteries are connected in parallel, a single battery with an internal short will cause all the batteries to discharge. Disconnect all battery cables; then recharge and test each battery separately.

ALTERNATOR DISASSEMBLY AND TESTING

If testing determines that the alternator has failed, the alternator can be disassembled for additional testing to determine if it will be more economical to repair or replace the alternator. Many shops no longer repair alternators and instead replace defective alternators with new or remanufactured units. This is unfortunate because alternators are fairly simple machines and are

Performing a Parasitic Current Draw Test

PHOTO SEQUENCE 2

PS2-1 Switch off all electrical loads on the truck and place the key switch in the off position. If possible, disable the dome lamp from illuminating when a door is opened to permit access to the fuse block if fuse block is located inside the cab.

PS2-2 Place a clamp-on current probe around the negative battery cable. On trucks with more than one negative battery cable, it will be necessary to disconnect all but one negative battery cable to cause all current to flow through one cable only. If the clamp-on ammeter indicates that current is more than 5A, something has probably been left on such as lights. Make sure everything on the truck has been switched off before proceeding. This DMM shown here is indicating a current draw of approximately 11.5A.

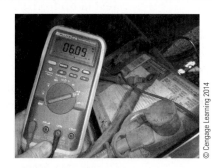

PS2-3 If the clamp-on ammeter indicates that parasitic current draw is less than 5A, disconnect the negative cable from the negative battery terminal. Connect the DMM direct-reading ammeter in series with the negative battery cable and negative terminal. The reading on the DMM may indicate a few amps as the various electronic modules on the truck reset due to the battery disconnection and reconnection. Do not switch on any load with the ammeter in place. Doing so may damage the ammeter or blow the fuse in the ammeter.

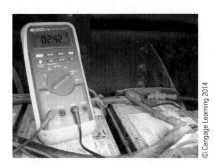

PS2-4 Watch the value displayed on the DMM ammeter as the various modules on the truck power down. The value displayed on the DMM after a couple of minutes is the parasitic load current. The parasitic load current on this vehicle is shown to be about 2.42A. This is a high level of parasitic load current and would cause the truck to have to be jump-started on a regular basis and have very short battery life.

PS2-5 Using the circuit diagram as a guide, remove the applicable circuit protection devices one by one (reinstalling each time) until the circuit that is drawing an excessive amount of current is located as evidenced by the ammeter reading dropping substantially.

PS2-6 The major circuit that is causing the current draw has been found and the ammeter indicates that parasitic load current has dropped to 180mA with the circuit protection device for the offending circuit removed. This is still a somewhat high amount of parasitic load current, but much lower than the 2.42A measured previously.

Component	Test Connection	Normal Reading	If Reading Was	Trouble Is
Rotor	Ohmmeter from slip ring to rotor shaft	Infinite resistance	Very low	Grounded
	Test lamp from slip ring to shaft	No light	Lamp lights	Grounded
	Test lamp across slip rings	Lamp lights	No light	Open
Stator	Ohmmeter from any stator lead to frame	Infinite resistance	Very low	Grounded
	Test lamp from lead to frame	No light	Lamp lights	Grounded
	Ohmmeter across any pair of leads	Less than 1/2Ω	Any very high reading	Open
Diodes	Ohmmeter across diode, then reverse leads	Low reading one way; high reading other way	Both readings low Both readings high	Shorted Open
	12V test lamp across diode, then reverse leads	Lamp lights one way, but not other way	No light either way Lamp lights both ways	Open Shorted

© Cengage Learning 2014

Figure 7-51 Guidelines for alternator component testing.

made up of only a few components. Photo Sequence 3 illustrates the disassembly and reassembly of a Delco Remy 22 SI alternator.

After disassembly, each component is cleaned with a soft cloth and individual components are inspected and tested. The bearings are checked for noise and roughness and replaced if necessary. Inspect the rotor shaft surface that contacts the bearings contact for damage. Many of the other components in the alternator are tested using an ohmmeter and a powered test lamp. The chart in **Figure 7-51** shows the test connections and the expected results for each major component.

Testing the Rotor

The rotor contains the slip rings and the field windings. The rotor field windings are typically enamel- or varnish-coated copper wire. The coating acts as an insulator. If a rotor has been shorted, the rotor windings may have become overheated, causing the insulating coating to burn away and resulting in wire-to-wire shorts or shorts to ground. A rotor that shows signs of overheating must be replaced.

The resistance measured between the slip rings indicates the rotor winding resistance (**Figure 7-52**). Typical winding resistance should be somewhere

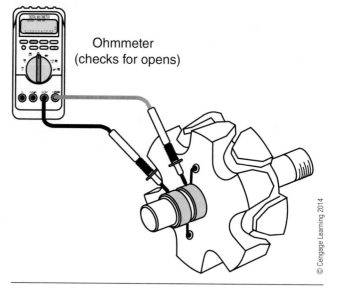

Ohmmeter (checks for opens)

© Cengage Learning 2014

Figure 7-52 Testing for rotor field winding resistance.

between 1.5 ohms and 6 ohms, depending on the OEM's specifications. Readings that are much lower or higher indicate a wire-to-wire short or an open field winding circuit requiring rotor replacement.

The rotor field windings are also tested for a short-to-ground condition, as shown in **Figure 7-53**. The resistance between the slip ring and the rotor shaft should be infinite.

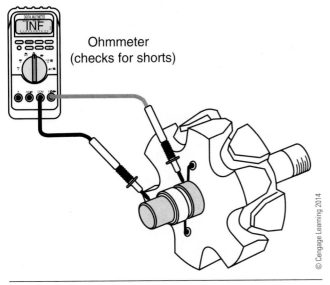

Figure 7-53 Testing for a short to ground field winding.

Testing the Stator

Like the rotor field windings, the stator is made of varnish- or enamel-coated copper wire. Excessive heat can damage the coating, causing wire-to-wire shorts and shorts to the stator frame. Replace any stator that has signs of overheating. The stator is tested for open circuits between the phases using an ohmmeter (**Figure 7-54**). The resistance between all three phases should be near 0 ohm. An open phase will cause an infinite resistance or very high resistance.

The stator is also tested for windings that are shorted to the frame, as shown in **Figure 7-55**. The ohmmeter should indicate infinite resistance. Otherwise, replace the stator or alternator.

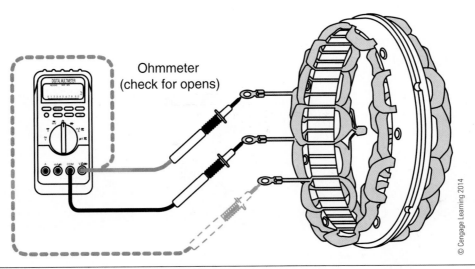

Figure 7-54 Testing the stator for open circuits.

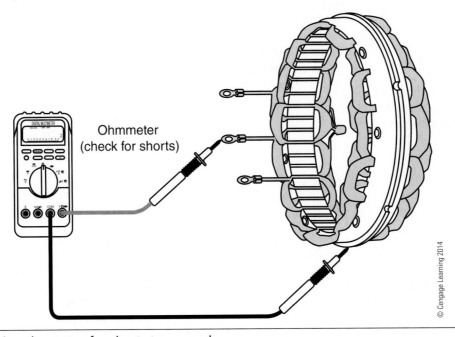

Figure 7-55 Testing the stator for shorts to ground.

PHOTO SEQUENCE 3 Delco Remy 22 SI Alternator Disassembly and Reassembly

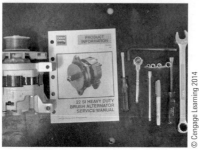

PS3-1 Assemble the tools and technical information that will be necessary to disassemble the alternator. This example is a Delco Remy 22 SI brush-type alternator. Mark the alternator end covers and stator for proper reassembly.

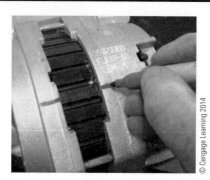

PS3-2 Mark the alternator end covers and stator for proper reassembly.

PS3-3 Remove the four bolts used to attach the two end covers.

PS3-4 Carefully pry the front end cover away from the stator using a small pry bar. This 22 SI alternator has four notches where the pry bar can be inserted. Be careful not to nick the stator wiring insulation with the pry bar. Do not attempt to pry the rear cover away from the stator at this time.

PS3-5 Pull the front end cover with rotor attached away from the alternator rear housing and stator. A snapping sound will be heard as the spring-loaded brushes extend when the slip rings on the rotor are pulled away from the rear housing.

PS3-6 Remove the three stator connections from the diode rectifier bridge. The stator connections on this 22 SI alternator use ring terminals and the diode bridge has three threaded studs, making removal an easy task. Other alternators may have the stator leads soldered to the rectifier bridge and must be desoldered to permit stator removal.

PS3-7 Pry the stator assembly out of the rear housing in the same way that the front cover was removed. Lift the stator away from the rear housing, being careful not to bend the stator wiring where it connects to the rectifier bridge, and set the stator aside.

PS3-8 Loosen the fastener used to retain the diode trio assembly to the voltage regulator.

PS3-9 Pay close attention to the length of each fastener that is removed and any insulating sleeves or insulating washers that are present. It is critical that fasteners and insulating washers be reinstalled in their correct location. The fastener that connects the voltage regulator to the diode trio output is insulated by a plastic sleeve as shown.

PS3-10 The diode trio on the 22 SI alternator also contains an auto start assembly. Remove the two fasteners that attach the diode trio assembly to the rectifier bridge. Note that one fastener is insulated and is connected to the positive heat sink; the other fastener is not insulated and is connected to the negative heat sink.

PS3-11 Carefully lift the diode trio and auto start assembly out of the alternator rear housing and set aside. The auto start assembly output supplements the rotor residual magnetism.

PS3-12 Remove the nut that attaches the alternator B+ output stud to the voltage regulator.

PHOTO SEQUENCE 3 (Continued)

PS3-13 Remove the alternator B+ stud by pulling the stud through the rear housing. Note the insulating sleeve, which isolates the B+ stud from the grounded rear housing.

PS3-16 Lift the voltage regulator out of the alternator housing and set aside.

PS3-14 Remove the two fasteners used to retain the brush holder assembly to the rear cover. Note that one fastener is insulated; the other fastener is not insulated. Lift the brush holder assembly out of the alternator rear housing and set aside.

PS3-17 Remove the R terminal strap and R terminal and set aside. The strap connects the R terminal to the rectifier bridge connection. The R terminal has an insulation sleeve and insulating washers, which insulate the R terminal from the grounded rear housing.

PS3-15 Remove the remaining noninsulated fastener that attaches the voltage regulator to the rear housing.

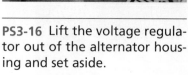

PS3-18 Remove the remaining noninsulated fastener, which attaches the negative heat sink to the rear housing.

PHOTO SEQUENCE 3 **(Continued)**

PS3-19 Lift the rectifier bridge assembly out of the rear housing and set aside.

PS3-20 The rear roller bearing can be pressed out and replaced if necessary. The rotor can also be pressed out of the front cover for access to the front bearing if necessary. When reassembling the alternator, lubricate the bearings with the specified grease.

PS3-21 Reassembly is basically the opposite of disassembly with some exceptions. Clean the slip rings using 600 grit sandpaper or 500 grain polishing cloth. Wipe down all components with a soft cloth. To reassemble the brush holder, press the brushes all the way into the holder. A small drill bit or stiff wire is then passed through small holes in the brushes and brush holder. This keeps the brushes depressed while the alternator is reassembled. A small hole in the rear housing permits the drill bit to pass to the outside of the rear housing.

PS3-22 Make sure that the correct fasteners are used when reassembling the alternator and torque fasteners to the OEM specified value. The insulation sleeve must be present on any component that is not grounded to prevent a short to ground. The insulation sleeves and washers around the B+ terminal, R terminal, and I terminal, if applicable, must be reinstalled so that the terminals are not grounded.

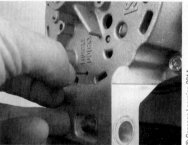

PS3-23 After the alternator is reassembled, remove the drill bit or stiff wire that is keeping the brushes depressed. You should hear two snapping sounds as each brush makes contact with its corresponding slip ring.

Testing the Diodes

Common rectifier diodes should only permit current to flow in one direction and should block the flow of current in the opposite direction. The diode test setting on a DMM or analog ohmmeter is utilized to test the rectifier diode assembly, as shown in **Figure 7-56**. The rectifier bridge typically has three positive diodes and three negative diodes. Current should only flow through the diodes in the appropriate direction. Replace the rectifier bridge if defective.

The field diode assembly or diode trio is also tested using a DMM on the diode test setting, as shown in **Figure 7-57**. Current should only flow in one direction through all three diodes.

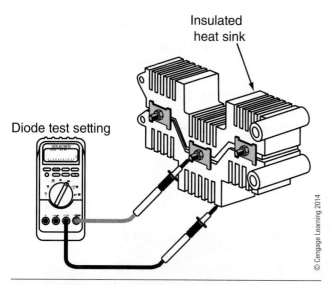

Figure 7-56 Testing the rectifier diodes.

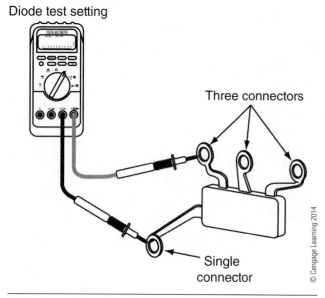

Figure 7-57 Testing the field diode or diode trio.

Testing the Brushes

The length of the brushes decreases with use. Many alternator problems are caused by worn brushes. The springs behind the brushes cannot maintain good contact between the brush and slip ring when the brushes are worn, resulting in too low of a rotor field current and low alternator output. Brushes are typically replaced during any alternator repair.

Testing the Voltage Regulator

Bench testing the voltage regulator is not typically performed. The on-vehicle full-field test that bypasses the regulator should determine if the voltage regulator is the cause of the low alternator output. An alternator with a high voltage output may be caused by a defective regulator or an internal short to ground somewhere in the field circuit.

ALTERNATOR REASSEMBLY

Reassembly of the alternator is basically the reverse order of disassembly. However, the specific OEM instructions must be followed. These instructions may include information on bearing lubrication and the installation of a special compound behind the voltage regulator or rectifier for proper heat transfer.

Pay close attention to any insulating washers such as those used to insulate the positive output terminal and the voltage regulator. Leaving these out or installing them at the wrong location can result in a short-to-ground condition or an unregulated output voltage.

EXTRA FOR EXPERTS

This section provides more information related to alternating current and alternators.

RMS Voltage

Most people would expect that the maximum or peak amplitude of the voltage for a typical wall socket would be 120V instead of 170V. The 120V designation of North American power standard indicates an RMS voltage level. RMS stands for root mean square, which is a math term. To simplify, the RMS voltage of an AC waveform is the DC equivalent voltage with regard to the heating ability. A resistor connected to a 120V AC RMS voltage source would dissipate the same amount of heat as an identical resistor connected to a 120V DC source. Part of the time, the AC voltage is much less than 120V, while at other times it is above 120V. The polarity of the voltage changes, but the heat given off by the resistor is independent of which direction current is flowing in the resistor. Current flow in either direction through a resistor will cause heat to be dissipated by the resistor.

All DMMs in the AC volts setting display the RMS value of a sinusoidal waveform. Using the DC voltage setting of a DMM to measure a sinusoidal waveform will cause the average value of the waveform to be displayed, which is 0V. Half of the waveform is positive; the other half of the waveform is negative, so the average is zero.

Electronic Voltage Regulators

A simplified schematic for an internal electronic voltage regulator is shown in **Figure 7-58**. The transistor identified as TR1 is the large high-power bipolar transistor that switches the field current off and on (pulse width modulation) to control the alternator output voltage. The base of the TR1 transistor is controlled by the transistor identified as TR2. The circuit is designed so that when TR2 is switched on, TR1 is switched off and vice versa. This occurs because when TR2 is switched on, it effectively shorts out the voltage at the base of TR1, causing TR1 to switch off.

The resistors identified as R2 and R3 form a series voltage divider circuit. The battery terminal voltage is divided proportionally across these two resistors per the rule of series circuits. The zener diode identified as

D2 is the heart of the voltage regulator circuit. Recall that a zener diode will not conduct electric current in the reverse-biased direction until the voltage across the diode exceeds the zener voltage rating of the diode.

An increase in alternator output voltage causes the voltage dropped across resistor R3 to increase accordingly. When the voltage drop across R3 rises to the zener voltage rating of D2, the zener diode D2 conducts current in the reverse-biased direction. When D2 switches on in the reverse-biased direction, the base of TR2 is supplied with a positive voltage. This causes transistor TR2 to switch on, which results in TR1 switching off. TR1 is now interrupting the rotor field current, resulting in a drop in the alternator output voltage. A drop in output voltage causes the voltage across R3 to decrease, resulting in zener diode D2 no longer conducting. With D2 not conducting, zero voltage is supplied to the base of TR2. This causes TR2 to switch off, permitting TR1 to switch back on again to conduct field current. The field current restores the rotor magnetic field and the alternator output voltage once again increases. This process is repeated several hundred times per second as needed to maintain a constant alternator output voltage of about 14.2V.

The voltage regulator controls the duty cycle of the rotor field through PWM. Each time a high current load such as the headlamps is switched on, the output voltage at the alternator drops momentarily as a portion of the alternator output current is diverted to the load. The voltage regulator circuit responds to the drop in the alternator output voltage by increasing the duty

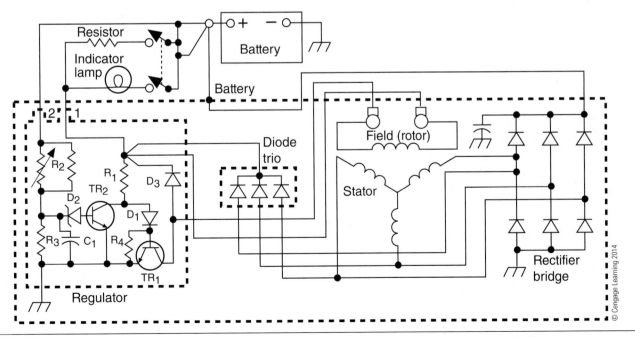

Figure 7-58 Simplified electronic voltage regulator schematic—zener diode D2 is the primary component.

cycle of the rotor field current, thus regulating the alternator output voltage.

Notice that the resistor R2 in **Figure 7-58** is shown as a variable resistor. The resistance of R2 increases as the temperature decreases (NTC). This change in resistance requires the alternator output voltage to increase before the zener diode D2 will conduct. The result is that the alternator output voltage will increase as alternator temperature decreases. Likewise, alternator output voltage also decreases when the temperature increases. This type of voltage regulator is known as a temperature-compensating voltage regulator. The increase in alternator output voltage in cold temperatures recharges the battery more efficiently. The decrease in alternator output voltage in hot weather prevents overcharging the battery. This temperature-compensated voltage regulator is utilized in some trucks, although many modern trucks have a non-temperature-compensated voltage regulator.

The simplified voltage regulator was shown for a three-wire-type system with a charging indicator lamp. A single-wire-type system supplies the connection to the voltage regulator for voltage sense inside the alternator instead of using a direct connection to the battery-positive terminal.

The regulated charging voltage of some alternators can be adjusted to customize the charging voltage to the truck's application to improve battery life. The adjustable regulator may be in the form of an externally mounted removable resistor cap on some Delco Remy alternators; it can be installed in one of four different positions to provide four different alternator output voltages. This resistor cap adds a series resistance to the regulator voltage divider to increase the alternator output voltage.

Summary

- The charging system supplies the energy to recharge the truck's batteries.

- The alternator produces a three-phase alternating current through the principles of electromagnetic induction. The alternating current is rectified to a direct current.

- The rotor is an electromagnet that produces a rotating magnetic field inside the stationary windings called the stator. The magnetic lines of force produced by the rotor cut through the stator windings to induce a sine waveform voltage in the stator windings.

- The three-phase stator windings may be connected in a delta or wye (Y) configuration. Either configuration produces a three-phase alternating current waveform.

- The means of making electrical contact with the rotor field windings is through slip rings and brushes.

- The alternator output voltage is controlled by regulating the current flow through the rotor field windings. The voltage regulator uses PWM to control the field winding current level.

- The rectifier uses a series of diodes to transform the negative half of the sine waveform into a positive waveform. The pulsating positive polarity waveform is smoothed by a capacitor. The small amount of pulsation that remains in the waveform is called ripple voltage.

- Testing the charging system includes testing the batteries, making a visual inspection, and measuring charging system circuit resistance. Circuit resistance testing is performed by measuring the voltage dropped on the charging circuit with rated alternator current flowing through the charging circuit.

- Parasitic current draw is the key-off current draw of the truck's electrical system. Parasitic current draw is measured using an ammeter. The source of the current draw can be determined by removing fuses one by one.

Suggested Internet Searches

Try these web sites for more information on alternators and charging system components:
http://www.delcoremy.com
http://www.prestolite.com (Leece-Neville alternators)
http://www.cooperindustries.com
http://www.vanner.com

Review Questions

1. Which is a true statement?

   A. Moving a conductor parallel to magnetic lines of force in the direction of the arrows on the magnetic lines of force causes a voltage to be induced in the conductor.

   B. Moving a conductor parallel to magnetic lines of force in the opposite direction to the arrows on the magnetic lines of force causes a voltage to be induced in a conductor.

   C. Moving a conductor so that magnetic lines of force are being cut or interrupted causes a voltage to be induced in the conductor.

   D. Voltage cannot be induced in a copper conductor because copper cannot be magnetized.

2. A waveform repeats itself 60 times per second. What is the frequency of the waveform?

   A. 120 hertz

   B. 1 hertz

   C. 3600 hertz

   D. 60 hertz

3. Why are slip rings in an alternator necessary?

   A. They permit the stator to rotate.

   B. They provide a high resistance connection to the stator windings.

   C. They prevent a delta from forming.

   D. They permit current to flow through a rotating component called the rotor.

4. Technician A says a delta-connected stator produces a three-phase voltage. Technician B says that a Y-connected stator produces a three-phase voltage. Who is correct?

   A. A only

   B. B only

   C. Both A and B

   D. Neither A nor B

5. A rectifier diode bridge is used to perform which operation in an alternator?

   A. Convert DC into AC

   B. Regulate voltage output

   C. Bridge the gap between the stator and the rotor

   D. Convert or rectify the negative half of a sine wave into the positive half of a sine wave

6. The batteries have been accidentally hooked up backward on a truck. Which component has probably been damaged and for what reason?

   A. The regulator because it has conducted all of the reverse current

   B. The rectifier because it has been reverse biased

   C. The stator because it has been shorted to ground

   D. The rectifier because it has been forward biased with almost no resistance in the circuit

7. One diode in the diode trio has opened. How would this probably impact the alternator's performance?

   A. Increased alternator output voltage

   B. Increased alternator output current

   C. No expected change in the alternator's performance

   D. Decreased maximum alternator output current

8. Which of these places has the highest voltage in a properly working truck electrical system with the engine running and the batteries charging?

A. Across battery terminals

B. Across cranking motor B+ and cranking motor ground

C. Across the alternator plus output stud and alternator ground stud

D. All of the above would be the exact same voltage

9. A truck that is only operated on weekdays must be jump-started every Monday morning but manages to start Tuesday through Friday unassisted. Which is the most likely cause?

A. Loose alternator drive belt

B. Parasitic loads

C. Too high of a charging voltage

D. Excessive charging circuit resistance

10. An alternator has no output until a technician performs a full-field test. When the alternator is full-fielded, the alternator's output voltage reaches 16V with more than 100A of current flowing out of the alternator. What does this probably indicate?

A. Defective internal regulator

B. Worn brushes

C. Loose alternator drive belt

D. Defective stator, rotor, or bridge rectifier

11. A diode trio is being tested using a DMM in the diode test setting. The DMM indicates an open circuit in both directions for all three diodes. What would have been the likely output for this alternator?

A. Very high output voltage

B. High charging current

C. Normal output current

D. No output voltage or current

12. The stator windings in an alternator are being tested with an ohmmeter. The resistance measured between each of the three windings is nearly 0 ohms. What does this indicate?

A. The stator windings do not have an open circuit.

B. The stator windings are shorted to the stator frame.

C. The stator windings are open.

D. The stator windings are magnetized.

13. Which of the following could cause the charge indicator warning lamp to illuminate on a truck?

A. Broken alternator drive belt

B. Worn alternator brushes

C. Shorted to ground circuit between the alternator I terminal and the indicator lamp

D. All of the above

14. What is the purpose of full-fielding an alternator?

A. Restoring residual magnetism in the rotor

B. Restoring residual magnetism in the stator

C. Testing the alternator

D. All of the above

15. Where are the brushes located in a brushless alternator?

A. Plastic brush holder at the back of the alternator

B. Beneath the voltage regulator

C. Inside the rotor

D. None of the above

CHAPTER

8 DC Motors

Learning Objectives

After studying this chapter, you should be able to:

- List the main components of a typical starter (cranking) motor.
- Describe how interacting magnetic fields cause the armature in an electric motor to rotate.
- Explain why a starter motor draws less current as motor speed increases.
- List the advantages and disadvantages of a gear-reduction starter motor.
- Measure cranking circuit resistance using the voltage drop method.
- Troubleshoot the cause of a no-crank problem.
- Disassemble a starter motor, test the internal components, and reassemble.
- Perform a rapid assessment of a truck's electrical system.
- Explain how rotational direction is reversed with a permanent magnet motor.

Key Terms

armature

brushed DC motor

commutation

commutator

compound-wound

drive

electronic control unit (ECU)

field coil

gear-reduction

growler

milling

one-way clutch

overrunning clutch

permanent magnet

pinion gear

pole shoe

positive engagement

ring gear

series-wound

shunt-wound

solenoid

tooth abutment

INTRODUCTION

DC motors are used for several purposes on modern trucks. Accessories such as windshield wipers, heating and air conditioning, and some hydraulic ABS systems utilize a DC motor. The largest DC motor found on most trucks is the engine starter (cranking) motor.

Electric motors convert electrical energy into mechanical energy. Electric motors perform the opposite function of generators, which convert mechanical energy into electrical energy, as illustrated in **Figure 8-1**. Like generators, electric motors operate on the principles of electromagnetism. Almost all of the motors found on a truck also use brushes to make electrical contact with the

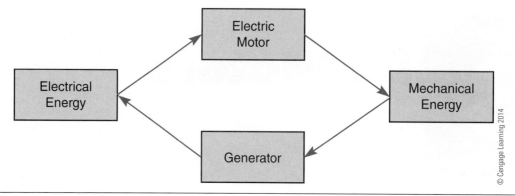

Figure 8-1 Electric motors and generators convert one form of energy into another.

rotating motor elements. Therefore, these types of motors are referred to as **brushed DC motors**.

BRUSHED DC MOTOR BASICS

Electric motors operate on the principles of magnetism. A brief review of some of the principles of magnetism will assist you in understanding how a brushed DC motor operates.

Magnetism Review

The arrows on invisible magnetic lines of force indicate that the lines start at the north magnetic pole and end at the south magnetic pole. The arrows on the lines of force indicate the direction that a compass needle would point if placed in the magnetic field. The magnetic lines of force do not actually flow or rotate.

The magnetic field between the poles of a horseshoe magnet is illustrated by the magnetic lines of force shown in **Figure 8-2**. When electric current passes through a conductor, magnetic lines of force surround the conductor. The magnetic lines of force appear as concentric (same center) circles that appear to rotate about the conductor, as shown in **Figure 8-2**. Note that if the direction of current flow through the conductor were reversed from that shown, the direction of the arrows on the magnetic lines of force would also be in the opposite direction from that shown. The right-hand rule, explained in **Chapter 3**, can be used to determine the direction of the arrows associated with the magnetic field surrounding the conductor. Knowing the direction of the arrows on the magnetic lines of force makes it possible to understand how magnetic lines of force from more than one source interact with each other.

If a current-carrying conductor were placed in the magnetic field between the poles of a horseshoe magnet, the magnetic field caused by the current flow through the conductor would distort the magnetic field of the magnet, as shown in **Figure 8-3A**. The magnetic

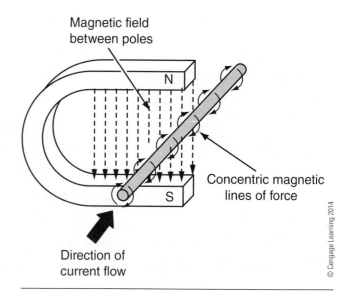

Figure 8-2 Current-carrying conductor outside of magnetic field caused by horseshoe magnet; no interaction between magnetic fields.

field surrounding the conductor is "rotating" in a clockwise direction because of the direction of current flow through the conductor, per the right-hand rule. This causes the magnetic lines of force, on the left side of the conductor in **Figure 8-3A** (due to the current flow), to point in the same direction (up) as the magnetic lines of force caused by the magnet. This leads to an important fact regarding magnetic fields:

Important Facts: Two magnetic fields that interact with each other combine to form a single magnetic field. If the arrows on the magnetic lines of force of both magnetic fields are pointing in the same direction, the resulting magnetic field is strengthened. If the arrows on the magnetic lines of force are pointing in opposite directions, the resulting magnetic field is weakened.

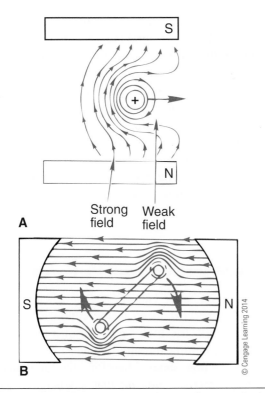

A

B

Strong field   Weak field

Figure 8-3 (A) Current-carrying conductor placed in magnetic field causes an interaction between magnetic fields; conductor is compelled to move from strong magnetic field to weak field. (B) Current-carrying conductor formed into a loop is compelled to rotate around its axis to move from strong field to weak field.

Figure 8-3A illustrates that the magnetic lines of force caused by the magnet and the magnetic lines of force caused by the current flow through the conductor have arrows that are pointing in the same general direction. These two separate magnetic fields combine to form a stronger magnetic field. At the same time, the magnetic lines of force on the right side of the conductor, caused by the current flow through the conductor, have arrows that are pointing in the opposite direction from the magnetic lines of force caused by the horseshoe magnet. This has the effect of canceling out a portion of the combined magnetic field on the right side of the conductor. This canceling out effect causes the combined magnetic field on the right side of the conductor to be a weaker magnetic field than the combined magnetic field on the left side of the conductor.

Recall from the study of magnetism in **Chapter 3** the following important fact:

Important Facts: Conductors that are carrying current are compelled (want) to move out of a stronger magnetic field into a weaker magnetic field.

The magnetic lines of force caused by the north and south poles of the horseshoe magnet act like rubber bands that flip the conductor from the strong field on the left side of the conductor to the weaker magnetic field on the right side of the conductor in this example.

When a current-carrying conductor is formed into a loop and placed in a magnetic field, as shown in **Figure 8-3B**, the loop will rotate to move from the strong magnetic field to the weak magnetic field. The interaction of the magnetic field caused by current flow through a conductor formed into a loop that is free to rotate and the magnetic field caused by a stationary magnet is the basis of brushed DC motor theory.

A Simple Electric Motor

A simple electric motor is shown in **Figure 8-4**. This motor looks similar to the introductory alternator discussed in **Chapter 7**. The conductive loop that rotates in an electric motor is called an **armature**. The armature is connected to a ring that is similar to alternator rotor slip rings. However, the motor slip ring in this simple motor is split into two pieces like a donut that is broken into two pieces. This two-piece slip ring causes current to flow through one-half of the split ring, through the loop of wire, to the other half of the split ring. This split slip ring in a motor is referred to as a **commutator**. Each piece of the commutator is called a commutator segment. Spring-loaded brushes make electrical contact with the commutator segments. The brushes and commutator segments permit the armature to rotate yet still make electrical contact similar to the brushes and slip rings in an alternator rotor.

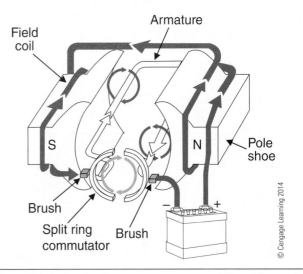

Figure 8-4 Brushed DC motor; current flow through armature reverses directions every 180 degrees of rotation.

The two magnets shown surrounding the armature in **Figure 8-4** are electromagnets called **pole shoes**. Pole shoes are made of iron or similar material that can be easily magnetized. Loops of copper conductor called the **field coils** or field windings are wrapped around the pole shoes. The field coil is wrapped in such a direction to cause the half of the pole shoe facing the armature to become a north pole magnet, as shown in **Figure 8-5**. The pole shoe on the left side of this illustration has the field coils wrapped such that the side of the pole shoe facing the armature becomes a south pole magnet. The pole shoes cause a magnetic field, as shown in **Figure 8-5**.

The simple motor shown in **Figure 8-4** has the field coils wired in series with the armature loop by way of the brushes and commutator segments. To observe this series connection between the field coils and the armature, trace the path of conventional current in the simple motor shown in **Figure 8-4**. Conventional current flows from the battery's positive terminal through the right side field coil, through the left side field coil, and on to the left side brush. Current then flows through the left brush to the split ring commutator and through the armature loop. From the armature loop, current flows through the other side of the split ring commutator, through the right side brush, and back to battery-negative terminal.

The magnetic field produced by the pole shoes interacts with the magnetic field surrounding the armature loop due to current flow through the loop; this interaction causes the loop to rotate a portion of a revolution as the armature moves from a strong magnetic field into a weaker magnetic field.

The design of the armature and the spacing of the commutator segments cause the current flow through the armature to reverse just as the loop escapes the strong magnetic field. Reversing the current flow through the armature loop causes the direction of the magnetic field set up around the armature to reverse as well. Only the current flow through the armature changes direction. The current through the field coils always flows in the same direction in brushed DC motors. This continuous magnetic field reversal in the armature at just the right time keeps the armature positioned in a strong magnetic field. This causes the armature loop to rotate continuously in an attempt to escape from this strong magnetic field. This process is referred to as **commutation** and is discussed in detail in the Extra for Experts section at the end of this chapter.

Counter-Electromagnetic Force

When a conductor cuts through magnetic lines of force, a voltage is induced in the conductor. The armature of an electric motor consists of a looped conductor that is cutting through magnetic lines of force created by the pole shoes. Therefore, the fundamentals of electromagnetism indicate that a voltage must be induced in the armature loop as it rotates through the magnetic lines of force set up by the pole shoes. This voltage induced in the armature loop as it begins to rotate has a polarity such that it acts as a series-opposing voltage to the battery-supplied voltage. This series-opposing voltage causes the current flowing through the armature to decrease, similar to the three-battery flashlight example in **Chapter 5** in which one battery is installed upside down. This induced voltage in the armature loop that opposes the battery voltage is referred to as counter-electromagnetic force (CEMF) and was introduced in **Chapter 3**. CEMF may also be referred to as back emf.

As discussed in **Chapter 7**, one way to increase the level of voltage produced by electromagnetic induction is to increase the speed with which the conductor cuts through the magnetic lines of force. As the motor begins to rotate faster, the speed with which the armature loop passes through the magnetic field set up by the pole shoes increases. This increase in rotational speed causes the amount of CEMF induced in the armature loop to also increase as more magnetic lines of force are being cut per second by the conductor.

The CEMF produced by the motor is a series-opposing voltage to the battery voltage. Therefore, an increase in CEMF as the motor speed increases causes a proportional decrease in the current drawn by the motor.

The concept of CEMF is discussed in greater detail in the Extra for Experts section at the end of this chapter.

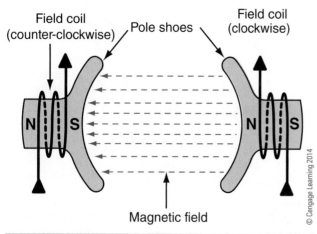

Field coil (counter-clockwise)   Pole shoes   Field coil (clockwise)

Magnetic field

© Cengage Learning 2014

Figure 8-5 Magnetic field developed by pole shoes.

Improving the Simple Electric Motor

The single-loop armature is good for illustrating the principles of DC brushed motors but does not produce a practical motor. The amount of torque developed by the armature loop is highest at the starting point shown in **Figure 8-4** and the point directly opposite this starting point (180 degrees). At the point where current reversal thorough the armature loop takes place (commutation), the armature is producing very little turning torque. Putting a load on this simple motor, such as requiring it to crank an engine, would cause the motor to stall at one of these commutation points.

To improve the simple motor design, several armature loops or windings could be added with equal spacing and the number of commutator segments would also be increased accordingly, as shown in **Figure 8-6**. Thus, each loop would cause two peak torque points per rotation, providing a relatively smooth torque output from the motor. As one loop undergoes commutation, another loop may be undergoing maximum torque, which keeps the motor rotating continuously. Additional pairs of pole shoes could also be added to the motor to further increase the torque developed by the motor. Four pole shoes are common in truck starter motors, along with 30 or more commutator segments.

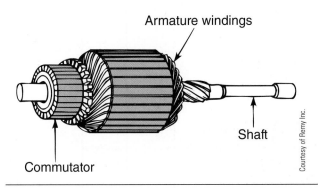

Figure 8-6 Armature windings and commutator segments.

Like the stator frame used on the alternator to concentrate magnetic lines of force, most motors use an iron or steel frame to provide a low-reluctance path for magnetic lines of force from the pole shoes to flow through. This frame also provides a means to retain the pole shoes.

The armature is made more effective by using a series of thin steel laminations to direct and concentrate magnetic lines of force. These laminations are stacked like the laminations used in alternator stators and prevent energy losses known as eddy currents from flowing in the armature. The laminations are slotted to hold the armature conductor loops. The laminations, conductors, and commutator are all pressed onto the armature shaft as shown in **Figure 8-7**.

STARTER MOTORS

Starter motors are brushed DC motors. A cutaway view of a starter motor is shown in **Figure 8-8**. The starter motor must provide a significant amount of power to crank a diesel engine. Starter motors are rated in watts or horsepower. Truck starter motors are similar to automobile starter motors except a truck starter motor for a 16-liter diesel engine may develop 11 hp (8.5 kW), which is perhaps 10 times that of a smaller automobile starter motor. This explains why most large trucks have four parallel-connected batteries with a total of perhaps 4000 cold cranking amps (CCA), which is also 10 times that of a small automobile battery.

Starter Motor Current Draw

The truck starter motor may draw 3000A or more initially, which drops to 1000A while the engine is cranking at its maximum speed. The current decreases as the motor begins rotating to crank the engine because of the CEMF produced by the rotating armature.

An example of starter motor current draw plotted with respect to time while cranking a Cummins 15-liter

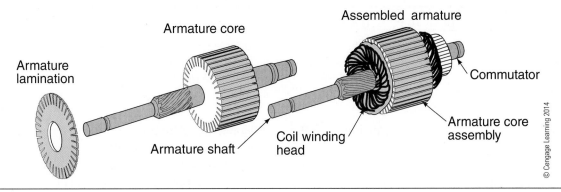

Figure 8-7 Armature components.

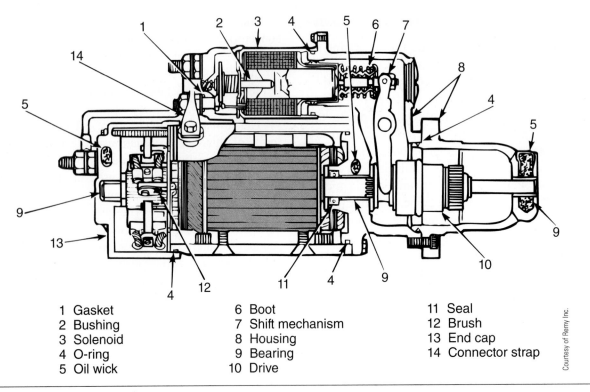

Figure 8-8 Cutaway view of a starter motor.

1 Gasket
2 Bushing
3 Solenoid
4 O-ring
5 Oil wick

6 Boot
7 Shift mechanism
8 Housing
9 Bearing
10 Drive

11 Seal
12 Brush
13 End cap
14 Connector strap

Courtesy of Remy Inc.

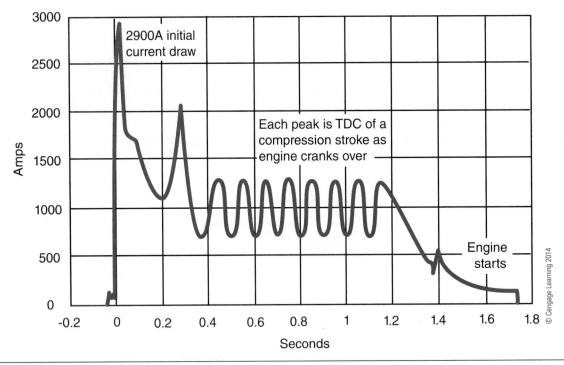

Figure 8-9 Oscilloscope trace of current drawn by starter motor; 500A per division vertical axis, 200 ms (0.2 sec) per division horizontal axis.

diesel engine at 75°F (24°C) is shown in **Figure 8-9**. Notice that when the starter motor is first supplied with voltage at 0 seconds, the current drawn by the motor rises rapidly to approximately 2900A. A stationary motor is basically just a coil of large-gauge copper wire, so the resistance from the motor B+ terminal to ground is only a couple of thousandths of an ohm (milliohms). As the motor armature begins to rotate, the current drawn by the motor drops substantially because of the CEMF produced by the armature.

The sine waveform that develops when the current drawn by the motor is plotted with respect to time as shown in **Figure 8-9** is caused by engine compression. Each time a piston approaches top dead center (TDC) of the compression stroke, the starter motor speed slows as engine load increases due to cylinder compression. The reduction in starter motor speed reduces the CEMF produced by the starter motor, which increases current drawn by the motor. This is illustrated by each of the current peaks in Figure 8-9 (not including the initial current draw peak). The compressed air in the cylinder then assists the starter motor to rotate the crankshaft, thus reducing motor load, increasing motor speed, and reducing the current drawn by the motor until the next compression stroke. Note that the average current drawn by the starter motor is approximately 1000A while the engine is cranking at full speed.

When the engine starts, the current drawn by the starter motor drops to approximately 100A for the brief period of time that the key switch is still held in the cranking position. This reduction in current drawn by the starter motor occurs because the motor is running at maximum or no-load speed. The corresponding increase in CEMF generated by the starter motor is causing this reduction in current drawn by the starter motor.

Starter Motor Fundamentals

The simple motor introduced in the previous section had a single pair of pole shoes. Most truck starter motors have two or three pairs of iron pole shoes, like that shown in **Figure 8-10**. The field coil winding that is used to magnetize the pole shoe is a ribbon of insulated copper. **Figure 8-11** illustrates a field coil winding assembly with brushes for a four-pole starter motor. The pole shoes with field coils are attached to the inside of an iron frame or housing (**Figure 8-12**). The housing provides a low-reluctance path for magnetic lines of force.

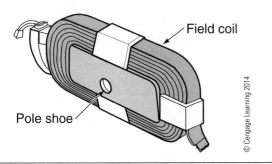

Figure 8-10 Pole shoe with field coil; current flow through field coil creates an electromagnet.

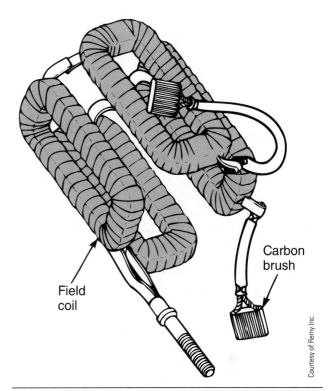

Figure 8-11 Four insulated field coils with brushes.

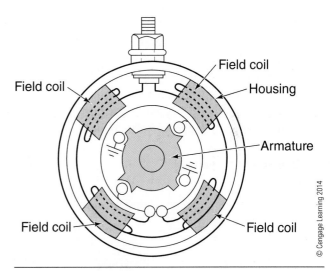

Figure 8-12 Pole shoes and field coils installed in iron housing.

A four-pole starter motor is shown in **Figure 8-13**. Unlike the two-pole simple motor introduced earlier, the north and south pole magnets are not situated directly across from each other. The field coil windings are wound such that alternating north and south pole magnets are formed. The resulting magnetic fields are illustrated in **Figure 8-13**. The result is rotation of the armature as it continually moves from a strong magnetic field to a weak magnetic field.

The simple motor introduced in the previous section had the field coils connected in series with the

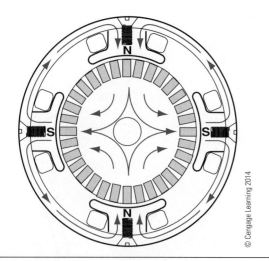

Figure 8-13 Interaction of magnetic fields results in armature rotation.

armature loop. The motor shown in **Figure 8-14** is such a motor with four field coils and two sets of brushes and is called a **series-wound** motor. Any current that flows through the field coils must also flow through the armature. As the load on the motor increases, the current increases, as does the torque developed by the motor. Most heavy-duty starter motors are series-wound motors.

Some small DC motors may have the field coils wired in parallel with the armature. This is called a **shunt-wound** motor and is shown in **Figure 8-15**. Shunt-wound motors are designed to limit the speed of the motor under lightly loaded conditions. However, the torque output of a shunt-wound motor does not

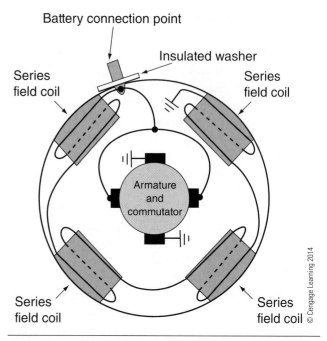

Figure 8-15 Shunt-wound motor.

increase by very much as the load placed on the motor is increased, so they are not typically used as starter motors.

Some light-duty starter motors are a combination of a series-wound and a shunt-wound motor called a **compound-wound** motor. Compound-wound motors may have three field coils connected in series with the armature windings and have one field coil that is connected in parallel with the series-connected windings, as shown in **Figure 8-16**.

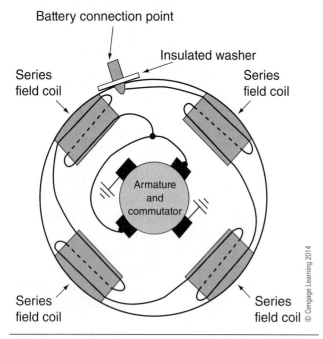

Figure 8-14 Series-wound motor.

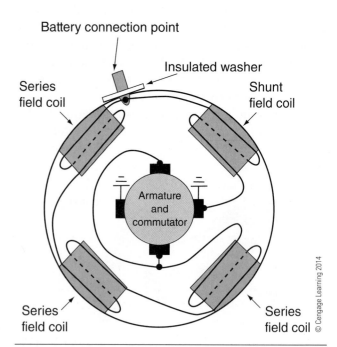

Figure 8-16 Compound motor.

Starter Motor Drives

A small-diameter **pinion gear** is connected to the starter motor's armature shaft and acts as the output gear for the starter motor. The pinion gear is designed to mesh with a large-diameter **ring gear**. The ring gear is part of the engine flywheel or flex-plate. The ratio of the number of teeth on the ring gear to the number of teeth on the pinion gear is perhaps 15:1. This means that the starter motor must rotate 15 revolutions to cause the engine to rotate one revolution. Typical warm diesel engine cranking speed may be 250 rpm, indicating that the starter motor is rotating at 3750 rpm.

When the engine starts, it may idle at perhaps 700 rpm. The ring gear rotates at the same speed as the engine because it is attached to the engine crankshaft, so the ring gear is also rotating at 700 rpm at low idle. This indicates that when the engine starts, the ratio of ring gear to pinion gear teeth would cause the pinion gear to rotate 15 times faster than the ring gear. Therefore, if the starter motor pinion gear were firmly attached to the armature shaft and always remained engaged with the ring gear, the starter motor armature would have to rotate at 10,500 rpm at low idle speed and 30,000 rpm at an engine speed of 2000 rpm. These high rotational speeds would cause the starter motor armature to disintegrate.

To prevent this damage to the starter motor due to excess rotational speed, a starter motor **drive** mechanism is used. The drive mechanism permits the pinion gear to disengage from the ring gear after the engine starts by a sliding action parallel with the armature shaft. The drive mechanism also permits the pinion gear to be locked to the armature shaft in one direction and free to rotate on the armature shaft in the other direction. This type of action is referred to as a **one-way clutch** or **overrunning clutch**. This one-way clutch prevents destruction of the armature because of rapid acceleration due to high rotational speed of the ring gear, which results from the brief period of time when the key may still be held in the crank position and the engine has just started. The armature shaft via the pinion gear can drive the ring gear, but the ring gear cannot drive the armature shaft in the direction of engine rotation due to the one-way clutch. This is similar to a bicycle pedal crank in which your leg muscles are like the starter motor. The pedal crank can drive the rear wheel via the chain, but the rear wheel cannot drive the pedal crank. Otherwise, going down a hill would require you to pedal the bicycle very fast and you could never coast.

The drive mechanism also permits the pinion gear to slide into mesh with the ring gear before the motor begins to rotate rapidly. Otherwise, sliding a rotating pinion gear into mesh with the stationary engine ring gear causes damage to both gears, as will be explained later.

> **Tech Tip:** Most starter motor drives of the past were manufactured by the Bendix® Company. Therefore, the term *bendix* is often used to describe a starter motor drive.

Starter motors use a **solenoid** to engage the drive mechanism. A solenoid is an electromagnetic device used to cause some remote mechanical movement. A solenoid typically contains a spring that opposes the magnetic attraction created by current flow through the coil (**Figure 8-17**). When current flows through the coil, the armature is pulled into the solenoid due to the magnetic field created by the current flow (**Figure 8-18**).

Most diesel engine starter motor drives are the fully enclosed shift lever type shown in **Figure 8-19**. A solenoid assembly mounted atop the motor housing is used to move the shift lever to engage the drive assembly. A spring keeps the drive assembly in the disengaged position so that the pinion gear is not meshed with the ring gear. When current flows through the solenoid windings, a magnetic field is developed that causes the soft iron plunger to be drawn

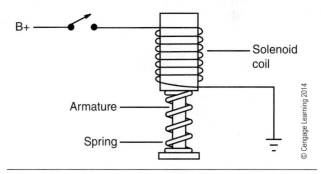

Figure 8-17 Solenoid with coil not energized.

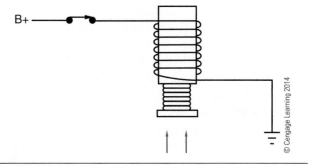

Figure 8-18 Solenoid with coil energized.

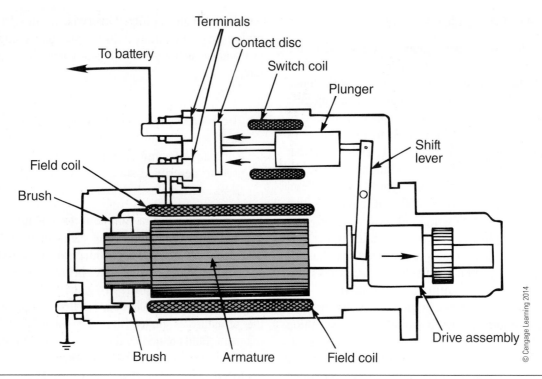

Figure 8-19 Shift-lever-type drive.

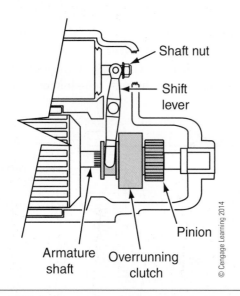

Figure 8-20 Drive mechanism.

into the magnetic field. This movement is translated to the pivoting shift lever, which compresses the spring and slides the drive assembly with pinion gear into mesh with the ring gear (**Figure 8-20**).

The solenoid assembly has another job besides pulling on the shift linkage to engage the drive assembly. The solenoid plunger assembly also contains a switch contact disc on the end opposite the shift lever. The solenoid plunger movement causes the high-current switch contact disc to make contact with the terminals and supply power to the starter motor (**Figure 8-19**).

If the solenoid plunger stoke is complete, meaning the pinion gear is fully engaged with the ring gear, the contact disc provides an electrical connection with the terminals and power is supplied to the starter motor windings.

Truck starter motor drives are typically **positive engagement** type. *Positive engagement* means that the motor windings will not be supplied with voltage until the pinion gear is fully engaged with the ring gear. In order for the motor to begin rotating, the contact disc in **Figure 8-19** must make contact with the terminals to provide voltage to the starter motor windings. This prevents the motor from rotating until the pinion is positively (fully) engaged with the ring gear. A helix or spiral machined in the armature shaft and drive collar causes the drive to rotate slightly as the drive assembly slides on the armature shaft. This slow pinion gear rotation along with tapered pinion gear teeth minimizes the likelihood of **tooth abutment** occurring. *Tooth abutment* refers to the pinion gear being in such a position that the pinion gear teeth crash into the ring gear teeth when the solenoid is energized, thus preventing engagement of the pinion gear with the ring gear.

Pinion gear to ring gear tooth abutment typically does not occur very often. However, should tooth abutment occur, the solenoid plunger will not achieve full stroke so the starter motor will not be powered. The key switch must be released and an attempt made at cranking the engine again. Typically, tooth abutment will not occur a second time in a row because only a small amount of rotational movement of the pinion gear is necessary to clear tooth abutment.

If the motor were to rotate under power with tooth abutment occurring, the pinion gear would act as a machinist mill and quickly destroy both the pinion gear and the ring gear. Such damage to the gears is known as ring gear or pinion gear **milling** and results in costly damage.

You should be aware of the action of a positive engagement type of drive because at some point you will probably encounter a tooth abutment while attempting to crank a diesel engine. You will know that tooth abutment has occurred because holding the key switch in the crank position causes just a single clunk noise with no cranking. The clunk noise is the result of the solenoid pulling the pinion gear into contact with the ring gear, and no cranking occurs until you release the key switch and attempt to crank again.

As mentioned, one of the tasks of the drive assembly is to permit the pinion to drive the ring gear but not permit the ring gear to drive the pinion gear. Most diesel engine starter motors use a sprag-clutch or roller-clutch type of drive to permit the one-way transfer of torque from the pinion gear to the ring gear. The sprag clutch and roller clutch are both one-way clutches, or overrunning clutches as they are sometimes known. A roller type of clutch is shown in **Figure 8-21**. The roller clutch is designed to permit a rotation in one direction yet lock the two elements shown in the other direction. The spring-loaded rollers act as wedges to lock the two elements.

Engine Crank Interlocks

The starter solenoid can exceed 60A (as high as 300A on gear reduction starters as discussed in the next section). As indicated, a relay or magnetic switch is used to switch this 60A of current to the starter motor solenoid. The control coil of this relay is typically supplied voltage by the key switch in the crank position. Federal Motor Vehicle Safety Standards (FMVSS) in the United States require that the engine on vehicles with automatic transmissions may only be cranked with the shift selector in the park or neutral position. The automatic transmission **electronic control unit (ECU)** may supply a ground or high side (positive) signal via a low side or high side driver when the transmission selector is in park or neutral for use in the crank control relay circuit, as shown in **Figure 8-22**.

Additionally, some OEMs prevent the engine from cranking on trucks equipped with manual transmissions unless the clutch pedal is depressed. The control relay coil for the starter motor solenoid may have a clutch switch or a manual transmission neutral switch wired in series with the key switch crank position output. The clutch switch may instead interrupt a path to ground for the control relay coil. On trucks with automated mechanical transmissions, the transmission ECU typically supplies both the high side (positive voltage) and ground for a crank-inhibit relay.

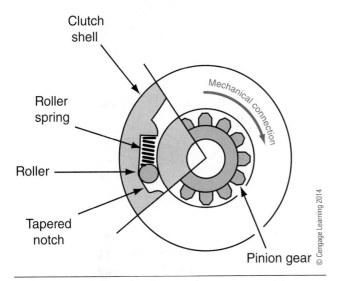

Figure 8-21 Roller clutch permits one-way drive.

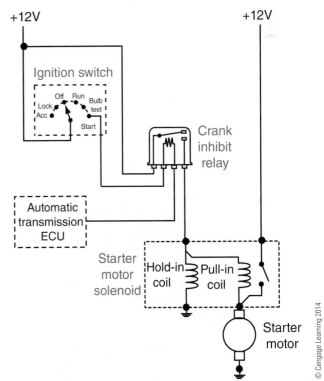

Figure 8-22 Crank inhibit circuit.

There is really no set standard indicating how a starter motor control circuit is designed from truck to truck. Until you learn how each starter motor control circuit is designed, a circuit diagram is an absolute necessity when troubleshooting a no-crank problem.

Thermal Protection

Starter motors should be operated for no more than 30 seconds at a time. Two or more minutes between cranking attempts must be provided to permit the starter motor to cool. Diesel engines should typically start in much less than 30 seconds, but cold ambient temperatures and restarting a diesel engine that has run out of fuel are two exceptions. It is often tempting to keep cranking an engine longer than 30 seconds when it seems the engine is just about to start in cold weather. However, the high temperatures caused by the large amount of current drawn by the starter motor can result in expensive damage to the starter motor, battery cables, and batteries.

To minimize the chances of damaging the starter motor due to excessive temperatures, some diesel starter motors have provisions for a thermal overcrank protection (OCP) switch. This normally closed switch is threaded into the starter motor to sense internal motor temperature and is designed to open if the starter motor becomes too hot. This switch is only designed to handle a few amps of current, so the thermal overcrank switch is typically wired in series with the crank control relay coil or magnetic switch coil. Thus, if the starter motor becomes too hot from an extended cranking cycle, the overcrank switch opens, which disables the starter motor for several minutes until the motor cools and the overcrank switch closes again.

GEAR-REDUCTION STARTER MOTOR

Up until the last few years, the typical diesel starter motor was a direct drive motor, like those discussed up to this point. The pinion gear was driven directly from the armature shaft of the motor without any intermediate gearing. A **gear-reduction** starter motor uses a motor that is physically much smaller than a comparable direct drive starter motor. However, this smaller motor is designed to rotate at a speed that is much greater than a direct drive motor. The output of the small motor is coupled to a gear set such as the planetary gear set shown in **Figure 8-23**. Planetary gear sets are found in automatic transmissions. The gear set is used to reduce the output shaft speed, thereby increasing the starter drive torque.

Gear-reduction starter motors are gaining popularity in the truck industry and have been utilized in

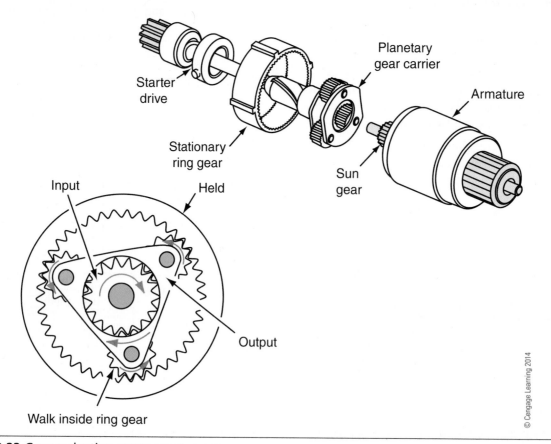

Figure 8-23 Gear-reduction starter motor cutaway.

automobiles for some time. Gear-reduction starter motors are now standard equipment for many heavy- and medium-duty diesel engines. Gear-reduction starter motors are also available as replacement starter motors for many older engines originally equipped with direct drive motors.

Advantages of Gear-Reduction Starter Motors

One advantage of a gear-reduction starter motor over a direct drive motor is the reduced mass (weight) of the gear-reduction motor.

Another advantage of using a gear-reduction starter motor is that an increased engine cranking speed over TDC of each engine compression stroke can be obtained, even though the average cranking speed remains about the same as a comparable direct drive motor. This increase in speed over TDC is said to improve the cold start performance of a diesel engine.

Gear-reduction starter motors are also more efficient than direct drive motors. This means that less of the electrical energy that is being converted to mechanical energy is lost as heat. One starter motor manufacturer indicates up to a 40 percent decrease in starter motor current with the gear-reduction motor compared to a comparable direct drive motor. This means less current is also flowing through the cranking circuit, so the available voltage at the starter motor terminals is increased compared to direct drive starters.

Most gear-reduction starter motors are soft-start motors. The soft-start design causes the armature to rotate slowly as the drive with attached pinion gear is sliding toward the ring gear. This slow rotation of the pinion gear provides a greater likelihood that the pinion gear is fully engaged with the ring gear before fully cranking power is applied, which reduces ring and pinion gear milling. However, the soft-start design also causes most gear-reduction starter motors to draw much more current through the motor's solenoid terminal compared to most direct drive starter motors.

Disadvantages of Gear-Reduction Starter Motors

The solenoid windings on most large direct drive starter motors draw about 60A of current initially, which drops off rapidly to perhaps 20A. See the Extra for Experts section regarding solenoid hold-in and pull-in windings for an explanation. By comparison, the solenoid windings on a gear-reduction starter motor may initially draw 300A. This is much more current than a typical relay can handle. Therefore, most gear-reduction starter motors require a separate magnetic switch (mag switch) to switch the 300A of solenoid

current. The magnetic switch is typically installed directly to the gear-reduction starter motor.

Because of the high level of solenoid current with gear-reduction starter motors, replacing a direct drive motor with a gear-reduction motor may require a magnetic switch to be installed if the truck previously used only a standard relay to control the starter motor solenoid. Otherwise, the truck's cranking solenoid relay may be quickly destroyed by the 300A of solenoid current. Some starter motor manufacturers supply the magnetic switch with new replacement motors.

DIAGNOSIS OF THE CRANKING SYSTEM

By far, the truck's starter motor requires more electrical power than any other electrical system component. Supplying adequate voltage to the starter motor is essential for reliable engine starts. Before any testing of the cranking system is performed, the batteries must be individually tested and recharged or replaced if necessary. If the batteries are not being properly recharged by the truck's charging system, this charging system problem should be addressed first before testing the cranking system. In many cases, a cranking system complaint such as slow cranking is the result of a charging system problem causing an inadequate state of charge of the batteries.

Cranking Circuit Resistance

The resistance of the starter motor battery cables and the resistance of the starter motor windings form a series circuit when the motor solenoid is closed during engine crank. The rules of series circuits indicate that the voltage supplied by the battery will be divided and drop proportionally to the series resistances.

The resistance of the cables between the starter motor and the battery is very low because the cables are made of large-diameter copper. The resistance of the positive and negative starter motor battery circuit together is typically designed to be less than $1m\Omega$ (0.001 Ω or 1/1000th of an ohm). The $1m\Omega$ of total resistance for the cranking circuit may not seem like much, until the average current of 1000A drawn by a typical starter motor is considered. Ohm's law indicates that 1000A of current flowing through $1m\Omega$ of resistance results in a voltage drop of 1V on the total cranking circuit. It has been estimated that a 1V decrease in voltage measured at the starter motor terminals results in a decrease of about 28 rpm in cold engine cranking speed. Thus, adding only 0.001Ω $(1m\Omega)$ of additional series resistance to the cranking circuit may decrease the cold cranking engine speed enough to prevent the engine from starting.

Excessive resistance in the starter motor supply circuit may be caused by loose or corroded connections. This corrosion may even be located between the strands of copper wire and the cable terminal. Cables that are undersized because of relocated battery boxes are another cause of high resistance in the cranking circuit. Adding additional length to the cranking circuit increases the cranking circuit resistance.

Depending on battery box location (length of cables), some trucks may require two parallel-connected 4-0 cables for the positive circuit and another pair of 4-0 cables for the negative circuit to achieve this 1mΩ value. It is also common for OEMs to use the frame rail as part of the negative circuit. A cable connected between the battery's negative terminal and frame rail and another cable connected between the starter motor ground stud and frame causes the frame to act as a negative battery cable. On some medium-duty trucks, the low resistance of the frame rail alone is sufficient to act as the negative circuit. Heavy trucks may use one or more 4-0 negative battery cables and use the parallel-connected frame rail as well for the negative starter motor circuit.

Measuring the resistance of the starter motor supply circuit cannot be performed using a standard DMM ohmmeter. The resistance value of this circuit is just too small. The resistance of the entire cranking circuit is much less than the resistance of the test leads used on a DMM. Additionally, poor connections in high-current circuits often do not show up as a problem until a large amount of current flows through the connection. For these reasons, the voltage dropped on the cranking circuit battery cables while conducting a known (typically 500A) steady amount of current is measured instead of directly measuring the circuit resistance, as shown in **Figure 8-24**. Once again, the

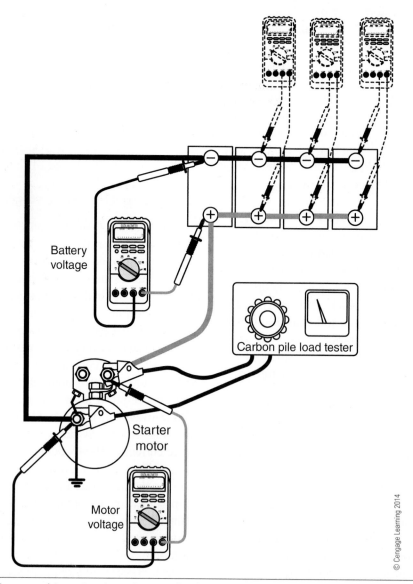

© Cengage Learning 2014

Figure 8-24 Measuring cranking circuit resistance by loading to 500A and measuring voltage drop on circuit.

batteries must be tested and recharged if necessary before performing any cranking system testing.

To cause a steady 500A of current to flow through the cranking circuit, a carbon pile load tester is used along with a digital voltmeter and a clamp-on ammeter. A method for testing the cranking circuit resistance is shown in **Figure 8-24**.

Details for this procedure are as follows:

1. Connect the carbon pile across the starter motor B+ junction and ground terminal.
2. Connect a DMM voltmeter across the battery's positive and negative terminals.
3. Briefly load the carbon pile to 500A and obtain battery terminal voltage measurements.
4. Connect the DMM voltmeter to the starter motor B+ and ground terminals. Do not connect the voltmeter to the carbon pile clamps because this will cause a measurement error.
5. Briefly load the carbon pile to 500A and obtain starter motor terminal voltage measurements.
6. Subtract the loaded starter motor terminal voltage from the loaded battery terminal voltage measurements. The result is the amount of voltage that is dropped on the positive and negative cranking circuit battery cables.

Alternatively, two DMMs can be used, along with an assistant to measure the battery terminal voltage and the starter motor terminal voltage simultaneously with 500A of current flowing in the cranking circuit.

The permissible voltage drop on the entire cranking circuit varies between OEMs. However, a total cranking circuit voltage drop of 0.5V with 500A flowing indicates a cranking circuit resistance of 1mΩ (0.001Ω). A total cranking circuit (positive and negative cables) voltage drop of 0.5V or less is usually acceptable for medium-rated diesel engines. Large-displacement heavy-duty diesels may require a cranking circuit total voltage drop of 0.375V (375mV) or less with 500A flowing. This equates to a cranking circuit resistance of 750μΩ.

If the cranking circuit voltage drop is excessive, the next step is to determine which cable, positive or negative, is the source of the excessive resistance. This test is shown in **Figure 8-25**.

Steps for performing this test are as follows:

1. Connect the carbon pile across the starter motor B+ junction and ground terminal.
2. Connect a DMM voltmeter between the battery-positive terminal and the starter motor positive terminal. Do not connect the voltmeter to the carbon pile clamps.

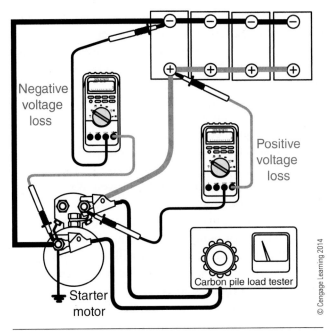

Figure 8-25 Finding the source of the high cranking circuit resistance.

3. Briefly load the carbon pile to 500A and obtain the positive circuit voltage drop measurement from the voltmeter.
4. Connect a DMM voltmeter between the battery-negative terminal and the starter motor negative terminal. Do not connect the voltmeter to the carbon pile clamps.
5. Briefly load the carbon pile to 500A and obtain the negative circuit voltage drop measurement from the voltmeter.

The positive and negative circuit voltage drops should each be approximately one-half the maximum allowable voltage drop for the entire cranking circuit. Values higher than this indicate excessive resistance in that circuit.

All of the preceding examples for cranking circuit voltage drop are specified for a 12V direct-drive starter motor. A comparable 24V starter motor has less current flow in the starter motor circuit than a 12V system, so the permissible voltage drops are increased for 24V starter motors. See the OEM's information for the permissible voltage drops with a 24V starter motor or for gear reduction starter motors.

If a carbon pile load tester is not available, an alternative means of testing the cranking circuit voltage drop is to disable the fuel system to prevent the engine from starting and crank the engine while measuring voltage drops. Consult the specific OEM's service information for the proper way to disable the fuel system to permit the engine to crank without starting. This engine cranking method is much more difficult to accurately test voltage drop across sections of the engine cranking circuit because the current draw is not a steady value due to engine compression, as was illustrated in **Figure 8-9**.

Tech Tip: Unfortunately, some shops are not as well equipped as others so some improvised diagnostic techniques may be required throughout your career. However, as a student technician it is important that you obtain a good understanding of the theory and concepts behind testing and diagnosis of a problem in a training environment where you should have all the necessary tools. This will assist you in developing your own diagnostic techniques in the future using the available tools.

If cranking circuit voltage drops are above the acceptable values, the cause of the excessive resistance must be located. Often, corroded or loose connections are the cause of the excessive resistance. **Figure 8-26** illustrates the measurement of the voltage dropped across the battery's positive terminal and the battery cable while a carbon pile is connected across the starter motor connections or the engine is cranking with the fuel system disabled. Cleaning the various terminals in the cranking circuit, including any frame rail to cable terminal junctions, will often substantially reduce starter motor circuit voltage drop. In some instances, the battery box may have been relocated farther away from the starter motor by a body builder and additional parallel cables need to be added to reduce circuit resistance to make up for the increased cable length. A parallel battery cable connection inadvertently may have been left disconnected the last time a

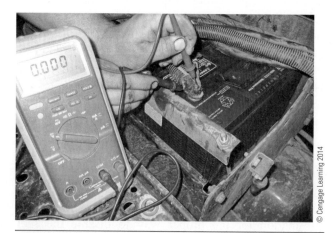

Figure 8-26 Testing voltage drop across battery terminals.

major repair was performed on the truck, causing an increase in cranking circuit resistance.

Another possible source of cranking circuit resistance is the high-current motor solenoid switch contacts. This switch contact resistance was not included in the cranking circuit voltage drop using the steady 500A load. On most starter motors, the output of the solenoid switch is accessible where it connects to the positive motor windings. To measure the solenoid switch voltage drop, connect a DMM across the solenoid switch terminals and crank the engine. The voltmeter will fluctuate some, but the voltage drop across the solenoid switch contacts should not exceed 200mV (0.2V) while the engine is cranking. The solenoid assembly containing the high-current switch is typically not serviceable and must be replaced as a unit if defective.

If cranking circuit resistance is acceptable as determined by the voltage drop method, the cause of a slow cranking speed may be an internal starter motor problem. In the real world of truck repair, time is usually very important. The fastest way of ruling out a starter motor problem is to replace the starter motor with a known good motor. Prior to replacing the motor though, verify that excessive cranking circuit resistance is not the cause of the problem and that the battery system is in good order. Also, take current draw and engine cranking speed measurements before replacing the motor. Compare these measurements after replacing the motor.

If an engine still cranks slowly after starter motor replacement, as indicated by the current draw and cranking speed measurements, the cause of the cranking problem may be the engine itself or by components driven by the engine. Excessive friction between rotating components in the engine can cause slow cranking speeds, as well as high external parasitic loads such as hydraulic pumps or automatic transmission problems. In some cases, it may be necessary to disconnect the engine from all external loads to determine if the slow cranking problem is caused by something outside of the engine.

Starter Motor Control Circuit Tests

The starter motor supply circuit voltage drop discussed previously is just one test. Several other voltage drop tests of the cranking system may be required based on the problem. Many of these tests deal with the starter motor solenoid control circuit. The starter motor solenoid for a conventional direct-drive starter motor may draw 60A or more when both the hold-in and pull-in windings are energized to shift the starter drive into mesh with the ring gear. The control circuit

0W-30   5W-30   10W-30   15W-40   30

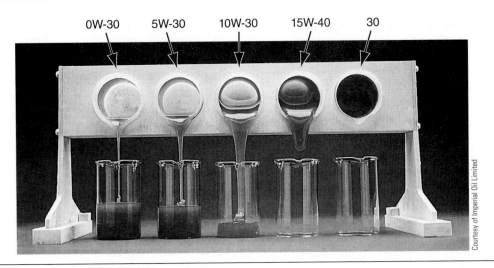

Figure 8-27 Cold engine oil viscosity demonstration.

for gear-reduction starter motors may draw 300A or more. Testing this circuit may also require the use of a carbon pile to place a specific load on a circuit or component such as a magnetic switch while the applicable voltage drop is measured. The acceptable value of voltage drop varies with the circuit and the OEM. Again, the most common cause of high resistance is corroded terminals at connection points throughout the circuit.

Other tests of the starter motor control circuit involve the use of a voltmeter to measure the various voltage drops on circuits when the magnetic switch or relay is energized. The voltmeter is moved throughout the circuit to test for excessive resistance or open circuits. OEM circuit diagrams and troubleshooting information should be consulted for these tests because safety interlock devices such as a clutch switch may be used to prevent the engine from cranking unless the clutch is depressed.

Engine Oil and Cold Weather Cranking

Having the correct viscosity of engine oil is very important for cold weather starts, as illustrated by the cold weather flow ability of oil shown in **Figure 8-27**. Many failures to crank fast enough or long enough to start a diesel engine in cold weather have been falsely blamed on batteries or the starter motor, when the real problem was the wrong engine oil viscosity for the ambient temperature. The difference in cranking speed at $0°F$ ($-18°C$) between 10W-30 oil and 15W-40 oil can be significant. Most modern diesel engines require a cranking speed of at least 125 rpm or more to start. Many fleets in moderate climates use the same viscosity of oil year-round (often SAE 15W-40) even though the owner's manual may indicate that 10W-30

or lighter oil is necessary for colder temperatures, as shown in **Figure 8-28**. Severe cold spells in these moderate climates may result in a rash of no-start complaints because of engine oil viscosity causing the engine to crank too slowly to start.

Like most accidents, there is often not just one root cause but rather a series of events or marginal conditions that cause a failure. The wrong viscosity of engine oil alone may not prevent a diesel engine from starting when the temperature drops to $10°F$ ($-12°C$) on a cold Monday morning. However, a minor parasitic current draw decreasing the battery state of charge combined with a layer of corrosion on the battery cables along with the wrong viscosity of engine oil have an end result of an engine that cranks too slowly

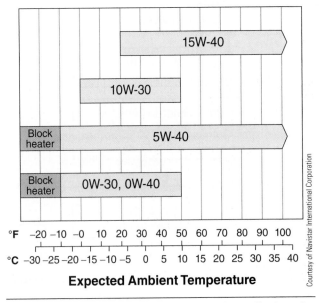

Figure 8-28 Recommended SAE engine oil viscosity grades for post-2007 engines.

to start. A good troubleshooter learns to evaluate all of the clues.

Diagnosing a Sample No-Crank Problem

The troubleshooting procedure introduced in **Chapter 4** will be used to troubleshoot a sample no-crank complaint. This procedure is shown in **Figure 8-29**.

A medium-duty truck with an automatic transmission has been towed to your shop. The work order indicates "engine will not crank." The words *start* and *crank* indicate two very different things to a technician, even though they may be interchanged by persons outside the trade. *Start* means that the engine runs or rotates under its own power. *Crank* means that the starter motor is causing the engine to rotate. You turn the key switch to the crank position and determine that the engine does not crank. This is good news for you as a technician because the problem is not an intermittent problem. You have verified the concern, which is the first step in the troubleshooting procedure.

Step 2 on the chart is to perform preliminary checks. Because this truck has an automatic transmission with a cable-operated control linkage, you cycle the shifter throughout each position while holding the key switch in the crank position. The engine does not attempt to crank in any of these positions. You also listen very carefully when the key switch is in the crank position to determine if you hear any solenoid click. When the key is switched to the crank position, no relay clicking or any other noises are heard. However, the headlamps are bright and everything else in the electrical system seems to work as normal, even when you attempt to crank the engine. This indicates that the no-crank problem is not caused by discharged batteries. A visual inspection, which is a part of preliminary checks, does not reveal any obvious problems.

Step 3 on the chart is to refer to service information. This requires the OEM circuit diagram or troubleshooting guide to be consulted, especially if you are not familiar with the electrical system on a particular model of truck.

The diagram in **Figure 8-30** indicates that a magnetic switch is used to supply power to the starter motor solenoid. The circuit diagram indicates that the high side of the magnetic switch coil is provided with a positive voltage from the key switch in the crank position. The low side of the magnetic switch coil is connected to a neutral start switch on the transmission linkage. The neutral start switch provides a ground only when the transmission is in neutral.

BASIC ELECTRICAL/ELECTRONIC DIAGNOSTIC PROCEDURE FLOWCHART

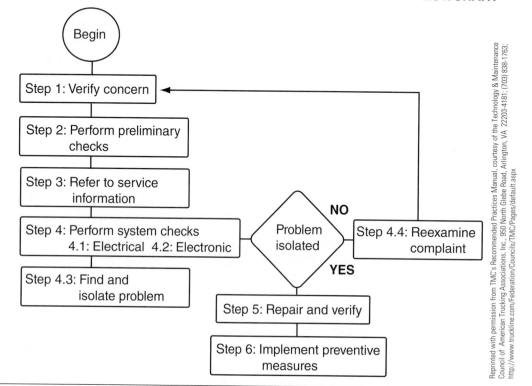

Figure 8-29 Diagnostic flowchart.

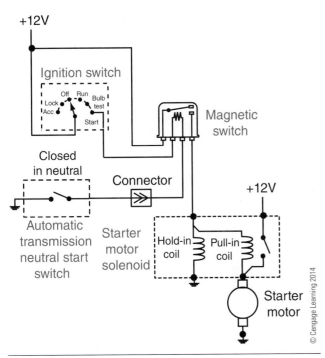

Figure 8-30 Cranking circuit diagram.

Therefore, two things are necessary for the magnetic switch on this particular truck to close:

- The neutral start switch must provide a ground to the magnetic switch control coil low side when the transmission is in neutral.
- The key switch must provide +12V to the magnetic switch control coil high side with the key switch in the crank position.

The next step, step 4, is to perform system checks. This is much easier to do now that you know how the system is supposed to operate.

Looking at the circuit diagram, you observe that it would be very useful to know if the magnetic switch was being energized with the key switch in the crank position. The magnetic switch is very easy to access on this truck, so that makes it an ideal place to start the troubleshooting. An assistant will be necessary for this task. Touching the magnetic switch to feel for a click while an assistant attempts to crank the engine indicates that the magnetic switch is not clicking. The cause of the magnetic switch not being energized on this particular truck may be as follows:

- The high side control of the magnetic switch coil supplied by the key switch is not present at the magnetic switch.
- The low side control of the magnetic switch coil (ground) supplied by the neutral start switch is not present at the magnetic switch.
- The magnetic switch is defective.

Step 4.3 is to find and isolate the problem. Using a voltmeter, you determine that there are +12V referenced to chassis ground at the high side of the magnetic switch coil with the key switch in the crank position, as shown in **Figure 8-31**. This rules out a problem with the high side of the magnetic switch control circuit. You also determine that there are +12V at the low side of the magnetic switch coil referenced to ground when attempts are made to crank the engine. This indicates that the low side of the magnetic switch is not being grounded. If the low side of the magnetic switch coil were being grounded by the neutral switch, the voltmeter should have indicated 0V. This +12V measurement at the low side of the magnetic switch coil referenced to ground with the key switch in the crank position also indicates continuity through the magnetic switch coil, so the magnetic switch is likely not the source of the no-crank problem.

The trouble has now been isolated to a problem with the low side of the magnetic switch coil circuit.

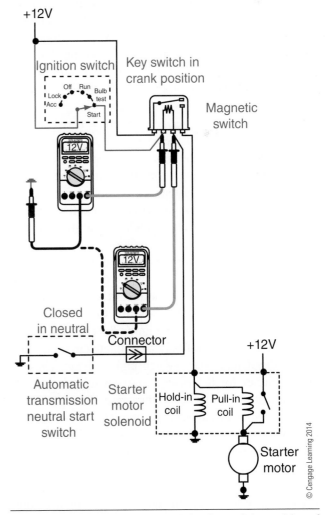

Figure 8-31 DMM measuring high and low side of magnetic switch coil during crank.

The circuit diagram indicates that there is a connector set between the magnetic switch low side and the neutral start switch. Where to go next in the diagnosis is largely dependent on what is easiest to access and what is most prone to fail. Getting to the connector to determine if the neutral start signal is present will be difficult on this particular truck. There is also not much to fail at that location anyway. A visual inspection indicates that the connector set is mated, so you decide to find an easier place to check for now. On the other hand, the transmission neutral start switch is easy to access and because it is a switch with moving components, it is more prone to failure than the connector set. Therefore, you disconnect the connector for the neutral start switch and inspect the terminals for corrosion, pushed out terminals, and other problems. The wiring harness terminals and the terminals on the neutral start switch all appear to be okay.

The circuit diagram indicates that there is a chassis ground connection on one side of the neutral start switch and a circuit to the magnetic switch coil on the other side of the connector. Several steps could be performed next to locate the source of the problem. One of the easiest steps is to verify that ground is present at the neutral start switch wiring harness connector, using a DMM to measure continuity to chassis ground. This test indicates that there is a good ground at the neutral start switch wiring harness connector. Next, you verify continuity between the neutral start switch and the magnetic switch coil low side. This could be done using an ohmmeter, but this would require connecting a long test lead all the way to the magnetic switch. Instead, you could have an assistant attempt to crank the engine with the neutral switch connector disconnected while you measure the voltage between the two terminals of the neutral start switch connector. If the circuit between the neutral start switch and the magnetic switch coil is intact and the ground circuit for the neutral start switch is intact, you should measure +12V across the two terminals of this connector, as shown in **Figure 8-32**. In doing so, you verify that there is likely a complete circuit all the way to the neutral start switch and an intact chassis ground circuit for the neutral start switch. It appears that the neutral start switch is the cause of the problem. Because you have isolated the problem to the neutral start switch, you can go to step 5, repair and verify. A new or known-good neutral switch is connected to the wiring harness before the old switch is removed from the vehicle. The engine now cranks and starts with the new switch connected to the neutral start switch wiring harness connector when the switch is closed (simulated neutral position). However, before removing the old

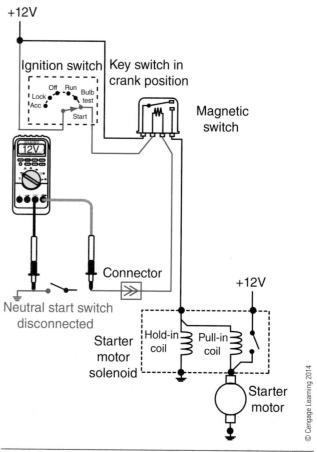

Figure 8-32 Voltage measurements at neutral start switch.

switch, disconnect the new switch and reconnect the wiring harness neutral start switch connector to the old switch just to verify that in moving things around, you have not just temporarily corrected an intermittent wiring problem. With the old switch reconnected to the wiring harness, the engine will not crank again. With an assistant holding the switch in the crank position, you wiggle the wiring and still the engine does not crank. Your earlier test in which you moved the shifter from position to position while trying to crank should rule out a shifter cable or switch adjustment problem, so it appears that a failed neutral start switch is the cause of the no-crank problem.

Determining the root cause of the problem is desired. However, in this case the neutral start switch is a sealed component with no serviceable components. Replacing the neutral start switch with a new switch takes two minutes. The engine now cranks when the key is placed in the crank position.

It is very important to verify that everything works correctly after the repair. The engine cranks and starts. What else needs to be done? Recall that a truck with automatic transmission should only crank with the transmission in park or neutral. In checking for crank inhibit in ranges other than neutral, you find that the truck cranks with the shift selector just at the edge of the reverse range. The engine also does not crank when the shifter is pulled to the bottom end of the neutral position. An adjustment on the shifter cable linkage or neutral start switch is necessary to correct this problem. Now the engine will only crank with the shift lever in neutral.

This shows the importance of checking your work. Giving the truck back to the owner with the neutral switch out of adjustment might have resulted in another tow-in for a no-crank problem. Permitting the engine to crank with the transmission in reverse could have resulted in an accident.

Step 6, the last step, is to implement preventive measures. A quick check of the rest of the electrical system indicates no other problems. However, you find some other problems with the truck and bring these to the attention of the service writer.

STARTER MOTOR DISASSEMBLY AND TESTING

In the past, one of the common tasks of a truck technician was to rebuild or recondition starter motors. Unfortunately, these skills are being lost because most shops just replace questionable starter motors instead of repairing or reconditioning them. The high cost of replacement components (if even available) and labor typically makes it more economical to replace a starter motor that has an internal problem. Even though most technicians will probably never disassemble a starter motor outside of the classroom, having some knowledge of the internal workings of the motor will make you a better troubleshooter.

After removing the starter motor from the engine but prior to disassembly, no-load rpm and current draw measurements are obtained and compared against the manufacturer's specifications for the specific motor. The test setup is shown in **Figure 8-33**. This testing reveals problems such as excessive friction or high internal motor resistance. For example, higher than specified current draw at a lower than specified no-load speed can be an indication of excessive friction caused by tight starter motor bearings. Think about this; it should make sense. Tight bearings decrease the motor speed, which increases the motor current draw. Problems with the windings, such as a wire-to-wire short in the armature windings or a grounded field coil, could also cause high-current draw with lower than specified no-load speed. These defects alter the magnetic field interactions, resulting in poor motor performance.

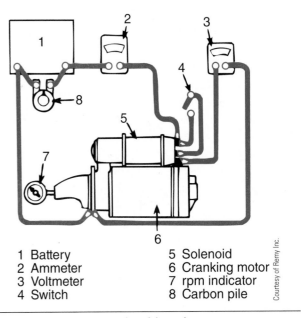

1 Battery
2 Ammeter
3 Voltmeter
4 Switch
5 Solenoid
6 Cranking motor
7 rpm indicator
8 Carbon pile

Courtesy of Remy Inc.

Figure 8-33 Starter no-load bench-test setup.

A motor with a lower than specified no-load speed and a lower than specified current draw indicates excessive internal motor resistance. This may be caused by poor contact between the brushes and commutator bars due to burning. This should also make sense because the excessive resistance limits the current drawn by the motor, which weakens the magnetic fields and thus reduces no-load speed.

Tech Tip: Often, the technician is given a list of possible causes for a problem in a repair manual such as those listed previously for the slow cranking speed. You may find it useful to evaluate why each possible cause listed could cause the problem. No list of potential causes can provide every possible cause for a problem. This type of exercise helps to sharpen your troubleshooting skills as you learn to formulate your own list of possible causes. Having this ability helps you to figure out problems when other technicians are stumped.

After the no-load speed and current tests, the motor is disassembled and each component tested to a service standard. Disassembly of a Delco Remy 37-MT direct-drive starter motor, used on many medium-duty diesels, is illustrated in Photo Sequence 4. The external components of this starter motor are shown in **Figure 8-34**.

After disassembly, clean the parts using a soft cloth. The components are then inspected and tested.

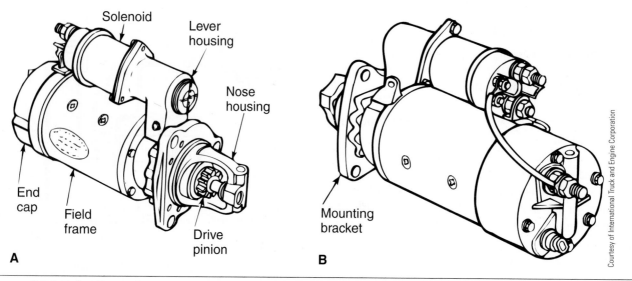

Figure 8-34 Delco Remy 37-MT starter motor external components.

Field Coil Testing

Field coil testing involves using an ohmmeter to test for open or shorted-to-ground field coils as shown in **Figure 8-35** and **Figure 8-36**. Not all field coils are connected in the same manner, so specific OEM information must be consulted to determine how the field coils are connected. Typically, any problem found with the field coils results in motor replacement.

Armature Testing and Service

A tool called a **growler** (**Figure 8-37**) is used to check for shorted armature windings. The growler causes an alternating magnetic field to be developed in

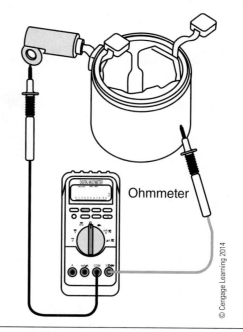

Figure 8-36 Testing for a shorted-to-ground field coil.

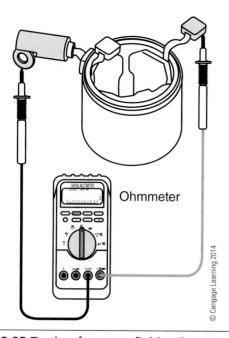

Figure 8-35 Testing for open field coils.

Figure 8-37 Growler used to test armature for shorts.

PHOTO SEQUENCE 4
Delco Remy 37-MT Starter Motor Disassembly

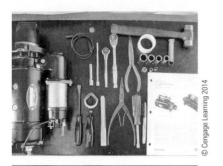

PS4-1 Assemble the tools and technical information that will be necessary to disassemble the starter motor. This example is a Delco Remy 37-MT starter motor.

PS4-4 Remove the solenoid shaft nut using an air tool if necessary. This nut is typically a self-locking type of nut that causes the solenoid shaft to rotate, preventing removal of the nut with hand tools.

PS4-7 Remove the solenoid assembly from the shift lever housing and set aside.

PS4-2 Mark each section of the motor that will be disassembled to aid in reassembly.

PS4-5 Remove the strap between the solenoid M terminal (output terminal) and the motor windings and set aside.

PS4-8 Loosen the recessed fasteners that attach the nose housing to the field frame. Note that the nose housing is designed to be installed or clocked in several different positions so that the motor can be used on several different engines. Make sure that the nose housing is marked in relationship to the field frame so that the nose housing can be reassembled in the same location.

PS4-3 Remove the plug from the shift lever housing at the drive end of the motor.

PS4-6 Remove the two fasteners that attach the solenoid to the motor shift lever housing.

PS4-9 Pull the nose housing away from the field frame and the armature shaft. It may be necessary to tap the housing flange with a soft-faced hammer to dislodge it from the field frame.

PHOTO SEQUENCE 4 (Continued)

PS4-10 Place a 5/8-in.-deep socket over the armature shaft and tap the socket with a soft-faced hammer to drive the retainer washer toward the pinion gear to expose the snap ring.

PS4-11 Remove the snap ring on the armature shaft. Note that the drive assembly cannot be removed until after the armature assembly is removed.

PS4-12 Remove the four fasteners that attach the end cap to the field frame and remove the large nut on the motor ground terminal stud. Remove the lock washer and insulation washers for the ground stud and set aside. Note the order of the ground stud washers so that they can be reinstalled in the same order.

PS4-13 Tap the motor end cap away from the field frame using a soft-faced hammer or small pry bar. Carefully pull the end cap away from the field frame. Note that the ground stud will remain attached to the field frame and will pass through the hole in the end cap.

PS4-14 This motor has four brushes. Two are positive brushes and two are negative brushes. The negative brushes are attached to the motor ground stud that passed through the end cap. The positive brushes are connected to the field windings. The springs hold the brushes tight against the commutator.

PS4-15 Remove the fastener that attaches each of the two positive brushes to the field windings. Move the field winding connections out of the way enough to permit the brush holder assembly to be removed.

(Continued)

PHOTO SEQUENCE 4 **(Continued)**

PS4-16 Carefully slide the brush holder assembly away from the field frame and set aside.

PS4-17 Pull the armature assembly out of the field frame and set the armature aside.

PS4-18 With the armature shaft removed, the drive assembly can be removed from the shift lever by rotating the drive slightly as shown.

PS4-19 Loosen the pair of fasteners that attach one of the pole shoes to the field frame. Do not let the pole shoe fall free inside the field coils. Reach inside the motor as the fasteners are removed to keep the pole shoe from falling and possibly damaging the field coil insulation.

PS4-20 Remove the pole shoes from the field frame and repeat for each pole shoe.

PS4-21 Pull out the grommet that insulates the field windings from the field frame and carefully push the field winding connector inward so that it clears the field frame.

PS4-22 Carefully lift the field windings from the field frame. Note the position of any insulation sheet between windings and field frame so that it can be reinstalled in the same location.

the armature while a thin piece of metal like a hacksaw blade is passed over the armature as the armature is rotated, as shown in **Figure 8-38**. The blade will vibrate when a short is found. A shorted armature winding usually requires armature replacement.

An ohmmeter can be used to check for grounded armature windings. The meter should indicate an open circuit between any commutator segment and the armature shaft (**Figure 8-39**). Alternatively, a special 120V AC test light is used to test for grounded armature windings. The test light uses a standard household light bulb that illuminates if an armature winding is grounded.

Testing for open circuits in armature or field coils can be performed using a standard DMM ohmmeter. There should be near 0Ω of resistance between each commutator segment (**Figure 8-40**). The reason for such low resistance is that there is only a loop of heavy-gauge copper wire between each commutator segment.

Wire-to-wire shorts, open circuits, or shorted-to-ground circuits in the field or armature require replacement of the field or armature.

Brushes

Brushes decrease in length as the brush wears during normal use. The length of brushes is measured and compared to the OEM's specifications, as shown in **Figure 8-41**. If the brush length is less than the service specification, the brushes are replaced.

Drive Assembly

The motor drive is not serviceable on most motors and should not be lubricated or cleaned with solvent unless specified by the OEM. Motor drive problems include slipping or locked one-way clutches. These conditions require drive replacement.

Solenoid

The starter motor solenoid is serviced as an assembly. Solenoid problems include open hold-in or pull-in

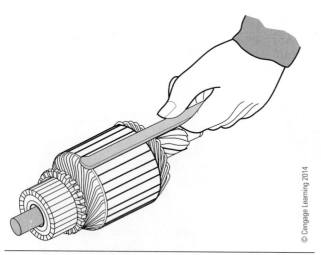

Figure 8-38 Blade vibrates and a growling noise is heard when armature is rotated to a shorted area.

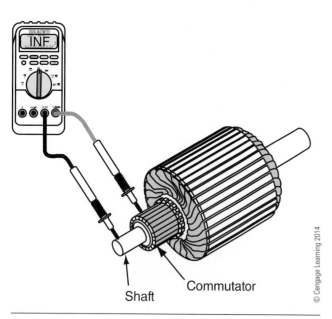

Figure 8-39 Testing armature for shorted-to-ground windings.

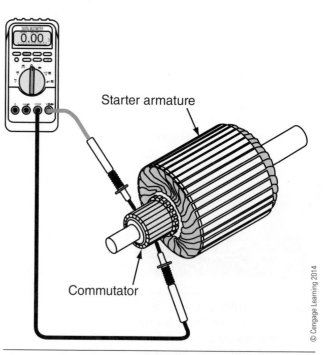

Figure 8-40 Testing armature for open circuits.

Brush holder side

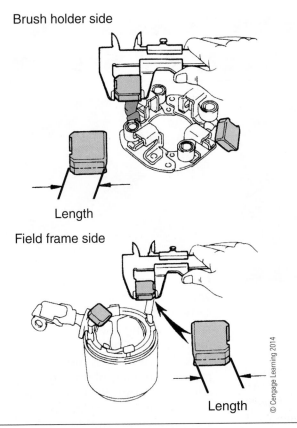

Length

Field frame side

Length

Figure 8-41 Measuring length of the brushes.

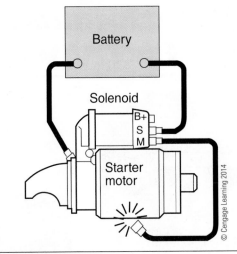

Figure 8-42 Energizing the starter motor solenoid to measure pinion gear clearance.

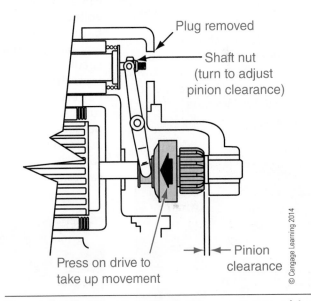

Plug removed

Shaft nut (turn to adjust pinion clearance)

Press on drive to take up movement

Pinion clearance

Figure 8-43 Checking the pinion gear to drive housing clearance.

windings and high-current contact switches with excessive resistance.

STARTER MOTOR REASSEMBLY

Starter motor reassembly involves reversal of the disassembly process, but specific OEM information should be consulted for details on motor reassembly.

Some truck starter motors require that the pinion gear to drive housing clearance be measured and adjusted to the OEM's specifications, if necessary. To measure this clearance, the motor solenoid must be electrically energized to cause the drive to be shifted into the cranking position. The motor cannot be rotating during this measurement, so a special procedure must be followed to energize only the solenoid without causing current to flow through the armature windings. An example of the electrical connections necessary for this procedure is shown in **Figure 8-42**, but specific OEM instructions should be consulted. The distance between the pinion gear and the retainer is then measured using a feeler gauge (**Figure 8-43**). Adjustment to the clearance can be made on many truck starter motors by an adjustment nut on the end of the solenoid shaft.

RAPID ASSESSMENT OF VEHICLE CHARGING AND CRANKING SYSTEM

Now that batteries, alternators, and starter motors have all been discussed, a method of performing a rapid test of the system can be addressed. This rapid test is used to determine the general condition of the charging and cranking system, not to diagnose a specific problem. This test can be performed as part of a preventive maintenance program or used when checking a customer's truck that was brought in for other problems. This procedure is outlined in Photo Sequence 5.

PHOTO SEQUENCE 5 Performing a Rapid Assessment of a 12V Vehicle Electrical System

PS5-1 Test the battery state of charge by measuring the open circuit voltage. Switch the headlamps on for 1 minute to remove any surface charge. Next, disconnect the ground cables from each battery. Use a DMM DC voltmeter to measure the open circuit voltage of each battery.

PS5-2 Record the open circuit voltage measurement for each battery. Recharge and retest any battery that has an open circuit voltage measurement less than 12.4V, assuming battery is at room temperature. Apply temperature compensation to all specifications as required if batteries are cold. Reconnect the battery cables.

PS5-3 Test the cranking voltage. Disable the fuel system to prevent the engine from starting per the OEM instructions. Connect a DMM DC voltmeter across the starter motor terminals, observing the correct polarity. Next, continuously crank the engine for 15 seconds. The DMM should indicate a voltage above 9.6V while the engine is cranking. Load test batteries and measure cranking circuit voltage drops if less than 9.6V.

PS5-4 Test the charging system voltage. Re-enable the fuel system to permit engine to start. Connect a DMM DC voltmeter across the battery terminals. Start the engine and run at 75 percent of its rated speed for 2 minutes. Switch on all electrical accessories. Record the battery terminal voltage.

PS5-5 If the battery terminal voltage measurement is less than 13.5V, a problem with the alternator, alternator drive mechanism, voltage regulator, or electrical connections is indicated.

PS5-6 If the battery terminal voltage measurement is greater than 14.5V with the truck at room temperature, a problem with the voltage regulator or the electrical connections is indicated.

OTHER BRUSHED DC MOTORS

In addition to the starter motor, many other motors throughout the truck are brushed DC motors. These are used for applications such as power windows, windshield wipers, and HVAC blower motors.

These motors do not have to supply nearly as much torque as the starter motor. To simplify the motors and reduce cost, most of these smaller motors have pole shoes that are **permanent magnets**. A permanent magnet is designed to retain its magnetic properties after initial magnetizing. Because the pole shoes are made from permanent magnets, no field coils are necessary. The only current flow through the motor is via the armature windings, as shown in **Figure 8-44**. Otherwise, the operation of permanent magnet motors is very similar to the starter motors described earlier. In fact, many small automobile starter motors use permanent magnets for the pole shoes.

Reversing Motor Rotation

One of the requirements of many permanent magnet motors, such as a power window motor, is reversibility. Reversing the direction of motor rotation requires a change in the direction of current flow through the motor armature. This means reversing the polarity of the voltage at the motor terminals. Because the voltage polarity changes, neither brush is connected directly to ground. The motor must have two terminals that are both isolated from chassis ground. Power window circuits are designed to reverse the polarity of the voltage applied to the two terminals of a power window motor armature based on the power window switch position, as shown in **Figure 8-45**.

EXTRA FOR EXPERTS

This section covers additional information related to electric motors.

Commutation

To illustrate the magnetic field interaction between the armature and the pole shoe magnets, imagine looking into the cross section of a single armature winding, as shown in **Figure 8-46**. If you could view from the vantage point shown in **Figure 8-46**, you would observe two circles representing the cross section of the armature winding. Because in this example current is flowing through the armature from the left side commutator segment to the right side commutator segment, the arrow or dart indicates that conventional current enters the left side circle and leaves the right side circle, as shown **Figure 8-46**. The armature winding is shown so that it is positioned in the middle of the commutator segments and lying flat through the centerline of the pole shoes in the right side of **Figure 8-46**. This is done for illustration

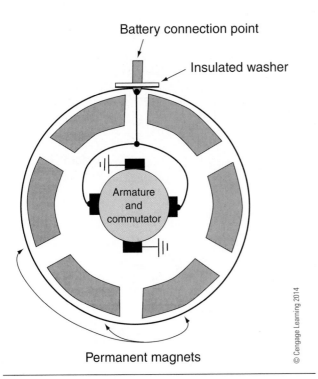

Figure 8-44 Permanent magnet motor.

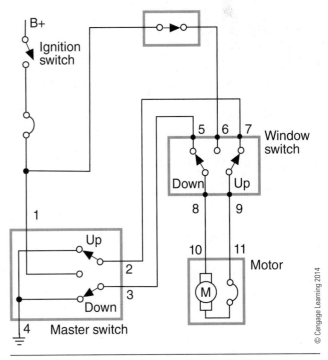

Figure 8-45 Power window circuit. Switch reverses voltage to the motor to raise or lower window.

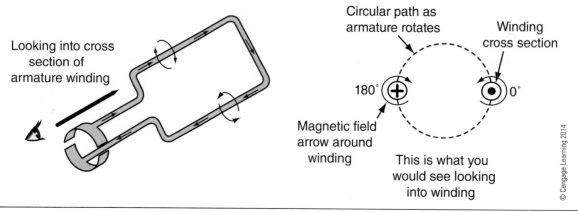

Figure 8-46 View looking into a single-loop armature.

purposes only. The loop could be started at any position on a real motor. This initial starting position will be the reference point for the remainder of this section, so it is considered the 0 degrees of rotation point as shown in **Figure 8-46**.

When the simple motor introduced earlier in this chapter in **Figure 8-4** is connected to a battery, current flows through the field coils and armature winding and back to the battery. The field current causes the pole pieces to become electromagnets with the polarity shown in **Figure 8-4**. The electromagnets cause magnetic lines of force with arrows indicating the lines "travel" from the north pole of the pole shoe on the right to the south pole of the pole shoe on the left. At the same time, the armature current causes a magnetic

field to surround the armature winding with magnetic lines of force "rotating" around the conductor in the direction shown in **Figure 8-4**, as indicated by the right-hand rule. An interaction between the two magnetic fields occurs as shown between the horseshoe magnet and conductor shown in **Figure 8-3A**.

The magnetic field interaction causes the armature loop to rotate in a clockwise direction from the 0-degree starting position as the current-carrying armature moves from an area where the magnetic field is strong to an area where the magnetic field is weak (**Figure 8-47A**). The arrows shown in **Figure 8-47** illustrate the direction of the force or torque to which the armature winding is subjected because of the magnetic field interactions. This force causes the armature loop to rotate. One of

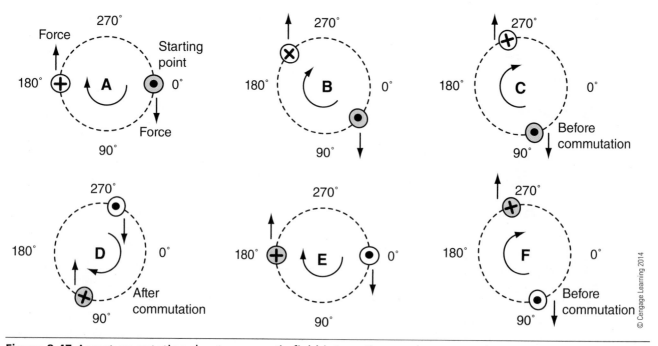

Figure 8-47 Armature rotating due to magnetic field interactions and commutation.

the cross sections of the armature conductor is shaded so that this cross section may be followed as the armature loop rotates. The torque imparted on the armature loop is due to the interactions of the magnetic fields causing the loop to move out of a strong magnetic field into a weaker magnetic field.

When the armature loop reaches a position 90 degrees clockwise from the starting position, the armature would stop rotating if the direction of current flow through the armature did not reverse directions. This is because any further rotation in either the clockwise or the counterclockwise direction would cause the armature to move into an area where the magnetic field is stronger. The 90-degree position shown in **Figure 8-47C** is as far away from the strong magnetic field that the armature can rotate as shown by the force arrows. The force arrows show the direction of the force placed on the armature loop by the magnetic field interaction. The force arrows are pointing away from each other because of the interactions of the magnetic fields. The armature cannot get further away from the strong magnetic field than at this 90-degree point. Any rotation past the 90-degree point without a reversal of current flow through the armature would cause the interacting magnetic fields to force the loop back to the 90-degree point.

A motor that only rotates 90 degrees and then stops is not of much use. However, if the direction of current flow through the armature were reversed just as the motor passed through the 90-degree point as shown in **Figure 8-47C** and **Figure 8-47D**, the loop would then continue to rotate clockwise another half-turn to the 270-degree position instead of stalling at the 90-degree point. The inertia of the rotating armature is necessary to prevent a stall at the 90-degree point. Once the current flow reverses through the loop due to commutation, the 270-degree point is where the armature is compelled to go, as shown by change in direction of the force arrows (**Figure 8-47D**). The action of the commutator causing current flow to reverse directions in the armature winding is called commutation. The force arrows indicating the torque imparted on the armature are now pointing in the opposite direction because of the change in the armature current flow direction at the commutation point. The direction of current flow through the armature is reversed just as the armature reaches the 90-degree point by the design of the split ring commutator. Notice that the direction of current flow has changed after commutation as shown by the dot and cross in the cross section of the armature loop, which indicate the point and tail of the current flow dart as explained in **Chapter 3**.

As the brushes pass over one section of the commutator to the other section of the commutator at this 90-degree point, the direction of current flow through the armature changes. The change in direction in armature current causes the arrows on the magnetic lines of force surrounding the conductor to reverse directions, as predicted by the right-hand rule. The strong and weak magnetic fields around the conductor have now reversed sides because the pole shoe magnetic field has stayed the same. Only the direction of the current flow in the armature has changed directions. The direction of current flow through the field coils remains unchanged by the reversal of current through the armature, so the magnetic field created by the field coil electromagnets also remains unchanged throughout commutation.

This reversal in the direction of current through the armature winding and the inertia of the rotating armature causes the armature to rotate past the 90-degree point an additional half-turn to the 270-degree point, as shown in **Figure 8-47F**. At this point, the brushes once again pass from one commutator segment to the next, causing the current flow through the armature to reverse again to the original direction. This causes the armature to rotate through the 0-degree starting position to the 90-degree point once again. The armature will rotate continuously in the direction shown because the commutator design causes the current through the armature to reverse directions at just the right moment to maintain armature rotation.

Tech Tip: The commutation process can be thought of being like a dog that is chasing its own tail. No matter how fast the dog runs around in a tight circle, it is never able to get the elusive tail. Just at the point when the dog almost reaches the tail, it moves away from it. The rotating armature loop keeps trying to move from a strong field to a weak field. Just as the loop gets to the weak field, the direction of current flow changes through the loop, which causes the loop to rotate another 180 degrees.

Additional CEMF Information

Brushed DC motors are very similar in construction to DC generators. DC generators are now obsolete but were used for many years as the voltage generator on automobiles and trucks prior to alternators. In fact, many smaller engines of the past used a single motor/

generator device. This dual-function device acted as a motor to crank the engine and became the DC generator after the engine was running.

The CEMF is an induced voltage in the armature windings with a polarity that acts as a series-opposing voltage source to the battery voltage as shown in **Figure 8-48**. The CEMF is shown as the circular object within the motor. The winding resistance is shown as the resistor. It is not possible to separate the winding resistance and the CEMF. Doing so is done only for illustration purposes.

Because the battery and motor are in series, the battery voltage and the motor CEMF are series-opposing voltages. Therefore, the only current that will flow through the motor windings, which are shown as the resistor in **Figure 8-48**, is the difference between the battery voltage and the CEMF voltage. For this reason, an increase in CEMF means a decrease in current drawn by the motor.

When the motor armature is not rotating, there is no CEMF being generated so the amount of current that flows depends upon the winding resistance. Because

the windings of a starter motor are just heavy-gauge copper wire, the resistance is very low as shown by the 0.01Ω resistor in **Figure 8-48**. With no CEMF being generated by the stationary armature, 1200A of current flows unopposed through the motor. When the motor is rotating at full speed, the example shown at the bottom of **Figure 8-48** indicates that 11V of CEMF is being generated. This series-opposing voltage results in only the difference of the 12V supplied by the batteries and the 11V of CEMF to be applied to the motor winding resistance. Therefore, an effective voltage of only 1V is supplied across the 0.01Ω winding resistance with the motor rotating at full speed. This results in only 100A of current flow with the motor running at full speed. The resistance of the motor has not changed, but the motor is drawing significantly less current when rotating than when stationary. Thus, an increase in motor speed means a decrease in motor current, as illustrated by the plots shown in **Figure 8-49**.

The CEMF must always be at least slightly less than battery voltage level that caused the armature to rotate in the first place. If the CEMF were equal to the battery voltage, then zero current would flow through the motor windings after the motor was brought up to operating speed. In theory, the battery could then be disconnected after the motor was brought up to speed and the motor would continue to rotate forever. Unfortunately, this perpetual motion is not possible due to losses from friction and conductor resistance.

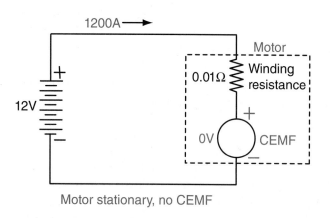

Motor stationary, no CEMF

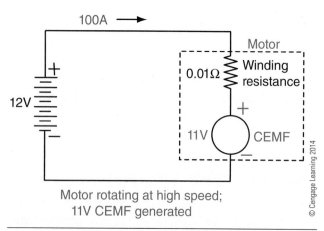

Motor rotating at high speed;
11V CEMF generated

Figure 8-48 CEMF with motor stationary (top) and motor rotating at full speed (bottom) and the effect on current drawn by the motor.

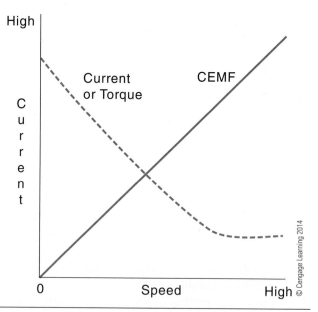

Figure 8-49 Motor speed versus CEMF generated and motor speed versus motor current draw.

It may seem strange that the current drawn by the motor decreases as the motor speed increases. It is almost like getting more work and paying for it with less energy. However, even though the current decreases with motor speed, the torque produced by the motor also decreases with motor speed. The torque imparted to the armature by the magnetic field interaction is highest at low speeds where the motor current is also highest. **Figure 8-49** represents current and torque with the same dashed line because both are dependent upon the speed of the motor.

Armature rotation produces CEMF, which reduces current drawn by the motor and reduces the torque produced by the motor. The high-torque production at low-speed characteristic of electric motors make them ideal for diesel engine starter motors, where a large amount of torque is necessary to begin to rotate a high-compression engine.

Starter Motor Solenoid Windings

Most starter motor solenoids have two sets of windings to create the magnetic field necessary to cause the solenoid plunger to move, as shown in **Figure 8-50**. One set of windings called the pull-in coil has much larger diameter wire than the other winding and draws much more current, thus creating a stronger magnetic field. The other winding, called the hold-in coil, draws much less current, thus creating a weaker magnetic field. In **Figure 8-50**, the high side of both the pull-in and hold-in coil are supplied with +12V when the ignition switch is placed in the crank

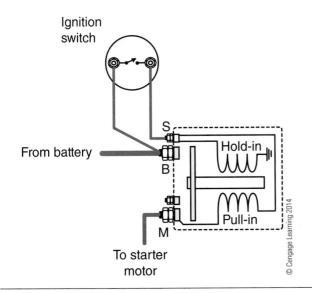

Figure 8-50 Starter solenoid hold-in and pull-in windings.

position. The other end of the hold-in coil is connected directly to ground. However, the other end of the pull-in coil is connected to terminal M, which is the high-current switch output. Terminal M is connected to the starter motor field coil and armature windings, which are connected to ground. The resistance of the armature and field coil windings is very low (much less than 1Ω), so it is as though the pull-in winding is also connected to ground even though it is really in series with the extremely low-resistance armature windings and field coils.

A relay or magnetic switch is typically used to minimize the current through the actual key switch crank position contacts, but one is not shown in **Figure 8-50** to simplify the circuit. When the pull-in and hold-in coil windings are supplied with current, both windings work together to create a very strong magnetic field to pull the solenoid plunger into the coils. The pull-in solenoid current is also flowing through the starter motor field coils and armature on its way to ground. The solenoid plunger movement causes the clutch lever to pivot and engage the drive assembly with the ring gear. The movement also causes the high-current solenoid switch contacts to close.

When the high-current switch in the solenoid is closed, +12V is supplied to the field coils and armature windings through terminal M. The current flow through these windings causes the motor to begin to rotate. Notice that with the solenoid high-current switch closed, the pull-in coil essentially has battery voltage at both ends of the coil. Because there is no difference in the voltage across the pull-in coil, all current flow through the pull-in coil ceases, which eliminates the magnetic field generated by the pull-in coil. However, current is still flowing through the hold-in coil. The magnetic field created by the hold-in coil is sufficient to keep the solenoid plunger "held-in," thus the name *hold-in coil*.

The hold-in coil alone does not generate a strong enough magnetic field to pull the solenoid plunger into the coil. Because the strength of a magnetic field decreases at a nonlinear rate with distance, a very strong magnetic field is necessary to cause the "distant" solenoid plunger to be pulled into the coil. This is the job of the pull-in coil, thus the name. By using this two-winding technique, the approximately 55A of current that would have been drawn by the pull-in coil can be used by the motor to develop torque instead of generating a magnetic field around the pull-in coil, which is no longer needed. The current drawn by the hold-in coil is typically about one-tenth that drawn by the pull-in coil.

When the engine starts and the key switch is released to the run position, current will briefly flow from the still-closed contact disc M terminal through the pull-in coil and hold-in coil to ground. The direction of current flow through the pull-in coil is now opposite to that of the hold-in coil. The result is opposing magnetic fields that cancel each other out, which causes the solenoid switch to rapidly open and interrupt the flow of current through the starter motor.

Summary

- Electric motors convert electrical energy into mechanical energy. Most electric motors used on a truck are brushed DC-type motors.

- A brushed DC motor has spring-loaded brushes that make contact with the commutator segments. The commutator segments are attached to loops of wire that make up the armature assembly. The armature is the rotating component in the motor.

- The pole shoes are the stationary electromagnets bolted to the motor frame. Field coils surround the pole shoes. Current flow through the field coils causes the pole shoes to be magnetized. This sets up a stationary magnetic field. The stationary magnetic field interacts with the magnetic field surrounding the armature windings. The interaction causes areas of weak and strong magnetic field inside the motor. The armature rotates to escape the areas of strong magnetic field.

- The commutation process describes the reversal of current flow through the armature windings at just the right time to keep the armature in a location of strong magnetic field. This current reversal causes the armature to continually rotate in an attempt to escape this strong magnetic field.

- Counter-electromagnetic force (CEMF) is the voltage that is induced in the armature windings as they pass through the magnetic field set up by the pole shoes. The CEMF acts as a series-opposing voltage to the battery voltage. The CEMF increases as motor speed increases. This causes the current drawn by a starter motor to decrease as the motor speed increases. The highest level of starter motor current draw occurs when the motor is stationary.

- The starter motor drive assembly contains the pinion gear. The drive assembly causes the pinion gear to be meshed with the engine ring gear when the motor solenoid is energized. The drive assembly contains a one-way clutch that permits the starter motor to drive the engine but prevents the engine from driving the starter motor.

- A positive engagement type of starter motor is designed not to rotate until the pinion gear is fully in mesh with the ring gear. This reduces the likelihood of ring gear milling.

- The starter motor solenoid causes the drive with pinion gear to slide into mesh with the ring gear and also causes the high-current contacts for the starter motor to close.

- Cranking circuit resistance is determined by causing a steady known amount of current to flow through the battery cables using a carbon pile resistor. The voltage dropped on the cables with the known current flowing is used to determine if cranking circuit resistance is acceptable. Low cranking circuit resistance is vital for proper engine cranking speed.

- Many smaller motors found on a truck are permanent magnet motors. The term permanent magnet refers to the pole shoes, which are constructed of material that has been magnetized. The direction of these motors can be reversed by changing the direction of current flow through the armature windings through motor voltage polarity reversal.

Suggested Internet Searches

Try these web sites for more information on starter motors:
http://www.delcoremy.com (Delco brand)
http://www.prestolite.com (Leece-Neville brand)

Review Questions

1. Which of the following is a true statement?

    A. A magnetic field surrounds a piece of wire that is conducting electric current, and the right-hand rule indicates the direction of the arrows on these magnetic lines of force.

    B. The arrows drawn on magnetic lines of force indicate the direction the north end of a compass needle would point if placed in the magnetic field.

    C. When two magnetic fields interact with each other, magnetic lines of force with arrows pointing in the same direction combine to form a stronger magnetic field.

    D. All the above are true statements.

2. What is the difference between slip rings used in an AC generator and a commutator used in a brushed DC motor?

    A. Slip rings provide an electrical connection for a rotating component, and a commutator does not rotate.

    B. Slip rings use brushes, and the commutator does not require brushes.

    C. Slip rings are continuous, and a commutator is broken up into segments.

    D. Slip rings and commutators are identical.

3. What is the main purpose of the field coils in a DC motor?

    A. Create a stationary magnetic field

    B. Create a magnetic field in the armature

    C. Create a CEMF

    D. Reverse the polarity in the armature winding just as commutation occurs

4. Technician A says that a typical starter motor draws more current as the motor speed increases. Technician B says that attempting to crank an engine with seized main bearings would cause a very low cranking current. Who is correct?

    A. A only

    B. B only

    C. Both A and B

    D. Neither A nor B

5. A rapid clunking sound coming from the starter motor area can be heard when attempting to crank an engine. The starter motor can be heard to rotate slightly but drop out and then crank again repeatedly with the key in the crank position. The vehicle batteries are tested and are fully charged. What is the most likely cause of this problem?

    A. Open circuit in the solenoid pull-in winding

    B. Burned motor commutator segments and brushes

    C. Open circuit in the solenoid hold-in winding

    D. Defective neutral safety switch

6. Which is true regarding a gear-reduction starter motor?

    A. Gear-reduction starter motors draw significantly more current than direct drive motors while cranking the engine.

    B. Gear-reduction starter motors rotate at a much lower speed than direct drive motors.

    C. The solenoid S terminal of a gear-reduction motor is typically connected directly to the key switch.

    D. Gear-reduction motors are typically smaller and weigh less than comparable direct drive motors.

7. An engine cranks slowly. Which is probably not a cause of the complaint?

   A. Incorrect engine oil viscosity for the climate

   B. Too much cranking circuit resistance

   C. Ring gear damage

   D. High battery internal resistance

8. A new truck operator indicates that occasionally the engine will not crank when the key is held in the crank position. The operator states that he hears a single clunk. Releasing the key and cranking again always causes the engine to crank. What is the likely cause?

   A. Normal tooth abutment with positive-engagement starter motors

   B. Intermittent open circuit in the starter motor control circuit

   C. Intermittent open circuit in the solenoid pull-in winding

   D. Batteries have a low state of charge

9. Cranking circuit resistance is best measured using which tool(s)?

   A. DMM ohmmeter

   B. Carbon pile resistor, clamp-on ammeter, and DMM voltmeters

   C. Growler

   D. Battery charger, carbon pile resistor, and clamp-on ammeter

10. An engine will not crank. Which is *not* a likely cause of this condition?

   A. Blown fuse in the positive cable between the battery-positive terminal and the starter motor B+ terminal

   B. Automatic transmission neutral start switch out of adjustment or defective

   C. Defective manual transmission clutch switch

   D. Engine ECM battery supply circuit is disconnected

11. A power window motor operates in one direction but not the other direction. Which is the most likely cause of this complaint?

   A. Worn brushes

   B. Defective permanent magnets

   C. Loss of residual magnetism in the armature

   D. Defective power window switch

12. Which component is not typically found in a starter motor?

   A. Drive assembly

   B. Brushes

   C. Diodes

   D. Armature

13. The resistance between each of the commutator segments of an armature is very low. What does this indicate?

   A. The armature windings are not open.

   B. The armature windings are shorted to the armature shaft.

   C. The armature windings are internally shorted together.

   D. The armature must be replaced.

14. A growler is used for what purpose?

   A. Testing pole shoes

   B. Testing solenoid windings

   C. Testing for armature windings that are shorted together

   D. Testing hacksaw blades

15. A field coil winding has become shorted to the field frame. What would be the expected result?

   A. High level of parasitic load current

   B. Low charging voltage

   C. High current draw when attempting to crank the engine

   D. Field coils are connected to ground by design, so there is no problem

9 Lighting Systems

Learning Objectives

After studying this chapter, you should be able to:

- List the various lights used on a truck and indicate their location and color.

- Describe why an incandescent lamp draws more current when first switched on than when the lamp is illuminated at full brilliance.

- Explain how the various types of turn signal flashers operate.

- Describe the difference between separate stop/turn lamps and combination stop/turn lamps.

- Troubleshoot a rear lighting problem on a truck with combination stop/turn lamps.

- Troubleshoot a trailer lighting problem.

- List the advantages and disadvantages of LED lighting.

Key Terms

back-feed

Canadian Motor Vehicle Safety
    Standard (CMVSS)

clearance lamps

combination stop/turn

composite headlamps

conspicuity

daytime running lights (DRL)

dual-filament lamp

Federal Motor Vehicle Safety
    Standard (FMVSS)

feedback

halogen

high intensity discharge (HID)

identification lamps

incandescent lamps

license lamp

marker lamps

parking lamps

sealed-beam

separate stop/turn

tail lamps

three-bar light

xenon

INTRODUCTION

Exterior lighting repairs on trucks and trailers are some of the most common electrical repairs performed by a truck technician. Lighting problems are typically not too difficult to diagnose and are a good place for a troubleshooter-in-training to get some practical troubleshooting experience. **Figure 9-1** illustrates some of the lighting found on a modern truck.

REQUIREMENTS

You may have noticed that the front turn signal lamps on your car and on every other vehicle on the

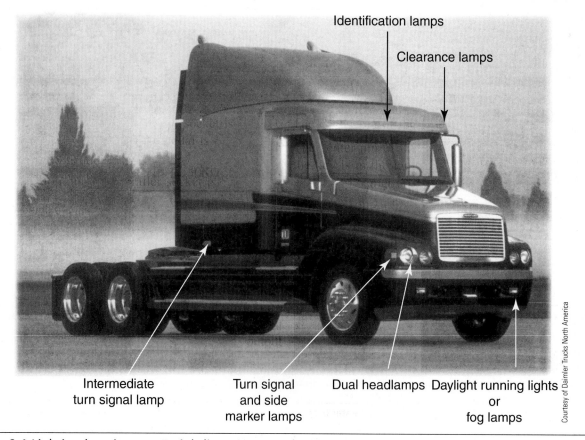

Identification lamps

Clearance lamps

Intermediate
turn signal lamp

Turn signal
and side
marker lamps

Dual headlamps Daylight running lights
or
fog lamps

Courtesy of Daimler Trucks North America

Figure 9-1 Lighting locations on Freightliner Century Class tractor.

road are amber (yellow). The stop lamps at the rear of the vehicle are always red. Cars and trucks in the United States are subject to **Federal Motor Vehicle Safety Standard (FMVSS) 108** and Canadian vehicles are subject to **Canadian Motor Vehicle Safety Standard (CMVSS) 108**. These standards define vehicle exterior lighting requirements, such as that front turn signals must be amber while stop lamps must be red on all motor vehicles. FMVSS and CMVSS 108 also define the placement of reflectors and reflective tape. Reflective material is known as **conspicuity**.

An overview of FMVSS and CMVSS 108 is shown in **Figure 9-2**. This chart references lights displayed in the pictorial shown in **Figure 9-3**. The chart indicates how many of each type of light are necessary, along with the required color. Most lights at the front or center of the truck or trailer must be amber. Most lights at the rear of the truck or trailer must be red. Turn signals at the rear of the vehicle or trailer may be amber or red. However, if the rear turn signal lamp also functions as a stop lamp, then it must be red. This kind of information is important for a truck technician to be aware of so that a truck operator is not fined for non-compliance of applicable regulations.

In addition to FMVSS 108, there are Federal Motor Carrier Safety Administration regulations (FMCSA) of

which technicians should be aware related to vehicle lighting in the United States. Some states or provinces may also have additional requirements.

Exterior Lighting Requirements

The chart in **Figure 9-2** mentions several exterior lights besides the headlamps. Collectively, these may be referred to as running lights. These lights are all typically illuminated when the headlamp switch is in the park or headlamp positions, or there may be individual switches to control some of the lamps. These exterior lights are as follows:

- **Marker lamps** are located on the sides of the truck or trailer.
- **Identification lamps** are three lamps that are centered on the front of the truck at the top edge of the cab, at the rear on the top edge of the box of a van truck, or at the rear of the top of the trailer at the center. The identification lamps are often called a **three-bar light** because the three lamps are sometimes attached to a single assembly.
- **Clearance lamps** are located at the front and rear of the vehicle at the topmost corners.
- **Tail lamps** are located at the rear of the vehicle or trailer.

Light	Location	Height	No. of Lamps	Color
		Required on All Vehicles		
1. Headlamp (not required on trailer)	Front	22"–54"	2 or 4	White
2. Tail lamp	Rear	15"–72"	2 or more	Red
3. Turn signal lamp[3]	Front	15"–83"	2 or more	Amber
	Rear[1]	15"–83"	2 or more	Red or amber
4. Hazard lamp[3] (same lamp as stop or turn signal)	Front	15"–83"	2 or more	Amber
	Rear	15"–83"	2 or more	Red or amber
5. Stop lamp[3]	Rear	15"–72"	2 or more	Red
6. License plate lamp	Rear, at license plate	No requirements	1 or more	White
7. Side marker lamp	Side near front	15" minimum	1 each side	Amber
	Side near rear (not required on truck-tractor)	15" minimum	1 each side	Red
8. Back-up lamp (not required on trailer)	Rear	No requirements	1 or more	White
9. Rear reflector	Rear	15"–60"	2 or more	Red
10. Side reflector	Side near front	15"–60"	1 each side	Amber
	Side near rear (not required on truck-tractor)	15"–60"	1 each side	Red
11. Intermediate side marker (if overall length is 30' or greater)	Side near center (not required on truck-tractor)	15" minimum[2]	1 each side	Amber
12. Intermediate side reflector (if vehicle overall length is 30' or greater)	Side near center (not required on truck-tractor)	15"–60"	1 each side	Amber
13. Parking lamp (only if vehicle is less than 80" wide)	Front	15"–72"	2 or more	Amber or white
		Required on Vehicles 80" or Wider		
14. Identification lamp	Front, spaced 6"–12" on center (not required on trailer)	As high as practicable[5]	3	Amber
	Rear (not required on truck-tractor)	As high as practicable	3	Red
15. Clearance lamp	Front, at widest point	As high as practicable[6]	2	Amber
	Rear, at widest point (not required on truck tractor)	As high as practicable[4]	2	Red
16. Conspicuity tape				Red/white

[1]Not required on truck-tractor if front turn signals are double-faced and visible from the rear.
[2]60" maximum on rear of trailers 80" or wider.
[3]Front identification lamps can be mounted on either the top of the body or on the cab's roof above the windshield.
[4]Rear clearance lamps need not meet the requirement that they be located as high as practicable when the rear identification lamps are mounted at the extreme end of the vehicle.
[5]Original equipment requires a minimum of 11.625 in^2 (75 cm^2) of effective, projected luminous lens area on vehicles 80" or wider.
[6]Front clearance lamp can be lower on straight trucks if it is at the widest part of the vehicle.

NOTE:
- All heights are measured from ground level.
- All lamps in pairs must be symmetrical and located as far apart as practicable.
- Auxiliary lighting cannot be mounted on the cab windshield pillar, and must not interfere with required lighting.
- Any two or more lamps and reflectors can be combined as long as visibility requirements are met, with the following conditions.
  1. Stop lamp and turn signal lamp can be combined if the stop lamp is deactivated when the turn signal is turned on.
  2. Tail lamp and clearance lamp cannot be combined optically. Use two separate lamps.

Lighting regulations referenced are FMVSS 108 and CMVSS 108 requirements. Any specific state of provincial regulations have not been included. Each lamp must be secured permanently to a fixed portion of the vehicle, parallel or perpendicular to the vehicle's centerline and not obstructed by any part of the vehicle. Auxiliary device is installed to meet required visibility.

© Cengage Learning 2014

Figure 9-2 Motor vehicle lighting requirements for the United States.

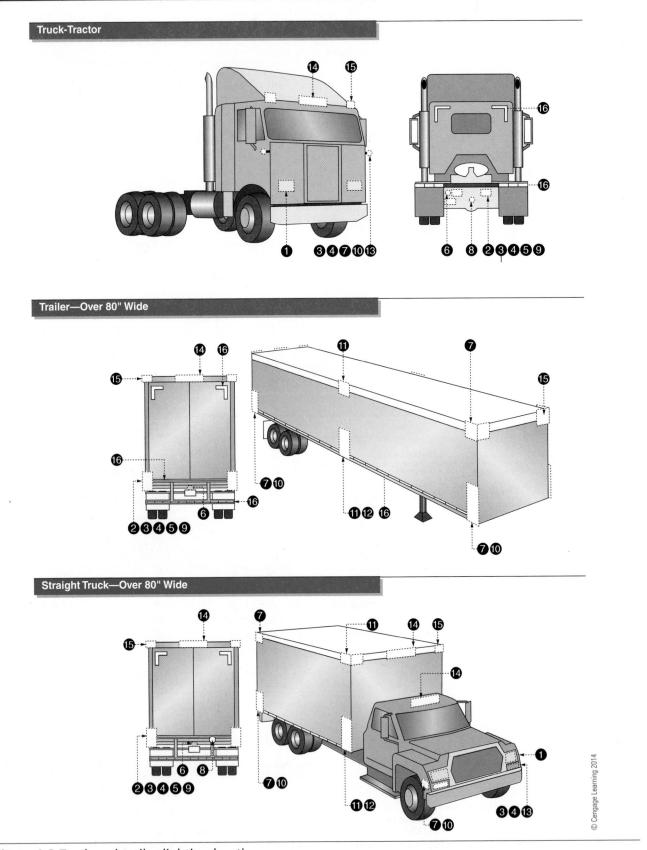

Figure 9-3 Truck and trailer lighting locations.

- **Parking lamps** are located at the front of the truck.
- The **license lamp** is located above the truck and trailer rear license plates.

INCANDESCENT LAMPS

Incandescent lamps are basically made of a piece of thin tungsten wire inside a sealed glass container. Incandescent lamps are often called light bulbs and the tungsten wire is called a filament. When electric current passes through the thin filament, the resistance of the filament causes heat to be generated. The heat results in the filament incandescing, which results in visible light being given off by the filament. The sealed glass container is a vacuum, or contains an inert gas such as **halogen**, to prevent the filament from melting due to the heat.

Automotive incandescent lamps are classified as miniature lamps. The term *miniature* indicates that the voltage rating of the lamps is low compared to household lamps.

Lamp Ratings

Incandescent lamps have several ratings. One of the most important lamp ratings is the design voltage rating. The design voltage rating defines the terminal voltage at which all other lamp characteristics are measured because changing the voltage changes the current and the amount of light given off by the lamp. This design voltage rating is approximately 12V or 24V for most truck lighting. Most truck electrical systems have 12V lighting systems, even though the starter motor may be 24V on some trucks. However, some trucks in South America and most in Europe and other parts of the world have 24V lighting systems. Military vehicles typically also have 24V lighting systems. The design voltage rating of the bulb must match the truck lighting system voltage. Installing a lamp designed for a 24V system in a 12V system will result in a significant decrease in light output because a 24V lamp has more resistance than a 12V lamp. Installing a 12V lamp in a 24V lighting system will result in a very bright lamp because of too much current flow through the filament, which will probably burn out after a few seconds of operation.

Lamps have a rating that indicates the intensity of the light emitted from the lamp (brightness). This rating is called the mean spherical candela (MSCD). Candela (CD) is also known as candlepower. This MSCD rating is dependent upon the voltage supplied to the lamp. A small decrease in voltage can result in a large decrease in light output. Therefore, a poor electrical connection that adds series resistance and a voltage drop to the lamp circuit will cause the light output to decrease accordingly.

Lamps may also have an amperage rating. This rating indicates the approximate amount of current the lamp will draw when supplied with the design voltage. A change in voltage supplied to the lamp will of course affect the current drawn by the lamp as predicted by Ohm's law. This lamp amperage rating is good information for the technician because it can be used in troubleshooting. For example, a circuit breaker that supplies the trailer marker lights may trip because too many marker lights have been added to the trailer. By counting the number of marker lights with a known amperage rating, you can determine what the expected current draw should be based on the number of lamps on the trailer.

The power rating of a lamp in watts is an indication of the amount of heat the lamp will dissipate. Substituting a lamp with one that has a higher power rating than the light socket is designed for can cause damage to the light socket or the plastic lens cover.

Automotive lamps are identified by an industry standard numbering system. It is important that lamps be replaced with the OEM's recommended lamp. Lamps with the same standard number, regardless of brand, should have the same ratings. Just because a lamp may fit into the socket and illuminate does not mean the lamp is correct for the application. Installing the wrong lamp may cause damage to the lamp socket or light lens or may cause the light not to be bright enough for the intended purpose.

Because trucks are driven many more miles and subject lamps to higher levels of vibration than automobiles do, special lamps for trucks have been developed. These longer-life lamps are identified by a fleet truck type of numbering system, as shown in **Figure 9-4**. This chart shows the fleet truck number along with the standard automotive number that each fleet truck type of lamp replaces.

Most automotive lamps have a clear glass globe so that they emit a white light. The translucent lens cover is colored either red or amber to provide color to the light. However, some lamps may have amber-painted glass globes or translucent amber glass globes. The translucent amber glass is called natural amber. Such lamps are identified in the numbering system with an A (painted amber) or NA (natural amber) suffix. Amber-colored lamps are designed to be used with clear lenses. Using an amber-colored lamp with a colored lens may produce a color of light that does not meet applicable regulations.

Miniature Service	Fleet Truck Types	Replaces These Lamps	Design			Rated Avg. Life @ Design Volts	Filament Type	Bulb Type	Base Type
			Volts	Amps	Candle Power				
Instrument Indicator Lamps	161	184	14.0	0.19	1	4000	C-2F	T-3 1/4	Wedge
	1445	182, 53, 53X	14.4	0.135	0.7	2000	C-2V	G-3 1/2	Min. Bay
Instrument Indicator, Identification, and Clearance Marker Lamps	293	1895, 57, 57X	14.0	0.33	2	7500	C-2F	G-4 1/2	Min. Bay
	1893	1889, 1891, 1898	14.0	0.33	2	7500	C-2F	T-3 1/4	Min. Bay
	194	158, 195	14.0	0.27	2	2500	C-2F	T-3 1/4	Wedge
	193*	194, 195	14.0	0.33	2	15000	C-2F	T-3 1/4	Wedge
Licence Plate Marker Lamps	97	67, 1155	13.5	0.69	4	5000	C-2V	G-6	S C Bay
	68	1178, 96	13.5	0.59	4	5000	C-2R	G-6	D C Bay
Turn Signal Lamps	199	1073, 1156	12.8	2.25	32	1200	C-6	S-8	S C Bay
Turn Signal Lamps	3156		12.8	2.1	32	1200	C-6	S-8	S F Wedge
Stop/Tail	198	1034, 1157	12.8	2.25	32	1200	C-6	S-8	D C Index
			14.0	0.59	3	5000	C-6		
	3057	2457, 2358	12.8	2.1	32	1200	C-6	S-8	D F Wedge
			14.0	0.48	2	5000	C-6		
	3157	2458, 2359	12.8	2.1	32	1200	C-6	S-8	D F Wedge
			14.0	0.59	3	5000	C-6		
	90	632, 99	13.0	0.58	6	750	C-2R	G-6	D C Bay
Dome Light	105	1003, 93, 1093	12.8	1.07	12	500	C-6	B-6	S C Bay
	631	89, 98	14.0	0.63	6	1000	2C-2R	G-6	S C Bay
	1004	104	12.8	0.94	15	200	C-6	B-6	D C Bay
	211-2	211, 211-1	12.8	0.97	12	1000	C-8	T-3	Min. Cap
	212-2	212, 212-1	13.5	0.74	6	2000	C-8	T-3	Min. Cap
*Operates hotter than #194—not recommended for instrument panels.									

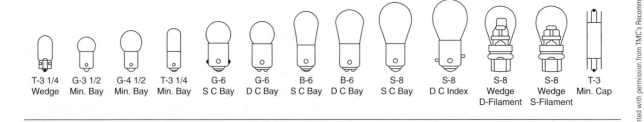

| T-3 1/4 Wedge | G-3 1/2 Min. Bay | G-4 1/2 Min. Bay | T-3 1/4 Min. Bay | G-6 S C Bay | G-6 D C Bay | B-6 S C Bay | B-6 D C Bay | S-8 S C Bay | S-8 D C Index | S-8 Wedge D-Filament | S-8 Wedge S-Filament | T-3 Min. Cap |

RP 127B-2

Reprinted with permission from TMC's Recommended Practices Manual, courtesy of the Technology & Maintenance Council of American Trucking Associations, Inc., 950 North Glebe Road, Arlington, VA 22203-4181; (703) 838-1763; http://www.truckline.com/Federation/Councils/TMC/Pages/default.aspx

Figure 9-4 Miniature lamp specifications.

Over time, the heat of incandescence causes the tungsten from the filament to deposit on the inside of the glass bulb. The coating of tungsten causes the glass to darken, which reduces light output. The loss of tungsten in the filament also causes the tungsten filament to become thinner the longer the lamp is in service, thus increasing filament resistance. This increase in resistance also causes less light output. The thinner filament will eventually burn open or fracture due to vibration, requiring lamp replacement.

Incandescent Lamp Terminals

Most lamps have a brass base that acts as the ground or negative terminal for the lamp. One end of the filament is connected inside the glass to the lamp base. The other end of the filament is typically connected to a soft metal terminal at the bottom of the lamp. The metal base typically has indexing pins known as bayonets that are used to retain the lamp in the lamp housing and provide electrical contact to the grounded lamp housing (**Figure 9-5**). The lamp is twisted into place in the lamp housing. The soft metal terminal at the base of most lamps makes contact with the positive terminal inside the socket.

Other types of lamps may just have exposed wires at the base of the lamp and are designed to be pushed straight into the lamp housing. The base of these lamps is made of a plastic insulating material or glass. The lamp housing contains terminals that are designed to contact the exposed wires on the lamp base, as shown in **Figure 9-6**.

Many lighting manufacturers recommend the addition of a special nonconductive anti-corrosion compound to incandescent lamp terminals and connectors to seal connections from water. This reduces corrosion of terminals. This compound is similar in texture to heavy grease. Do not substitute common grease for the special compound because some types of grease may attack plastic connectors and cause other problems.

Dual-Filament Lamps

It is common for one single light assembly to serve more than one purpose. For example, a single light assembly at the left rear of the truck with a red lens and a single lamp may serve as a tail lamp, a stop lamp, and a left turn signal lamp. This is accomplished by using a **dual-filament lamp**. One filament acts as

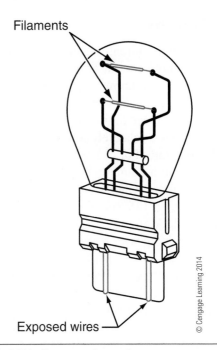

Figure 9-6 Wedge-type lamp base.

the tail lamp, and the other filament acts as the stop lamp. The stop lamp filament is flashed to create a turn signal lamp.

A dual-filament lamp typically has three terminals (**Figure 9-7**). One end of both filaments is connected together inside the lamp. This common connection is connected to the lamp's grounded metal base. A second terminal is connected to the tail lamp filament and the third terminal is connected to the stop lamp filament. The resistance of the tail lamp filament is higher than the resistance of the stop lamp filament. This causes the stop lamp filament to be much brighter and draw more current than the tail lamp filament. The indexing pins or bayonets are offset to ensure the lamp is inserted in the wiring harness lamp socket in the proper orientation and to prevent a single-filament lamp from inadvertently being installed in a dual-filament socket and vice-versa.

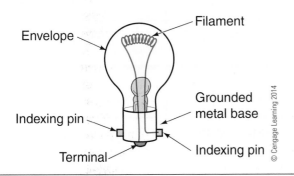

Figure 9-5 Indexing pins on lamp base used to retain lamp in socket and provide ground connection.

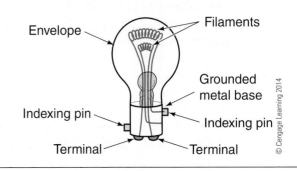

Figure 9-7 Dual-filament lamp; two filaments share a common ground. Offset indexing pins prevent incorrect installation.

Surge Current

The initial high current drawn by an incandescent lamp causes a surge current or inrush current that can be several times the normal operating current. This occurs because the filament in most incandescent lamps is merely a piece of tungsten steel. The tungsten lamp filament has a relatively low value of resistance at room temperature. For example, the cold resistance of the filament in a common stop/turn signal lamp is approximately 0.9Ω. Ohm's law would indicate that when supplied by a 12V source, the lamp should draw approximately 13A of current. The current drawn by a single stop/turn lamp with respect to time is shown in **Figure 9-8**. When voltage is first supplied to the lamp, the current drawn by the lamp will rapidly rise to approximately 13A. However, as the filament heats up, the resistance of the filament rises dramatically because tungsten has a positive temperature coefficient (PTC), meaning the resistance of the tungsten filament increases as the temperature of the filament increases. When the filament heats to normal operating temperature, the stop/turn lamp will only draw approximately 2A of current when supplied by a 12V source, as shown in **Figure 9-8**. This current level of 2A will remain steady until the voltage supplied to the lamp is interrupted. Ohm's law indicates that the hot resistance of the lamp filament is approximately 6Ω when the filament is hot. It would not be possible to measure this 6Ω using an ohmmeter, though, because an ohmmeter must only be used when a device is not powered.

Tech Tip: You should be aware of cold-filament resistance and surge current because a circuit that appears to be shorted to ground when tested with an ohmmeter may be due to the cold resistance of the tungsten lamp filaments. This is especially true when the resistance to ground of several parallel-connected lamps is measured. For example, the total resistance of four parallel-connected 1Ω cold-resistance lamp filaments is only 0.25Ω. Such a small value of resistance when measured with a DMM appears to be a short to ground.

HEADLAMPS

Headlamps are high-power sealed incandescent lamps that work in conjunction with a specially shaped polished reflector and prisms in the lens to focus light in a specific direction (**Figure 9-9**).

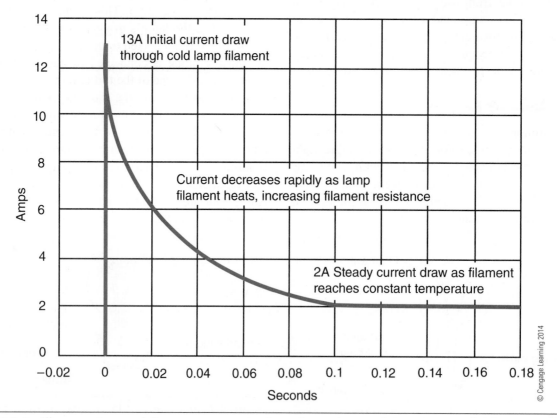

Figure 9-8 Oscilloscope trace of the surge current of one incandescent lamp, 2A per division vertical, 20 ms (0.02 sec) per division horizontal axis.

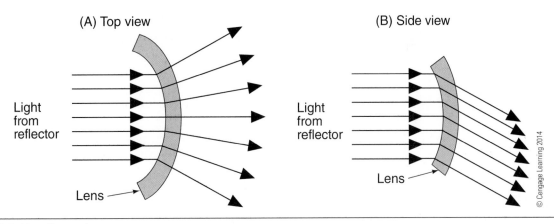

Figure 9-9 Headlamp lens.

There are two main types of headlamps:

- Sealed-beam headlamps
- Composite headlamp assemblies

Sealed-Beam Headlamps

There are two classifications of **sealed-beam** headlamps—conventional and halogen. Conventional sealed-beam headlamps are round or rectangular light assemblies that contain a tungsten filament in a sealed housing. Conventional sealed-beam headlamps are not used much in modern vehicles and have been mostly replaced by halogen sealed beams. Halogen sealed beams are also round or rectangular light assemblies but contain a sealed lamp with a quartz globe in a glass or plastic housing. The housing is also sealed and contains the reflector and lens (**Figure 9-10**). Both conventional and halogen sealed-beam headlamps are not serviceable and are replaced as an assembly.

The filament in a halogen sealed beam is much brighter than conventional sealed-beam headlamps, indicating that the filament is also much hotter. Such high

temperatures in a standard lamp would cause the tungsten to deposit rapidly inside the glass globe, resulting in significantly decreased light output as well as much decreased life. The halogen gas inside the sealed lamp causes the tungsten to be re-deposited on the filament instead of depositing on the interior of the glass. This extends the life of the lamp while keeping the glass clear, thus maintaining a higher illumination level throughout the life of the lamp.

Standard and halogen sealed beams may have the same physical dimensions and the same terminal connections. However, the two types of headlamps should not be mixed on the same vehicle.

Trucks with sealed-beam headlamps may have two or four headlamps at the front of the vehicle. If the truck has two sealed beams, then each sealed beam contains both a high- and a low-beam filament. These dual-filament lamps share a common ground, so this type of sealed beam has three terminals. The two filaments are placed within the lamp such that the angle of the emitted light changes based on which filament is illuminated (**Figure 9-11**). Only one filament in the same lamp assembly is illuminated at a time.

If the truck has four sealed beams, then two of the headlamps have both a high- and a low-beam filament as described for the two-lamp system, and the other two headlamps have only one filament and only serve as high-beam headlamps. The headlamps with two filaments, which act as the low-beam headlamps, are mounted on top when the headlamps are stacked vertically or on the outboard ends when the headlamps are arranged horizontally. With the four-headlamp system, two lamps are illuminated in low beam and four lamps are illuminated in high beam.

Composite Headlamps

Composite headlamps are not restricted to round or rectangular shapes, permitting the OEM to design

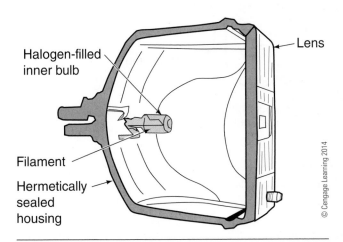

Figure 9-10 Halogen sealed beam.

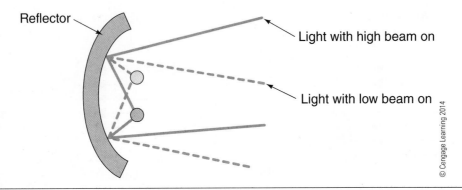

Figure 9-11 Headlamp reflector directs the light.

the front of the vehicle for improved aerodynamics and styling. Most new trucks use composite headlamps, as do most new automobiles. Composite headlamps use common replaceable halogen-filled lamps, like that shown in **Figure 9-12**. This permits replacement of just the lamp should one burn out (**Figure 9-13**). The lamp housing contains the reflector and lens. The lamp housing is sealed with some composite headlamps or it may be vented on some models. The lamp housing is typically not replaced unless it is damaged or the lens becomes discolored.

Some composite headlamps may use one halogen lamp per housing with dual filaments that acts as both high beam and low beam. Other composite headlamps may use two separate halogen lamps per housing, or use four composite headlamp assemblies.

The halogen lamps used with composite headlamps get very hot during use. Never touch a halogen lamp globe, even if cool (**Figure 9-14**). The oil from your fingers causes hot spots on the glass that may cause the lamp to fail prematurely.

1. Glass bulb
2. Low-beam filament with cap
3. High-beam filament
4. Lamp base
5. Electrical connections

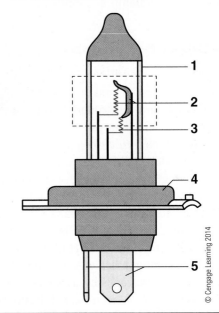

Figure 9-12 Halogen lamp.

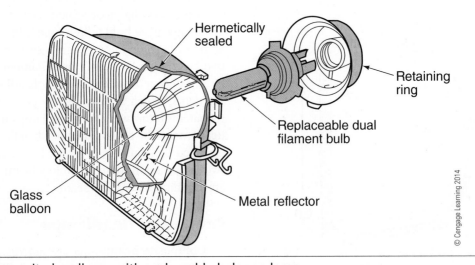

Figure 9-13 Composite headlamp with replaceable halogen lamp.

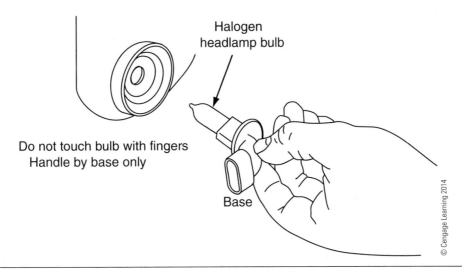

Figure 9-14 Handling a halogen lamp.

CAUTION *Halogen lamps contain pressurized gas. Handle halogen lamps with care to avoid severe cuts. The glass globe on any incandescent lamp can also break during attempts to remove the lamp from the socket and can cause cuts to the hands.*

Other Headlamp Technology

Like many high-end automobiles, **xenon** headlamps are now available for trucks. These are known as **high intensity discharge (HID)** lamps. Xenon is a chemical element and, like halogen, is a noble gas. A noble gas is an element with eight electrons in the valence band, making the element very stable and nonreactive. HID lamps have considerably higher light output compared to incandescent lamps with lower power consumption. The light color appears blue, but is actually close to natural sunlight in color.

HID lamps produce light in a manner similar to an arc welder. An arc is struck between two electrodes in a discharge chamber containing xenon, as shown in **Figure 9-15**. Up to 20kV is supplied by an electronic ballast to strike the arc. Metallic salts are vaporized by the heat and the lamps become metal halide lamps. The voltage is then automatically decreased to 85V.

Special optics are required for HID lamps and FMVSS 108 limits the installed height of HID lamps to 33.5 in. (85 cm) or lower from the ground. Therefore, you should never retrofit HID headlamps to trucks not originally equipped with HID headlamps. Doing so may actually be illegal in some areas.

Light-emitting diode (LED) headlamps are now being offered by at least one truck lighting

1. Glass capsule with UV shield
2. Electrical lead
3. Discharge chamber
4. Electrodes
5. Lamp base
6. Electrical connections

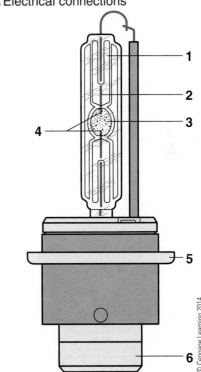

Figure 9-15 HID xenon headlamp components.

manufacturer as replacements for round sealed-beam headlamps and will likely be available in the near future as replacements for virtually all incandescent headlamps. These LED headlamps are claimed to last 50 times longer than standard incandescent sealed-beam headlamps. LED headlamps will likely become the norm over the next few years as the cost decreases and the technology improves.

Headlamp Switches

Headlamp switches are typically multi-terminal ganged switches. The headlamp switch may control the running lamps and panel illumination dimming as well. Headlamp switches are typically supplied with unswitched battery power to permit the headlamps and running lamps to operate with the key switch in the off position.

The circuit supplying the headlamp portion of the headlamp switch is often protected by an automatic reset type of circuit breaker. Some headlamp switches may contain an internal automatic reset type of circuit breaker. In the event of an intermittent short to ground in the headlamp circuit, the circuit breaker will automatically reclose to attempt to restore headlamp illumination. The headlamp circuit is also typically not shared by any other system to minimize the chances for a short to ground or overload of the headlamp circuit by other devices. Losing headlamps on an 80,000-lb (36,287 kg) tractor and trailer combination on a mountain road on a dark night is not a good situation.

WARNING Never *add any additional electrical loads to the headlamp circuit, such as driving lamps. The extra current may lead to headlamp circuit failure.*

The headlamp switch shown in **Figure 9-16** is in the off position, as indicated by the movable switch contacts designated by arrows. The switch contacts are ganged, meaning that they both move together. The dotted line between the movable switch contacts indicates a nonconductive mechanical connection, which causes the switch contacts to move in unison. With the headlamp switch in the off position, zero current flows from the battery-positive connections, shown as terminal numbers 1 and 5, to the lamp outputs, shown as terminal numbers 2, 3, and 4 in **Figure 9-16**.

With the headlamp switch in the park position, the shaded lines in **Figure 9-17** indicate that current flows from battery positive through the fuse to terminal 5 of the headlamp switch. Current then flows through the closed park switch contact to the running lamp output terminal 3 and out to the running lamps. Current also flows through the instrument panel illumination rheostat and out to the instrument panel lamps via terminal 4. Rotating the panel illumination rheostat increases or decreases the series resistance in the panel illumination circuit, resulting in dimmer or brighter panel illumination.

With the headlamp switch in the headlamp position, as the shaded lines in **Figure 9-18** illustrate, current flows from battery positive to terminals 1 and 5 of the headlamp switch. The current flow through the park lamp circuit is the same as in the park lamp position. Other than that, a parallel switch contact in the park switch circuit is being used. Current also

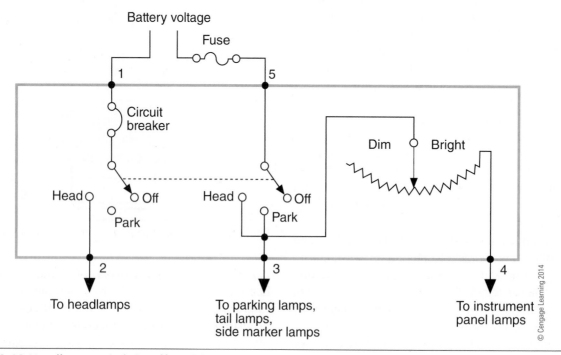

Figure 9-16 Headlamp switch in off position.

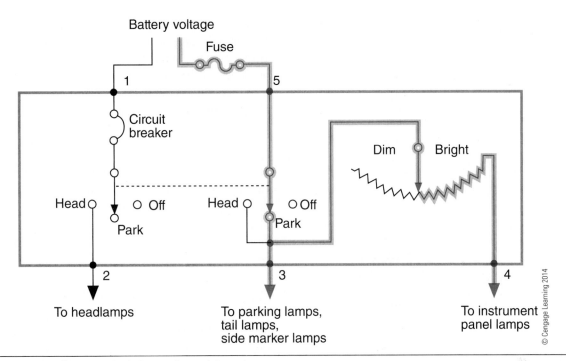

Figure 9-17 Headlamp switch in park position.

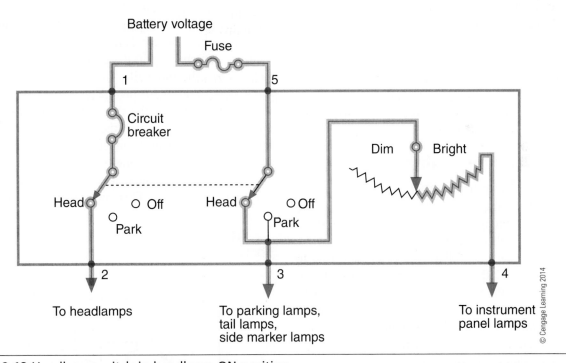

Figure 9-18 Headlamp switch in headlamp ON position.

flows from terminal 1 through the circuit breaker, and through the closed headlamp switch contact to terminal 2. Terminal 2 would typically be connected to the SPDT headlamp dimmer switch.

The dimmer switch is used to select the high beams or low beams. The dimmer switch may carry the full headlamp current, such as the floor-mounted type shown in **Figure 9-19**. The SPDT dimmer switch

directs current supplied by the headlamp switch to the high-beam or low-beam headlamps.

Most modern trucks do not use a floor-mounted dimmer switch. Instead, a turn signal lever-mounted dimmer switch or other multifunction switch is used, similar to that found in most automobiles. This type of switch does not typically carry the headlamp current. Instead, the dimmer switch controls a relay or the

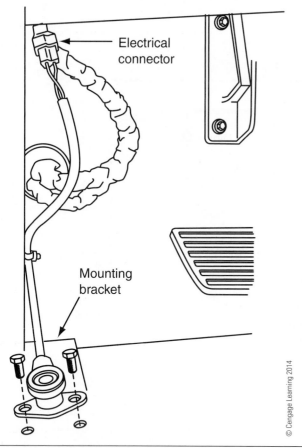

Figure 9-19 Floor-mounted SPDT dimmer switch.

switch is an input to a body control module, as will be explained in **Chapter 13**.

Daytime Running Lights

Daytime running lights (DRL) are pairs of white, yellow, or amber lamps at the front of the truck that are designed to be illuminated anytime the truck is in operation and the headlamps are off. DRL are required for trucks sold in Canada as well as many E.U. countries. In the United States, DRL is optional, but FMVSS 108 defines the limits of the intensity and the location of DRL.

Some OEMs supply a decreased voltage (PWM) from a dedicated electronic DRL module to the existing low-beam headlamps or the fog lamps to act as DRL. When the headlamps are switched to the on position, the DRL module switches off so as not to interfere with the headlamps. Other OEMs may have dedicated lamps at the front of the truck, which act as DRL, or they may be combined with the front turn signal lamps.

EXTERIOR LIGHTING COMPONENTS

Trucks have stop lamps, turn signal lamps, tail lamps, and back-up lamps like passenger cars. Much of the exterior lighting is controlled by body control modules on some trucks. However, the electro-mechanical components discussed in this section are still used on many modern trucks. Body control modules will be discussed in **Chapter 13**.

Flashers

Flashers are used to cause turn signal or hazard warning lamps (four-ways) to flash. Some trucks may use one flasher for both the turn signals and the hazard warning lamps. Other trucks may use two separate flashers. One flasher controls the turn signals, and the other flasher controls the hazard warning lamps. There are three main technologies used in flashers: bimetallic, hybrid, and solid state.

Bimetallic Flashers. Bimetallic thermal flashers use a heating element to cause a bimetallic strip to make or break a set of electric contacts as shown in **Figure 9-20**. This is very similar to the action of an automatic reset circuit breaker. The schematic symbol for an electric heating element is a squared zigzag line like that shown in **Figure 9-20**. The flasher is placed in series with the circuit supplying current to the lights to be flashed. When current flows through the flasher and heating element, the heat generated from the current flow causes the bimetallic strip to bend, which opens the contacts and interrupts the path for current flow to the lights. The bimetallic strip rapidly cools because no current is flowing through the heater with the contacts open. This causes the contacts to close again to start the flashing cycle once more.

Flashers are designed to flash at a rate of 70 to 110 times per minute. This rate may slow considerably

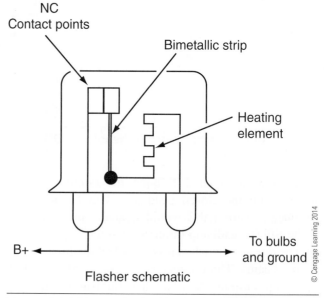

Figure 9-20 Thermal-type bimetalic flasher.

in very cold weather because of the bimetallic strip. The flash rate is also greatly affected by electrical system voltage. The flash rate decreases as system voltage decreases because of the corresponding reduction in flasher heating element current.

The flasher's opening and closing action also causes the familiar clicking sound to provide an audible indication that turn signals or hazard flashers are active.

There are several different classifications of bimetallic flashers even though most have the same two-terminal footprint. Bimetallic flashers are typically designed to flash a specific number of lamps. The flasher may also provide an indication of lamp outage by changing the flash rate when a lamp circuit is open, as with a burned-out bulb. For example, a turn signal flasher may be designed to only flash two lamps, one at the front and one at the rear of the truck. The two lamps in parallel cause a specific amount of current to flow through the flasher heater. An open circuit in one of the lamp circuits causes a reduction in current flow through the flasher heater. This reduction in current may cause the bimetallic strip to not heat up enough to snap the flasher contacts open or cause the flasher contacts to open at a greatly reduced rate. For example, a left front turn signal lamp that is burned out could cause the flasher to flash very slowly during left turns, while the right turn signal with both bulbs intact permits the flasher to flash at a normal rate. This change to the flash rate alerts the truck operator that there is a problem with the left turn signal lamps. This type of lamp outage indication flasher is typically found in automobiles and light trucks. Medium-duty and heavy-duty trucks do not typically use a lamp outage type of flasher because, unlike the case with passenger vehicles, there is no specific FMVSS 108 requirement for lamp outage indication on most trucks.

If a lamp outage type of flasher designed for two lamps were installed in a truck with three turn lamps (front, side, rear), the flasher would flash at a very fast rate because of the increase in current through the heater due to the extra turn signal lamp in parallel. You may have experienced this fast turn signal flash rate if you connected a trailer to your car or truck and it was equipped with a lamp outage indication type of flasher. The additional turn signal lamp in the trailer causes the flasher rate to increase.

Some bimetallic flashers may be designed for turn signals only. These types of flashers have a specific lamp rating of 2 to 5 lamps. The flasher rating indicates how many turn signal lamps the flasher is expected to flash. Other bimetallic flashers may be designed for both turn signals and hazard warning lamps. These flashers have a rating that indicates both the number of turn signals and the number of hazard lamps the flasher is designed to control. For example, a 2/4 lamp flasher is rated for 2 turn signal lamps and 4 hazard lamps.

Hybrid Flashers. Hybrid flashers contain an electronic circuit used to control an internal electro-mechanical relay. The electronic circuit provides a precise flasher rate that remains constant with changes in temperature or system voltage. Hybrid flashers generally have a longer service life than bimetallic flashers due to the lack of a heating element. The long service life of hybrid flashers makes them ideal for truck electrical systems.

Hybrid flashers may also have lamp outage detection. The flasher monitors lamp current and causes the flash rate to increase if one of the turn signal lamps is burned out or a lamp circuit is otherwise open. This is unlike the bimetallic flasher, which causes the flasher rate to decrease or stop flashing altogether with an open lamp circuit or a burned-out lamp.

Solid-State Flashers. Solid-state flashers typically contain a power MOSFET transistor acting as a switch and have no moving contacts. An electronic circuit controls the MOSFET to provide precise flasher rates in all conditions. These flashers are more expensive than hybrid flashers but have a very long service life.

Stop Lamp Switches

Stop lamps, or brake lamps as they are often referred to, are located at the rear of the truck or trailer and illuminate when the truck operator depresses the brake pedal.

Trucks with air brakes without provision for trailer air brakes (no tractor protection valve) typically have two air-pressure-activated stop lamp switches. These switches are normally open pressure switches designed to close with 2 to 6 psi (14 to 41 kPa) of application pressure. One switch is located in the primary air system downstream of the foot valve; the other switch is located in the secondary air system downstream of the foot valve. The two switches are typically wired in parallel so that brake pressure in either the primary or secondary air system downstream of the foot valve causes the switches to close. Two switches are used so that in case either the primary or secondary air system has lost pressure, the stop lamps will still illuminate.

Trucks with electronically controlled diesel engines may also have dedicated stop lamp switches that act as an input to the electronic engine control module; these are not used to control the stop lamps. These switches are often normally closed and open when brake application pressure is present. These switches are used by the engine control module to disable cruise control with service brake application. These switches may also have gold switch contact material because of the very low level of current that is being switched. As discussed in **Chapter 4**, special switches are necessary for controlling very low levels of current. Using a standard switch in place of the specified switch will cause the switch contacts to develop a layer of oxidation over time, which will make it appear that the switch is open when it is actually closed.

Trucks with air brakes that have provisions for trailer air brakes may only have a single normally open stop lamp switch located in trailer brake air supply circuit. This air circuit is designed so that it receives the greater of either the primary or secondary air application pressure. Thus, only one stop lamp switch is necessary for these trucks because pressure in either the primary or secondary apply circuit will be detected by the stop lamp switch.

Trucks with hydraulic brakes typically use a mechanically operated normally closed switch mounted to the brake pedal linkage (**Figure 9-21**). Depressing the brake pedal causes the switch to close.

Turn Signals, Hazard Lamps, and Stop Lamps

You have probably noticed that on many passenger cars and light trucks, the rear turn signals, hazard lights, and stop lamps all use one single lamp on each side of the vehicle. In this case, the rear turn signal lamp is red. This type of light is called a **combination stop/turn** lamp. Depressing the brake pedal causes both rear combination lamps to illuminate. Switching on the left turn signal with the brake pedal depressed causes the left rear combination lamp to flash and the right combination lamp to illuminate steadily for as long as the brake pedal is depressed.

Some passenger cars have a separate rear turn signal lamp that is often amber in color. This turn signal lamp also functions as the hazard lamp. The two rear stop lamps on these vehicles are separate from the rear turn signals and are red. When the stop lamps are contained in different lamp housings than the rear turn signal lamps, the lamps are called **separate stop/turn** lamps. The stop lamps and the turn signal lamps function independently of each other with separate stop/turn lamps.

Most trucks in North America have combination stop/turn lamps at the rear of the truck or tractor, while most large trailers have separate stop/turn lamps at the rear of the trailer. European- or Asian-built trucks often have separate stop/turn lamps at the rear of the truck.

Tech Tip: It is important to note that left and right side on a vehicle always refers to the truck operator's left and right when seated in the driver's seat. This is opposite of left and right as viewed facing the front of the truck.

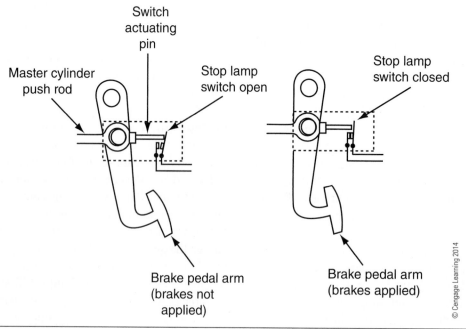

Figure 9-21 Hydraulic brake stop lamp switch.

One Lamp Filament, Three Functions

With combination stop/turn lamps, the turn signal switch design provides the ability for one lamp filament to perform three different functions: turn, stop, and hazard flashers. The turn signal switch is a multi-terminal ganged switch with several switch contacts. The turn signal switch has connection to the turn signal flasher, the hazard flasher, the stop lamp switch, front turn signal lamps, and rear combination lamps. As you may imagine, this switch is somewhat complex. However, the circuit diagram makes it easy to explain the operation of this switch.

A simplified circuit diagram for a rear combination stop/turn lighting system, less the hazard circuit, is shown in **Figure 9-22**. There are four movable switch contacts in this switch, one for each of the four turn signal lamps on the truck. The left side turn signal switch contacts are at the top of the schematic for the switch, and the right side turn signal switch contacts are at the bottom of the schematic for the switch. The left and right turn signal movable switch contacts are each ganged together, as shown by the dashed line. Remember that this dashed line indicates a mechanical connection, not an electrical connection. There is no continuity between the left rear and left front or right rear and right front turn signal lamps with the turn signals in the off position.

The switch shown in **Figure 9-22** is drawn so that the turn signal switch is in the center or off position. Notice that in the off position, there is continuity between the left and right rear combination lamps through the stop lamp switch connection. Terminals 1, 2,

and 3 of this switch are common when the turn signal switch is in this no-turn or off position.

If the brake pedal were depressed with the turn signal switch in the off position, the stop lamp switch would close (**Figure 9-23**). This causes current to flow into terminal 2 of the turn signal switch and out to the left rear combination lamp through terminal 1 and out to the right rear combination lamp through terminal 3, as shown by the colored lines. This causes both rear combination lamps to illuminate steadily. Notice that there is no current flow to the front turn signal lamps. Front turn signal lamps do not illuminate with the brake pedal depressed.

If the brake pedal were no longer depressed and the turn signal switch were placed in the left turn position, the ganged left turn contacts would move to the positions shown in **Figure 9-24**. This causes current to flow through the flasher heater and flasher contact to terminal 5 of the turn signal switch. Current then flows out to the left front turn signal lamp through terminal 4 and left rear combination lamp through terminal 1. Note that the left and right rear combination lamps are no longer electrically connected because the left turn signal switch has opened the common connection between the two rear lamps.

The current flow through the heater in the flasher causes the flasher contacts to open and then close again repeatedly to flash the left turn signal lamps at the front and rear of the vehicle.

Placing the turn signal switch in the right turn position would cause the same things to occur, except on the right-hand side of the vehicle.

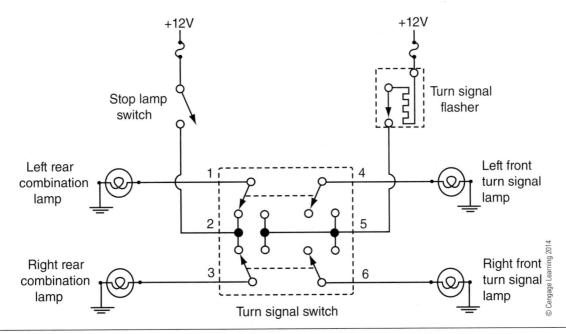

Figure 9-22 Turn signal with stop lamp switch open and no-turn position (off).

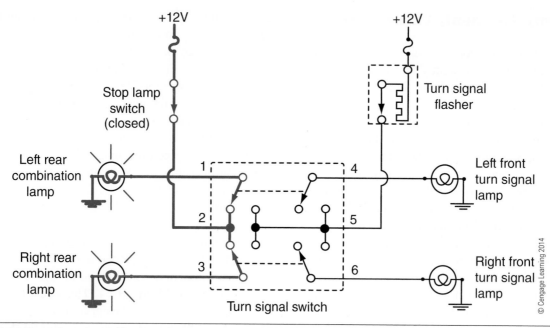

Figure 9-23 Stop lamp switch closed, turn signal switch off, both rear combination lamps illuminate.

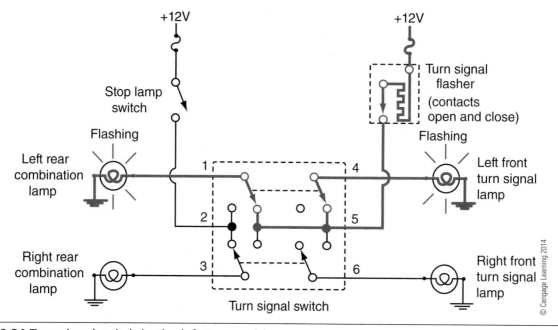

Figure 9-24 Turn signal switch in the left turn position.

If the brake pedal were depressed with the turn signal switch in the left turn position, there would be no change in the illumination of the left side lamps at the front and rear of the vehicle. The left side lamps at the front and rear would continue to flash. However, the depressed brake pedal would cause current to flow to the right rear combination lamp (**Figure 9-25**), resulting in the right rear combination lamp illuminating steadily while the left rear combination lamp and left front turn signal flash. The same thing would occur but on the right side if the turn signal switch were in the right turn position with the brake pedal depressed.

Hazard Lamps

The hazard switch (four-ways) is not shown in the simplified turn signal switches in the previous illustrations. The hazard switch typically places all four turn signal outputs in common while breaking the connection to the turn signal flasher. The separate hazard flasher then provides power to all four turn signal lamps. The heater in the hazard flasher causes the hazard flasher contact to open and close to flash all four turn signal lamps at the same time.

Some trucks are wired so that pressing the brake pedal with hazard flashers on causes the rear combination stop/turn lamps to come on steadily (stop

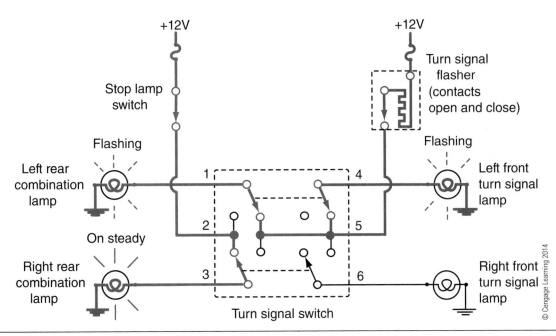

Figure 9-25 Left turn and stop lamp circuit.

overrides hazards). Some vehicles are instead wired so that pressing the brake pedal with the hazard lamps on causes the rear combination stop/turn lamps to continue to flash (hazards override stop).

Adding Tail Lamps to the Simplified Circuit

The rear combination stop/turn lamps typically also serve as the tail lamps for the truck. A dual-filament lamp provides this functionality. The dual-filament lamp has three terminals. One terminal is connected to the stop/turn filament, one terminal is connected to the tail lamp filament, and the other terminal is a common ground for both filaments. The front turn signal lamps are also typically dual-filament lamps and are used as park lamps. The park lamp switch (part of the headlamp switch) and the associated circuits have been added to the simple turn signal circuit (**Figure 9-26**). The addition of the running lamp circuits to this

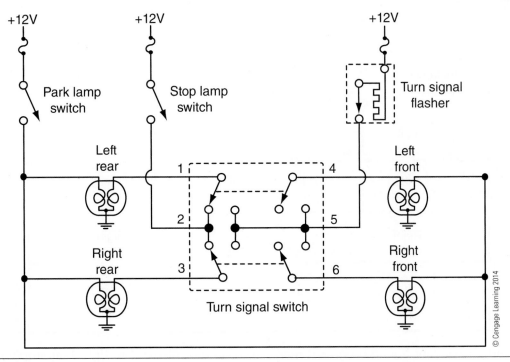

Figure 9-26 Adding headlamp switch and dual-filament lamps to simple turn signal switch.

schematic has no effect on the turn signal and stop lamp system, unless something goes wrong.

Strange Behavior with Combination Stop/Turn Lamps

The electrical connection inside the turn signal switch between the left and right rear combination stop/turn lamps with the turn signal switch in the off position can cause some seemingly strange behavior when the ground is open for one of the rear dual-filament lamps. The two lamp filaments in a dual-filament lamp used as a stop/turn and running lamp have different resistance values. The filament used for stop/turn lamps has about one-quarter the resistance of the filament used for running lamps (tail lamp or park lamp). This causes the stop/turn filament to be much brighter than the running lamp filament when both are supplied with 12V.

If the common ground for the combination stop/turn lamp is lost due to corrosion or a broken wire, the two filaments in a single combination lamp are no longer independent. The dual filaments in a single lamp can become series resistors (**Figure 9-27**) when the common ground connection is lost and only one of the filaments is supplied with voltage.

The ground for the left rear combination lamp has been opened in the circuit shown in **Figure 9-28**. Some strange behavior can occur when the common ground connection for a dual-filament lamp is open. For example,

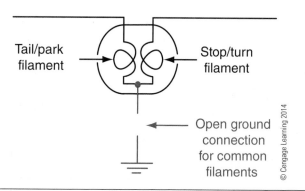

Figure 9-27 Filaments are in series when dual-filament lamp ground circuit is open.

switching on the park lamps without any other lights on still causes the left rear tail lamp to illuminate even though the ground for the left rear combination lamp is opened. However, the level of illumination is much dimmer than in the right rear tail lamp. You may think that the left rear tail lamp would not illuminate at all if the left rear ground connection were opened. However, current flows through the left rear tail lamp filament and "finds" or seeks a ground through the left stop/turn filament and through the right rear stop/turn filament to ground (**Figure 9-28**).

Notice that the current through the left rear stop/turn filament is flowing in a direction that is opposite of normal. This type of reverse current flow is often referred to as a **back-feed** or **feedback**. The path of this current flow is through the low-resistance left rear

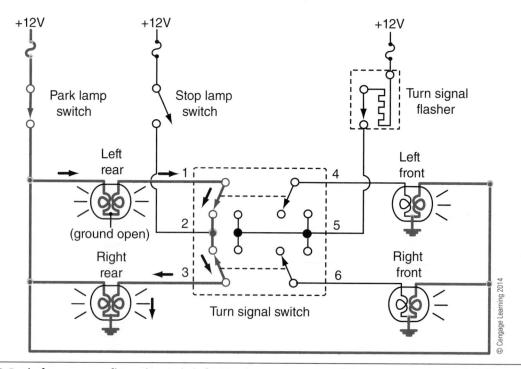

Figure 9-28 Path for current flow through left rear lamp stop/turn filament to right rear stop/turn filament with left rear ground open and park lamps on.

stop/turn filament, through the turn signal switch stop lamp connection, and out to the right rear stop/turn filament to ground, as shown by the arrows in **Figure 9-28**. The stop/turn filaments in both rear lamps will glow due to this current flow.

If the brake pedal were depressed with the park lamps on and with the left rear combination lamp ground still open, the left rear park lamp filament would no longer glow because battery voltage (+12V) would then be placed on both the left rear park lamp filament and the left rear stop/turn filament (**Figure 9-29**). With zero voltage differential across the lamp filaments, there is zero current flow through either left rear filament. The stop lamp filament in the left rear lamp would also not glow because of +12V present at both left combination lamp filaments. The right rear combination lamp would be unaffected by the loss of ground for the left rear combination lamp and would illuminate as expected.

If the park lamps were then switched off and the brake pedal were depressed with the left rear combination lamp ground still open, the left rear stop/turn filament would glow dimly, the left rear tail lamp filament would also glow dimly, as would all other running lamps on the truck. This occurs because the current supplied to the left rear stop/turn filament flows through the left rear tail lamp filament and "finds" a path to ground through the other parallel-connected running lamps on the truck, as shown in

Figure 9-30. Placing the turn signal switch in the left turn position with the running lights off will cause the same thing, except all the running lights will flash dimly. This may also affect the flash rate with thermal-type flashers.

Problems such as those described previously may seem difficult to troubleshoot, even though all are caused by the simple problem of an open ground at the left rear combination stop/turn lamp.

An open ground for the left front turn/park lamp will also cause all the running lights on the truck to flash dimly when the turn signal switch is in the left turn position. However, the lack of the common stop lamp connection in the front turn lamp circuit isolates this problem otherwise.

Another problem that sometimes occurs with dual-filament lamps is a short circuit between the two filaments inside the lamp. This failure may also occur if the soft metal terminals at the bottom of the dual-filament lamp make contact with each other or a terminal-to-terminal short occurs between the stop/turn and running light terminals in the lamp socket. This type of failure also causes some strange behavior similar to the open ground condition described earlier. For example, a short of the left rear stop/turn filament to the tail lamp filament with turn signals off and park lamps on causes both the right rear and left rear stop/turn filaments to illuminate brightly, as though the brake pedal had been depressed (**Figure 9-31**).

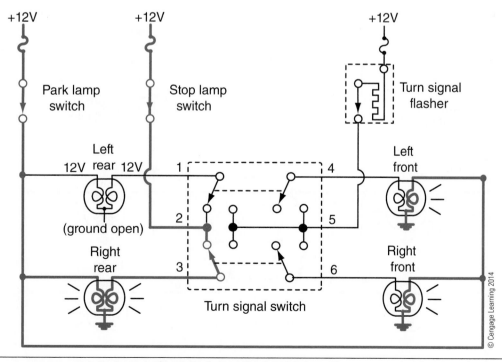

Figure 9-29 Left rear combination lamp ground open. Park lamps are on, stop lamps are on, and neither left rear filament is illuminated.

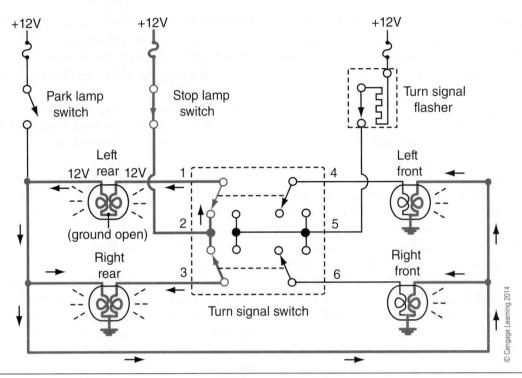

Figure 9-30 Left rear combination lamp ground is open. Park lamps are off, stop lamps are on, all running lights illuminate dimly when stop switch is closed.

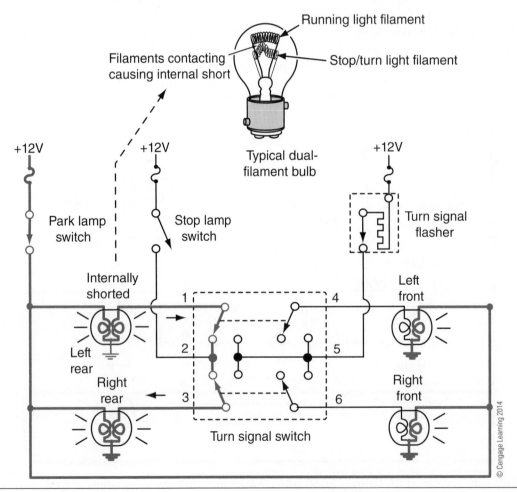

Figure 9-31 Internal filament short of left rear lamp filaments, park lamp switch on, brake not depressed.

If the park lamps were switched off and the brake pedal were then depressed, the internal short of the left rear filaments would cause all the running lamps on the truck to illuminate at less than normal intensity for as long as the brake pedal is depressed (**Figure 9-32**).

Placing the turn signal switch in the left turn position with the park lamps off will also cause all running lamps to flash with the left turn signal lamps with shorted lamp filaments. However, placing the turn signal switch in the right turn position with running lamps off will only cause the right turn signal filaments to flash and the running lamps will remain off. By using the circuit diagram and observing the behavior of the lights under various conditions, you may be able to determine the left rear combination lamp as the source of the problem without ever having to use a DMM to troubleshoot.

Tech Tip: Lighting problems that seem to make no sense are often caused a by dual-filament lamp that has a poor ground or a short between the filaments.

Back-Up Lamps

Back-up lamps or reverse lamps are located at the rear of the truck and are white. Back-up lamps are designed to illuminate only with the transmission in reverse and the key in the ignition position. Manual transmissions typically use a switch mounted in the transmission to control the back-up lamps. The switch is placed in series with the ignition supply and the back-up lamps. Placing the transmission in reverse causes the switch to close.

Automatic transmissions may use a mechanical switch on the transmission linkage, such as the neutral-start back-up (NSBU) lamp switch found on some Allison Transmission® 2000 series transmissions, to directly control the back-up lamps. Other automatic transmissions, such as the Allison World Transmission and Allison 4th Generation Controls 2000 and 3000 series, use a low side driver in the transmission ECU to control the back-up lamps. The transmission ECU energizes the low side driver when the transmission has attained reverse range. This means that the engine must be running for the back-up lamps to operate on these transmissions. The low side driver controls a relay that supplies current to the back-up lamps.

TRAILER LIGHTING

The voltage source for lighting on trailers is supplied by the tractor. The electrical connection between the tractor and trailer on modern trucks in North America is provided through a seven-wire trailer electrical cable with a plug on either end. The trailer

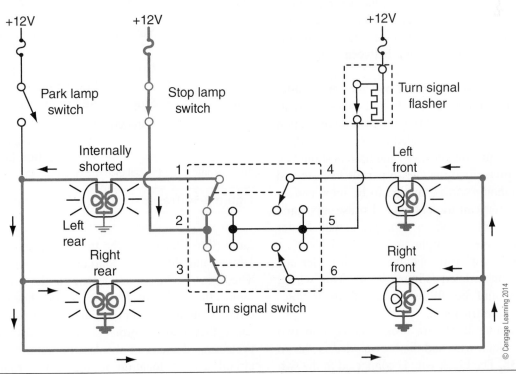

Figure 9-32 Internal filament short of left rear lamp filaments, brake depressed, park lamp switch not on.

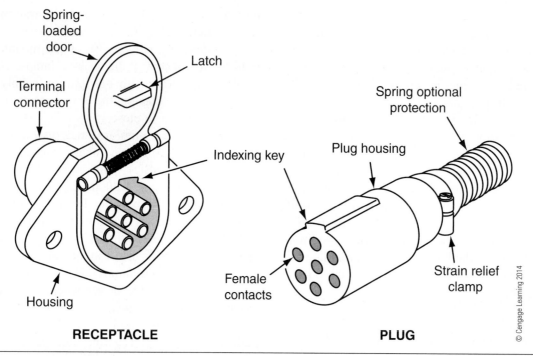

Figure 9-33 SAE J560 trailer electrical receptacle and plug.

cable is like a seven-wire electrical cord but has the same plug on both ends. This trailer wiring cable is typically coiled to permit movement between the truck and trailer, such as when turning.

Trailer Connections

Both the tractor and trailer have a seven-terminal trailer receptacle that accepts the trailer electrical cable plug (**Figure 9-33**). Double and triple trailers have trailer receptacles at both the front and rear of the trailer to provide interconnection between trailers and dollies.

Because most tractors do not pull the same trailer all the time, the trailer receptacle and cable are standardized in North America. The trailer interconnection is defined by SAE recommended practice J560. Each of the seven circuits on all trucks and trailers should be wired in the prescribed manner to permit tractors and trailers throughout North America to be interchangeable. South American trailers may also use this same convention or may use a European-style trailer connector and have 24V lighting.

The color of wiring insulation for each trailer circuit is also standardized (**Figure 9-34**). This wiring convention should be followed to aid in troubleshooting and identification.

Trailer running lights are broken down into two circuits—the black wire and the brown wire as shown in **Figure 9-35**. The black wire is connected to the two front marker and clearance lamps, the identification

SAE J560 STANDARDS

Conductor Identification	Wire Color	Lamp and Signal Circuits
Red	Red	Stop lamps and antilock devices
Blk	*Black	Clearance, side marker, & license plate lamps
Wht	White	Ground return to towing vehicle
Yel	Yellow	Left-hand turn signal & hazard signal lamps
Grn	Green	Right-hand turn signal & hazard signal lamps
Blu	Blue	ABS or auxiliary
Brn	*Brown	Tail, clearance, side marker lamps, & identification lamps

*It is recommended to balance the circuits as practicable.

Figure 9-34 Trailer standard wire insulation colors.

lamp (three-bar light), the two lower rear side marker lamps, and the center marker lamps (if applicable). The brown wire is connected to the rear clearance lamps, the tail lamps, and the license plate lamp. These circuits should be kept isolated from each other because they may be protected by two different circuit protection devices in the tractor. Otherwise, a short circuit at one location in the trailer running lamp wiring would disable all trailer running lamps.

Conductor No.	Color	Key	Lamp and Signal Circuits
1	White	⏚	Ground return to towing vehicle
2	Black	X	Side marker and identification lamps
3	Yellow	■	Left-hand turn signal and hazard signal
4	Red	▲	Stop lamps and antilock devices
5	Green	⬡	Right-hand turn signal and hazard signal
6	Brown	●	Tail, combined rear clearance, and license plate
7	Blue	AUX	Auxillary, optional lamps, dome, or ABS

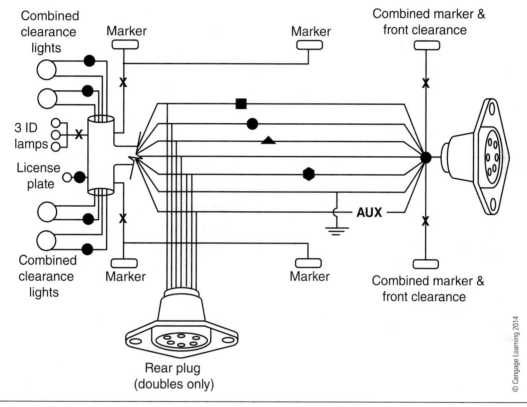

© Cengage Learning 2014

Figure 9-35 Trailer lighting wiring.

The layout and wire colors of the trailer plug and the trailer receptacle shown in **Figure 9-36** is something that you should memorize. The view shown is the mating end of the tractor or trailer's seven-way receptacle and the plug on the trailer cord's mating end. A test light, DMM, or a special trailer light test tool is used to test for voltage at the specific terminal to determine if the problem is with the truck or the trailer. Note the larger-diameter terminal at the top; this is the trailer ground terminal. Because all trailer light current must pass through the ground circuit, it uses a larger-diameter terminal than the other terminals and uses a larger-diameter wire than the other wires in the seven-way cable.

The tractor typically provides the circuit protection for trailer wiring. However, some trailers may have circuit breakers for trailer lighting circuits as well.

The length of the trailer wiring must be considered when replacing trailer wiring. Even though the rear trailer lamps may have the same ratings as the truck lamps, the long distance to the lamps can result in a substantial voltage drop on the wiring. This reduces the voltage at the trailer lamps. Wire gauge is especially important for double and triple trailers because the total circuit from the rear trailer light feed to the tractor can be more than 80 ft (24 m). The chart shown in **Figure 9-37** should be used to determine the wire gauge necessary for trailer wiring repair or replacement.

Trailer ABS Interconnection Cable

Trailers and dollies with air brakes manufactured for sale in the United States after March 1, 1998 must

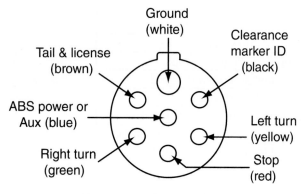

**VIEW LOOKING INTO TRACTOR
OR TRAILER RECEPTACLE MATING END**

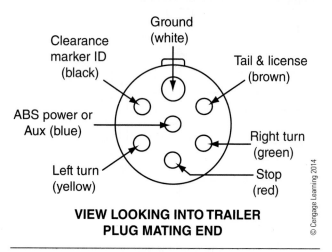

© Cengage Learning 2014

**VIEW LOOKING INTO TRAILER
PLUG MATING END**

Figure 9-36 Trailer socket and trailer cord connector layout looking into the mating ends of the connectors.

have antilock brakes (ABS). The trailer ABS electronic control unit (ECU) is located in the trailer and requires a power and ground circuit. Trailer ABS will be discussed in detail in later chapters. The center pin of the seven-way trailer lighting socket provides the primary means of supplying this trailer ABS power, while the trailer stop lamp circuit acts as an alternative supply. The trailer lighting ground circuit acts as the trailer ABS ECU ground circuit. To increase the voltage at the trailer ABS ECU, a special trailer interconnection cable that typically has a green outer jacket identified as ABS or SAE J2394 is to be used when the trailer has ABS. This marking indicates that the ground circuit, center pin circuit, and stop lamp circuit in the seven-way cable are of larger diameter than standard trailer interconnection cables to increase the voltage that is supplied to the trailer ABS ECU by minimizing the voltage drop on the cable. The SAE J2394 ABS type of trailer cable can be used to connect to trailers with or without ABS. However, only the SAE J2394 specification cable should

be used to connect the truck or tractor to trailers with ABS.

Trailer receptacles on the tractor that are designed for trailer ABS have the center pin powered at all times with the key switch in the ignition position. This receptacle is often identified by a green door and housing and may be further identified by white lettering indicating ABS. Only install the green receptacle on trucks, tractors, and trailers that are equipped for trailer ABS. Otherwise, a truck operator may assume that a non-ABS unit is equipped for trailer ABS.

Modular Trailer Wiring

Trailer wiring on most modern trailers is a modular, sealed wiring harness (**Figure 9-38**). Sections of the sealed harness can be individually replaced if damaged. The connectors on the sealed harness are molded from soft plastic and mate with other sealed sections of harness (**Figure 9-39**).

Sealed wiring harnesses may connect directly to sealed lamp assemblies. These sealed lamp assemblies are not serviceable and must be replaced if a lamp burns out. These sealed lamp assemblies are held in place by a rubber grommet and can be pushed out of the grommet for replacement. In very cold weather, it may be necessary to allow the grommet to warm to permit the removal of the lamp assembly.

LED LIGHTING TECHNOLOGY

LEDs have become increasingly popular for truck and trailer lighting. LED lamps use several individual LEDs wired in parallel. Details of the internal construction of an LED are shown in **Figure 9-40**. The actual diode portion of the LED is very small and the wire connecting the LED to the two internal terminals is a very small gauge. As indicated in **Chapter 6**, the current flow through an LED must be limited by a resistor, or these very small internal components will be destroyed. The light assembly contains the current-limiting resistor, so the leads of an LED light assembly can be connected directly to a 12V source like an incandescent light assembly. Unlike incandescent lamps, the polarity of the voltage supply must be observed. LED lamps typically have a series rectifier diode, to prevent damage to the LEDs if voltage polarity is accidentally reversed.

LED lights have several advantages over incandescent lamps, including:

- Improved life
- Instant on-time with no surge current
- Greater lighting efficiency

TOTAL LENGTH OF CABLE IN CIRCUIT FROM BATTERY TO MOST DISTANT LAMP

12-Volt System

Amperes (approx.)	Up to 30 Feet	40 Feet	50 Feet	60 Feet	80 Feet	100 Feet	120 Feet
	Gauge	Gauge	Gauge	Gauge	Gauge	Gauge	Gauge
1.0	14	14	14	14	14	14	14
2	14	14	14	14	14	14	14
3	14	14	14	14	14	14	12
4	14	14	14	14	14	12	12
5	14	14	14	14	12	12	12
6	14	14	14	14	12	12	10
7	14	14	12	12	12	10	10
8	14	12	12	12	10	10	10
10	14	12	12	12	10	10	8
11	14	12	12	10	10	8	8
12	14	12	12	10	10	8	8
15	12	12	10	10	8	8	8
18	12	10	10	8	8	8	8
20	12	10	10	8	8	8	6
25	12	10	8	8	8	6	6
30	12	10	8	8	8	6	6
35	12	10	8	8	6	6	6

Recommended minimum wire gauge for stop lamp and ground circuits.

	STOP LAMP (Red)	GROUND (White)
Single trailer (up to 50' length)	–12 Gauge	10 Gauge
Doubles trailer (2–28' lengths)	–12 Gauge	10 Gauge
Doubles trailer (2–40' lengths)	–10 Gauge	8 Gauge
Triples trailer (3–28' lengths)	–10 Gauge	8 Gauge

Compiled from SAE, TMC, and other sources.

© Cengage Learning 2014

Figure 9-37 Recommended wire gauges for trailer wiring; consult OEM information for trailers with ABS.

Advantages of LED Lighting

LED lights have a very long life expectancy—estimated at 100,000 hours or more—compared to comparable incandescent lamps, with 2000 to 15,000 hours of life. An LED light assembly may last the life of the vehicle. LEDs do not generate much heat and are not as susceptible to shock or vibration like the filament in an incandescent lamp. These two factors cause incandescent lamps to fail.

LEDs also have a very fast on-time. When voltage is supplied to the LED light assembly, the LEDs supply full light intensity almost instantly. By comparison, an incandescent turn signal or stop lamp may require 1/4 second or more to reach 90 percent of full lamp intensity. A driver following a truck with incandescent stop lamps will not know that the truck is braking until the stop lamp filaments are sufficiently heated. LED stop lamps provide at least an additional 20 ft (6 m) of stopping distance at highway speeds due to their nearly instant on-time. You may have noticed trucks with LED lights on the highway when the truck's turn signals are flashing. The difference in illumination time is very evident if the trailer has LED lights and the tractor has incandescent lamps. Even though both the trailer and tractor lights are being supplied with voltage at the same time, the LED lights illuminate much faster and have an almost flash-bulb type of effect because the intensity of the light output does not ramp-up like incandescent lamps.

88911—Rear sill harness with auxiliary red/blue/white dropouts

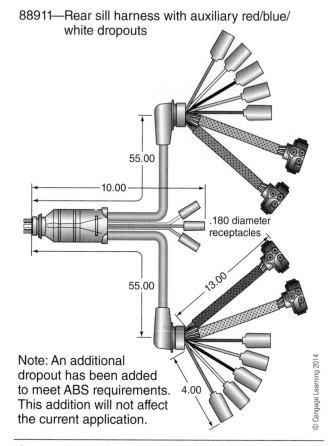

55.00

10.00

.180 diameter receptacles

55.00

13.00

4.00

Note: An additional dropout has been added to meet ABS requirements. This addition will not affect the current application.

Figure 9-38 Sealed wire harness.

LED lamps also have almost no surge current. Incandescent lamps can draw a significant amount of current as the lamp filament heats up. This surge current for four incandescent stop lamps can exceed 60A. This surge causes the voltage throughout the electrical system to dip briefly. Switching devices such as relays and flashers are stressed by these surge currents. LED lamps can also increase the life of these switching devices.

LED lights also maintain their brightness over the life of the lamp assembly. Incandescent lamps become dimmer with time due to the tungsten filament becoming thinner and depositing tungsten on the inside of the glass globe of the lamp.

LED lights operate much cooler than incandescent lamps yet provide the same light output. This indicates that LEDs are much more efficient than incandescent lamps because the heat generated by incandescent lamps is mostly wasted energy. LEDs therefore draw much less current than a comparable incandescent lamp. LEDs are much more efficient at creating light from electric current than are incandescent lamps. The chart shown in **Figure 9-41** compares the current drawn by incandescent lamps versus LED lamps. This decrease in current draw gives the LEDs several advantages over incandescent lamps.

INTERNAL GROUND

Description
1 6 Conductor Main Cable, 60'
2 Rear Sill Harness
3 Upper ID Harness, 192"
4 Rear Marker Lamp Harness, 24"
5 License Lamp Harness, 36"
6 Front M/C, Harness, 168"
7 Mid Turn Harness, Left Side, 420"
8 Mid Turn Harness, Right Side, 420"

CHASSIS GROUND

Description
1 6 Conductor Main Cable, 60'
2 Rear Sill Harness
3 Upper ID Harness, 236"
3a ID Harness
4 Rear M/C Jumper Harness, 30"
5 License Lamp Jumper Harness, 48"
6 Front M/C, Harness, 168"
6a M/C Jumper Harness, 24"
7 Mid Turn Harness, Left Side, 420"
8 Mid Turn Harness, Right Side, 420"
M/C and License Plug

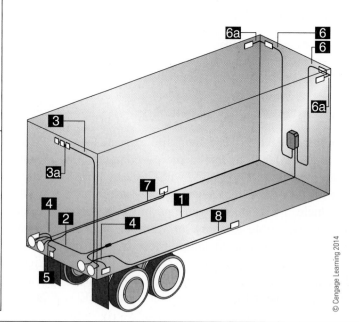

Figure 9-39 Trailer sealed harnesses layout.

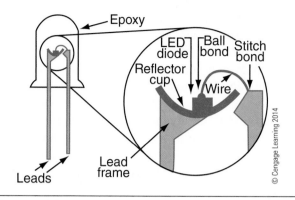

Figure 9-40 LED construction; a series resistance is necessary to limit current flow through the LED.

Lamp Type	Typical No. of Lamps	Current Draw (a) Incandescent*	Current Draw (a) LEDs*
Clearance	two	1.32	0.1
ID	three	1.98	0.15
Stop**	two	4.2	1.0
Tail	two	1.18	0.15
Turn**	two	4.2	1.0
Total current		12.88	2.4

* At rated voltage

** Can be combination stop/turn lamps

Figure 9-41 LED versus incandescent lamp current draw.

Because LEDs draw less current, less voltage is dropped on the wiring harness, providing a higher terminal voltage for an LED lamp compared to an incandescent lamp. This is very important for double or triple trailers, which may have 200 ft (61 m) of wire and frame ground between the alternator B+ feed to the rear of the last trailer and back to the alternator ground.

Trailers with ABS require an adequate voltage supply at each trailer ABS ECU. Using LED lights on the trailer reduces the current flowing through the trailer ground circuit. This reduces the voltage drop on the trailer ground circuit, which also acts as the ground for the trailer ABS ECU. Any voltage dropped on the trailer ground circuit reduces the voltage available to the trailer ABS ECU. Again, this is very important for double and triple trailers. LED lights significantly increase the available voltage at the trailer ABS ECUs during an ABS event with trailer lights illuminated.

Disadvantages of LEDs

There are some disadvantages of LEDs. The biggest disadvantage is initial cost. LED light assemblies cost significantly more than incandescent lamps. Because of this, LED lamps are prone to theft. To discourage theft, LED lamps may be attached with tamper-resistant fasteners that must be drilled out to replace the lamps, should they become damaged (common with trailer lighting due to loading dock damage).

The heat from incandescent lamps assists in keeping the light lens clear of snow and ice when driving in wintry conditions. Because of the high efficiency of LEDs, they generate very little heat. Therefore, it may be necessary to scrape snow and ice from LED lamps in winter driving conditions.

INTERIOR TRUCK LIGHTING

Lighting in the interior cab includes dome and courtesy lights, which illuminate when a door is opened. Trucks with sleeper cabs may have several interior lights, including florescent lamps. Other interior lighting includes instrument panel illumination.

Dome and Courtesy Lamps

The dome lamp is typically mounted in the front center of the cab interior on the roof. Courtesy lamps are typically mounted at the bottom of the instrument panel to illuminate the steps for entering or exiting the cab. Door sensing switches in the door A (front) pillar, B (rear) pillar, side of instrument panel, or incorporated into the door latch mechanism are used to control the dome and courtesy lamps. The dome and courtesy lamps are typically provided with full-time battery power (positive) on one side of the parallel-connected lamp filaments, while the other side of the filaments is connected to normally closed door switches. The closed door keeps the normally closed switches held in the open position. Opening the door causes the switch to close and provide a ground for the dome and courtesy lamps. The switch on the passenger door is typically wired in parallel with the driver's door switch so that opening either door causes all interior lamps to illuminate.

Many dome lamp assemblies have a switch to permit the dome lamp to be switched on with the doors closed. Some trucks may also incorporate the dome lamp switch with the panel illumination control. Rotating panel illumination past full illumination causes the dome lamp to illuminate.

Sleeper Cab Lighting

Trucks with sleeper cabs often have several other interior lights. These are mostly very simple and use switches located in the sleeper compartment to control the lights. Twelve-volt florescent lighting is commonly found in sleeper cabs.

Most sleeper cabs have luggage compartments. These compartments are the doors at the lower rear sides of the cab. These luggage compartments are typically lighted and make use of normally closed switches similar to those found on the truck's doors to provide a ground for the luggage compartment lamp.

Tech Tip: Luggage compartment lamps are often overlooked in the search for parasitic key-off load currents because there are no windows in the luggage compartment to determine if the luggage lamps are shutting off with the luggage doors closed.

Panel Illumination Lamps

Panel illumination includes the instrument cluster backlighting, radio backlighting, and various switch backlighting. Panel illumination lamps are typically incandescent lamps, but LEDs are common in individual switches as backlighting. The level or intensity of panel illumination is typically adjustable, like that on most passenger cars. Panel illumination lamps are typically provided with a full-time ground. The headlamp switch typically provides the panel illumination feed (**Figure 9-42**) with the headlamp switch in the park lamp position or headlamp position. A rheostat (variable resistor) connected in series with the panel illumination feed controls the current supplied to the parallel-connected panel illumination lamps. This rheostat is typically a wire-wound resistor.

Rheostats are prone to failure because they generate heat and depend on a mechanical contact between the wiper and the wire resistor. Over time, the wire may burn open or contact is lost between the wiper and wire. Many newer trucks use an electronic panel illumination device instead of a rheostat. The concept of pulse width modulation (PWM) was introduced in the study of voltage regulators. By cycling the panel illumination off and on at a rapid rate (perhaps 100 Hz) and controlling the duty cycle or on-time, the panel illumination lamps can be made to dim. If the voltage to the panel illumination lamps is only supplied 10 percent of the time and interrupted 90 percent of the time, the lamps will only reach partial intensity and will only glow dimly. Increasing the duty cycle so that the lamps are on 50 percent of the time and off 50 percent of the time will cause the panel illumination lamps to increase in intensity.

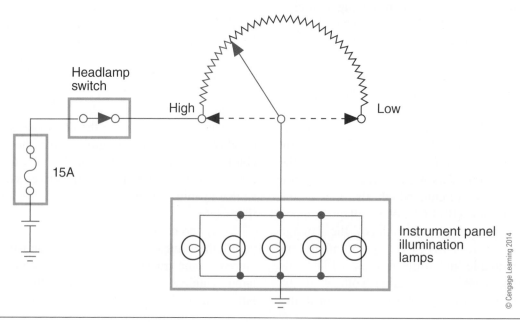

Figure 9-42 Instrument panel illumination.

Summary

- Truck lighting is subject to FMVSS 108 or CMVSS 108 and other government requirements. It is important to perform lighting repairs so that the truck does not violate the various lighting requirements.

- Incandescent lamps are lamps that contain a piece of thin metal wire such as tungsten. The sealed glass container does not contain oxygen, which permits the filament to glow white-hot (incandesce) without being consumed.

- Dual-filament lamps contain two filaments in one lamp globe. The two filaments share a common ground but have a separate voltage source for each filament. Dual-filament lamps are commonly used for turn signals and tail lamps, permitting one lamp to perform two or more functions.

- The resistance of an incandescent lamp increases significantly as the filament is heated, due to current flow. This causes a surge current when the lamp is first illuminated. The surge current of a typical incandescent lamp may be 10 times or more than the current drawn with the lamp at operating temperature.

- Trucks and trailers may have rear combination stop/turn lamps or separate stop/turn lamps.

- Trucks with rear combination stop/turn lamps can exhibit some seemingly strange behavior when a ground for a dual-filament lamp is opened. Other failures such as filaments that are shorted together can also cause problems that seem to be difficult to diagnose.

- Trailers receive voltage for lighting from the tractor through a seven-way trailer socket. This socket is standardized in North America and other parts of the world.

- Trailer sockets that are green and marked with the abbreviation ABS indicate that the center pin circuit is powered anytime the truck ignition is active. This center pin is used to supply the trailer ABS control module.

- LED lighting has several advantages over incandescent lamps, such as long life. Disadvantages of LED lighting include the higher initial cost.

Suggested Internet Searches

Try these web sites for more information on truck lighting:
http://www.grote.com
http://www.truck-lite.com
http://www.nhtsa.gov
http://www.fmcsa.dot.gov
http://www.gelighting.com
http://www.sylvaniaautocatalog.com

Review Questions

1. Which of the following is a true statement?

   A. A trucker can legally choose whatever color of marker lamps he wants for his truck.

   B. Stop lamps can be either amber or red.

   C. Front turn signals must be red.

   D. Rear turn signals can be amber if not also used as stop lamps.

2. An ohmmeter is used to measure the resistance of a headlamp. The ohmmeter indicates 0.5Ω. Ohm's law indicates that the headlamp lamp will draw approximately 28A at 14V. Since there are two headlamps wired in parallel, the total current draw of a pair of headlamps will be approximately 56A per Ohm's law. This does not cause a 20A type I circuit breaker used to protect the headlamp circuit to trip when the headlamps are switched on. Why?

   A. Ohm's law does not apply to incandescent lamp filaments because they are not resistors. The actual current drawn when the lamps are first switched on will be much less than 56A per Watt's law.

   B. The calculation is incorrect. Actual total current draw is 5.6A.

   C. The circuit breaker will probably trip because the headlamp lamps are apparently shorted internally.

   D. The two headlamps together will draw approximately 56A initially. However, the filament material has a positive temperature coefficient that will cause the current to drop off rapidly as the filaments heat.

3. A truck with air brakes has two pressure switches that are located in a delivery port of the foot valve. These switches are used to control the truck's stop lamps. Which best describes the way that these two switches are most likely being used?

   A. The two switches are wired in series so that there must be pressure in both the primary and secondary brake chambers to cause the stop lamps to illuminate.

   B. One switch is normally open, and the other switch is normally closed. The switches are wired in parallel. Brake application pressure causes the switches to change states and the stop lamps to illuminate.

   C. Both switches are normally closed. A drop in air pressure in either the primary or secondary air system causes the switches to open, which causes the stop lamps to illuminate.

   D. The two switches are normally open and are wired in parallel. Brake application pressure in the primary or secondary air system causes one or both of the switches to close and the stop lamps to illuminate.

4. A particular tractor has two rear lamp assemblies, each containing a single dual-filament incandescent lamp. The trailer has four rear lamp assemblies that each contains a single dual-filament incandescent lamp. Technician A says that the tractor probably has combination stop/turn lamps at the rear. Technician B says the trailer probably has separate stop/turn lamps at the rear. Who is correct?

   A. A only

   B. B only

   C. Both A and B

   D. Neither A nor B

5. The high-beam headlamps do not illuminate on a truck with a floor-mounted dimmer switch. Low-beam headlamps operate as expected. Which is the most likely cause?

   A. Open high-beam headlamp ground circuit

   B. Defective headlamp switch

   C. One high-beam headlamp burned out

   D. Defective headlamp dimmer switch

6. All of the lights on a trailer flash off or dim considerably when driving on rough roads and when turning. What is the most likely cause?

   A. Short circuit in the trailer ground.

   B. The connection at the center pin of the seven-way trailer socket is opening intermittently, causing a loss of voltage for all trailer lamps.

   C. An open circuit in the trailer lighting ground circuit is causing the trailer to have a ground only through the fifth wheel.

   D. The marker interrupt switch is interrupting all of the trailer lighting due to a defective switch contact.

7. The rear lights of a truck with combination stop/turn lamps have some very strange problems that seem to make no sense. Some lights are dim; some flash when they are not supposed to. What is a likely cause of such behavior?

    A. Front turn signal circuit is probably shorted to ground.

    B. A fuse is back-feeding or causing feedback.

    C. Electromagnetic interference, probably from a CB radio.

    D. Poor ground at a rear combination lamp.

8. Several lamps are burned out on a truck. What would be a likely cause?

    A. Poor ground causing excessive lamp resistance.

    B. Ground circuit with too low a resistance, causing lamp current to increase to a high level.

    C. Batteries have a high internal resistance.

    D. High charging system voltage.

9. Which of the following could reduce incandescent lamp light output?

    A. High resistance in lamp ground circuit.

    B. Reduction in current flowing through the lamp.

    C. Transfer of tungsten from the filament to the inside of the lamp globe.

    D. All of the above could reduce incandescent lamp output.

10. A trucker decides to add numerous incandescent marker lamps to his trailer. The 20A automatic reset circuit breaker in the trailer marker lamp circuit now cycles the trailer marker lamps off and on every few minutes. What should be done to resolve this problem?

    A. Increase the circuit breaker to 30A after measuring the marker lamp current and making sure it is less than 30A.

    B. Increase trailer ground circuit wire gauge size.

    C. Locate the intermittent open circuit that is causing the circuit breaker to trip.

    D. Recommend to the truck operator that he remove some of the trailer marker lamps because adding parallel loads causes total resistance to decrease and current draw to increase.

11. The circuit with dark green insulation in an SAE J560 seven-way trailer cord is used for what purpose?

    A. Trailer ABS power supply

    B. Trailer right turn

    C. Trailer left turn

    D. Trailer ground

12. The bayonets on an incandescent lamp are connected to which circuit on a negative-ground vehicle?

    A. Ground

    B. Turn signal

    C. Tail lamp

    D. Positive

13. One low-beam headlamp has burned out on a truck. What is the reason that the other low-beam headlamp is unaffected?

    A. The headlamps are wired in series.

    B. The low-beam filaments are isolated from ground.

    C. The headlamps are wired in parallel.

    D. They are dual-filament lamps.

14. Which lamps are not commonly found on trucks in North America?

   A. Front stop lamps

   B. Marker lamps

   C. Clearance lamps

   D. Identification lamps

15. What should a green trailer socket cover identified as ABS at the back of a tractor signify?

   A. The center pin of the trailer socket is powered when the brake pedal is depressed.

   B. The top pin of the trailer socket is grounded.

   C. The center pin of the trailer socket has a green wire.

   D. The center pin of the trailer socket is powered when the key is in the ignition position and is used to supply the trailer ABS ECU.

10 Electrical Accessories

Learning Objectives

After studying this chapter, you should be able to:

- Troubleshoot an electric horn circuit.
- Explain how a windshield-wiper motor system operates, including the wiper park feature.
- Diagnose the cause of a blower motor that does not operate.
- Troubleshoot the electrical portion of an air conditioning system.
- Explain the concept of a latched relay.
- Describe the difference between a normally open and normally closed air solenoid.
- Troubleshoot an on-off-type fan drive system.
- Diagnose an inoperative power window.
- Explain how a compression brake and exhaust brake provide braking force.
- Troubleshoot a Hydro-Max™ booster pump system.

Key Terms

clockspring

HVAC

Hydro-Max™

linear power module (LPM)

multifunction switch

resistor block

INTRODUCTION

Electrical accessories refer to electrical features that provide increased safety and operator comfort. These accessories include horns, windshield wipers, HVAC systems, and other features that are commonly found on automobiles. Other accessories not typically found on automobiles, such as engine brakes and on-off fan drives, are also covered.

Many of the components discussed in this chapter may be controlled by a body control module on some trucks, as will be discussed in **Chapter 13**.

HORNS

Trucks are well known for air horns. However, trucks are required to have electric horns. An electric horn has some similarities to both a speaker and a relay. Electric horns are electromagnetic devices. A typical electric horn is shown in **Figure 10-1**. The horn consists of an electromagnet, an armature, a diaphragm, and a set of normally closed switch contacts. The armature is attached to the diaphragm. The diaphragm is like the cone that vibrates in a speaker. When electric current flows through the coil, the electromagnet causes the armature to move. The

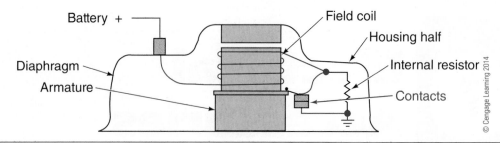

Figure 10-1 Electric horn components. Note closed contacts, which provide a path to ground for the field coil.

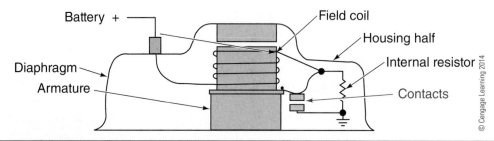

Figure 10-2 Armature movement due to magnetic field has physically opened the contacts to interrupt the flow of current through the field coil. Process of opening and closing contacts will repeat at a rapid rate.

armature movement causes the switch contacts to open, which interrupts the flow of current through the coil (**Figure 10-2**). With no current flow through the coil, a spring action causes the armature to return to the position shown in **Figure 10-1**. This also closes the switch contacts again, starting the process over again. The movement of the armature and diaphragm causes air to be displaced, resulting in formation of pressure pulsations or sound waves in the same way as in a speaker. The frequency of the sound is established by the number of times the diaphragm vibrates per second. This frequency is adjustable on some horns, but most modern horns have no provisions to adjust horn frequency. Most trucks use a pair of horns wired in parallel that are designed to have different sound frequencies to combine to create the sound of a chord.

A steering-wheel mounted normally open switch is typically used to control the electric horn. Because the steering wheel must be able to rotate several revolutions from the center, it is not possible to just route wires through the steering column to the steering-wheel switch. Doing so would result in wrapping the wires around the steering shaft every time the wheel was turned. Instead, a slip ring may be used to permit continuous electrical contact while the steering wheel is rotated (**Figure 10-3**). This is similar to the slip ring and brushes found in alternators.

Some trucks may have additional steering-wheel mounted switches, such as cruise control switches and air horn. It would require several different slip rings to control all of these features. Additionally, driver side air bags are now available on some trucks and the

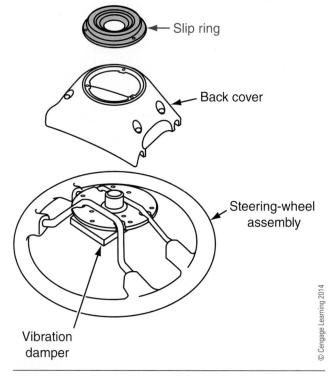

Figure 10-3 Slip ring provides electrical connection throughout steering-wheel rotation.

electrical connection for the air bag circuit requires a continuous electrical connection to prevent inadvertent air bag deployment. Some OEMs use **clocksprings** instead of slip rings, even if the truck is not equipped with air bags. A clockspring is similar to a measuring tape; clocksprings have been used in automobiles for several years for air bags and steering-wheel mounted switches. A clockspring consists of several thin copper circuits that are sandwiched between thin insulating tape materials. A clockspring is shown in **Figure 10-4**. The clockspring has two or more sets of electrical connectors so that it can be plugged into the stationary steering column wiring-harness connector and into the connector on the steering wheel that rotates with the steering wheel.

A clockspring requires some special attention to prevent breakage. Because the clockspring is like a measuring tape, it can only be rotated a specific number of times in either direction or the tape will be pulled tight inside the clockspring and may possibly break. This is typically not a problem because the vehicle steering linkage permits only a limited amount of steering-wheel revolutions. However, if the steering linkage is disconnected and someone rotates the steering wheel, the clockspring may break. In addition, if the steering wheel is rotated only one or two revolutions with the steering linkage disconnected, the clockspring will not be centered when the steering linkage is reconnected and may break the first time the steering wheel is fully rotated for a sharp turn.

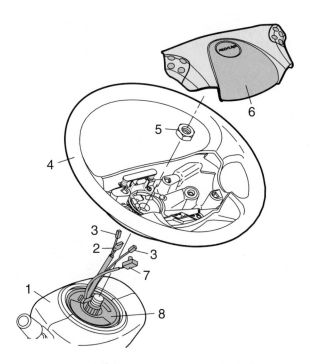

1. Steering Column	5. Steering Wheel Nut
2. Steering Wheel Switch Connector	6. Steering Wheel Cover Module
3. Horn Button Connectors	7. Air Bag Connector
4. Steering Wheel	8. Clockspring

© Cengage Learning 2014

Figure 10-4 Clockspring provides continuous rotation throughout steering-wheel rotation for several circuits.

> **CAUTION** *It is important to lock the steering wheel on trucks equipped with clocksprings when the steering linkage is disconnected to prevent clockspring damage. This is especially true on trucks equipped with air bags. Breakage of the clockspring may result in air bag deployment.*

Most trucks use a horn relay to control the electric horns instead of routing horn current through the horn switch in the steering wheel (**Figure 10-5**). The horn switch in the steering wheel typically provides a ground for the horn relay coil, while the high side of the relay coil is typically provided with battery voltage. This permits the horn to operate with the key switch in the off position.

Because a horn causes repeated, rapid interruption of current flow through a coil, a horn is potentially capable of generating very high levels of negative voltage spikes, as explained in the discussions on inductors in **Chapter 3**. Modern electric horns contain internal suppression of some type to minimize these potentially damaging voltage spikes. It is important to install only suppressed horns on modern trucks.

Air horns used on trucks may have the familiar lanyard chain, which is used to control a manual air valve that supplies compressed air to the air horn. Some trucks may instead have electrically controlled air horns. An electric solenoid is used to control air supplied to the remote-mounted air horn. Closing a switch causes the solenoid to energize and permit air to pass to the air horn.

WINDSHIELD WIPERS

In the past, truck windshield-wiper motors were often compressed air-operated devices. Most modern trucks use electric windshield-wiper motors, like passenger cars. Modern truck wiper motors are typically permanent magnet motors with two speeds. Permanent magnet motors require no field coils like those used in a starter motor. However, some trucks may instead have a three-speed wiper motor that uses field coils instead of permanent magnets.

Wiper motor assemblies typically contain a gear mechanism to increase the wiper arm torque (**Figure 10-6**). The armature shaft typically has a worm gear machined into the drive end. The worm gear arrangement is the reason that the wiper linkage and wiper arms are locked in position when the wiper motor is not powered. The motor linkage mechanism is designed to transform the continuous rotary motion of the wiper motor into the familiar back-and-forth motion of the wiper blades (**Figure 10-7**).

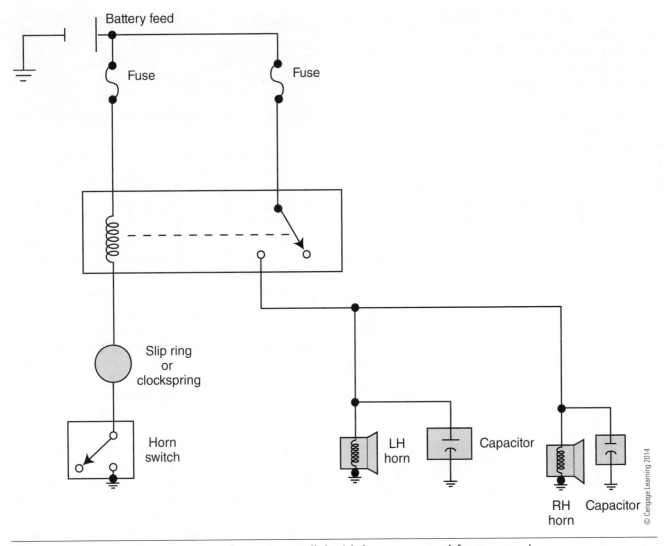

Figure 10-5 Horn circuit. Capacitors shown in parallel with horns are used for suppression.

Two-Speed Permanent Magnet Wiper Motors

A permanent magnet wiper motor can be made to have two speeds through the use of three brushes instead of two brushes. This type of motor has two different positive brushes and a single negative or common brush, as shown in **Figure 10-8**. Only one positive brush is provided with voltage at any given time. The placement of the positive brushes relative to the ground brush causes the motor to have two different speeds. Because of the placement of the high-speed brush, fewer armature windings are connected between the high-speed brush and the ground brush than are connected between the low-speed brush and the ground brush. The reduced number of armature windings used by the high-speed brush reduces the magnetism of the armature, resulting in a decrease in CEMF generated by the motor. The net result is an increase in motor speed when voltage is supplied to the high-speed brush compared to when voltage is supplied to the low-speed brush.

Because a motor is also a generator, a voltage will be generated at the high-speed brush when the wiper is operating in low-speed, and a voltage will be generated at the low-speed brush when the motor is operating in high-speed. This voltage can be much higher than the 12V electrical system. However, this is typically not a problem because the system is designed so that a completed circuit is only provided for one of the positive brushes (high or low) at a time.

Many wiper motors also contain an internal automatic reset-type circuit breaker or overload device. This internal circuit breaker is designed to open to protect the motor from damage should the wiper motor become stalled due to snow and ice on the windshield. The high level of current drawn by a stalled motor causes the temperature of the windings to increase. CEMF reduces the amount of current drawn by a motor, as explained in **Chapter 8**. A motor that is not rotating is not producing

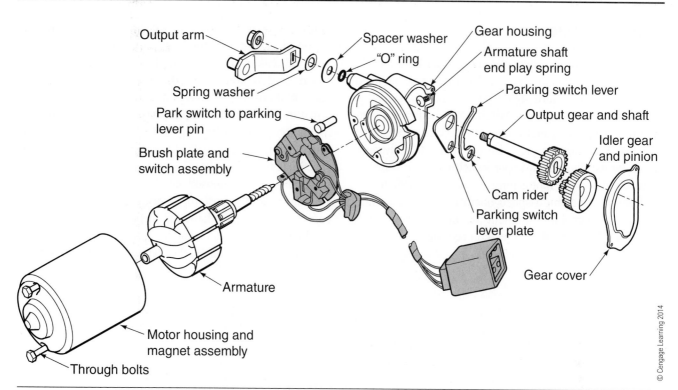

Output arm
Spacer washer
"O" ring
Gear housing
Armature shaft end play spring
Spring washer
Parking switch lever
Park switch to parking lever pin
Output gear and shaft
Idler gear and pinion
Brush plate and switch assembly
Cam rider
Parking switch lever plate
Gear cover
Armature
Motor housing and magnet assembly
Through bolts

© Cengage Learning 2014

Figure 10-6 Windshield-wiper motor internal components.

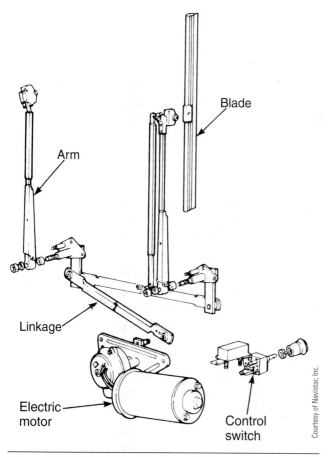

Blade
Arm
Linkage
Electric motor
Control switch

Courtesy of Navistar, Inc.

Figure 10-7 Windshield-wiper linkage and components.

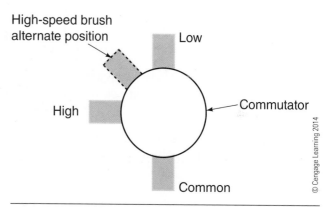

High-speed brush alternate position
Low
High
Commutator
Common

© Cengage Learning 2014

Figure 10-8 Two wiper motor positive brushes provide two wiper motor speeds.

any CEMF. Therefore, a stalled motor will draw a large amount of current and can be damaged due to excess temperature. Most trucks also use an automatic circuit breaker or other circuit protection device in the supply circuit for the wiper motor. The primary purpose of this external circuit breaker is to protect the wiring supplying the wiper motor. The internal wiper motor circuit breaker is typically designed to open before the truck's circuit breaker supplying the motor. It is important to never add additional loads to the wiper motor circuit. This could result in the wiper circuit protection device opening due to the additional electrical loads, causing the wipers to stop functioning.

Wiper Switch

Windshield-wiper motors are controlled by a switch. This switch is often incorporated onto a stalk-mounted switch, such as the turn signal stalk. This stalk switch is known as a **multifunction switch**. Other wiper switches may be instrument panel mounted rotary or sliding-type switches. Some trucks may also use relays that are controlled by the windshield-wiper switch instead of directly powering the wiper motor through the wiper switch.

Typical wiper high-speed operation using a switch to directly control a permanent magnet wiper motor is shown in **Figure 10-9**. The colored lines show the path of current flow through the ganged wiper switch.

With the switch in the high-speed position, current is supplied by an ignition (run) or accessory-powered circuit breaker. This current flows through the high-speed switch contact to the high-speed brush. The current then flows through the high-speed brush and armature windings to the grounded brush, causing the motor to operate at high speed. Note that the low-speed brush circuit is open with the wiper switch in the high-speed position. As indicated previously, a voltage will be generated on the unused brush when the motor is in operation, so the system is designed to only have a completed circuit available for one brush circuit at a time.

Typical low-speed operation is shown in **Figure 10-10**. Current flows through the wiper switch low-speed

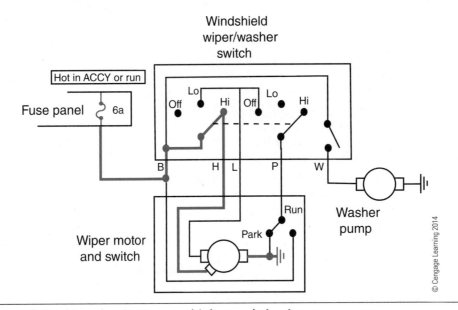

Figure 10-9 Current flow through wiper motor high-speed circuit.

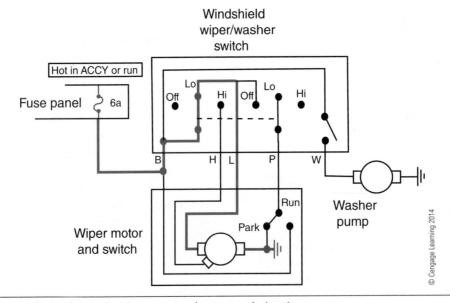

Figure 10-10 Current flow through wiper motor low-speed circuit.

contacts to the low-speed brush, through the armature windings to the grounded brush, causing the wiper motor to operate at low speed.

Wiper Park

Windshield-wiper systems are designed so that the wiper arms automatically return to the park position when the wiper switch is placed in the off position. This prevents the wiper arms from obstructing the operator's view when the wipers are not in use. To cause the wiper arms to park when the wipers are switched off, a park switch is incorporated into the wiper motor assembly. The park switch is designed to cause an interruption of current flow at the point in wiper travel when the wipers are in the parked position.

A typical wiper park methodology is shown in **Figure 10-11**. When the wipers are not in the parked position and the wiper control switch is placed in the off position, electric current will flow through the wiper park switch inside the wiper motor to the wiper control switch P (park) terminal, as shown by the colored lines in **Figure 10-11**. This causes the wiper low-speed brush to be powered, resulting in wiper low-speed operation until the wipers move to the parked position. In the parked position, the wiper park switch opens, due to mechanical linkage inside the wiper motor. The open park switch interrupts the flow of current to the wiper motor P terminal and low-speed brush. This causes the wiper motor to stop when the wiper arms have moved to the park position.

Most wiper motors are designed so that the internal wiper park switch is grounded with the wiper arms in

the park position. With the wiper arms in the parked position and the wiper switch in the off position, a low-resistance circuit exists between the wiper low-speed brush and the wiper ground brush through the wiper switch. A motor is also a generator because of the CEMF that is generated in the armature as it rotates through the magnetic field set up by the pole shoes, as explained in **Chapter 8**. When the switched-off wiper motor rotates to the park position, the voltage supplied to the motor is interrupted but the system still has inertia, which causes the motor to coast. The low-resistance path provided by the wiper switch between the wiper low-speed brush and ground brush causes the coasting wiper motor to stall quickly when the wipers reach the park position. The coasting motor is acting as a generator and the low-resistance path through the wiper switch acts as a short circuit across the terminals of the motor. The short circuit through the wiper park circuit causes the current flow in the armature caused by CEMF to reverse directions, resulting in the armature coming to a rapid stop at the park position. This dynamic braking action prevents the inertia of the wiper system from causing the wiper motor to skip through the park switch and continue operating, even though the wiper switch is in the off position. Keep the dynamic braking feature of wiper motors in mind when troubleshooting wiper systems.

The wiper park switch is typically supplied with ignition or accessory voltage and not direct battery voltage so that the wipers can be stopped in mid-travel by switching the key switch off with the wipers in motion. This permits ease of wiper blade replacement and other wiper system service.

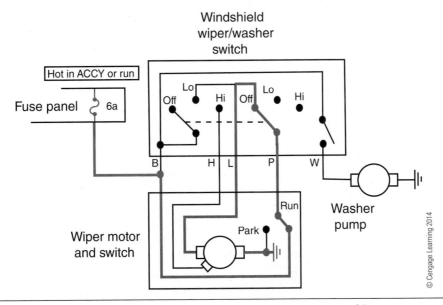

Figure 10-11 Current flow through wiper motor circuit while wipers are parking.

Intermittent Wipers

The term *intermittent wipers* (or interval wipers) describes a repeated single wipe of the windshield wipers with a time delay or dwell between wipes. Intermittent wipers are often obtained by supplying the low-speed brush with a brief pulse (e.g., one-half second) so that the wipers move off the park position. The wiper park system design causes the wipers to complete a full wipe cycle. An electronic controller provides this short-duration voltage pulse to move the wiper motor off the park position. The controller for intermittent wipers is sometimes built into the wiper switch, or is a separate control module. The wiper switch may provide six or more varying intermittent wiper delay periods. The wiper switch may contain a stepped resistor network or variable resistor, which provides a variable voltage to the wiper control module, indicating which position of the intermittent wiper switch the truck operator has selected (**Figure 10-12**).

HVAC SYSTEM

HVAC stands for heating, ventilation, and air conditioning. Most trucks in North America are equipped with air conditioning. A typical truck HVAC system is

shown in **Figure 10-13**. The main electrical components in a truck HVAC system include the blower motor system and the air conditioning electromagnetic clutch system. Some trucks may also use small electric motors and an electronic HVAC module to control the doors in the HVAC housing used to route the airflow through the ducts and to control temperature.

Blower Motor

The blower motor used in trucks is typically a permanent magnet brushed DC motor. The blower motor drives a squirrel-cage-type blower wheel (**Figure 10-14**). The airflow generated by the fan is routed through a heater core to produce warm air or through an evaporator coil to produce cold air. The heater core is a small radiator that uses engine coolant as the heat source. The evaporator coil is part of the air conditioning system.

Trucks with sleeper cabs typically have two separate HVAC systems due to the large interior size of these trucks. The HVAC system—including the blower motor, fan, heater core, and evaporator coil for the sleeper compartment—may be contained in a box assembly as shown in **Figure 10-14**. The HVAC system for the driver area is mounted in an assembly under the dash panel, similar to that found in a passenger car (**Figure 10-15**).

The blower motor speed is variable, with three or more speeds being typical. The speed of the blower motor is typically controlled by the use of a stepped resistor network referred to as a **resistor block** or assembly. The resistors are high-power wire-wound resistors (**Figure 10-16**). The resistors are placed in series with the blower motor, thus reducing the voltage supplied to the motor and resulting in a reduction in motor speed. A typical truck blower motor may draw 25A while running in high-speed, making it one of the highest electric power consuming devices in the cab of the truck. Due to the high current drawn by the blower motor, the resistor block gets very hot and is typically installed so that it is in the airflow of the blower motor to cool the resistors.

The thermal limiter shown with the resistor block in **Figure 10-16** is used to protect the resistors from overheating should the blower motor stall due to bearing failure or debris, or if the circuit between the blower motor and resistor block becomes shorted to ground. If either one of these conditions occurs and the blower motor switch is in the low- or medium-speed position, the fuse or circuit breaker protecting the blower motor circuit might not open or trip because the resistors in the resistor block are limiting

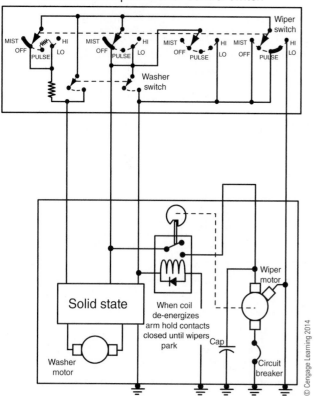

Intermittent wiper/washer control switch

Figure 10-12 Intermittent wiper system.

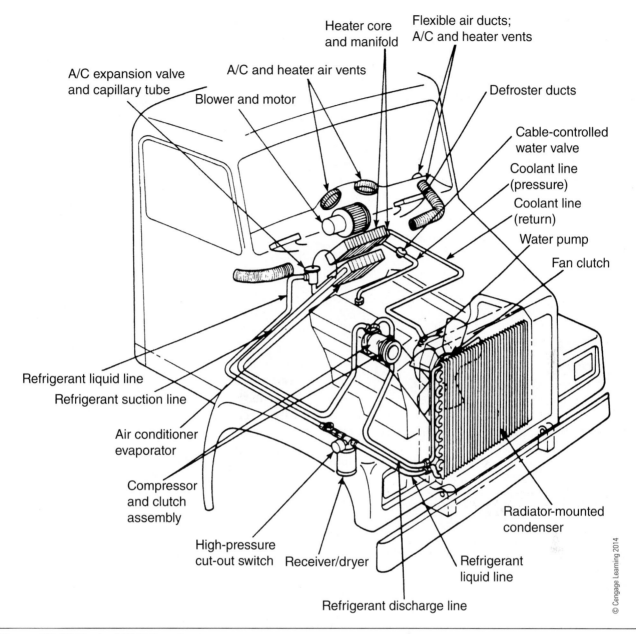

Figure 10-13 Typical HVAC system.

the current to a level below the CPD rating. However, the high level of current due to the short to ground or stalled motor can cause the resistors in the resistor block to overheat. On some trucks, the thermal limiter is designed to open the blower motor circuit like a CPD to prevent the resistors from overheating. On other trucks, the thermal limiter is designed to become a short circuit when overheated to cause the truck's CPD to open to prevent the resistors from overheating. In either case, it may be necessary to replace the resistor block as part of the repair in addition to repairing the original problem, such as a stalled blower motor.

The control circuit for a typical truck blower motor is shown in **Figure 10-17**. Note that the circuit is protected by a 30A fuse. The blower control switch has three motor speed positions. In the low-speed position, voltage is supplied to terminal 2 of the blower resistor. This places the two resistors in the resistor block in series with the blower motor, causing the blower motor to operate at a low speed. If you were to measure the voltage drop across the blower motor on a 12V system, you might find that only 7V was present with controls set for low speed indicating that about 5V was being dropped across terminals 2 and 3 of the blower resistor.

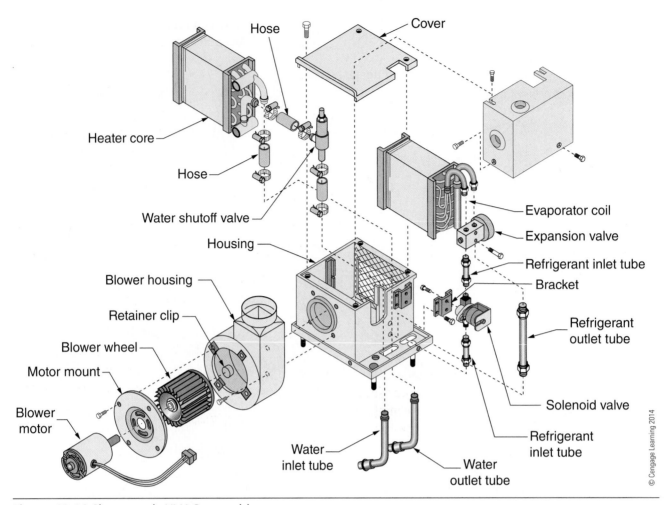

Figure 10-14 Sleeper cab HVAC assembly.

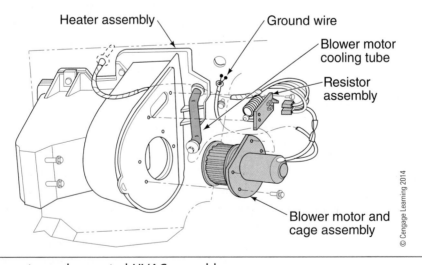

Figure 10-15 Instrument panel mounted HVAC assembly.

In the medium-speed position for the system shown in **Figure 10-17**, voltage is supplied to terminal 1 of the blower resistor. This places one resistor in series with the blower motor, causing a reduction in the series resistance resulting in a motor speed increase. The voltage drop across the blower motor might increase to 9V with the switch in the medium-speed position.

In the high-speed position, the resistor block is bypassed, resulting in no additional series resistance.

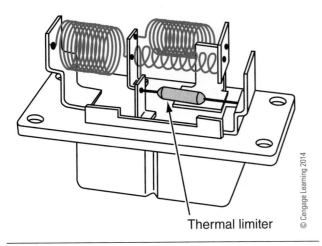

Thermal limiter

© Cengage Learning 2014

Figure 10-16 Resistor block used to control current through blower motor to reduce speed; many trucks use an electronic module to provide this function.

This causes the blower motor to operate at the highest speed and the voltage drop across the blower motor would increase to 12V.

Some modern trucks use an electronic module instead of a resistor network to control blower motor speed. This controller may use a PWM technique or use high-power transistors to control blower motor current. The system shown in **Figure 10-18** uses an electronic **linear power module (LPM)** to control blower motor speed. The LPM adds a series resistance to the ground circuit of the blower motor depending on the control voltage supplied by the HVAC control head. The resistance is obtained using transistors, not actual resistors like those found in a resistor block. The HVAC control head provides a 0V to 5V signal, depending on which of several possible fan speed positions is selected. An HVAC control head output voltage of 0V indicates fan off, and an output voltage of 5V indicates that maximum blower motor speed has been selected. Increasing voltage from the control head causes the LPM to increase motor speed by reducing the motor's ground resistance. Note that this system provides a constant battery voltage to the high side of the blower motor and the linear power module controls the blower motor ground to control blower motor speed. These electronically controlled blower motor systems have the advantage of offering more fan speed choices than resistor block systems.

Air Conditioning Clutch Control

The air conditioning (A/C) compressor clutch is an electromagnetic clutch. The clutch field coil is an electromagnet that causes a clutch hub, which is keyed to the compressor shaft, to be connected magnetically to the rotating belt-driven clutch pulley (**Figure 10-19**).

The compressor clutch is one of the larger electromagnets on a typical truck, drawing approximately 4A of current when energized.

Tech Tip: The A/C compressor clutch coil is typically diode suppressed, meaning a diode is in parallel with the clutch coil. The polarity of voltage supplied to the A/C clutch must be observed during any testing or troubleshooting to prevent damage to the suppression diode.

The climate control panel (control head), which contains the blower motor switch, mode control (defrost, panel, etc.), and temperature control, often supplies the high side for the air conditioning clutch, as shown in **Figure 10-17**. The air conditioning request switch may be a separate switch or may be incorporated into the mode control.

Typically, two pressure switches in the compressor clutch control circuit are used for A/C system protection. These switches may be placed in series with the compressor clutch on some trucks. If the two switches are wired in series with the compressor clutch, one of the pressure switches is a normally open switch that opens with system pressure less than approximately 20 psi (138 kPa). This low-pressure cut-out switch inhibits compressor operation if the system pressure is too low, indicating a loss of refrigerant in the system or a very low ambient temperature. Compressor operation is inhibited in either case to prevent damaging the compressor. A high-pressure cut-out switch is located in the high side of the A/C system. The high-pressure cut-out switch is normally closed and opens to inhibit compressor operation if the high side pressure is greater than approximately 375 psi (2.5 MPa). This indicates a condenser airflow problem or similar issue. Compressor operation is inhibited with either the high-pressure or the low-pressure cut-out switches open to prevent damage to the system. These two pressure switches are shown as being contained within the binary pressure switch in the typical A/C circuit illustrated in **Figure 10-17**.

Instead of the two pressure switches being connected in series, some trucks may have the low- and high-pressure cut-out switches wired in parallel with each other, as shown in **Figure 10-20**. The low-pressure safety switch shown is a normally closed switch that opens with pressures greater than approximately 20 psi (138 kPa). The high-pressure switch shown is a normally open switch that closes at pressures greater than 375 psi (2.5 MPa). Note that both of

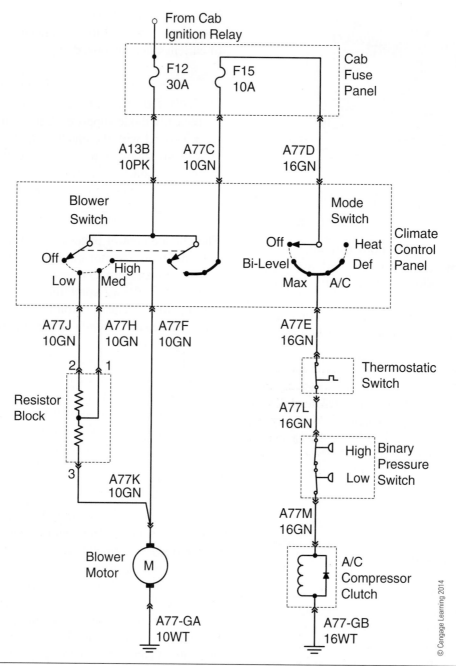

Figure 10-17 Typical HVAC electrical circuit. Note the switch ganged with the blower control switch, which powers the mode control switch.

these parallel-connected switches are opposite the normal states of the series-connected switches described in the previous section. Both switches in this parallel switch example would be open if low side pressure were greater than about 20 psi and the high side pressure were less than about 375 psi. With both of these switches open, the air conditioning relay would not be energized. Current sourced from the heater control switch would pass through the manual temperature control switch and on to the compressor clutch through the normally closed contacts of the air conditioning relay.

If either of the pressure switches shown in **Figure 10-20** closes, the air conditioning relay will be energized. The energized air conditioning relay interrupts the flow of current to the compressor clutch. Note also that energizing the air conditioning relay also causes the output of the normally open air conditioning relay contact (terminal 87) to be supplied to the high side of the air conditioning relay coil (terminal 86). If the manual temperature control switch is closed (refrigerant temperature greater than approximately 40°F (4°C)), then the air conditioning relay will remain energized even after both of the pressure

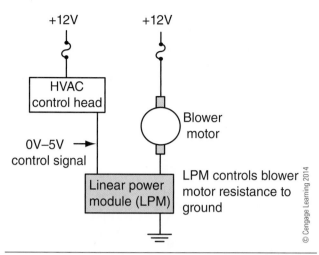

Figure 10-18 Electronic linear power module (LPM) used to control blower motor speed: 0V from control head is motor off, and 5V from control head is highest speed. Motor speed is directly proportional to control voltage.

switches are opened when the system pressure returns to an acceptable level. The output of the normally open terminal of the air conditioning relay (terminal 87) has latched the air conditioning relay in the energized state, which causes the compressor clutch to be de-energized. It will be necessary to interrupt the voltage supplied to the manual temperature control switch to de-energize the air conditioning relay to permit the compressor clutch to be energized once again. The voltage supplied to the manual temperature control would be interrupted when the ignition switch is cycled off or the air conditioning switch is cycled off.

This latching action of a relay is commonly used to prevent something from repeatedly cycling off and on due to some problem with the system. The latching action requires power to be cycled to "break" the latch. Without this latching action, the A/C compressor clutch could rapidly cycle if refrigerant is low. You should also be aware of this latching action of a relay when troubleshooting because inadvertent latching of relays can occur due to various wiring problems. Latched relays are covered in more detail in the Extra for Experts section at the end of this chapter.

Engine Fan Drive

Many trucks have an engine fan drive that is compressed air operated (**Figure 10-21**). These are also known as on-off fan clutches. A fan drive couples the engine fan to the fan pulley, which is driven by the engine accessory drive belt. The fan drive permits the fan to freewheel when it is not needed for engine cooling. The fan can also be forced on by the engine

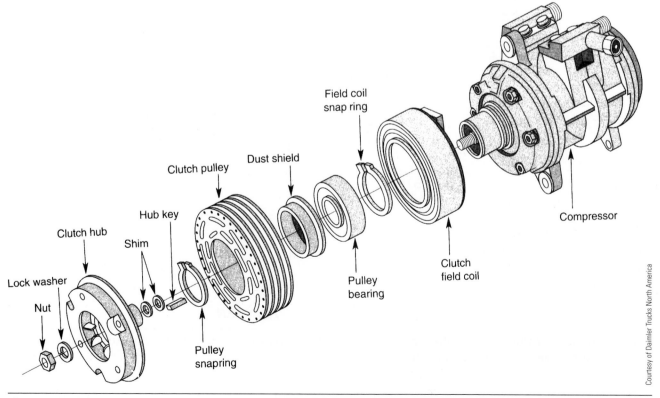

Figure 10-19 Compressor clutch assembly. Clutch field coil is one of the largest electromagnets found in a truck electrical system.

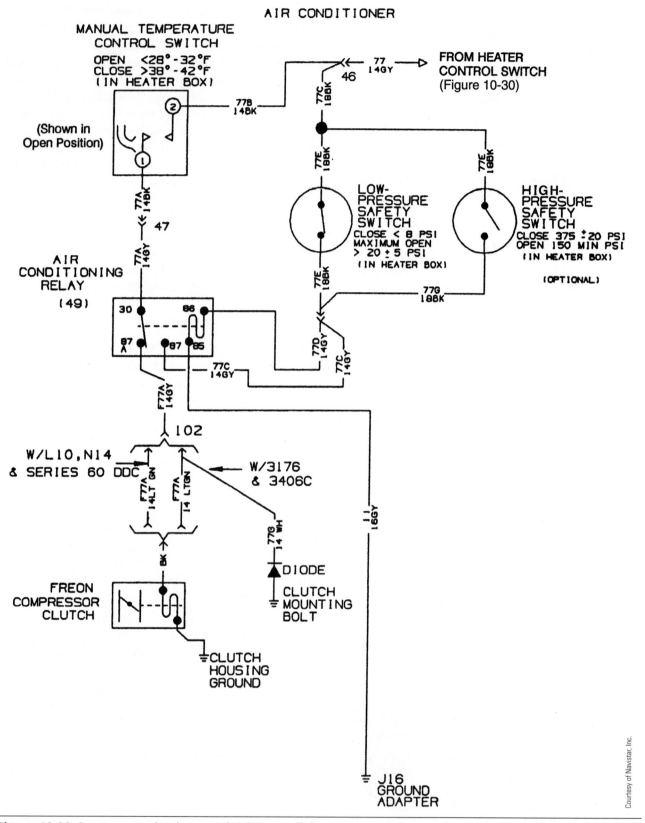

Figure 10-20 Compressor clutch control using parallel pressure switches and a relay. Note suppression diode connected in parallel with compressor clutch.

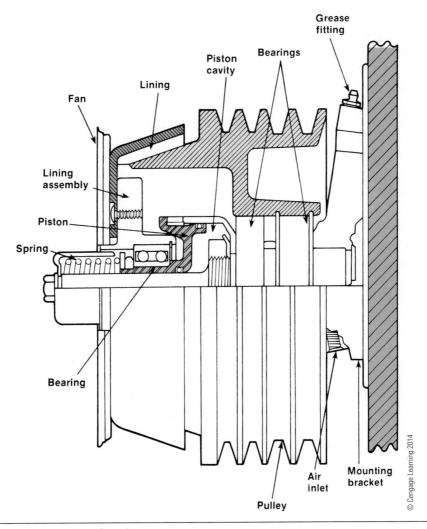

Figure 10-21 Air-operated engine fan drive.

ECM to act as a fan retarder, which is another form of engine braking. The airflow through the condenser provided by the engine fan is also necessary for low vehicle speed air conditioning performance. The compressed air supplied to the fan drive is controlled by an electric solenoid.

There are two main types of compressed air actuated fan drives: spring engaged and air engaged. Spring engaged fan drives use compressed air to release the fan clutch to permit the fan to freewheel. With no air pressure supplied to the fan drive, the fan is engaged by the springs in the drive. Air engaged fan drives use compressed air to engage the fan clutch. With no air pressure supplied to the fan drive, the fan is freewheeling.

On both types of fan drive systems, it is common to use an air solenoid valve to control the fan drive. These valves typically have three ports designated NO (normally open), NC (normally closed), and out or discharge port. This normally open and normally

closed terminology when discussing air or hydraulic solenoids and valves differs from that used with an electric relay or switch. When compressed air is routed to the normally open port of an air solenoid and the solenoid is *not* energized, air will flow through the solenoid valve to the discharge port. Energizing the solenoid will cause the normally open path through the valve to be blocked, thus preventing the flow of air from the normally open port to the discharge port. Additionally, the air that is trapped between the discharge line and the device being controlled is exhausted through the normally closed port of the solenoid. For this reason, you should not put a plug in the unused port of this type of solenoid.

The normally open terminology when dealing with air solenoids may seem to be opposite of the terminology used with electric relays. A normally open relay contact does not permit electric current to flow with the relay coil de-energized. A normally open air solenoid valve, on the other hand, does permit air to flow through to the discharge port if the solenoid coil is not energized. The term *open* in electricity and the term *open* in pneumatics mean different things. *Open* in pneumatics means that air *can* pass through, while *open* in electricity means that current *cannot* pass through. The term *closed* also means the opposite in electricity and pneumatics. *Closed* in pneumatics means that air *cannot* pass through while *closed* in electricity means current *can* pass through.

The electric solenoid valve that supplies an air fan drive may be controlled by switches, as shown in **Figure 10-22**. Note that there are three switches that control this fan solenoid valve. In the top drawing shown in **Figure 10-22**, the switches are all normally-closed-type switches and are wired in series. Note also that the fan drive in the top drawing of **Figure 10-22** is the type that uses air pressure to apply the fan clutch. The manual override switch is a dash-panel mounted switch that permits the operator to manually switch on

the fan drive. This optional fan override is no longer permitted on most engines due to exhaust emissions requirements related to engine coolant temperature. The A/C pressure switch monitors A/C high side pressure and opens at a predetermined pressure, typically near 275 psi (1.9 MPa). The thermal switch monitors engine coolant temperature and opens when engine coolant temperature reaches some value around 220°F (104°C). With all three of these switches closed, the solenoid valve will be energized, which blocks the flow of compressed air to the fan drive. The fan drive is freewheeling with all three switches closed. If any one of these three switches should open, the flow of current through the solenoid valve would be interrupted. This would cause compressed air to be supplied to the fan drive through the solenoid normally open port, causing the fan drive to engage and resulting in air being drawn over the condenser and radiator.

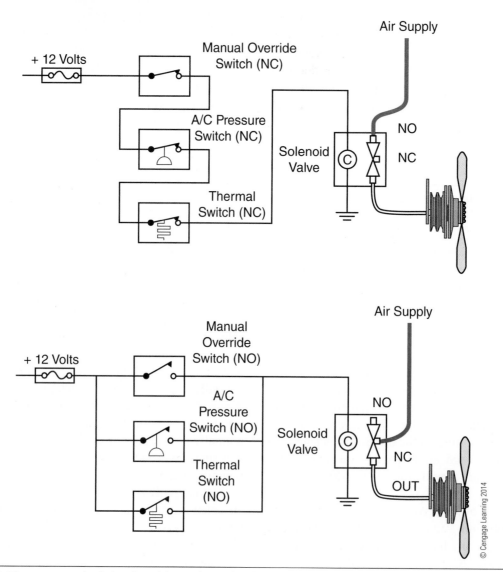

Figure 10-22 Direct control of the fan drive solenoid valve using pressure, temperature, and manual switches.

The system shown in the lower drawing of **Figure 10-22** is also the type of fan drive that requires air pressure to engage fan drive, but uses the normally closed port of the solenoid valve. The three switches used in this system are normally open switches that are wired in parallel. Closing any one of the three switches causes the solenoid to energize. Energizing the solenoid causes compressed air to flow to the fan drive to engage the fan clutch.

The fan solenoid on modern trucks is often controlled by the engine ECM. The various switches act as inputs to the ECM (**Figure 10-23**). The ECM controls a relay that in turn controls the fan solenoid valve. This type of control will be covered in more detail in later chapters.

Temperature and Mode Control

The control of the temperature and mode (defrost, floor, panel, and recirculation) doors in the HVAC housing is typically performed by simple cable control on older trucks. However, some modern trucks use electric motors contained in a gear-reduction package to control the position of the various doors in the HVAC assembly, similar to those found on passenger cars. This type of system requires an electronic module to control the door motors; the module is typically integrated into the HVAC control head assembly.

Some medium-duty trucks may use vacuum-controlled mode and temperature doors, like those commonly found on automobiles. The vacuum source for the HVAC control head on automobiles with gasoline engines is the intake manifold. However, the intake manifold pressure on a turbocharged diesel engine is typically never a negative pressure or vacuum. Therefore, a small electric motor is necessary to drive a vacuum pump to supply the vacuum source for the HVAC control head.

POWER WINDOWS

Power windows on trucks are similar to those used in passenger cars. Power windows typically use permanent magnet brushed DC-type motors. The motor drives the window regulator mechanism in a manner similar to the operation of the window crank on trucks without power windows. The regulator mechanism converts the rotary motion of the motor into the vertical movement of the window. Directional change of the motor is accomplished by reversing the polarity of the voltage applied to the window motor. Changing the direction of current flow through a brushed DC motor causes the motor to reverse directions. Both terminals of window motors are isolated from ground so that the polarity of the voltage supplied to the motor can be reversed, unlike

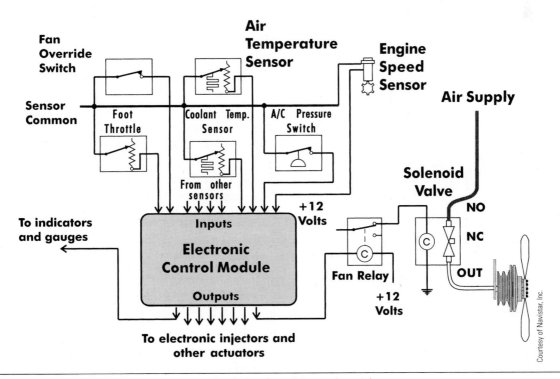

Figure 10-23 Engine ECM providing control of the fan drive solenoid.

wiper motors and blower motors, which always rotate in the same direction. The window control switches are designed to provide the appropriate polarity of voltage to the motor to cause the window to be raised or lowered.

Power window systems typically have a master switch on the driver's side of the truck that permits the truck operator to control all of the truck's power windows. Each door also typically has a power window switch to permit the passenger to control the window as well. Most trucks only have two doors, but some medium-duty trucks may have crew cabs with four doors. A simple power window circuit for a passenger-side power window is shown in **Figure 10-24**. This simple circuit shows a master switch at the driver side of the truck, a window switch at the passenger side of the truck, and the window motor for the passenger window. Power window switches are typically three-position switches with the center position being the normal position. Up is one direction of switch movement; down is the other direction of switch movement. All of the switches in this drawing are shown with the switches in the normal state, neither up nor down. Note that the switches are not ganged. The switch is designed so that only the up or the down switch changes state at any one time.

Note in **Figure 10-24** that ignition voltage is supplied to the master switch terminal 1 and the window

switch terminal 6. This drawing also shows that both of the motor terminals 10 and 11 are connected to ground by the normal state of both switches.

If the driver side switch for the passenger window were moved to the up position, the up switch in the master switch would make contact with the ignition voltage source provided by terminal 1. This would cause terminal 2 of the master switch to be supplied with +12V. Current would then flow from terminal 2 to terminal 7 of the window switch to terminal 11 of the motor. Because terminal 10 is grounded through the master switch terminal 3, the passenger side window motor will rotate in the up direction.

If the driver side switch for the passenger window were moved to the down position, the down switch in the master switch would make contact with the ignition voltage source provided by terminal 1. This would cause terminal 3 of the master switch to be supplied with +12V. Current would then flow from terminal 3 of the master switch to terminal 5 of the window switch, through the normally closed down switch to terminal 10 of the motor. Because terminal 11 of the motor is grounded through the master switch terminal 2 via the normally closed up switches, the passenger-side window motor will rotate in the down direction.

When the window motor reaches the stall position in the up or the down direction at the end of window travel, the current flow through the motor will increase due to the lack of a CEMF being generated by the stalled motor. Power window motors typically have an internal circuit breaker to prevent damage to the motor due to high current flow with the motor stalled. This would occur if the window switch were held in the up or down direction with the window at the end of travel.

A typical power window circuit for a medium-duty truck with a crew cab and four doors is shown in **Figure 10-25**. Current flow when the right rear window is being commanded to the down position using the right rear window switch is shown in **Figure 10-25**. Note that current flow is through the master switch contacts and the left rear window switch contacts.

A complaint about the right rear power window motor not operating in the down position when commanded by the master switch, yet operating correctly when commanded to the down position by the right rear door switch, would seem to be caused by a defective master switch. However, this complaint could be caused by either the master switch or the right rear switch because current must pass through the contacts of both switches.

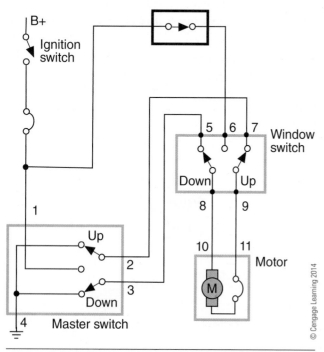

Figure 10-24 Simple power window circuit for a passenger-side window.

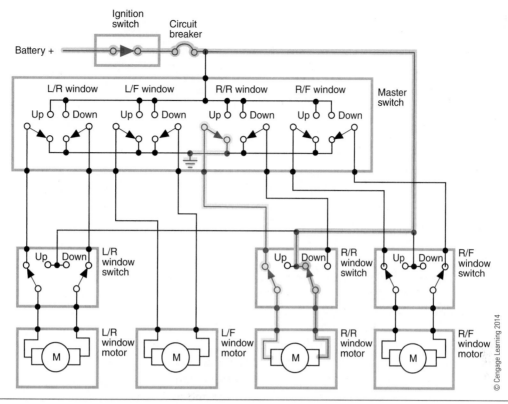

Figure 10-25 Current flow with the right rear window switch in the down position.

....

Tech Tip: The importance of using a circuit diagram to evaluate how the system operates and how you are going to solve the problem is vital for efficient troubleshooting.

....

MOTORIZED AND HEATED MIRRORS

Motorized mirrors are a popular option on heavy-duty trucks. Heated mirrors use an electric heater grid inside the mirror to keep ice from forming on the mirrors.

Motorized Mirrors

Motorized exterior mirrors with dual-axis control use two motors to control the horizontal and the vertical movement of the mirror. The motors are mounted inside a sealed motor drive assembly inside of the mirror housing. The polarity of voltage applied to the mirror motors is reversed to change the motor direction of rotation to bring the mirror glass in or out, up or down. The control switch for the

power mirrors can appear quite complex, as shown in **Figure 10-26**. For this reason, a function table is typically provided in the circuit diagram. This function table indicates what the polarity of the applicable circuit should be when the mirror switch is in each position. For example, the function table in **Figure 10-26** indicates that placing the mirror switch in the left-up position causes circuit 78R to have continuity with circuit 78C (+12V), while the same motion will cause circuit 78P to have continuity with circuit 78-GA (ground).

Some trucks may only control the horizontal motion of both the driver and passenger side mirror because this is the most important motion for large mirrors when backing up the truck. This simplifies the power mirror circuit, as shown in **Figure 10-27**. Note that the switches used to control the mirrors are ganged-type switches.

Heated Mirrors

Heated mirrors contain a small resistive heating element behind the mirror glass. Heated mirrors may be controlled by a switch in the dash panel or may be powered directly from the ignition so that the mirrors are heated at all times. The heated mirrors

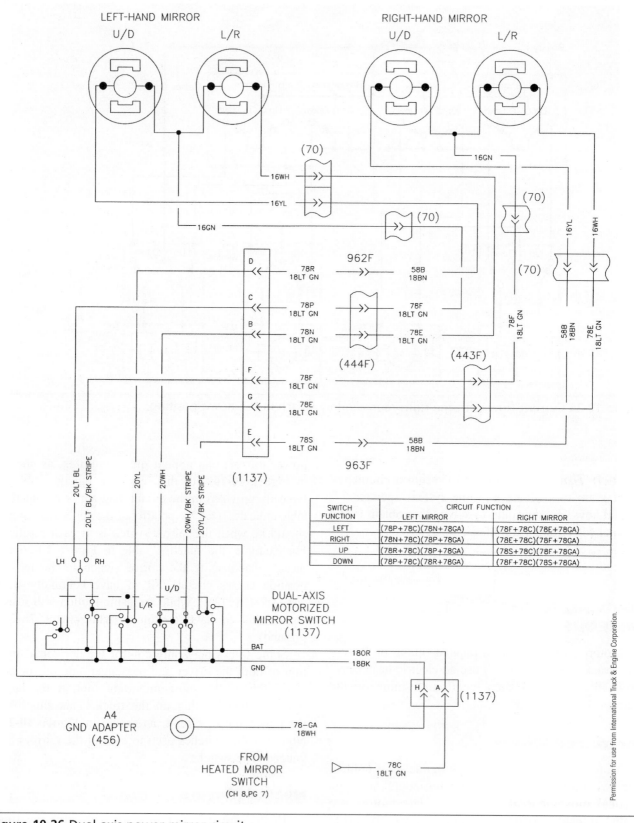

Figure 10-26 Dual-axis power mirror circuit.

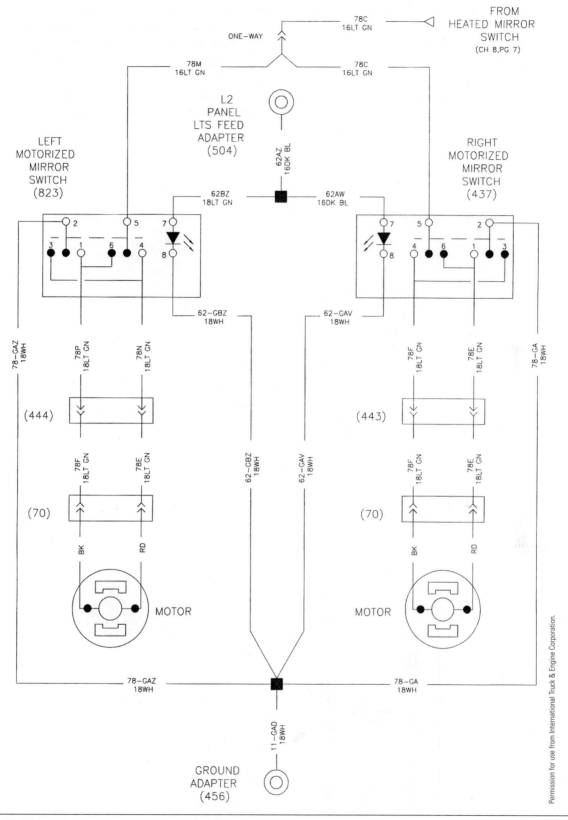

Figure 10-27 Single-axis power mirror circuit.

shown in **Figure 10-28** use a dash-mounted switch to control the mirror heat. Note the LEDs in the switch, which are utilized to provide an indication of mirror heat ON and to provide backlighting for panel illumination.

The mirror in the example shown in **Figure 10-28** also has a running lamp mounted to the mirror, which shares a common ground with the mirror heater. This type of information can be useful in troubleshooting. A complaint of the left side mirror heater not working along with an inoperative left mirror running light could indicate a problem with the common ground for the two electrical devices. Once again, you would not have this knowledge of a common ground for the two circuits without the information provided by the circuit diagram.

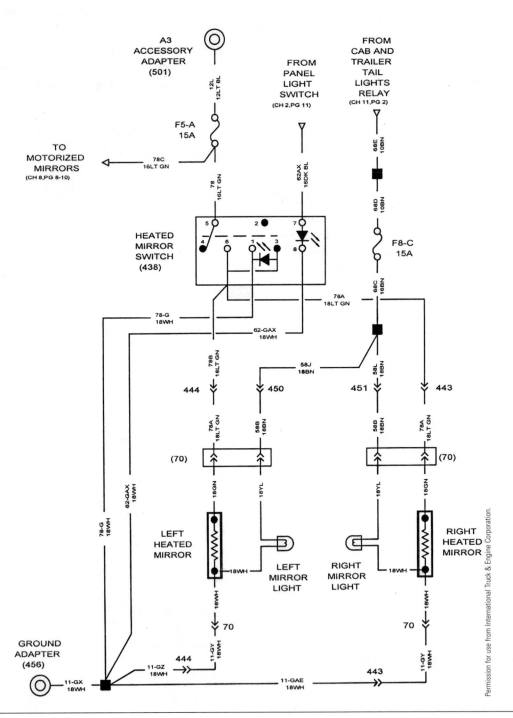

Figure 10-28 Heated mirrors circuit.

ENGINE BRAKE SYSTEMS

The terms *engine brake* or *engine retarder* describe devices used to slow the truck by using the powertrain in conjunction with the service brake system.

Compression Brake

A diesel engine compression brake operates by opening the exhaust valve as the piston nears TDC of the compression stroke while the truck is coasting. This causes the compressed air to exit out the exhaust instead of acting as a spring. A compression brake is commonly known as a Jake Brake™, which is the name of a popular compression brake system supplied by Jacobs Vehicle Systems™.

A schematic of a typical compression brake for a diesel engine with a mechanical fuel system is shown in **Figure 10-29**. The electrical portion includes a dash-mounted on-off switch, a fuel pump switch to sense that the accelerator is not depressed, and a clutch pedal switch. Opening any one of these three switches will disable the compression brake. With all three switches closed, the normally closed electric solenoid valve is energized to permit engine lubrication oil to enter the brake housing assembly to act as a hydraulic fluid. The motion of an adjacent cylinder's exhaust rocker lever is used to depress a master piston. This causes the slave piston to open the exhaust valve a small amount.

Note the diode in **Figure 10-29**. The diode is used as suppression for the solenoid valve.

The engine ECM on electronically controlled diesel engines typically provides control of the compression brake. The compression brake control switches act as inputs to the ECM.

Bleeder Brake

Another type of engine brake is known as a bleeder brake. This brake holds all the exhaust valves open by a small amount while another device restricts the exhaust. The exhaust restriction can be accomplished by closing down the vanes of a variable geometry turbocharger (VGT) or a movable throttle valve in the exhaust. This bleeder brake is not as effective at engine braking as a conventional compression brake, but the noise emitted is much less than a typical compression brake.

Exhaust Brake

An exhaust brake uses a throttle valve to restrict the exhaust near the exhaust manifold. Restricting the

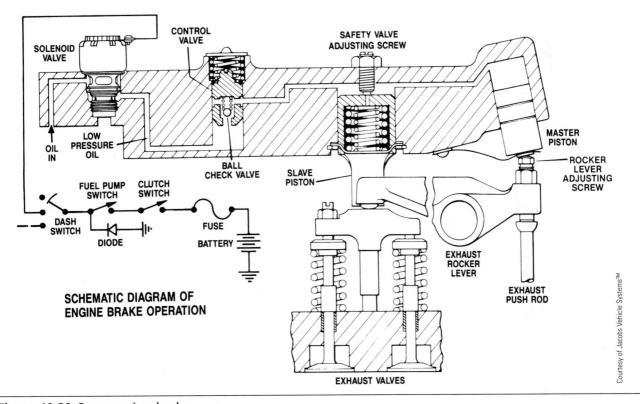

Figure 10-29 Compression brake.

exhaust during coasting results in a moderate level of engine braking. Exhaust brakes are not as effective as compression or bleeder brakes, but still provide a moderate level of engine braking. An air-powered cylinder that is controlled by an electric solenoid typically is used to actuate the exhaust brake, as shown in **Figure 10-30**.

Exhaust brakes typically have a dash-mounted on-off switch. On some electronically controlled diesel engines, the on-off switch acts as an input to the ECM, which energizes a relay to control the electric solenoid when conditions for exhaust brake actuation are met (**Figure 10-31**).

Some diesel engines with variable geometry turbochargers may use the turbocharger vanes as the means of restricting the exhaust. The engine ECM commands the turbocharger vanes to the closed position, which restricts the exhaust, causing the turbocharger to act as an exhaust brake.

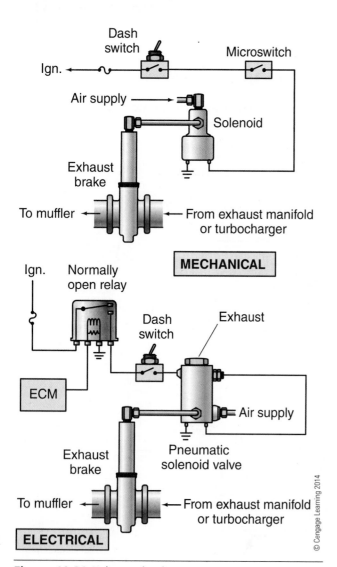

Figure 10-30 Exhaust brake.

HYDRAULIC BRAKE SYSTEM BOOSTER

Many medium-duty trucks have hydraulic brakes. Most modern medium-duty trucks with hydraulic brakes have disc brakes on all four wheels. The force supplied by the truck operator's foot on the brake pedal alone is not sufficient to generate the hydraulic brake system pressure at the calipers needed to stop a medium-duty truck in a reasonable distance. Passenger cars often use a vacuum booster to provide hydraulic brake assist. Diesel engines do not have intake manifold vacuum as a gasoline engine does, so the power steering system is typically used in trucks with hydraulic brakes to provide hydraulic brake assist. The standard truck hydraulic brake booster system is the Bosch **Hydro-Max™** system (**Figure 10-32**).

In the event of the loss of power steering fluid flow, such as with a stalled engine, the truck would be very difficult to stop without hydraulic assist. An electric motor and pump act as a back-up means of providing hydraulic assist when there is low power steering fluid flow. The back-up electric motor is typically designed to operate when there is low power steering flow and the key switch is in the run position. The back-up electric motor also typically operates if the brake pedal is depressed with the key switch in the off position.

The truck electrical circuit for a typical Hydro-Max system is shown in **Figure 10-33**. The flow switch is a normally closed switch that opens when power steering fluid is flowing through the booster. The flow switch closes if power steering fluid flow is below some flow rate, to provide a ground to the hydraulic brake relay. The high side of the hydraulic brake relay is provided with +12V from either ignition or from the brake pedal switch circuit. Therefore, the pump motor will only operate if the key is in the ignition position and the flow switch is closed, or if the brake pedal is depressed and the flow switch is closed, regardless of key switch position.

The diodes shown in **Figure 10-33** prevent backfeeds or feedback of the electrical system from occurring (see Chapter 9 for an explanation of backfeed). If the diodes were damaged such that they were internally shorted and permitted current flow in both directions, depressing the brake pedal with the key off would cause the fuse F3 circuit to be supplied through the hydraulic brake switch. The ignition system would be powered as though the key were in the run position in this circumstance. Additionally, the engine ECM would always detect the hydraulic brake switch as

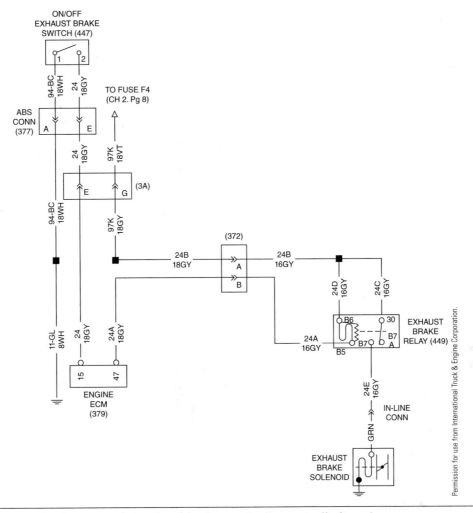

Figure 10-31 Exhaust brake control circuit with electronically controlled engine.

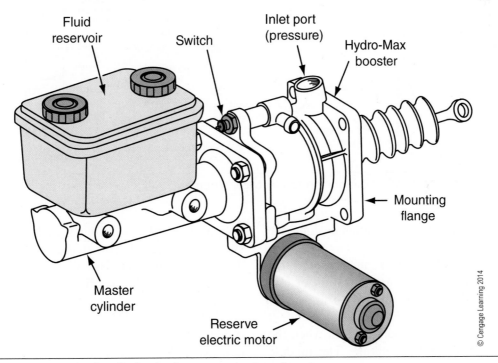

Figure 10-32 Hydro-Max hydraulic brake booster pump and motor.

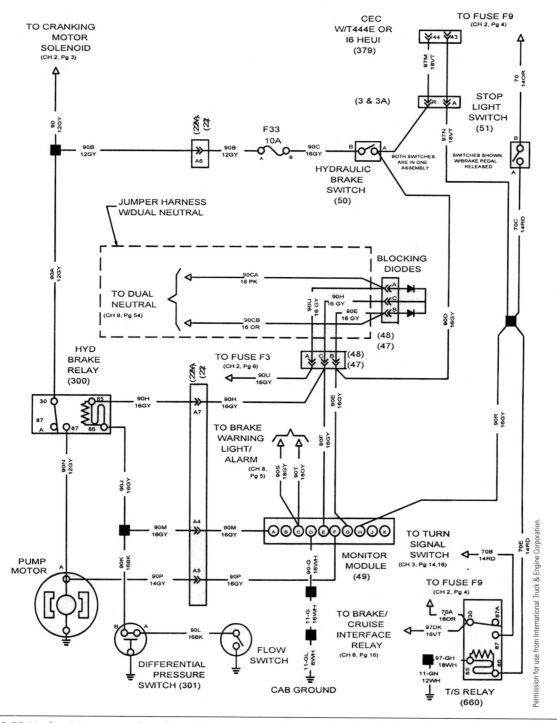

Figure 10-33 Hydro-Max control and monitoring electrical schematic.

closed with the key switch in the ignition position due to the shorted diodes.

Most trucks use a monitor module for the Hydro-Max system to provide an indication of Hydro-Max system failure to the operator. The monitor module provides a signal that is used to illuminate a warning lamp and sound an audible alarm when the Hydro-Max pump motor is operating. The module

also provides this warning signal if a closed flow switch is not detected when the key is first placed in the ignition position. This is necessary because the flow switch could be disconnected or otherwise faulted in the open circuit position, which would prevent the Hydro-Max motor from operating when needed. When the key switch is first switched to the ignition position, the engine is not running, so there should be zero power steering flow. Therefore, the flow switch should be closed when the key is first switched to ignition. After the engine starts, the flow switch should open. The Hydro-Max monitoring module verifies that the transition from flow switch closed to flow switch open occurs. If the flow switch is not closed with the key on and the engine not running, the monitoring module will provide a signal to illuminate a warning lamp and sound an audible alarm. The module also monitors the electric pump motor circuit continuity and causes the warning lamp and alarm to sound if the motor circuit is open.

The Hydro-Max monitoring module may be incorporated into the body systems controller on some trucks, as explained in **Chapter 13**.

EXTRA FOR EXPERTS

This section covers additional topics related to truck accessories.

Latched Relay

A relay utilized to latch off a load when some abnormal condition occurs is shown in **Figure 10-34**. The normally closed contacts of the relay (87A) are used to permit current to pass to the load from an on-off switch when no abnormal condition exists (**Figure 10-34A**). Note the wire that runs from the normally open relay terminal (87) to the high side of the relay coil (86). When an abnormal condition exists that causes the inhibit switch to close, the relay is energized (**Figure 10-34B**). This interrupts the path of current flow to the load. Should the abnormal condition clear itself, such as high temperature, the inhibit switch opens but the relay remains energized because the high side of the coil is being supplied by the normally open contact (**Figure 10-35C**). The relay has latched itself on and the coil will remain energized until the on-off switch is opened to reset the latched relay (**Figure 10-35D**). The on-off switch could then be placed back in the on position, causing the load to be powered once again.

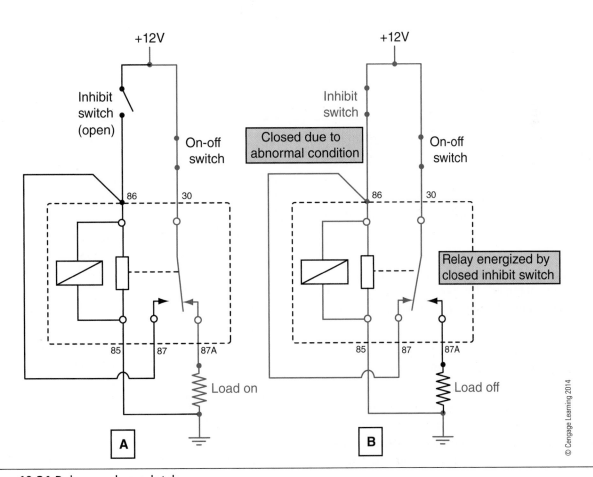

Figure 10-34 Relay used as a latch.

© Cengage Learning 2014

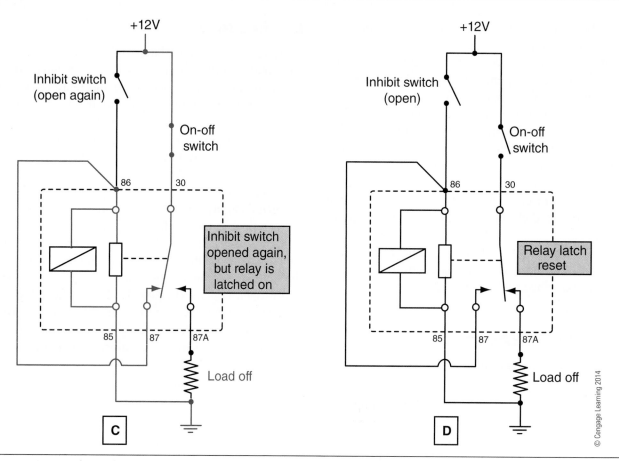

Figure 10-35 Relay is latched on until the on-off switch is cycled to break the latch.

Summary

- Electrical accessories refer to items that improve safety and operator comfort.

- Electric horns are electromagnetic devices. Current flow through the windings is interrupted repeatedly to cause a diaphragm to move up and down and emit a tone.

- Windshield-wiper systems typically use a two-speed permanent magnet brushed DC motor. The two speeds are obtained by positioning one of the brushes to change the magnetic field strength of the armature winding.

- The wiper motor system mechanical linkage causes the wipers to travel back and forth across the windshield.

- Intermittent wipers often use the wiper park system to provide one sweep of the wipers. A brief pulse is supplied to the parked motor, causing the wipers to move off the park position and return to the park position.

- The electrical portion of the HVAC system on a truck typically includes the blower motor and the A/C clutch. Blower motor speed is often controlled by switching in a series of resistors known as a resistor block.

- The engine fan drive on many trucks is an electrically controlled pneumatic system. The engine fan can be engaged when needed for the A/C system or when the engine coolant temperature increases.

- Power window motors typically use permanent magnet brushed DC motors. Reversing the current flow through the motor causes the window to be raised or lowered. The switches in the power window system control the direction of current flow through the power window motors.

- The brake booster system on a truck with hydraulic brakes may use an electric motor to drive a pump, should the engine stall.

Suggested Internet Searches

Try these web sites for more information on truck accessories:
http://www.tricoproducts.com
http://www.boschusa.com
http://www.hortonww.com
http://www.jakebrake.com

Review Questions

1. Which of the following is a true statement?

   A. A truck with a clockspring in the steering column must have the clockspring wound tightly before installation to prevent breakage.

   B. The slip ring used for providing electrical contact between steering column and steering wheel is like the commutator in a brushed DC motor.

   C. All trucks with air horns have the horn located on the roof of the cab and a lanyard hanging down from the cab roof controls the horn.

   D. Some trucks use two electric horns with different pitches that are connected in parallel.

2. The windshield wipers on a truck do not operate and stay in the park position. The wiper arms do not even move when a technician attempts to move the wiper arms by hand. What does this test indicate?

   A. The wiper linkage must be jammed or the wipers could be moved manually.

   B. The wiper motor gear is probably broken.

   C. This test does not indicate anything of relevance.

   D. The circuit breaker is probably tripped and needs to be reset.

3. The speed of a two-speed wiper motor is typically controlled through what method?

   A. Series resistors are switched to reduce motor current and motor speed.

   B. The motor supply voltage is PWM to reduce speed.

   C. The position of the low-speed brush compared to the position of the high-speed brush causes a change in the CEMF produced by the armature, resulting in a motor speed change.

   D. The number of motor field coils is increased in low speed compared to high speed, resulting in a change of motor speed.

4. Two technicians are discussing an electric windshield-wiper system. Technician A says that the wipers travel back and forth because the voltage supplied to the wiper motor reverses polarity at either end of the wiper travel. Technician B says that wipers that are frozen to the windshield in the winter typically cause the wiper battery supply fuse to blow due to excessive CEMF generated by the wiper motor. Who is correct?

   A. A only

   B. B only

   C. Both A and B

   D. Neither A nor B

5. Which is the most likely method of controlling the HVAC blower motor speed?

   A. Resistors are switched in series with the motor to reduce speed.

   B. The fan switch is actually a rheostat that adds series resistance to the blower motor circuit.

   C. The resistor block is placed in parallel with the motor to draw away a portion of the motor current and thus reduce motor speed.

   D. A saturated NPN transistor is used.

6. The power window motor does not operate on the passenger side of a truck. Which would be the most likely cause from the choices listed?

   A. The motor ground strap is broken or has a poor contact, causing high resistance.

   B. The motor field windings are open.

   C. The driver-side power window switch is defective.

   D. There is excessive CEMF in the passenger window motor.

7. A truck with a Hydro-Max system has a brake pressure indicator warning lamp illuminated in the instrument cluster with the engine running. The Hydro-Max pump motor does not operate when the key is in the ignition position with the engine not running. Of the choices listed, what is the most likely cause of the warning lamp being illuminated?

   A. Power steering fluid level is low

   B. Flow switch is disconnected

   C. Low air pressure

   D. Low brake fluid in master cylinder

8. A normally open air solenoid will permit air to pass through the solenoid to the device being controlled under which of the following conditions?

   A. Electrical connector for the solenoid is disconnected

   B. Fuse controlling the air solenoid is blown

   C. Solenoid is energized

   D. Both A and B

9. The fan drive remains engaged at all times on a truck with an on-off air-operated fan drive. The drive is a spring-applied type that uses air pressure to permit the fan to freewheel. What is a possible cause of this problem?

   A. Disconnected air line to the fan drive

   B. Open circuit in the fan control solenoid valve wiring

   C. Defective A/C pressure switch

   D. All of the above

10. The A/C clutch will not engage on a truck when air conditioning is selected. Which of the following is the *least* likely cause of this problem?

    A. Open high-pressure cut-out switch circuit

    B. Open low-pressure cut-out switch circuit

    C. Open A/C clutch suppression diode

    D. Low A/C refrigerant charge

11. What is the primary purpose of a resistor block for an HVAC blower motor?

    A. To protect the blower motor from a short to ground

    B. To add a series resistance to the blower motor to control blower motor speed

    C. To heat the inlet air

    D. To provide suppression for the blower motor windings

12. The windshield wipers on a certain truck may sometimes stop in the middle of the windshield instead of parking when the wiper switch is switched off. Otherwise, the wipers function correctly. Of the choices listed, what is the most likely cause?

    A. A tripped circuit breaker

    B. A defective wiper park switch inside the wiper motor

    C. Poor motor ground

    D. Defective wiper blades

13. A truck that uses a clockspring in the steering column to provide connection to the horn and cruise control switches in the steering wheel has an inoperative electric horn and inoperative cruise control after replacement of a steering gearbox. Of the choices listed, what is the most likely cause?

    A. Clockspring not reconnected after the repair

    B. Drag link installed backward

    C. Defective horn and defective cruise control switches

    D. Broken clockspring due to steering wheel rotation with gearbox removed

14. The right rear power window does not operate on a four-door medium-duty truck. Technician A says that a defective master switch at the driver's door could be the problem. Technician B says that a defect right rear door switch could be the problem. Who is correct?

    A. A only

    B. B only

    C. Both A and B

    D. Neither A nor B

15. An HVAC blower motor only runs when the control is set to the maximum fan speed. Technician A says that a defective resistor block could be the problem. Technician B says that an improperly sized circuit protection device for the HVAC blower motor could cause this. Who is correct?

    A. A only

    B. B only

    C. Both A and B

    D. Neither A nor B

CHAPTER
11
Sensors, Digital Electronics, and Multiplexing

Learning Objectives

After studying this chapter, you should be able to:

- List the various types of sensors used on a modern truck.
- Explain the differences between a digital and an analog signal.
- List the main logic gates and develop the truth table for each gate.
- Describe the types of memory commonly used in an electronic module.
- List the four main types of electronic module inputs.
- Explain the concept of a pull-up and pull-down resistor in an electronic module input circuit.
- Discuss the various forms of multiplexing used on a modern truck.

Key Terms

active-high

active-low

analog multiplexing

analog to digital (A/D) converter

backbone

binary

bit

bus

byte

controller area network (CAN)

data link

digital

EEPROM

electronic service tool (EST)

flash memory

floating

gate

Hall effect

idle validation switch (IVS)

impedance

integrated circuit

J1587

J1708

J1939

logic

message identification (MID)

microprocessor

multiplexing

non-volatile memory

parameter group number (PGN)

parameter identifier (PID)

pull-down resistor

pull-up resistor

random access memory (RAM)

reference voltage

resistance temperature detector (RTD)

sensor

shield

signal conditioning

source address

stub

suspect parameter number (SPN)   thermocouple   variable reluctance

terminating resistor   time-division multiplexing   volatile memory

thermistor   transfer function   Wheatstone bridge

INTRODUCTION

Sensors are an important part of a modern truck's electrical system. Sensors convert some physical property such as pressure or temperature into an electrical signal. Sensors are also known as transducers or sending units. The basic principles of some common sensors will be examined in this chapter.

This chapter will also introduce digital electronics, which is the basis of modern truck electronic control systems. Multiplexing is also presented.

AMPLIFIERS

An amplifier is utilized to increase the amplitude or strength of a low-level signal as shown in **Figure 11-1**. Amplifiers are also known as *amps*, which is not to be confused with the unit of measurement of electric current. An amplifier is typically drawn as a sideways triangle symbol, as illustrated in **Figure 11-2**. The input side of the amplifier is the base of the triangle and may be shown with one or two inputs. The output terminal is at the point of the triangle. The amplifier shown in **Figure 11-2** is called a differential amplifier. A differential amplifier amplifies the difference in the voltage between its two input terminals. This type of amplifier is also known as an operational amplifier or op amp. An operational amplifier is able to perform various mathematical operations such as addition, subtraction, multiplication, and even some calculus operations. A detailed study of operational amplifiers is outside the scope of this book, although they are quite fascinating devices. Some of the OEM-provided information in the drawings in this chapter show basic amplifier circuits. It is only important at this point to understand that an amplifier (amp) increases the amplitude of a low-level electrical signal. For more information on operational amplifiers, see the Internet links at the end of this chapter.

SENSORS

Humans have five senses: sight, hearing, smell, taste, and touch. These five senses permit your brain to process what is occurring in the world around you. The eyes, ears, nose, taste buds, and skin could be thought of as being sensors. Sensors are used by the various electronic modules on a modern truck to gather a variety of information.

Temperature Sensors

Your skin provides your brain an indication of temperature. Nerves carry an electrical signal from the skin to the brain, which you interpret as temperature. In a similar manner, temperature sensors throughout a modern truck electrical system inform the various electronic control modules the temperature of the engine coolant, the manifold air temperature, the air conditioning evaporator inlet temperature, and so forth. A modern truck may have more than 20 different temperature sensors of various types. The most common types of temperature sensors will be evaluated.

Thermistors. Thermistors are two-terminal temperature-measurement sensors that are typically made of semiconductor material. Thermistors are one of the most common types of temperature measurement sensors found on a modern truck. The resistance measured between the two terminals of the thermistor changes proportionally to the temperature being measured. Most thermistors are constructed of semiconductor material and have a negative temperature coefficient (NTC) as described in **Chapter 6**. You may recall from algebra that the graph of a line that points downhill from left to right has a negative slope. NTC indicates that the resistance of the thermistor decreases as the temperature being measured increases, as shown in **Figure 11-3**. This particular graph is the

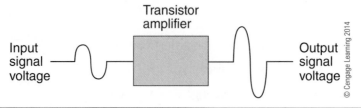

Figure 11-1 An amplifier increases the amplitude of a low-level signal.

Differential Amplifier

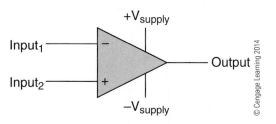

Figure 11-2 A differential amplifier intensifies the difference in voltage between its two inputs.

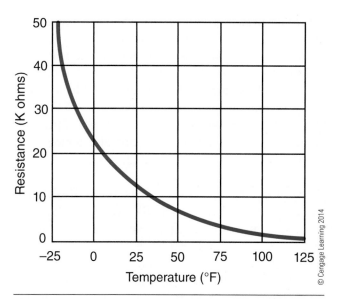

Figure 11-3 Air conditioning system thermistor resistance versus temperature.

temperature-resistance curve for an air conditioning evaporator temperature thermistor. The curved line indicates that the sensor output is nonlinear, so the change in resistance for a specific change in temperature is not constant.

The resistance versus temperature curve of NTC thermistors varies depending upon the expected temperature range that the thermistor will be measuring. The circuit shown in **Figure 11-4** is an engine coolant temperature (ECT) sensor circuit. The thermistor resistance versus temperature curve for this ECT sensor is different from the curve shown for the air conditioning sensor shown in **Figure 11-3**. This is because engine coolant and an air conditioning evaporator have very different normal operating temperature ranges. One disadvantage of thermistors is that because the

output is nonlinear, thermistors have a limited temperature range where the measurement system accuracy is acceptable. For example, an electronic module could not accurately determine the difference between $123°F$ ($51°C$) and $125°F$ ($52°C$) for the thermistor plot shown in **Figure 11-3** because the change in resistance of the thermistor may be less than 1Ω. At extremely low temperatures, an electronic module might not be

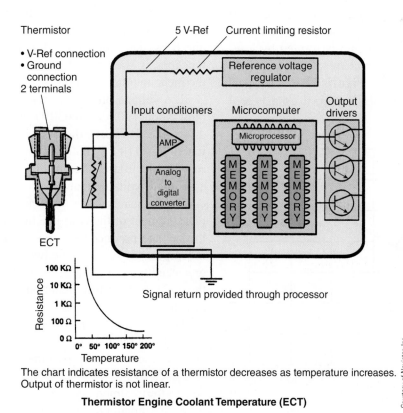

The chart indicates resistance of a thermistor decreases as temperature increases. Output of thermistor is not linear.

Thermistor Engine Coolant Temperature (ECT)

Figure 11-4 Engine coolant temperature (ECT) sensor circuit.

able to accurately determine the difference between an open circuit with infinite resistance caused by a disconnected wiring harness connector and the very high resistance of the thermistor at $-25°F$ ($-32°C$). However, the electronic module should be able to accurately determine the difference between $32°F$ ($0°C$) and $34°F$ ($1°C$), which is an important temperature range for an air conditioning evaporator temperature sensor.

Resistance Temperature Detectors. Resistance temperature detectors (RTD) are typically two-terminal temperature measurement sensors constructed of a thin metal wire. Metal has a positive temperature coefficient (PTC), meaning as the temperature of the metal increases, the resistance also increases. This is illustrated in **Figure 11-5**. The plotted line points uphill from left to right, indicating a positive slope. Notice that the plotted line is also straight or linear, unlike the curved nonlinear plot of the thermistor. This linear property makes RTDs very accurate over a wide temperature range. RTDs are used to measure temperatures up to about $1200°F$ ($650°C$), which makes them ideal for diesel exhaust aftertreatment system temperature measurements. Many RTDs are made from platinum, so they may also be known as platinum resistance thermometers (PRT).

Thermocouples. Thermocouples are temperature measurement sensors that are constructed from two dissimilar metals. A thermocouple is formed by joining two wires made of different metals such as iron and constantan, as shown in **Figure 11-6**. This particular combination of metals is known as a type J

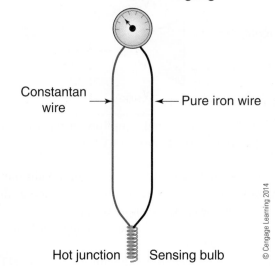

Cold junction with electronic module and gauge

Constantan wire → ← Pure iron wire

Hot junction   Sensing bulb

© Cengage Learning 2014

Figure 11-6 Type J thermocouple.

thermocouple. Both ends of the two wires are welded together to form two wire junctions. The junction where the temperature is being measured is referred to as the hot junction, while the reference junction with a known temperature is referred to as the cold junction. A phenomenon known as the Seebeck effect states that the difference in temperature between the two junctions of dissimilar metals produces a low-amplitude voltage that increases as the difference in temperature between the two junctions increases. The voltage produced by the temperature difference between the two junctions is used to determine the temperature of the thermocouple hot junction.

The voltage produced by a type J thermocouple is only about $30mV$ ($0.030V$) when the temperature difference between the hot junction and the cold junction is $1100°F$ ($600°C$). Therefore, thermocouples require electronic circuitry to amplify the very low amplitude voltage that is produced. The electronic circuitry must also perform an accurate temperature measurement of the cold junction temperature because the voltage produced is proportional to the difference in temperature between the two junctions. A thermistor or special type of diode is typically utilized to measure the cold temperature junction within the electronic module.

Thermocouples are often used to measure diesel exhaust gas temperatures in the exhaust manifold for an instrument called a pyrometer. Thermocouples are also used to measure the temperatures within the exhaust aftertreatment system on some engines.

A thermocouple temperature measurement tool is available that permits many high-end DMMs to be

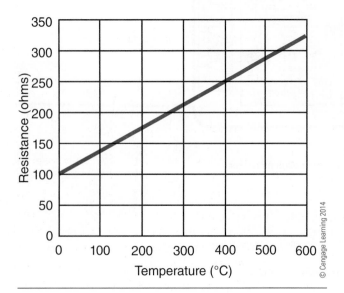

© Cengage Learning 2014

Figure 11-5 RTD temperature versus resistance.

used to measure temperature. These are very useful tools for troubleshooting problems with temperature sensors. For example, the temperature of the engine coolant indicated by the DMM thermocouple can be compared to the temperature indicated by the engine coolant temperature sensor. However, inaccuracy can occur if the coolant temperature sensor is measuring the temperature of the coolant in the block and the thermocouple is placed in the expansion tank. The sensor being tested and the thermocouple must be placed as close as possible to each other for an accurate temperature comparison.

Because a thermocouple operates on the principle of dissimilar metals producing a voltage, you should not attempt to repair thermocouple wire. Splice clips, solder, and butt splices used for wire repair will create a dissimilar metal junction resulting in temperature measurement error. Damaged thermocouple wire must be replaced. Many thermocouple temperature measurement devices, such as exhaust aftertreatment sensors, do not have any electrical connectors between the thermocouple and the electronic module. The two or three thermocouples for aftertreatment temperature measurement are typically integrated into a remote thermocouple electronic module assembly instead of connecting the thermocouple directly to the engine ECM. Damaged thermocouple wire may require replacement of the entire thermocouple and electronic thermocouple module assembly.

If the thermocouple wire insulation becomes overheated and melts, the two dissimilar conductors can make contact with each other to form an inadvertent second hot junction. This is a common failure with exhaust aftertreatment sensors should the sensor wiring not be correctly routed and clipped. The electronic thermocouple module may not be able to detect this failure electrically and will measure the temperature of this second hot junction instead of the actual hot junction temperature. In the case of exhaust aftertreatment, the difference in temperature between the real hot junction and the second false hot junction can be several hundred degrees, resulting in problems related to the aftertreatment system. Details on exhaust aftertreatment are discussed in **Chapter 14**.

Position and Rotational Speed Sensors

Modern trucks have a variety of sensors that are used to detect the position of some mechanical device. These can be as simple as the position of the accelerator (foot throttle) or as complex as the precise position of the crankshaft with the engine running.

These position measurement devices can also be used to determine the rotational speed of a component, such as the transmission output shaft speed. The vehicle speed can then be calculated by applying the rear axle ratio and tire revolutions per mile or per km value to the transmission output shaft rotational speed.

Potentiometers and Rheostats. Potentiometers and rheostats were described in **Chapter 6**. These simple resistive devices can be used to measure the angular position of the accelerator pedal, a fuel level float arm, or other device. A mechanical linkage is connected to the movable device and the wiper of the potentiometer or rheostat. In the case of a potentiometer used as an accelerator position sensor, depressing the accelerator causes the wiper to rotate. The corresponding change in voltage provided at the wiper terminal of the potentiometer indicates the accelerator position. To prevent unintentional acceleration should some wiring problem cause the accelerator position sensor voltage to increase without the wiper actually moving, many diesel engines incorporate an **idle validation switch (IVS)**. The switch closes or opens when the accelerator is off-idle to validate that the truck operator is depressing the accelerator. If a wiring problem such as a wire-to-wire short causes a voltage to be present at the ECM accelerator input and the ECM does not detect that the idle validation switch indicates an off-idle condition, the ECM will ignore the accelerator position sensor input signal from the potentiometer and limit engine speed to some value.

Some accelerator position sensors utilize two oppositely wired potentiometers to produce two different voltages. For example, one sensor voltage will increase from 1V to 4V when the accelerator is depressed; the other sensor voltage will decrease from 4V to 1V when the accelerator is depressed. If a wiring problem causes a voltage to be present at one of the ECM accelerator position sensor inputs and this voltage does not correspond to the voltage measured by the ECM for the other sensor, the ECM will ignore the inputs and limit the engine speed to some value.

Variable Reluctance Sensors. **Variable reluctance** sensors are basically miniature AC voltage generators. Variable reluctance sensors are also known as magnetic sensors or a variety of similar terms. A variable reluctance sensor consists of a coil of wire wrapped around a permanent magnet (**Figure 11-7**). A low-reluctance rotor, known as a target, timing disk, or a variety of similar terms, rotates past the tip of the sensor. The target has raised teeth like a gear. In some cases, the target is also an actual gear. The target is

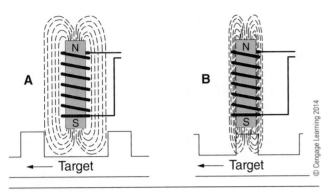

Figure 11-7 Variable reluctance sensor.

driven by the engine camshaft, crankshaft, transmission output shaft, axle shaft, or other device for which the rotational speed, and in many cases the position, is to be measured.

Reluctance is a property of a material that describes how easily magnetic lines of force can be directed through the material, as described in **Chapters 3 and 7**. Materials such as iron and steel have very low reluctance. A change in reluctance near the tip of the sensor causes the magnetic lines of force created by the permanent magnet in the sensor to be directed into the low-reluctance material as it passes in front of the sensor. This change in the shape of the magnetic field is the basic principle of variable reluctance sensors.

When a tooth of the target is not close to the permanent magnet in the sensor, the magnetic lines of force caused by the permanent magnet surround the magnet, as shown in **Figure 11-7A**. When a tooth of the target rotates so that it nears the tip of the permanent magnet, the magnetic lines of force are redirected from their normal paths and are concentrated inwardly by the presence of the tooth on the low-reluctance target (**Figure 11-7B**). The magnetic lines of force moving inwardly cut through the coil wrapped around the sensor magnet. Cutting magnetic lines of force through a coil of wire causes a voltage to be induced in the coil, as shown by the trace labeled e_g in **Figure 11-8**. In this illustration, the target is moving from right to left across the tip of the variable reluctance sensor, shown at the top of **Figure 11-8**.

When the tooth of the target is directly in front of the magnet contained within the variable reluctance sensor, the magnetic lines of force have stopped moving inwardly and are stationary. This causes the voltage induced in the coil to decrease to zero, as shown in **Figure 11-8**.

When the low-reluctance target tooth is moving away from the sensor, the magnetic lines of force surrounding the magnet begin to return outward to their original shape shown in **Figure 11-7A**. The movement

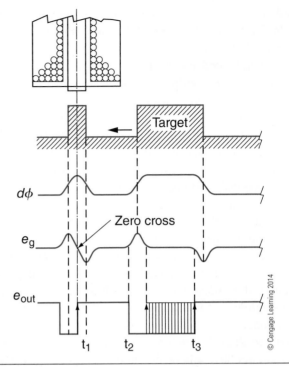

Figure 11-8 Variable reluctance sensor output voltage (e_g) with target moving from right to left.

of the magnetic field as it expands outwardly causes the magnetic lines of force to cut through the coil in the opposite direction, compared to when the target tooth was approaching the magnet. This outward expansion of the magnetic field causes a voltage to be induced in the coil of wire with a polarity opposite of the voltage induced when the magnetic field was moving inwardly, thus producing a negative voltage. Therefore, a complete sine waveform is produced as a tooth and valley of the target pass in front of the sensor, as shown by the e_g trace in **Figure 11-8**.

The frequency of the voltage waveform generated by the variable reluctance sensor indicates the rotational speed of the target. Truck antilock brake systems (ABS) use variable reluctance sensors to measure individual wheel speeds. A rotating tone ring or tone wheel with square-cut teeth similar to a gear passes close to the wheel speed sensor to act as a low-reluctance target. The tone wheel used in an ABS system may be machined or cast as webs in the backside of a brake rotor, or it may be a stamped metal ring attached to the wheel hub on drum brakes. The tone wheel passes in front of the stationary sensor as the wheel is rotated. The sensor output is an AC voltage with a frequency directly proportional to wheel speed. Details on ABS are discussed in **Chapter 15**.

Variable reluctance sensors may also be used as both speed and position sensors, such as engine camshaft position and crankshaft position. A target with 60 or more teeth is attached to the crankshaft and provides the engine ECM with information about engine speed as well as engine crankshaft position. This crankshaft target typically has a single tooth missing, corresponding to a specific crankshaft position. The missing tooth on the target causes a gap in the output waveform, as shown in **Figure 11-9**. The missing tooth is used as an initial starting point (i.e., cylinder #1 TDC). To know if the #1 cylinder is on the compression or exhaust stroke, a variable reluctance sensor on the camshaft may also be used in conjunction with the crankshaft position sensor. The camshaft target may have one tooth for each cylinder, plus one extra tooth or one larger tooth that corresponds to the missing tooth on the crankshaft target (see **Figure 11-14**). Alternatively, the camshaft target may be a single raised pin on the face of the camshaft gear indicating a reference position. Using the signals from both the camshaft and crankshaft position sensors, the engine ECM "knows" the crankshaft position to a high degree of accuracy when the engine is running. Crankshaft position is vital information for an electronically controlled diesel engine, as will be discussed in **Chapter 14**.

Some older electronically controlled diesel engines may only use a camshaft position sensor with a multiple-toothed target to determine crankshaft position.

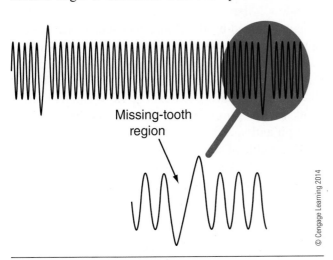

Figure 11-9 Variable reluctance crankshaft position sensor output voltage.

The amplitude of the voltage produced by the variable reluctance sensor must be high enough to be detected by the electronic module that is connected to the sensor. The amplitude of the voltage produced by variable reluctance sensors largely depends on two factors: the air gap and the rotational speed of the target. The air gap is the distance between the tip of the sensor and the tooth of the target. The larger the air gap, the lower the voltage produced by the sensor. This occurs because the variable reluctance sensor depends on a distortion of the magnetic field to generate a voltage. The closer the tooth of the tone wheel is to the tip of the sensor, the greater the distortion of the magnetic field and the more lines of force there are to cut through the coil of wire. Some variable reluctance sensors have adjustable air gaps. The air gap adjustment procedure varies, so consult the specific OEM's information for instructions.

An ABS wheel-speed sensor with too great an air gap will cause a problem in the ABS system because the amplitude of the signal is too low to be accurately measured. ABS systems can also detect loose wheel bearings or warped or bent tone wheels based on signal amplitude. These defects all may cause variation in the air gap throughout rotation or during cornering, which is detected as variation in the amplitude of the voltage produced by the sensor as the wheel is rotated (**Figure 11-10**). This variation in amplitude is also the fundamental of AM radio, which uses amplitude modulation (AM) to carry a signal.

Tech Tip: Proper wheel-bearing adjustment is very important on trucks with ABS because a loose wheel bearing can cause the ABS sensor to be pushed away from the tone wheel, resulting in an inconsistent air gap between the sensor and tone wheel. Just a small increase in air gap between the tip of the sensor and the tone wheel will result in a large decrease in the voltage produced by the sensor.

The other factor that determines the amplitude of the voltage produced by a variable reluctance sensor is the

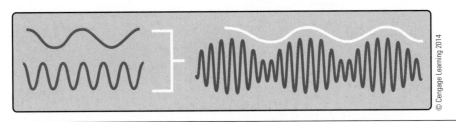

Figure 11-10 Amplitude modulation of wheel speed sensor signal with misadjusted wheel bearings.

speed at which the teeth of the tone wheel pass in front of the sensor. The faster the tone wheel passes in front of the sensor, the greater the amplitude of the voltage produced by the sensor. This occurs because of the principle of magnetic induction, according to which the more magnetic lines of force that cut through a conductor per second, the higher the amplitude of the voltage induced in the conductor. Peak ABS wheel speed sensor output voltage at 2 mph (3.2 km/h) may only be 1V or so, while at 60 mph (96 km/h) the same sensor output voltage may exceed 35V. For this reason, variable reluctance sensors are typically not used where slow movement must be detected. If the amplitude of the voltage signal produced by the sensor is too low, the electronic control module may not be able to accurately detect the frequency of the signal. This low speed-sensing limitation of variable reluctance sensors is not a factor with ABS systems because ABS is not active when vehicle speed drops below a few miles per hour.

For an engine crankshaft or camshaft position sensor, too large of an air gap can result in an engine no-start in extreme cold temperatures. The combination of large air gap and slow cold cranking speed results in the amplitude of the signal produced by the sensor to be too low for the engine ECM to recognize a valid pattern; thus, no fueling occurs.

Hall Effect Sensors. Passing a current through a thin layer of semiconductor material causes a difference in potential (voltage) to be developed between the edges of the semiconductor material when exposed to a magnetic field. The level of the voltage produced in the semiconductor material is directly proportional to the strength of this magnetic field. This principle is known as the **Hall effect** (**Figure 11-11**). The voltage produced due to the Hall effect is referred to as the Hall voltage.

The manner in which the Hall effect produces a voltage is very different from electromagnetic induction. Electromagnetic induction requires cutting magnetic lines of force. The Hall effect generates a voltage with no relative motion between the semiconductor material and the magnetic lines of force. The Hall voltage is generated by the presence of a magnetic field without cutting any magnetic lines of force. The Hall voltage can be a DC voltage, provided the magnetic field strength remains at a constant level.

The Hall effect can be used to produce a Hall effect switch or sensor. A Hall effect switch typically has three terminals and contains a permanent magnet to generate a magnetic field. Two terminals provide a supply voltage and a ground to power the electronics in the sensor. The third terminal is the sensor output. The Hall voltage is amplified inside the sensor and is used to switch a transistor on and off to indicate the presence of the magnetic field. The example shown in **Figure 11-12** is a Hall effect switch-type camshaft position sensor that uses a switching transistor to make or break a path to ground. This switching action all takes place with no physical contact between the Hall effect sensor and the metal target. The timing disk is installed to the front of the camshaft. Note in this example that the timing disk has one "window" that is wider than the others are. The signal produced by the Hall effect sensor would be a square waveform with one longer off pulse when this wider window passes in front of the Hall effect sensor. The engine ECM uses this wider pulse to detect the camshaft position. The ECM can then infer crankshaft position and speed based on the camshaft position and speed.

The width of the magnetic target or tooth that is passed near the Hall effect sensor influences the length of time the signal is generated by the Hall effect sensor

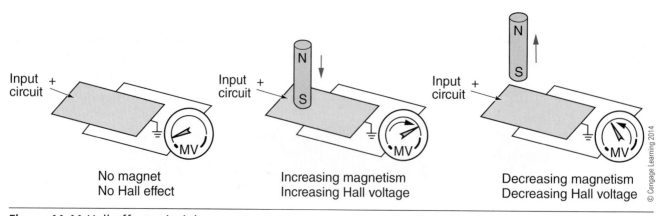

No magnet
No Hall effect

Increasing magnetism
Increasing Hall voltage

Decreasing magnetism
Decreasing Hall voltage

© Cengage Learning 2014

Figure 11-11 Hall effect principle.

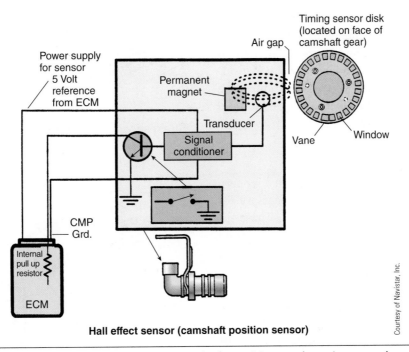

Hall effect sensor (camshaft position sensor)

Courtesy of Navistar, Inc.

Figure 11-12 Hall effect switch used to measure camshaft position and engine speed.

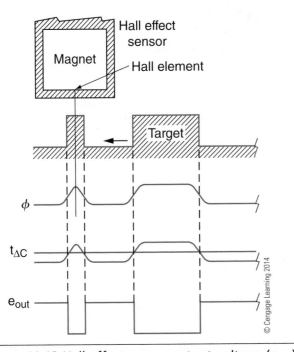

© Cengage Learning 2014

Figure 11-13 Hall effect sensor output voltage (e_{out}) when target is moving from right to left.

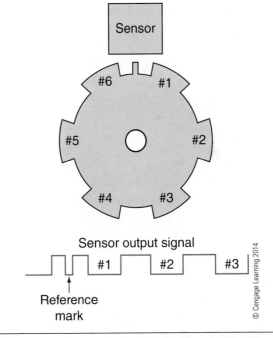

© Cengage Learning 2014

Figure 11-14 Camshaft target for Hall effect or variable reluctance sensor.

(**Figure 11-13**). The bottom trace in **Figure 11-13** shows the sensor output voltage, designated as e_{out}, corresponding to the position of the target as it moves from right to left across the tip of the Hall effect sensor.

Figure 11-14 illustrates a typical camshaft target, which could be used with a Hall effect switch or a variable reluctance camshaft position sensor. The small

extra tooth shown in **Figure 11-14** creates a reference mark in the sensor output signal, which lets the engine ECM "know" that the next large camshaft target tooth detected corresponds to cylinder #1.

As discussed in previous chapters, the magnetic field surrounding a current-carrying conductor is directly proportional to the current flowing through a conductor. Therefore, the Hall effect can be used to determine the amount of current flowing through a conductor by measuring the magnetic field caused by the current flow. This is how a clamp-on DC current probe operates. The voltage produced by the Hall effect sensor contained in the clamp-on current probe corresponds to the strength of the magnetic field generated by the current flow.

Pressure Sensors

There are a variety of pressure sensors on a modern truck. A number of pressure sensors are necessary for the diesel exhaust emission control system. These include intake manifold pressure, exhaust manifold pressure, and diesel particulate filter differential (delta) pressure. Pressure sensors may also be used to calculate flow, as in the case of an exhaust gas recirculation (EGR) differential pressure sensor, which measures the pressure drop across a venturi. This is analogous to measuring the voltage drop across a known value of resistance to determine the current flow.

Potentiometric Pressure Sensors. Potentiometric pressure sensors are simple potentiometers where the change of length of a Bourdon tube within the sensor causes the potentiometer wiper to move corresponding to the pressure being measured (**Figure 11-15**). The result is an output voltage that is proportional to the pressure being measured. Potentiometric sensors may be found on some older trucks as sensors (sending units) for instrumentation.

Strain Gauge Pressure Sensors. A strain gauge is a thin flexible material imprinted with a zig-zag pattern of an electrically conductive material, as shown in **Figure 11-16**. When subjected to force, the flexible material stretches resulting in an increase or decrease in the length of the conductive track, as well as a

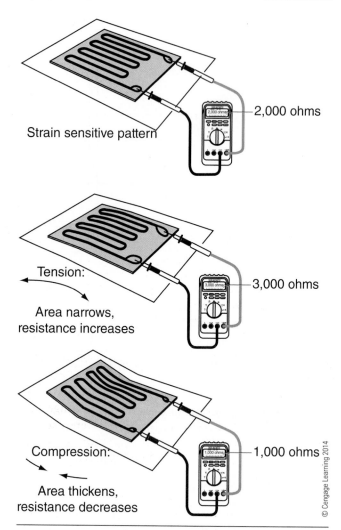

Figure 11-16 Strain gauge resistance change due to deformation.

corresponding narrowing or thickening of the conductive material. The change in length and thickness causes a small change in the resistance of the conductive track. The strain gauge is attached to a flexible diaphragm inside a sensor. Strain gauge type pressure sensors are constructed so that a reference pressure chamber with a known pressure is situated on one side of the diaphragm, while the other side of the diaphragm is subjected to the pressure being measured. As the pressure being measured increases, the strain gauge deforms with the diaphragm resulting in a change of resistance. A strain gauge pressure sensor that makes use of the change in resistance of a semiconductor material is known as a piezoresistive sensor.

The change in resistance in a strain gauge pressure sensor is typically converted to a voltage by an arrangement called a **Wheatstone bridge**. A Wheatstone bridge is a diamond-shaped arrangement of three fixed resistors of known value, plus a variable resistance identified as R_S in **Figure 11-17**. This is really just a

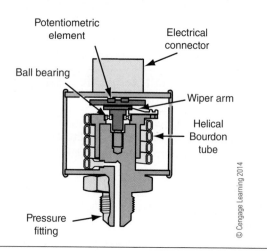

Figure 11-15 Potentiometric pressure sensor.

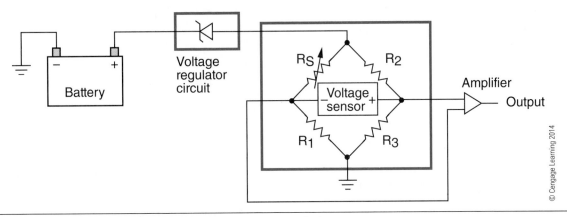

Figure 11-17 Wheatstone bridge used with strain gauge (R_S).

series-parallel circuit with a differential voltage measurement obtained between the two points shown, so do not let the diamond shape of the resistor arrangement be intimidating. The Wheatstone bridge is contained within the sensor. As the value of R_S (the strain gauge) changes when pressure is increased, the voltage measured by the sensor will change accordingly. The difference in voltage between the two terminals of the Wheatstone bridge caused by a pressure change may only be a few millivolts. However, a differential amplifier contained within the sensor amplifies the difference in voltage across the terminals of the Wheatstone bridge to a usable level.

Strain gauge type pressure sensors are typically three-terminal sensors. A fixed supply voltage, such as +5V, a ground circuit, and a sensor output signal indicating the measured pressure make up these three terminals. The reference pressure chamber is typically a pure vacuum, making the sensor an absolute pressure sensor. However, the chamber may be vented to atmospheric pressure on some sensors, making the sensor a gauge pressure type sensor. In the case of a differential pressure sensor, such as the sensor used to measure the diesel particulate filter (DPF) differential (delta) pressure, one side of the strain gauge diaphragm is exposed to the DPF inlet pressure, and the other side of the diaphragm is exposed to the DPF outlet pressure. The sensor output voltage is proportional to the differential pressure measured across the DPF. This differential pressure can be used by the ECM to determine the amount of soot in the DPF, as will be explained in **Chapter 14**.

Variable Capacitance Pressure Sensors. Variable capacitance pressure sensors, also known as capacitance discharge sensors, are three-terminal sensors that measure pressure through a change in value of an internal capacitor. Recall that capacitance is measured in units of farads. One of the factors that determine the

value of capacitance is the distance between the two plates of a capacitor. In a variable capacitance sensor, one plate of a capacitor is fixed while the other plate is attached to a flexible diaphragm as shown in **Figure 11-18**. One side of the moveable plate is exposed to a reference pressure, such as a pure vacuum. The other side of the movable plate is exposed to the sensed pressure. As the sensed pressure increases, the distance between the movable capacitor plate and the fixed capacitor plate decreases resulting in an increase in the capacitance value. If the reference pressure is a vacuum, then the sensor is an absolute pressure sensor. This indicates that this sensor can measure a pressure less than atmospheric pressure.

An electronic circuit within the sensor converts the capacitance change into a corresponding voltage. In **Figure 11-18**, the sensor is supplied with a 5V **reference voltage** and ground. The output voltage at the sensor signal terminal of the sensor is some percentage of the 5V reference voltage. For example, this sensor may have an output voltage of 0.5V at the signal terminal when the sensor is measuring atmospheric pressure and 4.5V when measuring 80 psi (550 kPa). If the pressure being measured were less than atmospheric pressure, the output voltage would be less than 0.5V for this particular sensor. The output voltage of the sensor is typically linear and increases at a rate directly proportionally to the pressure being measured (output voltage increases as the pressure being measured increases).

Instead of a DC output voltage, some variable capacitance sensors may provide a pulse width modulation (PWM) output signal where the duty cycle (on-time) of the PWM signal is proportional to the pressure being measured. Electronic circuitry within the sensor generates the variable PWM signal.

Piezoelectric Pressure Sensors. Piezoelectric describes a phenomenon whereby a voltage is produced

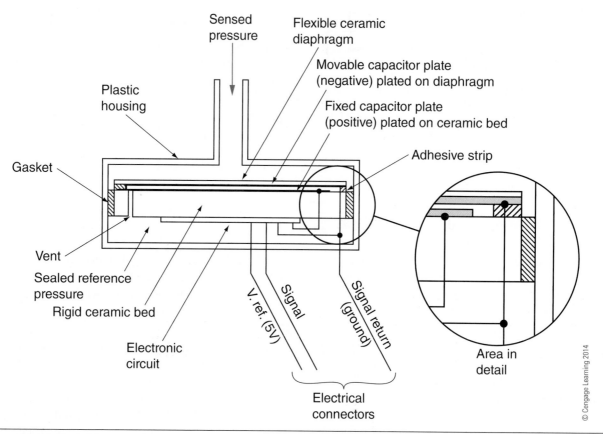

Figure 11-18 Variable capacitance MAP sensor.

when certain crystals, like quartz, are subjected to mechanical stress. In other words, squeeze a rock and produce a voltage. Piezoelectric pressure sensors measure a change in pressure by means of a quartz sensing element, as shown in **Figure 11-19**. Note that piezoelectric pressure sensors do not measure a steady

or static pressure, only a change in pressure. Piezoelectric pressure sensors currently have minimal usage on modern trucks. However, diesel fuel injectors or glow plugs someday may contain a piezoelectric pressure sensor to measure the cylinder firing pressures. This will provide an indication of misfire for exhaust emissions compliance as well as more precise fueling and timing control.

The sensor shown in **Figure 11-19** operates on the same principles as an accelerometer, a sensor used to measure acceleration. A rapid decrease in speed (velocity) is a negative acceleration while an abrupt change in the direction of travel is a lateral acceleration. Accelerometers are used for air bag system crash detection and advanced ABS systems as will be discussed in **Chapter 15**.

The piezoelectric effect also works in the reverse; apply a voltage to a quartz crystal and the crystal will expand. The latest generation of high-pressure common rail diesel fuel injectors makes use of this principle, which will be discussed in **Chapter 14**.

Chemical Composition and Property Sensors

There are a variety of other specialized sensors used on a modern truck. These include sensors that

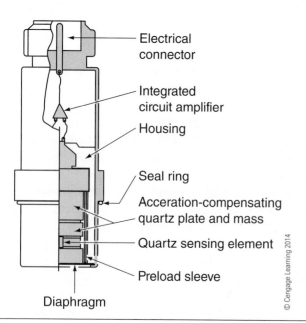

Figure 11-19 Piezoelectric pressure sensor.

determine the composition of the exhaust gas, such as the amount of oxygen and oxides of nitrogen (NO_x) that are present. Other sensors measure the quality of the diesel exhaust fluid (DEF) or the engine oil condition. The internal operation of these sensors is outside the scope of this book, but additional information can be found by researching the Internet links at the end of this chapter. Future sensor technology related to diesel exhaust emissions control includes ammonia (NH_3) and soot sensors. Diesel exhaust emissions will be covered in **Chapter 14**.

One simple property sensor is the water in fuel (WIF) sensor found at the bottom of the fuel water separator or primary fuel filter housing. This sensor typically consists of two exposed electrodes with a parallel-connected resistor. An electrical current is passed through the sensor. Diesel fuel and water have different resistances. If sufficient water is present, the difference in resistance is detected and a warning lamp is illuminated to alert the operator to drain the fuel water separator.

DIGITAL ELECTRONICS

The term digital is often used in marketing the latest device. Something that is digital is typically advertised as being superior to something with older technology that is not digital. The term analog (analogue) is applied to the outdated device. Television signals stopped being broadcast in analog format in 2009 in the United States and were replaced with a digital signal.

One of the simplest comparisons between analog and digital is the difference between a light switch and a variable light dimmer control. The light switch can be in only one of two states: on or off. By comparison, a light dimmer control can provide a continuous range of illumination levels, from off to dim to fully illuminated.

Another comparison between analog and digital is the measurement and display of time. A conventional mechanical clock or wristwatch with hands that rotate continuously is an analog clock. An electronic clock that displays time with digits, such as with LEDs, is a digital clock. The hands on an analog clock have a continuous motion and are never stopped at any particular number on the dial for any period of time. The motion of the minute and hour hands may be slow, but they are never stopped at one specific value on the dial for any period of time. By comparison, a digital clock that displays the time of day to the nearest minute displays the same numerical value for the current time for a period of 1 minute. Going a step

further, the second hand on a mechanical clock is never stopped at a particular number on the dial. The second hand moves in a continuous arc. However, a digital clock that indicates the time of day to the nearest second shows the same value of time for a period of 1 second, even though time is actually continuous and does not stop for 1 second as indicated by the digital clock.

Analog versus Digital as Related to Music

If a microphone was placed in front of a guitar and a single string on the guitar was plucked, the microphone would convert the sound waves into an electrical signal. For the most part, the guitar string vibrates at one main or fundamental frequency based on how tight the string is tensioned and the weight (mass) of the string. An oscilloscope could be used to display an amplified voltage waveform representing the string vibration and the sound that is produced (**Figure 11-20**).

If several strings were played at the same time, such as in a chord, each string would vibrate at a particular frequency. The microphone would convert these combined vibrations into an electrical signal representing the sound produced by the guitar. The oscilloscope would display this single waveform as a complex waveform that appears to have a pattern, but not like a sine waveform. Playing other chords after the first chord and adding other instruments to form a band would cause the signal produced by the microphone to be a waveform similar to that shown in

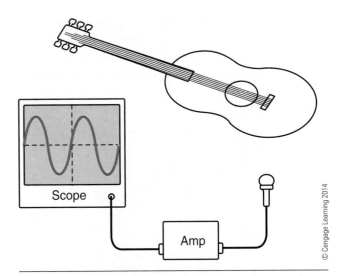

Figure 11-20 Plucking a single guitar string produces a fundamental frequency shown on the oscilloscope, plus harmonic frequencies.

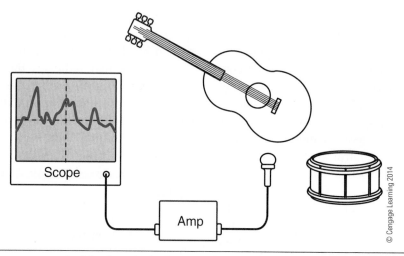

Figure 11-21 Waveform generated by band playing music.

Figure 11-21. The waveform looks like a random waveform without any pattern. However, it does represent in electrical form the sound that is present at the microphone.

In the past, the signal generated by the microphone could have been recorded to a vinyl phonograph record. Groves in the record were cut with bumps that matched the waveform of the signal being recorded. After the vinyl record is formed, the record could have been played on a record player (provided any are still in existence). A needle on the record player follows this groove. The movement of the needle over the bumps is transformed back into an electrical signal, which is amplified and used to drive a speaker. The speaker changes the amplified electrical signal back into sound waves. The sound emitted from the speaker is very close to the original signal that was recorded.

The signal produced by the microphone varies continuously and is called an analog signal (**Figure 11-22**). The phonographic recording device attempts to record this analog signal onto the vinyl record. A cassette tape recorder also attempts to record the analog signal from a microphone onto magnetic tape.

Compact discs (CDs) and MP3 devices have replaced records and cassette tapes. The information on a compact disc, such as music, is stored in a numerical format. This numerical format is referred to as digital. The analog signal recorded from the music shown in **Figure 11-23** could be converted into a digital signal. This analog signal is broken into equal time segments illustrated by the vertical lines. The amplitude of the signal is measured or sampled at each time point shown by the dots in **Figure 11-23**. Sampling indicates that the signal is not being continuously monitored or recorded. The amplitude of the signal is just being

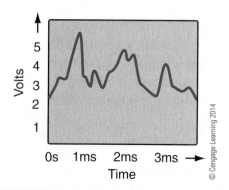

Figure 11-22 Analog signal produced by microphone.

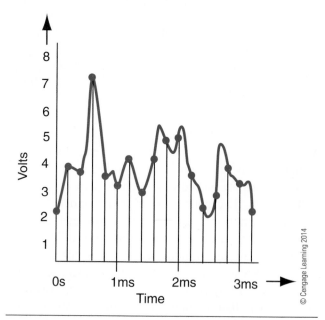

Figure 11-23 Analog signal divided into equal time increments.

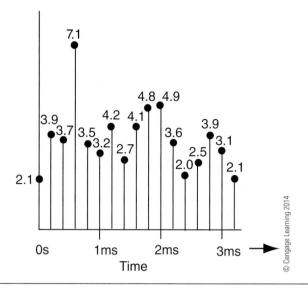

Figure 11-24 Amplitude measured at each time interval.

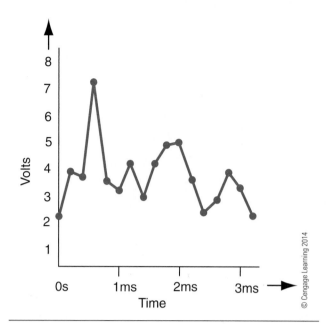

Figure 11-26 Re-created waveform from numerical information.

measured at prescribed time intervals as shown in **Figure 11-24**. Changes in the analog signal that occur between the sampling points are not detected.

This boxy-looking waveform could easily be recorded by storing the approximate amplitude (height) of the signal at each time interval, as shown in the spreadsheet in **Figure 11-25**. The boxy-looking waveform could then easily be re-created based on this information. The waveform shown in **Figure 11-26** has been re-created from the numerical information. Note that the re-created waveform does not look

Time (ms)	Volts
0	2.1
0.2	3.9
0.4	3.7
0.6	7.1
0.8	3.5
1	3.2
1.2	4.2
1.4	2.7
1.6	4.1
1.8	4.8
2	4.9
2.2	3.6
2.4	2
2.6	2.5
2.8	3.9
3	3.1
3.2	2.1

Figure 11-25 Waveform stored in numerical format.

exactly like the original analog waveform that was created by the musical instruments. Playing this re-created waveform through a speaker would sound something like the original music, but would lack clarity or fidelity. Some of the information in the original analog signal has been lost through sampling.

To make the signal less boxy-looking and thus make the music sound much better, the time interval between samples can be decreased such that the human ear cannot easily detect the transitions between time intervals. By reducing the time between samples of the signal amplitude (more samples per second), the analog signal created by the microphone is more accurately recorded. The time between samples has been decreased by one-half in **Figure 11-27**. The waveform re-created from this more frequently sampled numerical data is much closer to the original analog waveform, as shown in **Figure 11-28**.

Increasing the sampling rate (reducing time between samples) permits a waveform to be re-created from the stored information that is much closer to the original analog signal. However, this also requires more data to be stored because there are more time points with the associated signal amplitude to record. For example, CD audio typically has been sampled at a rate of 44.1k samples per second.

Old movie cameras using photographic film are similar to a digital signal. The movie camera attempts to record motion by taking a series of still pictures in rapid succession. Motion, such as someone walking, appears very jerky in old movies. As movie camera technology progressed, the time between each picture

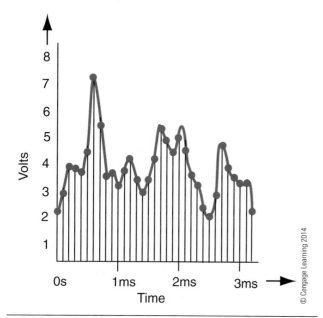

Figure 11-27 Increased sample rate (less time between samples).

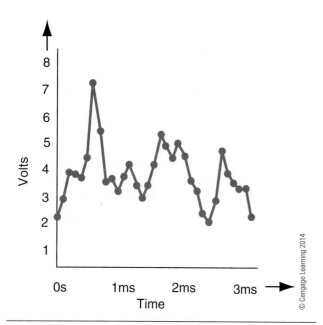

Figure 11-28 Re-created waveform is now much closer to original analog waveform.

decreased, which improved the quality of the movie. Eventually, motion in a movie appears to be lifelike because your eyes are not capable of detecting the time between frames of the movie.

Each individual frame or picture taken by the movie camera is like the sampling of a digital signal. Increasing the number of pictures taken per second by a movie camera results in a more lifelike movie. In the same way, increasing the number of digital samples per second (increasing the sample rate) results in a more accurate representation of the original analog

audio signal. At some point, your ears may not able to detect the difference between a digitally re-created analog signal such as that produced by a CD player and a true analog signal such as that produced by musical instruments at a live concert.

Digital Numbering System

Digital information is not stored or manipulated in the standard base 10 numbering system that you have used all your life. A base 10 numbering system uses combinations of the numbers 0 to 9 to generate all numbers. These familiar base 10 numbers of 0 to 9 are known as decimal numbers. There are 10 different numbers: 0, 1, 2, 3, 4, 5, 6, 7, 8, 9 in the decimal numbering system. Instead of decimal numbers, a base 2 or **binary** numbering system is used when working with digital information. Only the numbers 0 and 1 are used in a base 2 or binary numbering system. All numbers in the binary system are represented by combinations of a 0 or a 1. For example, the decimal number 9 is equal to 1001 in binary.

Many calculators have the ability to convert from base 10 to the binary number system, but converting from base 10 to base 2 is really not that important for a truck technician. However, those who are interested in computers probably already have some knowledge of the binary numbering system.

The reason that a binary numbering system is used in digital electronics is that there are only two possible digits, 0 and 1, in the binary numbering system. These two digits are combined to represent any decimal number. The two possible digits of binary work well in digital electronics because a switch also only has two states, open or closed. A light that is controlled by a switch can also only have two different states, either off or on. There is no in-between state for a light switch in which the light controlled by the switch is only partially on or partially off.

Instead of a switch used to control a light, a group of switches wired in parallel and connected to a 1V battery can be used to represent a binary number. With the switch closed, a voltmeter would indicate a value of 1V across a resistor, as shown in **Figure 11-29**. With the switch open, the voltage across the resistor would be 0V. The 1V represents a binary 1 while the 0V represents a binary 0.

The decimal number 9 is represented as 1001 in binary and could be represented in electrical form by using four parallel-connected switches, as shown in **Figure 11-30**. Closing the first and last switches and leaving the two middle switches open would cause the voltage dropped across the resistors to represent the

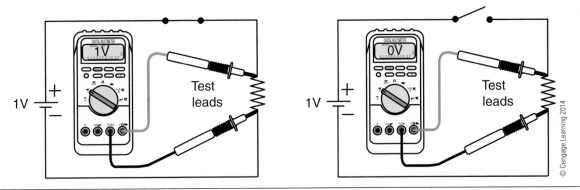

Figure 11-29 Voltage across resistors with 1V source, open and closed switches.

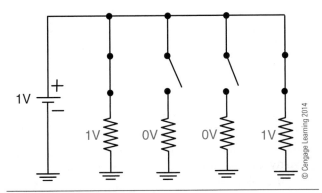

Figure 11-30 Four switches used to represent the number 9 in binary format.

decimal number 9. This pattern of 1V or 0V could be translated into the number 9 by using four switches. Larger decimal numbers could be represented by adding more switches and resistors.

Each of the digits in a binary number is called a **bit**. A bit is a single binary character and is either a 1 or a 0. You have probably heard the word *bit* used in relation to computers. The more bits there are in a binary number, the larger the value of the decimal number that can be represented. Another term often used when talking about computers is **byte**. A byte is a group of eight bits. The largest binary number in a single byte is 11111111, which equates to 255 in ordinary decimal numbers. The largest decimal number that can be represented by two bytes (16 bits) is sixteen 1s, which equates to 65,535 in decimal numbers.

You are probably familiar with the term byte used with computers when referring to information storage capacity. A CD disk is capable of storing about 682 megabytes of information. This means that 682 million bytes, each consisting of eight individual bits, can be stored on the disk. Each of these bits is either a 1 or a 0. Computer hard drives commonly hold several gigabytes (billions of bytes) of information. Computer hard drives use magnetism to record information onto the magnetic disk material. Areas of the disk that are magnetized in one manner represent a binary 1, while those magnetized in another manner represent a binary 0.

Logic Gates

Electronics can be divided into two general categories: analog and digital. Prior to this chapter, everything in this book has dealt with analog systems. Resistors, transistors, and other components are used in analog electronics. Ohm's law, Kirchhoff's laws, and other rules are used to predict or explain how an analog electric or electronic circuit will respond.

In digital electronics, calculating voltage drops, measuring current, and handling other tasks performed with an analog system are not necessary. Digital electronics is mostly only concerned with two different voltage levels. Voltages below some value are deemed as a 0 (logic 0), while voltages above some other value are deemed as a 1 (logic 1).

The focus of digital electronics is **logic**. Logic is the use of correct or valid reasoning to come to a conclusion. A decision based solely on logic would typically not include any emotion. In digital electronics, logical decisions are made by devices called **gates**. These gates are physical electronic components that are composed of transistors and other hardware. However, the hardware is not the important thing here. The important thing in digital electronics is the decision that is made by the hardware (gate).

The terms *input* and *output* are used quite frequently when dealing with personal computers. Inputs to personal computers are devices like the keyboard and mouse. Outputs of personal computers are the display screen and the speakers. Logic gates have one or more inputs and one output. The inputs of logic gates are voltages corresponding to logic 1 or logic 0 levels. Because these gates are logical, they will always make the correct decision. The decision that a logic gate makes is "what will my output be, a 1 or a 0?"

In analog electronics, symbols are used to represent hardware devices such as transistors, resistors, capacitors, inductors, and other devices. In digital electronics, symbols are used to represent the logical decisions that the hardware (gates) will make.

There are several types of logic gates. Each logic gate is designed to make a specific decision when supplied at its inputs with combinations of logic 1 voltage levels and logic 0 voltage levels.

The first type of gate that will be examined is an AND gate. To assist in explaining the operation of an AND gate, a simple electric circuit will be evaluated. The circuit shown in **Figure 11-31** consists of two switches wired in series with a battery and a light bulb. The light bulb is the output, and the two switches are the input. In order to cause the light bulb to illuminate, both switch A and switch B must be closed (on). Closing just one switch while leaving the other switch open will result in the light bulb not illuminating. In other words, switch A AND switch B must be closed for the light bulb to illuminate. Each of the possible switch states and the corresponding light bulb output could be listed in a table, as shown in **Figure 11-32**.

Because a switch is a digital device and can only be on or off, this table could be modified so that open switches are identified as 0, while closed switches are identified as 1. Additionally, because the light bulb can only be on or off, the outcome can be identified as 1 for ON and 0 for OFF, as shown in **Figure 11-33**. This type of table is referred to as a truth table.

Switch A	Switch B	Out
0	0	0
0	1	0
1	0	0
1	1	1

Figure 11-33 Status of Switch A and Switch B truth table.

The schematic symbol for an AND gate and the truth table for the AND gate are shown in **Figure 11-34**. This particular AND gate has two inputs and one output. The truth table indicates what outcome or decision the AND gate will make when each possible combination of logic level (1 or 0) is present at the inputs A and B of the AND gate. An AND gate could have more than two inputs, resulting in a larger truth table.

Another common logic gate is the OR gate. The function of an OR gate is represented by the simple electric circuit shown in **Figure 11-35**. This circuit has two switches that are in parallel. Closing either switch will cause the light to illuminate. Closing both switches will also cause the light to illuminate. It is necessary to open both switches to cause the light to not be illuminated. Thus, closing either switch A OR switch B will cause the light to illuminate. The schematic symbol and truth table for an OR gate with two inputs are shown in **Figure 11-36**.

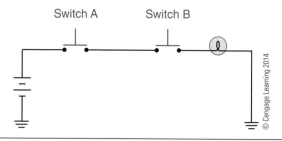

Figure 11-31 AND gate constructed of two switches in series.

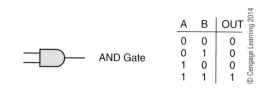

Figure 11-34 AND gate symbol and truth table.

Switch A	Switch B	Out
Off	Off	Off
Off	On	Off
On	Off	Off
On	On	On

Figure 11-32 Status of Switch A and Switch B and resulting outcome of light bulb.

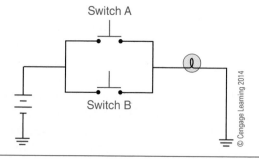

Figure 11-35 OR gate constructed of two parallel switches.

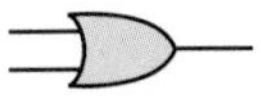

	A	B	OUT
OR Gate	0	0	0
	0	1	1
	1	0	1
	1	1	1

© Cengage Learning 2014

Figure 11-36 OR gate symbol and truth table.

Tech Tip: The OR gate symbol looks somewhat like a shovel, which could be used in mining ore (one way to remember the meaning of the shape).

A NOT gate is another logic gate. In speech, placing the word *not* in front of a word makes the outcome of the sentence have the opposite meaning. A NOT gate is also known as an inverter. Because there are only two possible logic states, a NOT gate inverts the logic level at its input. An input of 1 at a NOT gate results in an output of 0, and an input of 0 at a NOT gate results in an output of 1. The schematic symbol for a NOT gate is shown in **Figure 11-37**. Note the small circle at the tip of the arrow in the schematic symbol.

The NOT gate and the AND gate can be combined together to form a NAND gate. The schematic symbol and truth table for a NAND gate are shown in **Figure 11-38**. The small circle is taken from the circle in the NOT gate. Note that the output of the truth table for a NAND gate is opposite that of an AND gate.

The NOT gate and the OR gate can be combined together to form a NOR gate. The output of the truth table for a NOR gate is the opposite of an OR gate (**Figure 11-39**).

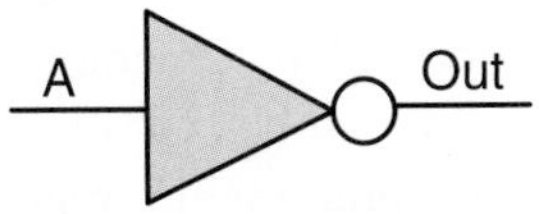

	A	Out
NOT Gate	0	1
	1	0

© Cengage Learning 2014

Figure 11-37 NOT gate and truth table.

	A	B	OUT
NAND Gate	0	0	1
	0	1	1
	1	0	1
	1	1	0

© Cengage Learning 2014

Figure 11-38 NAND gate and truth table.

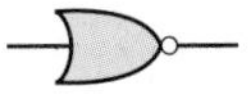

	A	B	OUT
NOR Gate	0	0	1
	0	1	0
	1	0	0
	1	1	0

© Cengage Learning 2014

Figure 11-39 NOR gate and truth table.

Microprocessors

One of the major breakthroughs in modern technology that occurred during the mid-twentieth century was the invention of the **integrated circuit**. An integrated circuit is a circuit in which the transistors, diodes, resistors, and other components are formed during the manufacturing process of the device. Integrated circuits are commonly known as chips (**Figure 11-40**). Each of the transistors, diodes, and resistors in the integrated circuit is microscopically small. These components are not just small individual or discrete components like diodes and transistors; instead, the components are formed from layers of silicon made of P-type and N-type material. The logic gates discussed in the previous section are typically made from transistors that are formed in an integrated circuit. The manufacturing processes for integrated circuits have become so refined that integrated circuits are now very reliable and relatively low cost. The small size of the components formed in an integrated circuit also reduces the electric power requirements of the device.

Modern computers contain a device called a processor. The processor is the main component in the computer. The processor is only capable of working with binary numbers (1s and 0s). The term **microprocessor** is often used as a generic reference to a digital processor. The microprocessor is not capable of "thinking" or "knowing," but these terms are often used to describe what is occurring inside the microprocessor. The microprocessor just processes information based on its programming instructions. Any decisions that the microprocessor makes are the outcomes of how the microprocessor is programmed or instructed to respond. These program instructions use the status of the microprocessor inputs in the decision-making process.

The microprocessor is an integrated circuit and may contain millions of transistors. Microprocessors are complex devices that are beyond the scope of this book. However, it is not necessary to fully understand how a microprocessor operates to troubleshoot a system that is microprocessor controlled. Most modern trucks contain a variety of electronic modules that are

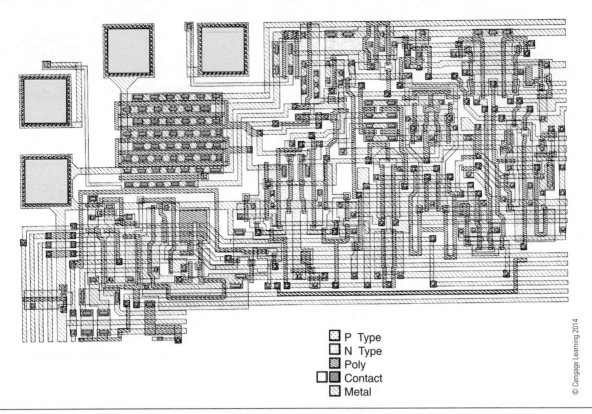

P Type
N Type
Poly
Contact
Metal

© Cengage Learning 2014

Figure 11-40 Integrated circuit.

microprocessor based. The function of some of these modules will be covered in later chapters.

Memory

You have probably heard the term *memory* used frequently in reference to personal computers. Memory is basically just a place where digital information is stored. Memory is binary, meaning that a memory location is either a 1 or a 0. Many types of memory are integrated into truck electronic modules. These types of memory can be broken down into two main categories: **volatile memory** and **non-volatile memory**.

Volatile memory is memory that is lost or reset when power is taken away from the electronic module. The **random access memory (RAM)** of a personal computer is volatile memory. If the power has ever gone out while you were working or playing a game on a personal computer, you probably lost a portion of your work or had to start your game over again. Your work or play is being written to RAM until you save the information to the hard drive. Volatile memory such as RAM is used by the microprocessor to temporarily store information in a way that is similar to writing on a chalkboard with a piece of chalk. Disconnecting the power to the electronic module causes the information stored in RAM to be lost in a way that is similar to erasing the chalkboard.

Non-volatile memory is memory that is retained through a disconnection of electric power to the electronic module. Non-volatile memory is used to retain the microprocessor's programing instructions. A personal computer contains a hard drive on which the operating system and program files are stored. Disconnecting the power to the personal computer has no effect on the data stored on the hard drive because it is stored magnetically. Truck electronic modules do not have a hard drive to store the microprocessor's program instructions. Instead, the program instructions are stored in electronic non-volatile memory. This is similar to the type of memory used in some MP3 players and most other consumer electronic devices.

Most non-volatile memory can be erased and rewritten with new information. Non-volatile memory used in truck electronic modules that can be rewritten is typically electrically erasable, programmable, read only memory (**EEPROM**). Some types of EEPROM are designed to be erased and rewritten many times. This type of EEPROM is used in truck electronic modules to store information that changes frequently. For example, the engine ECM may maintain several pieces of information such as total engine hours and total accumulated vehicle mileage (odometer). When the key is switched off, the engine ECM may update this information stored in EEPROM. This is like saving a file that you have been working on to the hard drive of a personal computer before shutting off the computer.

Flash memory is a type of EEPROM that is used to retain the major portion of module programming instructions. Flash memory differs from standard EEPROM in that large chunks or sections of flash memory must be erased at the same time, unlike standard EEPROM that can be erased and rewritten in small sections or bytes. The type of flash memory used in electronic modules typically has a limited number of write cycles so it is not used to store information that changes frequently. The terms *flash* or *reflash* are often used to describe the operation of reprogramming the flash memory section on an electronic module, such as the engine ECM. OEMs often find problems with the original programing instructions, such as an engine performance problem under specific or unique operating conditions. To resolve these problems, a revised version of software is released by the OEM and the ECM is reprogrammed or flashed with the revised software.

All types of memory store information in a digital format. This information storage is accomplished by storing an electric charge in a floating gate of a very small field effect transistor (FET), similar to storing a charge in a capacitor. If a charge is present in the floating gate, the FET cannot be switched on, indicating one bit with logic state 0. If a charge is not present in the floating gate, the FET can be switched on, indicating one bit with logic state 1. The charge can be stored in the floating gate almost indefinitely, even when the power to the device is interrupted. To erase the memory, a voltage is applied that causes the electric charge stored in the floating gates to be discharged. The theory of operation of some types of flash memory are explained by the Heisenberg uncertainty principle and quantum tunneling; therefore, the physics of flash memory is definitely outside the scope of this book.

Inputs and Outputs

The five human senses act as inputs to the human version of a processor known as the brain. This human processor was pre-programmed with some specific instructions for survival such as breathing. Other programming was added or modified by life experiences. Touching a hot stove causes the touch sensors (via electrical impulses) to provide information to the human's processor. The processor responds by commanding the muscles (also via electrical impulses) to take the hand off the stove, NOW! This command to the muscles is an output of the human processor.

Like a human brain, microprocessors also have inputs and outputs. Microprocessors evaluate the information taken from their inputs and use this information to make decisions based on the microprocessor's programming. The decisions made by the microprocessor cause the microprocessor to control the state of its outputs accordingly.

Microprocessor Inputs and Outputs. All inputs to a modern microprocessor are digital (1 or 0). All outputs from the microprocessor are also digital. However, an electronic module that contains a microprocessor can have both analog inputs and analog outputs, as will be explained later in this section.

Most of the sensors addressed earlier in this chapter provide a variable signal corresponding to the temperature, pressure, or other quantity being measured. Because microprocessors can only accept digital information, the analog signals provided by sensors must be converted to the digital (binary) equivalent. An integrated circuit called an **analog to digital (A/D) converter** is used to convert the analog voltage measured at the input of the electronic module into a digital signal. The analog voltage at an electronic module input at a particular instant in time is measured by the A/D converter and assigned a binary (digital) value, as illustrated in **Figure 11-41**.

In another example, the battery voltage level is a necessary piece of information for the decision making of many electronic modules, such as the engine ECM. An analog value of 12V measured at the battery voltage input of the electronic module must first be converted to the digital or binary equivalent of 12. The binary number for 12 is 1100. The A/D converter would change the 12V measured at the input to a binary value of 1100. The microprocessor can work with this binary number and "know" through its programming that a binary value of 1100 means that there is 12V present at the electronic module battery voltage input. Increasing the voltage at the input to 13V causes the A/D converter to supply the microprocessor with 1101, the binary equivalent of 13. However, if the battery voltage were 12.4V, it would be necessary to round down to 12V because this simple conversion from analog to digital is only capable of 1V steps.

Because it is necessary for the microprocessor to resolve the truck's battery voltage level more accurately than 1V steps, the A/D converter could be designed to provide a single bit for each 1/10 of a volt present at the input. Therefore, a value of 12.1V measured at the A/D converter input is equal to 121 bits when each bit is designed to be worth 1/10 of a volt. This 121 would equate to 1111001 in binary. Note that

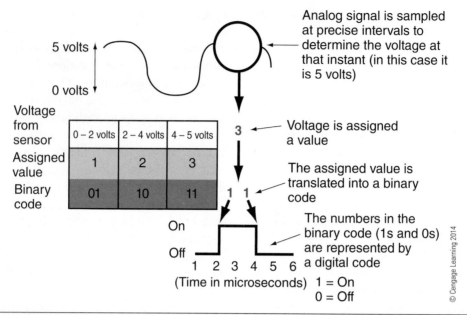

Figure 11-41 Analog to digital conversion.

it is necessary to have more bits in order to represent 12.1V in binary than 12V. A principle of digital electronics is that in order to measure something more accurately, more bits are necessary. This is called resolution or volts per bit. An actual A/D converter may be described as an 8-bit device, meaning there are 11111111 binary or 255 possible steps into which the analog voltage being converted could be divided. Typically, the analog input voltage is scaled or reduced by voltage divider circuits within the electronic module before being supplied to the A/D converter. This scaling limits the maximum voltage to the A/D converter to some value, such as 5V. The A/D converter uses this maximum voltage as a reference voltage. Therefore, an 8-bit A/D converter would have a resolution of 19.6mV per bit for a 5V reference voltage.

The other factor besides sample rate that impacts the quality of digitally recorded music is the resolution of the amplitude of the original analog signal being converted. For example, CD audio typically has 16-bit resolution of the signal amplitude or 65,536 possible voltage steps.

The output of the 1s and 0s from the A/D converter can be serial or parallel. *Serial* means the A/D converter will produce one bit after the other. For example, the A/D converter may output each bit for a period of 1 μsec per bit in a very slow system (by today's standards). This means that it would take 8 μsec for the A/D converter to produce the value of an 8-bit binary number. The microprocessor could place these 8-bits in a memory storage location as they were received

from the A/D converter (**Figure 11-42**). Only two wires or conductive paths between the A/D converter and the microprocessor are necessary for this serial method (one path being the common ground).

If you own a personal computer, you probably know the clock speed of the processor used in the computer. Clock speeds in the gigahertz region are common in PCs. The clock in a processor generates a continuous pulse or square wave. This clock pulse is used to cause the magic of digital processing to occur. The clock speed refers to the frequency of this pulse. The higher the frequency of the clock pulse, the faster

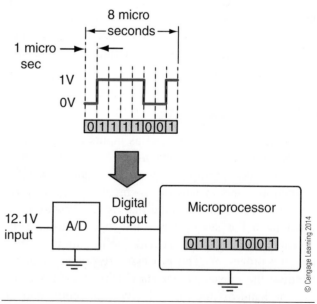

Figure 11-42 Serial processing of a digital signal.

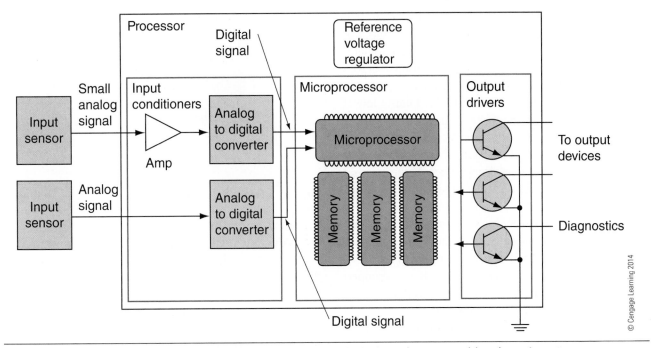

Figure 11-43 Analog signal is conditioned, converted to digital, and processed by the microprocessor.

the processor is capable of processing digital information and performing instructions.

An overview of A/D converters, microprocessors, and memory is shown in **Figure 11-43**. Analog signals supplied by sensors are amplified and conditioned if necessary. **Signal conditioning** refers to the manipulation of the signal so that it is properly prepared for the next stage of the process. The scaling of the analog input voltage so that it is always less than 5V at the actual input of the A/D converter described previously is an example of signal conditioning. Filtering or smoothing of the analog voltage using a capacitor is also a part of signal conditioning. After the analog signal is conditioned, the analog signal is converted to a serial digital signal. This digital signal is an input to the microprocessor. The microprocessor places this digital information into RAM memory for temporary use. The microprocessor then accesses its programming instructions from flash memory to determine what should be done next, based on the information obtained from the sensor.

Electronic Module Input Circuitry. Electronic modules used in a truck electrical system such as the engine ECM depend on input devices like sensors. An electronic module input device such as a sensor is connected to an input terminal of the electronic module. Some of the internal circuitry used for signal conditioning in a typical electronic module will be examined in this section. This circuitry is not as complicated as you might have imagined especially if

the circuit is simplified to its most important components to illustrate the principles.

If a 1kΩ resistor were connected to the positive terminal of a 12V battery at one end and left disconnected or **floating** at the other end, a voltmeter would indicate 12V between the end of the resistor that is floating, as shown in **Figure 11-44**. The reason for this is that the resistor and the voltmeter form a series voltage divider circuit. The resistor has 1kΩ of resistance; the voltmeter has approximately 10MΩ of internal resistance (see the Extra for Experts section on Additional DMM Information in **Chapter 2** for an

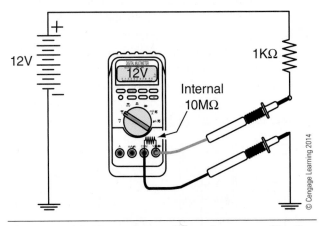

Figure 11-44 Measuring voltage between floating 1k ohm resistor connected to B+ and ground with DMM voltmeter with 10M ohm internal meter resistance. Almost all voltage is dropped across the meter's internal resistance.

explanation). The laws of series circuits indicate that nearly all of the voltage is dropped across the 10MΩ of internal resistance inside the voltmeter. The voltmeter is merely displaying the voltage that is dropped across its 10MΩ of internal resistance, which is effectively 12V. This is an important concept. Use Ohm's law if necessary to calculate the voltage dropped across the 10MΩ of internal voltmeter resistance and the voltage dropped across the 1kΩ resistor. You will find that nearly all of the voltage is dropped across the internal resistance of the voltmeter, thus the DMM reading of approximately 12V.

If the 1kΩ resistor were instead connected to ground at one end and left floating at the other end, a voltmeter would indicate 0V across the resistor (**Figure 11-45**). This is because there is zero voltage being dropped across the resistor. The voltmeter is in parallel with the resistor so the voltmeter indicates that 0V is dropped across the resistor. This may seem obvious, but the concept is important.

The circuitry contained inside an electronic module that is connected to the input terminals of the module can be simplified to a resistor and a voltmeter. This resistor may be connected to a positive voltage inside the module like the example shown in **Figure 11-44** for some types of inputs. Alternatively, this resistor may be connected to ground inside the module like the example shown in **Figure 11-45** for other types of inputs. The A/D converter is like the DMM voltmeter shown in the previous examples. The A/D converter is connected to the module input terminal, as shown in **Figure 11-46**. This terminal is where the vehicle wiring harness would provide connection to the sensor or switch that is acting as an input device to the electronic module.

The A/D converter measures the voltage between the input terminal and ground. In the example shown in **Figure 11-46**, nothing external to the electronic module is connected to the input terminal so the input

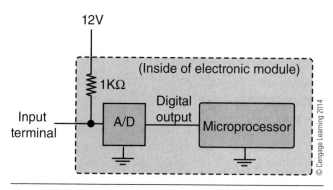

Figure 11-46 An A/D converter measures voltage like a DMM voltmeter; nearly all of the 12V is dropped across the very high internal resistance of the A/D converter. The A/D converter measures approximately 12V and converts this to the digital equivalent.

terminal is effectively floating. The internal resistance of the A/D converter is very high, like a DMM voltmeter. If the top of the resistor shown in **Figure 11-46** is connected to +12V, then the A/D converter would measure about +12V referenced to ground. This is very similar to the circuit shown in **Figure 11-44** except the high internal DMM voltmeter resistance is the internal resistance of the A/D converter. Compared to the 1kΩ resistor, the internal resistance of the A/D converter is 10 Mega Ω or more. Therefore, nearly all the voltage, 12V in this case, will be dropped across the high internal resistance of the A/D converter, causing the A/D converter to read a value of 12V for the voltage value at this input. A millivolt or so is also dropped across the 1kΩ resistor, but this is a negligible amount in this case.

The 1kΩ resistor shown in **Figure 11-46** is called a **pull-up resistor**. The resistor pulls the input terminal up to some voltage level when nothing external to the module is connected to the input terminal. In this example, the input is "pulled-up" or elevated to battery voltage. With nothing external to the electronic module connected to the input terminal, this particular input is pulled-up to 12V through the 1kΩ resistor inside of the module. This concept will be discussed in more detail later in this chapter.

Adding External Devices to the Input. If a variable resistor-type sensor that provides a path to ground were connected to the input terminal of the electronic module, the input terminal would no longer be floating and a series voltage-divider circuit would be formed. The 1kΩ pull-up resistor inside the module and the variable resistor outside the module are connected in series, as shown in **Figure 11-47**. This connection will cause current to flow from the +12V source, through the pull-up resistor and through the sensor resistance to

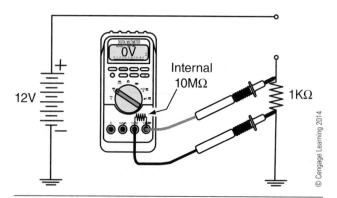

Figure 11-45 Measuring voltage across floating 1kΩ resistor connected to ground.

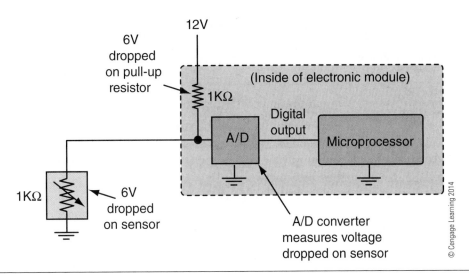

Figure 11-47 Connecting a variable resistor to the input terminal.

ground. If the sensor also has a resistance of 1kΩ, the pull-up resistor and the sensor would each have 6V dropped across them as the 12V supplied by the battery divides equally. The A/D converter is measuring the voltage between the input terminal and ground. Therefore, the A/D converter is connected in parallel with the sensor. This means that the A/D converter is actually measuring the voltage that is dropped across the variable-resistor type sensor because parallel-connected devices must always have the same voltage dropped across them, per the rules of parallel circuits.

Do not let the fact that one of the resistors is inside of an electronic module and the other resistor, the variable resistance of the sensor, is outside of the module cause confusion. With Ohm's law and the laws of series circuits, it does not really matter where the series-connected resistors are located. The fact that the pull-up resistor and the sensor resistance each have 6V dropped across them is just applied Ohm's law along with the rules of series circuits.

The A/D converter would measure the 6V that is dropped across the sensor, convert this analog voltage to a digital value, and provide the microprocessor with the binary equivalent of 6V. The microprocessor now "knows" that the voltage at this input terminal is 6V.

In reality, the internal resistance of the A/D converter is in parallel with the 1kΩ external variable resistor that is connected to the input. However, the internal resistance of the A/D converter is so high compared to the 1kΩ external variable resistor that the A/D converter's internal resistance can be ignored. This is a very important concept to understand. Placing a very high resistance in parallel with one of two much smaller resistors that are wired in series results in almost no difference in the voltage level that is dropped across either of the original series resistors.

To prove this concept, verify that the voltage displayed on the DMM voltmeter shown in **Figure 11-48** is correct (ignoring the internal resistance of the DMM). The first step is to calculate the equivalent

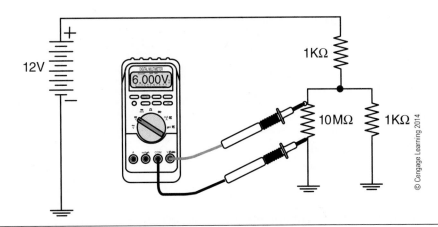

Figure 11-48 Series-parallel circuit with 1kΩ connected in parallel with 10MΩ, connected in series with 1kΩ.

resistance for the two parallel resistors, which is 1kΩ in parallel with 10MΩ. This results in a value of 999.9Ω for the equivalent resistance. This 999.9Ω is in series with the 1kΩ resistor, resulting in a total circuit resistance of 1999.9Ω. Ohm's law indicates that a total current of 0.0060003A (6.0003mA) will flow through the circuit. Multiply this current by the equivalent resistance for the two parallel-connected resistors to find the voltage dropped across the parallel resistors. Therefore, the voltage dropped across the parallel combination of the 1kΩ resistor and the 10MΩ resistors is 5.9997V. However, this is essentially 6V and most DMMs would indicate 6.000V when configured to display the highest possible precision.

Tech Tip: Knowing the level of precision or accuracy of measurements required to diagnose a problem comes with experience. Expecting too much precision in a voltage measurement during troubleshooting may result in needless replacement of components. Conversely, not being accurate or precise enough in your measurements when necessary may cause you to misdiagnose the problem. For example, there is no general rule on when a difference of 100mV is important and when 100mV can be ignored. However, most OEMs' troubleshooting information typically provides a range of acceptable readings, such as a voltage measurement, which should be observed.

Modifying the variable-resistance sensor so that its resistance is now 2kΩ will cause the voltage measured

at the module input terminal to become 8V, as shown in **Figure 11-49**. This is because the fixed-value pull-up resistor is 1kΩ and the variable-resistor sensor is 2kΩ and these two resistors are connected in series (ignoring the high internal resistance of the A/D converter). The A/D converter measures this 8V at the input terminal, converts the analog signal to a digital signal of equivalent value, and supplies the microprocessor with this information. The microprocessor now "knows" that the voltage at this input terminal is 8V.

If the variable-resistance type sensor, was a thermistor that is measuring engine coolant temperature, the microprocessor could be programmed to know that 6V at this input means that coolant temperature is 120°F (49°C), while 8V at this input means that coolant temperature is 60°F (16°C).

The 1kΩ pull-up resistor in **Figure 11-49** is used to form a series voltage divider network with the resistance of the sensor connected to the input terminal in this example. The pull-up resistor is said to cause the input to be pulled-up to +12V. With nothing external to the electronic module connected to the module's input terminal (**Figure 11-46**), the voltage measured at the floating input terminal referenced to ground is +12V. An electronic module input terminal that is internally connected to a pull-up resistor is inclined to be pulled-up to some positive voltage level until something outside the module, such as the resistance of a sensor providing a path to ground, causes the voltage measured at the input to decrease. The pull-up resistor inside the module is necessary because it forms a series voltage divider circuit with the device that is connected outside of the module to the input terminal such as a sensor. The pull-up resistor also limits the

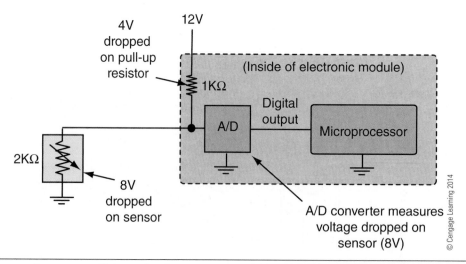

Figure 11-49 Sensor changes to 2kΩ, and voltage measured by A/D converter increases to 8V as voltage divides between pull-up resistor and sensor resistance.

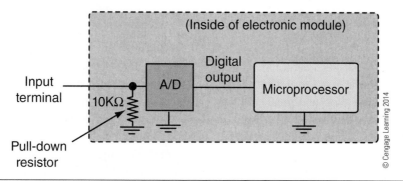

Figure 11-50 Pulled-down input.

amount of current that will flow if the input terminal is grounded and permits diagnosis of wiring failures, such as an open or shorted-to-ground circuit. This will be discussed in detail later in this chapter.

The internal input circuitry of an electronic module can also be constructed such that a resistor is connected between the input terminal and ground inside the module. This type of input is referred to as being pulled-down (**Figure 11-50**). The resistor inside the module is called a **pull-down resistor**. The resistor pulls the input circuit down to ground level voltage or 0V with nothing outside the electronic module such as a sensor connected to the input terminal. An input that is pulled-down is designed to be connected to a device outside of the module that provides a positive voltage to the module's input terminal. The pull-down resistor creates a series voltage divider circuit with a device that is connected to the input terminal outside of the module such as a sensor that supplies a positive output voltage signal.

For example, an air pressure sensor that provides an output signal voltage of 2V when measuring a pressure of 50 psi (345 kPa) could be connected to a pulled-down input terminal of a module as shown in

Figure 11-51. This would cause a voltage drop of 2V across the pull-down resistor. The A/D converter that is connected in parallel with the pull-down resistor would convert this voltage into the digital equivalent of 2V for use by the microprocessor. The microprocessor would be programmed to "know" that a voltage of 2V at this input means that air pressure is 50 psi. The programming instructions of the microprocessor, which converts an input voltage into a pressure, temperature, or other physical measurement, is called a **transfer function**. A transfer function is a math equation used to describe the relationship between the physical variable the sensor is measuring, such as pressure, and its output voltage. For example, if an air pressure sensor provides an output voltage of 2V at 50 psi (345 kPa) and an output voltage of 3.5V at 100 psi (690 kPa), the transfer function for the sensor in units of psi would be:

$$\textbf{Voltage} = \textbf{0.03} \times \textbf{Pressure} + \textbf{0.5}$$
$$\textbf{or}$$
$$\textbf{Pressure} = \textbf{33.333} \times (\textbf{Voltage} - \textbf{0.5})$$

Therefore, a voltage of 4.1V measured at the electronic module input would indicate the pressure is 120 psi

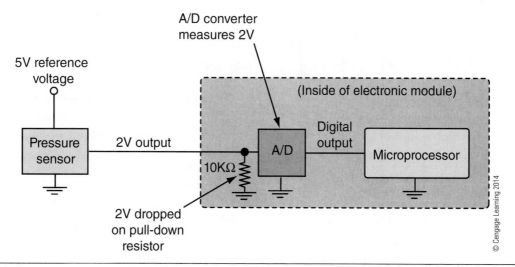

Figure 11-51 Air pressure sensor supplying 2V to pulled-down input.

(827 kPa). The transfer function for this sensor is plotted in the graph shown in **Figure 11-52**. This particular sensor has a linear output, as indicated by the straight line shown in the graph. If you have taken algebra, you may recognize the $y = mx + b$ format of this sample transfer function. For nonlinear sensors such as thermistors, the transfer function would be more complex and the plotted line would be curved, not straight.

Pull-down resistors ensure that the voltage measurement at an electronic module will be 0V when the input terminal is floating. Pull-down resistors also limit the amount of current that will flow if the electronic module input terminal is connected to a positive voltage source.

Digital and Analog Electronic Module Inputs. There are four basic types of electronic module inputs. Inputs can be classified as being pulled-up or pulled-down. The circuit shown in **Figure 11-46** is a pulled-up input. The circuit shown in **Figure 11-50** is a pulled-down input.

Electronic module inputs can also be classified as being analog type or digital type. The input circuit examples shown in **Figure 11-47** and **Figure 11-51** are analog inputs. The analog input voltage must be converted to the digital equivalent binary 1s and 0s by an A/D converter before being supplied to the microprocessor. For example, the A/D converter would transform the 2.0V analog value in **Figure 11-51** to 10100 binary, given a resolution of 0.1V per bit.

Electronic module digital inputs are inputs that are capable of only recognizing two different states or two different ranges of voltage. Digital inputs are typically

connected to a switch because a switch also only has two different states: closed or open. Voltages present at a digital input below some value are considered as logic 0 state while voltages above some value are considered logic 1 state. For example, a voltage below 1V measured at a module digital input terminal is typically considered logic 0 state while voltages above about 3V are considered logic 1 state. Voltages between 1V and 3V might not be "guaranteed" to always be evaluated as either logic 1 or logic 0 states, so the circuit is designed to stay away from voltages in this unknown region. Digital inputs do not require an A/D converter because the input is already digital (high or low, 1 or 0).

In reality, there are additional signal conditioning components to filter (smooth) the input voltage and additional voltage divider circuits inside the module to further reduce the voltage supplied to the microprocessor's digital input terminal. However, these devices are not so important for a basic understanding of electronic modules.

Electronic module digital inputs, like analog inputs, may also be classified as being pulled-down or pulled-up depending on the device that is connected to the input. For example, an electronic module digital input terminal that is connected to ignition voltage (+12V) that is used to determine if the key switch were in the ignition position or in the off position would be pulled-down. This is because when the key switch is in the off (open switch) position, the ignition voltage is 0V. The digital input would indicate logic 0 state with the key switch in the off position because of the connection to ground through the pull-down resistor (**Figure 11-53**, top). With the key switch in the ignition (switch closed) position, the voltage at the input would be ignition voltage or about 12V (**Figure 11-53**, bottom). The digital input would indicate logic 1 state, indicating that the key switch is in the ignition position. The microprocessor would use the status of this digital input in the decision-making processes involving the state of the key switch. A pulled-down digital input may also be referred to as an **active-high** input.

Electronic module pulled-up digital inputs would typically be connected to switches or other devices that provide a path to ground when closed, such as the park brake switch shown in **Figure 11-54**. With the park brake switch open, the digital input would be determined as being logic 1 state because of the voltage measured at the microprocessor input due to the pull-up resistor's connection to a 12V source. With the switch closed, the digital input would be determined as being logic 0 state because of the path to ground provided by the closed switch. The amount of current that flows in this path to ground is minimal because of

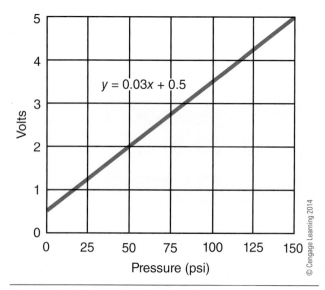

$$y = 0.03x + 0.5$$

Figure 11-52 Transfer function for pressure sensor.

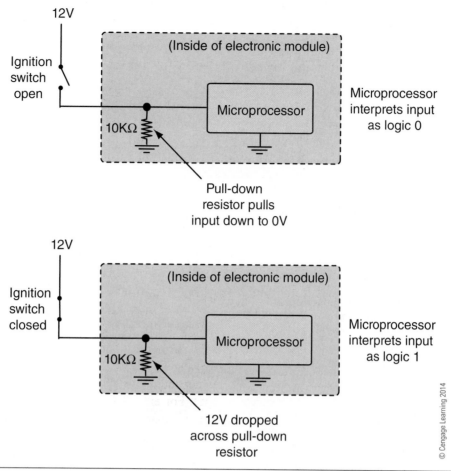

Figure 11-53 Pulled-down digital input measuring key off and key in ignition position.

the series resistance provided by the 10kΩ pull-up resistor. The switch and the pull-up resistor become series voltage divider circuits. The switch has near infinite resistance when open and near zero resistance when closed. The pull-up resistor causes the voltage at the input terminal to be nearly the supply voltage value (12V in this case) or nearly 0V depending on the switch being opened or closed. A pulled-up digital input may also be referred to as an **active-low** input.

Analog Input Examples. Analog inputs are capable of determining a specific voltage level at the input within a predefined range of voltages. Analog inputs are typically connected to a device that has a range of resistance or devices that supply a variable voltage. An example of a device that would be connected to an analog input is an electronically controlled diesel engine accelerator-position sensor. This sensor is often a potentiometer and is used to measure the amount the truck operator is depressing the accelerator pedal. Modern electronically controlled diesel engines are "drive by wire" with respect to the accelerator. There is no cable between the accelerator and the fuel system, which controls how much fuel is to be injected.

The output voltage of the accelerator position sensor is used to determine how much power the driver is requesting from the engine. Using a digital input for the accelerator would not be a good idea because the engine would respond with low idle-only fuel or maximum fueling, with nothing in between. This would make the truck very difficult to drive. Therefore, an accelerator-position sensor would be connected to an ECM analog input instead of an ECM digital input. An analog input would permit the ECM, to determine the accelerator position based on the voltage measured at the analog input as shown in **Figure 11-55**. This input is pulled-down because the output of the accelerator position sensor (APS) is a variable positive voltage. The microprocessor in the engine ECM could be programmed to "know" that a voltage of 1V means low idle, while a voltage of 4V means the accelerator is pressed to the floor.

Note the 5V reference voltage shown in **Figure 11-55**. In order for the ECM to accurately determine accelerator position, the voltage supplied to the accelerator position sensor must be precisely 5V. The accelerator position sensor shown in **Figure 11-55** is a potentiometer. Supplying the potentiometer with 6V instead of 5V

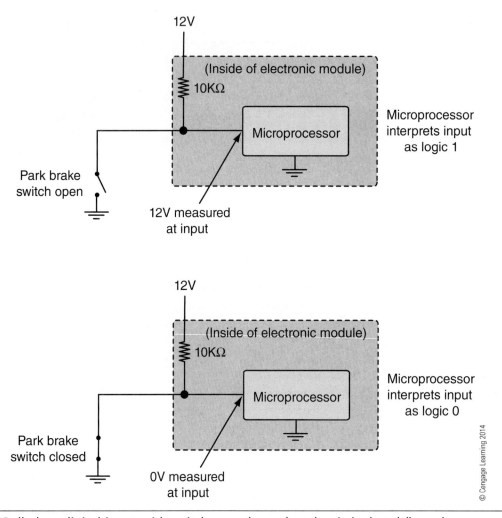

Figure 11-54 Pulled-up digital input with switch open (upper) and switch closed (lower).

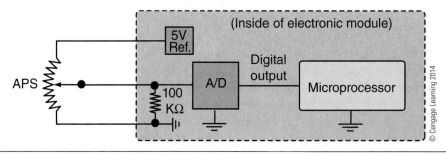

Figure 11-55 Accelerator position sensor connected to pulled-down analog input.

will cause the voltage measured at the ECM input terminal to be higher than it would be if the sensor were supplied with 5V as a reference voltage. This will cause the ECM to "think" that the accelerator is partially depressed, even though the truck operator does not have his foot on the accelerator. Conversely, if the reference were 4V instead of 5V, the ECM would not detect the accelerator being fully depressed when the truck operator is requesting maximum engine power. An electronic voltage regulator within the electronic module is responsible for supplying a precise sensor reference voltage (typically +5V) which is vital for accurate sensor measurements.

Tech Tip: Wiring harness problems with three-wire sensor reference voltage and ground (return) circuits can result in sensor measurement errors. High levels of circuit resistance caused by terminal corrosion can result in misdiagnosis and unnecessary component replacement.

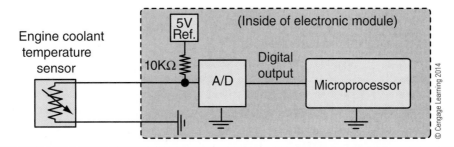

Figure 11-56 Engine coolant temperature sensor connected to pulled-up analog input.

An electronic module analog input could be designed to be pulled-up if the external device connected to the input terminal, such as a sensor, supplied a variable resistance path to ground. An example is a thermistor used to measure coolant temperature, such as the circuit shown in **Figure 11-56**. As the resistance of the thermistor decreases due to an increase in coolant temperature, the voltage measured at the input terminal of the module would also decrease accordingly. The microprocessor in the engine ECM can then apply the sensor transfer function information that is stored in the ECM program memory to determine the coolant temperature based on the voltage measured at its coolant temperature input terminal.

To summarize, there are four basic types of electronic module inputs:

- Pulled-up digital input
- Pulled-down digital input
- Pulled-up analog input
- Pulled-down analog input

Although a circuit diagram may not indicate what type of electronic module input the external circuit is connected to, you should be able to figure this out based on the type of device present (i.e., a switch or a sensor) and what is being supplied to the electronic module by the device (i.e., a positive voltage or a path to ground). Having this knowledge can be very useful in troubleshooting, as will be demonstrated in later chapters.

Electronic Module Digital Outputs. In **Chapter 6**, the concept of low side drivers and high side drivers was introduced. These drivers are transistors being used as an electronic switch. Low side drivers "sink" or provide a path to ground, meaning conventional current is flowing from the load into the electronic module. High side drivers "source" or provide a positive voltage, meaning conventional current is flowing out of the electronic module to the load. The control of the drivers is digital, meaning the driver is either off or on, like a switch. The programming instructions in the microprocessor cause the microprocessor to determine when to switch on and when to switch off each driver, based on the status of the microprocessor's inputs. One of several digital outputs of the microprocessor is used to act as the control signal for the driver, as shown in **Figure 11-57**. This driver shown is a low side driver. A digital 0 provided by microprocessor output #1 causes the low side driver *not* to provide a ground for the relay coil, similar to an open switch that could be used to control the relay. A digital value of 1 supplied by microprocessor output #1 causes the low side driver to switch on and sink a path to ground for the relay coil, similar to a closed switch. The decision to switch the low side driver on or off is based on the microprocessor's programming instructions. In this example, the relay for the on-off air fan clutch is being controlled by a low side driver. The microprocessor's programming instructions are designed to provide an output of 0 *not* to switch on the fan and a value of 1 to switch on the fan. The logic statement in **Figure 11-57** illustrates the microprocessor programming instructions for this output.

Electronic Module Analog Outputs. Microprocessors provide digital values as outputs. Many times, a digital value of 1 or 0 is fine if what is being controlled will be either on or off, such as a relay coil or a light. Otherwise, an integrated-circuit device called a digital to analog (D/A) converter is used to transform a digital value into an analog value. Analog outputs would be used by a CD player to produce the audio signal that drives the speakers to produce music.

A digital output can also be switched off and on rapidly by the microprocessor using PWM so that it has the appearance of being an analog output. For example, instrument panel illumination dimming can be controlled by switching a high side driver off and on several times per second using PWM to control the lamp intensity.

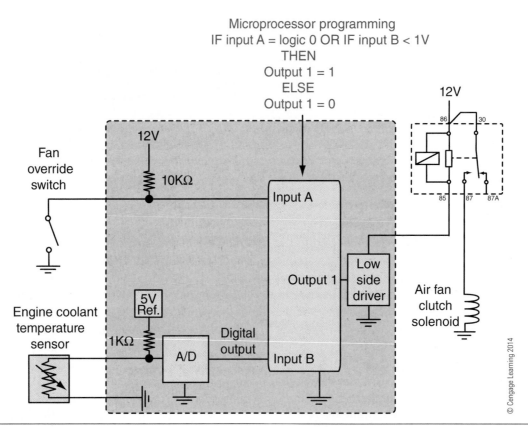

Microprocessor programming
IF input A = logic 0 OR IF input B < 1V
THEN
Output 1 = 1
ELSE
Output 1 = 0

Figure 11-57 Microprocessor controlling one of its outputs based on its input values and programming instructions.

MULTIPLEXING

The term **multiplexing** refers to methods used to combine more than one channel of information into a common signal path. Multiplexing is not a new idea, having been used in the communications industry for many years. There are many different types of multiplexing. The most common forms of multiplexing used in truck electrical systems are analog multiplexing and time-division multiplexing.

Analog Multiplexing

An example of **analog multiplexing** in truck electrical systems is the use of a group of switches wired in parallel that are connected to a single analog input of an electronic module. The example shown in **Figure 11-58** illustrates a pulled-up analog input that is connected to a group of parallel-connected switches. Each switch has a different value of series resistance and each switch corresponds to a different type of action. Therefore, closing only the switch connected to the 1kΩ resistor in **Figure 11-58** will cause the voltage measured at the analog input to change from 12V to 6V as the pull-up resistor in the electronic module and 1kΩ resistor in series with the switch form a series

voltage divider circuit. The microprocessor would respond to this voltage level at this input by causing some specific action to occur, based on its programming instructions. Opening the switch connected to the 1kΩ resistor and closing the switch connected to the 2kΩ resistor will cause the voltage level at the input to increase from 6V to 8V. The microprocessor could be programmed to cause some other action to occur due to the change in voltage measured at the input.

The single analog input is being used for four different purposes in the example shown in **Figure 11-58**. The microprocessor could cause four different actions to occur based on which switch has been closed. These actions could be something as simple as switching on four different lights or something more complicated. One use of analog multiplexing in modern trucks is the four steering-wheel mounted cruise control switches. The rotation of the steering wheel requires the use of a clockspring to provide electrical continuity as the steering wheel is rotated. By reducing the number of wires necessary to connect to four different cruise control switches (on, off, set, resume) from five to two (including the ground), the number of circuits in the clockspring can be reduced. This also reduces the number of wires in the wiring harness.

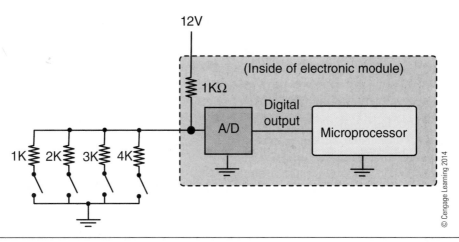

Figure 11-58 Analog multiplexing.

Time-Division Multiplexing

Time-division multiplexing used in modern trucks can be described as the time-sharing of a pair of wires. Like a vacation home time-share, several different electronic devices use the pair of wires to transmit their information at different times.

The simplest example of time-division multiplexing is the use of a pair of rotary switches that are rotated to the same position at the same time (synchronized). The example in **Figure 11-59** permits the output of four different variable resistance sensors to use one pair of wires to connect to four different analog inputs of an electronic module. Both of the rotary switches in this example are moved to position A. The microprocessor takes a reading on input A and the switches are then both moved to position B. The microprocessor then takes a reading on input B and the switches are moved to position C, then position D, and back to position A, and so forth. If the rotary switches are both rotated very quickly in unison with each other, then it almost appears that all four sensor values are being measured almost continuously by the microprocessor through the use of two wires. The pair of

wires is being time-shared by the sensors and the microprocessor inputs.

Rotary switches are not actually used in time-division multiplexing but are just being used to present the concept of time-division multiplexing.

Serial Data Communications. The serial and USB ports found on a PC are a form of serial data communication. The 1s and 0s are transmitted one after the other to communicate digital data between the PC and some other device plugged into a USB port.

The use of digital serial data communications is common in modern truck electrical systems to permit the various electronic modules on the truck to communicate with each other. Serial data communication is a form of time-division multiplexing. A pair of wires is typically used as the communication means. This pair of wires is often referred to as a **data link**. Even though this terminology is not technically correct, the term *data link* will be used in the remainder of this text for a serial communications network because the term has widespread usage in the industry to describe the pair of wires used to communicate information in a

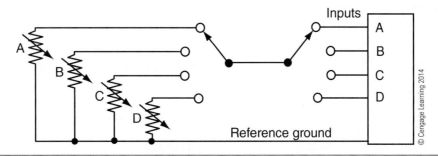

Figure 11-59 Time division multiplexing concept. Rotary switches with four sensors connected to four analog outputs. Switches are simultaneously placed in the same position for a brief period of time, permitting two wires to carry the information of four different inputs.

digital format. The various modules on the truck take turns using this common pair of wires connected to each module. Each module will transmit (place) 1s and 0s on this common data link while the other modules connected to the pair of wires receive or "listen" to this digital data. The 1s can be a voltage value measured across the two wires that is above some level, while the 0s can be a voltage value measured across the two wires that is below some level. The digital data is transmitted and received serially or one bit after the other.

Only one module at a time may transmit information on the data link, but all modules can receive this information at the same time. This is like being in a meeting with several people present. Only one person at a time can speak or no one will be able to understand what is being said. The digital data is several bits in length and includes information such as which module is sending the information, what type of information will be transmitted, and the actual information such as the engine oil pressure.

The engine ECM may be connected to the automatic transmission ECU and the ABS ECU on a truck through a serial communications data link. The engine ECM has direct access to a great deal of information such as engine oil pressure and coolant temperature obtained through its own hardwired sensors. This information can be useful to other modules in their decision-making processes. For example, the information that the engine ECM receives from the accelerator-position sensor is a necessary piece of information for the automatic transmission controller's decisions about when to make the next shift. Instead of hardwiring the accelerator-position sensor to both the engine and the automatic transmission control module, the engine ECM can share the accelerator position information with all the modules on the truck using a data link.

The information being transmitted on the data link may not be useful for all modules. Data that is not needed by a module will be ignored. This is like sitting in a meeting (or classroom) where information that you do not care about is being presented. You will probably not listen very attentively until the topic of the meeting shifts to something that pertains to you personally or the classroom instructor indicates that the material will definitely be on the test.

The digital information that is transmitted onto the data link in modern trucks is defined by a Society of Automotive Engineers (SAE) standard so that each electronic module, regardless of who manufactured the device, is able to understand the information being transmitted on the data link. The standard is like a common language that is understood by all modules.

SAE J1587/J1708 Standard

For many years, SAE **J1587** was used as the message content standard or protocol used in truck serial data communications. The J1587 messages have a bit transmission rate of 9600 bits per second (bps).

SAE **J1708** is the standard that defines the hardware, including the physical data link wiring used with the J1587 protocol. Together, the J1708 and J1587 standards are often referred to as the ATA data link. The J1708-defined data link consists of a twisted pair of copper wires covered with a standard wire insulation material. The wires are twisted together to reduce the effects of electromagnetic interference (EMI). EMI will be discussed in more detail in later chapters. The two data link wires are typically identified as + and − by most OEMs. However, neither wire is directly connected to a positive or negative voltage source.

An interim message protocol standard that also uses the SAE J1708 hardware is the SAE J1922 protocol. The J1922 protocol is similar to the J1587 protocol.

SAE J1587 Message Structure. Messages transmitted over J1587/J1708 first indicate the major system or device that is sending the information, called the **message identification (MID)**. SAE has assigned an industry-wide three-digit MID for each type of major system that is connected to the J1587/J1708 data link. For example, the engine ECM is MID 128 and the transmission ECU is MID 130.

Next, a **parameter identifier (PID)** is transmitted. A PID is an SAE assigned number indicating the specific information, such as PID 110 indicating engine coolant temperature. After the PID, the actual data such as the engine coolant temperature is transmitted in a coded format. After the PID and its data, additional PIDs and their associated data could be transmitted within the same message, up to maximum message length of 21 bytes. Any compatible electronic device throughout the industry that is connected to the J1587/J1708 data link is capable of decoding the MID, PID, and the actual data.

Following the MID and PIDs, a checksum value is transmitted. The checksum is used for error checking to ensure that all devices have received the data correctly. If a device indicates an error, the information is retransmitted.

SAE J1939 Standard

The J1587/J1708 serial communications standard, along with J1922, has mostly been replaced by SAE **J1939**. The bit transmission rate of J1939 is 250k or 500k bps compared to the 9600 bps rate of J1587/J1708.

This indicates that many more messages can be transmitted and received in the same amount of time with a J1939 data link system compared to J1587/J1708. The difference in data transmission rates between J1939 and J1587/J1708 is similar to the difference between the fastest available cable modem and dial-up Internet service.

J1939 Specification Cable. The relatively low data transmission rate of the J1587/J1708 data link does not require much special attention other than twisting the two conductors together. However, the high data transmission rate of the J1939 data link requires some special consideration to provide reliable communications. Unlike J1587/J1708, both the protocol and the hardware, such as the physical data link cable, are covered under sections of the same J1939 standard. Two different versions of cable are defined by SAE J1939. Both versions of cable use a twisted pair of copper wires but with a special plastic insulation. The special plastic insulation maintains a specific value of capacitance between the two wires of the data link. The two copper wires are like the two plates of a capacitor. The plastic insulation material is the dielectric between the plates of this capacitor. Twisting of the wires together limits electromagnetic interference and susceptibility, but it also maintains a specific distance between the two data link wires. Recall that the distance between the plates of a capacitor is one of the factors that determine the value of capacitance.

Together, the insulation material and the distance between the two conductors cause the capacitance between the two wires of the data link to be some specific value. This is important for reliable high-speed data communications. An example of high-frequency wiring with which you may be familiar is satellite or cable television coaxial cable. This coaxial cable has a center copper conductor surrounded by a special insulation material. A wire mesh surrounds this insulation material. The center copper conductor is one conductor; the wire mesh is the other conductor. The insulation material surrounding the center conductor is the dielectric between the two plates of the capacitor (the two conductors). This causes a specific amount of capacitance between the center conductor and the wire mesh. Again, a specific value of this capacitance is very important for good TV reception.

If you have ever tried to substitute plain insulated wire for this special coaxial cable, you know that the quality of the picture suffers. The reasons for this are beyond the scope of this book. However, you may have noticed that the coaxial cable is referred to as 75Ω cable. The resistance of the wire or the resistance between the two conductors is not actually 75Ω. This resistance value is describing what is known as the characteristic **impedance** of the cable.

Impedance is like resistance but also includes the effect that alternating current has on inductors and capacitors. Inductors and capacitors react differently to alternating current than they do to direct current due to the energy storage capabilities of inductors and capacitors. An ohmmeter cannot measure this impedance because an ohmmeter only supplies a known value of DC voltage and measures the amount of DC current passing through the unknown resistance, then calculates the resistance using Ohm's law. The resistance between the two conductors of the coaxial cable is infinitely high because the plastic insulation material is an insulator. However, the coaxial cable has some specific value of impedance. Impedance, like resistance, is also measured in ohms. This is the reason that coaxial cable has a 75Ω rating, even though the resistance between the two conductors is near infinity. The characteristic impedance for the transmission media (cable) becomes more important as the frequency of the signal increases.

Do not worry if you do not understand the concept of characteristic impedance. Just realize that you cannot connect your ohmmeter across the conductors of coaxial cable rated at 75Ω and measure a resistance of 75Ω. It may be helpful to think of impedance as being "AC resistance" and the resistance that you can measure with a DMM ohmmeter as being "DC resistance."

The cable used in a J1939 data link has a 120Ω characteristic impedance. This 120Ω characteristic impedance cannot be measured with an ohmmeter in the same way that the 75Ω characteristic impedance of coaxial cable cannot be measured using an ohmmeter.

Two different J1939 standards define the data link cable. The simplest type of cable is defined by SAE standard J1939/15. This standard describes a twisted pair of insulated wires with a special insulation material. The twisted pair is then covered with a thin plastic cable jacket. The jacket of cable that meets the requirements of SAE J1939/15 standard is marked with lettering indicating SAE J1939/15.

The other standard is SAE J1939/11. This standard also describes a twisted pair of wires. However, this standard includes a metal foil covering over the twisted pair called a **shield** (**Figure 11-60**). A third noninsulated copper wire contacting the shield acts as a drain wire. The drain wire is connected to ground at one point in the network through a capacitor, typically within one of the electronic modules. The whole assembly is also covered with a plastic jacket. The purpose of the shielding is to provide immunity to

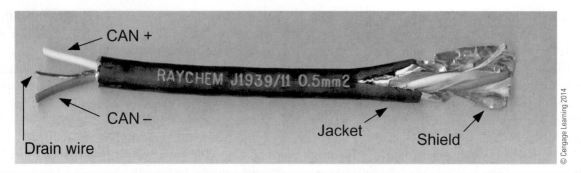

Figure 11-60 J1939/11 shielded cable.

electromagnetic interference. Simply put, the shielding prevents stray electric fields, such as those generated by a CB radio, from causing information on the data link to be corrupted. The jacket of cable that meets the standards of SAE J1939/11 is marked with lettering indicating SAE J1939/11.

Some OEMs use combinations of J1939/11 and J1939/15 on the same truck. The data link inside the cab may be J1939/15 while the data link outside the cab is J1939/11 shielded cable.

SAE J1939 cable is specified to have yellow insulation on one wire and green insulation on the other wire. The yellow wire is designated CAN+ or CAN HIGH while the green wire is designated CAN− or CAN LOW. CAN stands for controller area network, which is a type of communications network as explained later in this section.

Tech Tip: Because of the high rate of data communications transmitted over the J1939 data link, specific OEM instructions should be followed when repairs are performed on the data link wiring. Also, never substitute ordinary wire for J1939 data link cable because the plastic wire insulation material is not the same thickness or dielectric as J1939 specified cable and could result in communication problems.

Network Topology. Topology refers to the manner in which a computer network is physically laid out. The J1939 network is defined as a **bus** topology. The term *bus* is used to describe a single communications line shared by several devices. In the case of SAE J1939, a single data link cable (twisted pair of wires) is the shared communications line that routes throughout the vehicle as shown in **Figure 11-61**. This single cable is known as the **backbone**. Each module is connected to the backbone by a section of cable less than 1 m in length called a **stub**. The stubs are spliced into the backbone near each electronic module throughout the vehicle.

Terminating Resistors. You have probably had trouble understanding an announcer at an outdoor sporting event because of multiple loudspeakers at various distances from you causing the sound to arrive at your ears at different times. For similar reasons, the J1939 data link requires a 120Ω **terminating resistor** at each end of the backbone, or the use of special electronic terminating bias circuits. Terminating bias circuits currently have limited usage in on-road truck J1939 networks. Therefore, only terminating resistors will be addressed in this text. The purpose of the terminating resistor is to prevent signal reflections from interfering with the data communications. Reflections can be thought of as being like echoes. The terminating resistors are not the same as the 120Ω characteristic

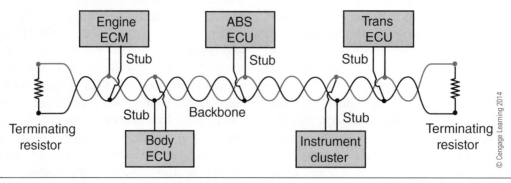

Figure 11-61 J1939 is classified as bus topology.

impedance of the data link but are real, ordinary resistors that can be measured using an ohmmeter. The resistors are connected across the CAN+ and CAN– circuits at the two ends of the J1939 data link backbone. These resistors absorb signal energy instead of permitting the signal energy to bounce back into the data link like an echo and corrupt the signal.

One important thing to know about the terminating resistors is that typically no communication between electronic devices can occur over a J1939 network without at least one of these terminating resistors being present across the data link, as illustrated in **Figure 11-62**. This illustration shows the voltage measured between CAN+ and CAN– using an oscilloscope. Time is shown on the horizontal axis and the voltage difference between CAN+ and CAN– is shown on the vertical axis. With no terminating resistor connected across the data link, the signal reflections result in no bus communications being possible. If just one terminating resistor is present, communication is possible, although there may be intermittent J1939 communications problems due to signal reflections.

Terminating resistors have a second function besides reducing signal reflections. Terminating resistors also provide an electrical load across CAN+ and CAN–, similar to a pull-down resistor as described in the previous section. The top oscilloscope trace shown in **Figure 11-62** indicates that the voltage difference between CAN+ and CAN– slowly decays to 0V with no terminating resistors present. Compare this to the lower oscilloscope trace in **Figure 11-62** with both terminating resistors present where the signal quickly drops to 0V. The slow signal decay to 0V due to the missing terminating resistors results in devices connected to the bus misinterpreting the signal.

Controller Area Network (CAN). A **controller area network (CAN)** describes a type of communications network in which there is no master controller. No single electronic module is responsible for controlling the network and there are no network controllers as there are in most computer networks. Instead, each electronic module in a CAN contains a communications chip called a CAN transceiver that is designed specifically to communicate with CAN transceivers in

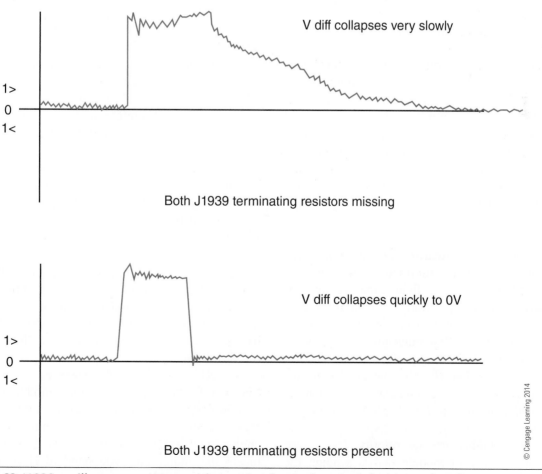

Figure 11-62 J1939 oscilloscope patterns with terminating resistors missing (upper) and present (lower).

other electronic modules that are connected to the same communications network. None of these modules is in charge of the network.

The CAN transceivers located in the various electronic modules on a truck—such as the engine ECM, the automatic transmission ECU, and the ABS ECU—all take turns using the J1939 data link to communicate specific information. This information includes vehicle speed, engine speed, coolant temperature, oil pressure, and many other pieces of information. When one module has finished using the data link to transmit a message, another module takes over and uses the same data link to transmit its message. In a modern truck even with a large amount of data link "traffic," the J1939 data link is sitting idle with no messages being communicated more than 50 percent of the time. The transfer of information over the J1939 data link occurs so fast that it appears that all of the modules are transmitting on the data link at the same time.

What is actually happening in a CAN network is that each module is patiently waiting a turn to begin transmitting its information. The high rate of data communications on J1939 means that the time that a module must wait for another module to finish communicating is very brief (microseconds). However, a method of determining which message is the most important if two or more modules begin to communicate at the same time (known as arbitration) is a part of the J1939 specification. If two controllers begin transmitting their messages at the exact same time, the message with higher importance is communicated first and the controller with a less important message will stop transmitting and will wait for another turn. These communication rules for determining message importance have been programmed into each CAN transceiver so no module needs to be in charge of the network.

SAE J1939 Message Structure. A J1939 message contains many pieces of information, as shown in **Figure 11-63**. The arbitration field at the beginning of the message contains the priority, which is used to determine which message is more important if two or more controllers begin communicating at the same time. The arbitration field may also contain the **parameter group number (PGN)**. The PGN indicates what type of data the message contains, such as engine speed related information (PGN 61444) or engine temperature related information (PGN 65262). Lower PGNs indicate a higher message priority. The **source address** for the type of module that is sending the information also comprises the arbitration field.

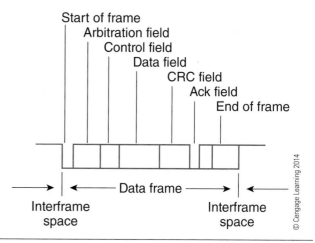

Figure 11-63 J1939 message format.

For example, the engine ECM is assigned a source address of 0 by the SAE J1939 specification while the automatic transmission ECU is assigned a source address of 3. Lower source address numbers indicate a higher priority of the devices; therefore, the engine ECM is assigned a source address of 0.

After the arbitration field, a control field indicating the length of the message is transmitted. After the control field, the actual data is transmitted. Typically, 64 bits of each message are the actual data contained in parameter packages known as a **suspect parameter number (SPN)**. An SPN is a number assigned by SAE to each parameter or measurable factor, such as SPN 110, engine coolant temperature, or SPN 175, engine oil temperature. Several related SPNs are grouped together as a PGN. Therefore, a J1939 message for PGN 65262 contains SPN 110 and SPN 175, along with four other SPNs related to engine temperature. A J1939 SPN is similar to a J1587 PID. Whenever possible, both utilize the same number for a common parameter, such as engine coolant temperature being represented by PID 110 in a J1587 message and SPN 110 in a J1939 message.

Lastly, J1939 messages contain a method of verifying that all electronic modules connected to the J1939 data link have received the same message called the cyclic redundancy check (CRC). If any module on the J1939 data link indicates that an error has occurred in receiving the message, the message is retransmitted.

Altogether, a typical J1939 message contains 128 bits. The time for one electronic control module connected to a 500k bps CAN bus to broadcast a typical 128-bit message, such as a message containing PGN 65262 engine temperature information, is 256 μsec. This means that more than 3900 standard J1939 messages per second could theoretically be transmitted and received by any electronic module that is connected to a 500k bps CAN bus. This large number of messages per second makes it appear that several messages are being sent at the same time, although each electronic module is just borrowing or time-sharing the two wires for 256 millionths of a second.

The actual 1s and 0s observed on a J1939 CAN data link are determined by the difference in voltage between the CAN+ and CAN− circuits. **Figure 11-64** illustrates what a two-channel (two-input) oscilloscope with one channel connected between CAN+ and ground and a second channel connected between CAN− and ground might indicate. The vertical axis in **Figure 11-64** indicates the voltage referenced to ground while the horizontal axis indicates time. If the difference in voltage between CAN+ and CAN− is approximately 0V, then the bit is logic 1 (recessive). If the difference in voltage between CAN+ and CAN− is approximately 2V, then the bit is logic 0 (dominant). This may seem backward, but all the devices connected to the J1939 data link know the rules. The binary value shown in **Figure 11-64** would be interpreted as 101. If this were a J1939 data link operating at 500k bps, then the time for one bit would be 2 µsec. Therefore, only 6 µsec of activity is shown across the horizontal axis in **Figure 11-64**.

The high rate of data communication possible over the J1939 data link has had a significant impact on modern truck electrical systems. The use of multiplexing has extended into several areas of the modern truck, as will be shown in the chapters that follow. Additionally, other types of communications networks utilized on modern trucks will be introduced.

Other Types of Digital Communication

In addition to the J1939 and J1708 data links, other more primitive forms of digital communication are commonly found in truck electrical systems. Up to this point, PWM has only been discussed as a means to control the average current by switching a DC voltage off and on to provide lamp dimming or alternator voltage regulation. PWM is also a form of digital communication. For example, on some engines the ECM controls the variable geometry turbocharger vane position by modification of the duty cycle of a single-wire PWM control signal supplied to the turbocharger actuator. This control signal is referenced to ground. A duty cycle of 10 percent of this control signal could indicate that the ECM is commanding the turbocharger actuator to drive the vane position to 10 percent of the maximum position. The PWM signal is only a means of communication in this case and is not a variable voltage supply for the device. Variable geometry turbochargers are discussed in **Chapter 14**.

The universal asynchronous receiver/transmitter (UART) is another type of simple digital communication. Asynchronous transmission refers to the transmission of data without the use of a digital pulse train called a clock signal between the two devices. Instead, start bits are used at the beginning of each digital message packet, which synchronizes the two devices. UART typically only requires a single wire with voltage levels being referenced to ground.

LIN (local interconnect network) is a serial communication protocol with a maximum transmission

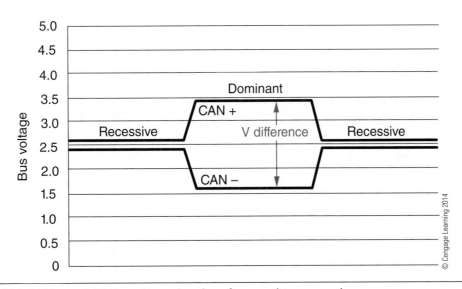

Figure 11-64 J1939 CAN+ and CAN− voltage levels referenced to ground.

rate of 20k bps. Unlike a CAN network, one device is the master device and initiates all communication while the other devices merely respond when commanded. A LIN network may be used as a means for the engine ECM, the master device, to receive information from remote sensor modules. Typically, only one wire is necessary for LIN communications, with voltage levels being referenced to ground.

SENT (single edge nibble transmission) is a low-cost one-way serial communication protocol used to permit sensors and other devices to send information to an ECM or ECU. The term *nibble* describes a group of four binary bits, or one-half of a byte.

Troubleshooting the Multiplexed Truck

In **Figure 11-65**, a two-channel oscilloscope trace of a J1939 message is shown. One oscilloscope channel is connected between CAN+ and ground, and the other oscilloscope channel is connected between CAN− and ground. The acknowledge bits shown in **Figure 11-65** indicate an end of the message. The amplitude of the acknowledge bit is greater than the other dominant bits in the message because all controllers connected to the bus are transmitting this bit at the same time indicating that they have all received the message. **Figure 11-65** illustrates a typical waveform pattern you would expect to see using a two-channel

oscilloscope to view activity on the J1939 data link. Note how the signal quickly rises and falls, unlike the signal shown in **Figure 11-62** where the terminating resistors were missing. Oscilloscopes can be useful tools in diagnosing electrical problems with the J1939 data link. More information on troubleshooting electrical problems with the J1939 data link such as shorts to ground or open circuits will be discussed in **Chapters 14 and 15**.

Although it might be possible to interpret each bit of a message displayed on an oscilloscope connected across the CAN+ and CAN− terminals to determine what information a module is communicating, this would be very difficult to perform and would be very time consuming. Therefore, the tool of choice when troubleshooting much of the electrical system on modern trucks is a PC and a special communication adapter device designed to permit the computer to be connected to the truck's data links as shown in **Figure 11-66**. The computer is connected to the truck's J1939 or J1587/J1708 data link via a standardized diagnostic connector on the truck. This permits the computer to interface or communicate with the various electronic modules on the truck. A 6-pin diagnostic connector shown in **Figure 11-67** is used on older trucks that only have a J1587/J1708 data link. Modern trucks with J1939 have a 9-pin diagnostic connector as shown in **Figure 11-68**. The color of this 9-pin diagnostic connector is specified as gray or black for 250k bps J1939 networks. A green-colored

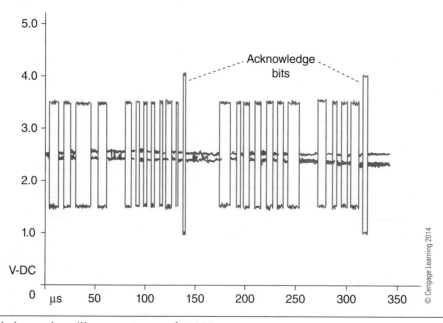

Figure 11-65 Dual-channel oscilloscope trace of J1939 messages.

Personal computer (PC)

Adapter cable

Adapter cable

Adapter cable

Communication adapter

Cable adapter

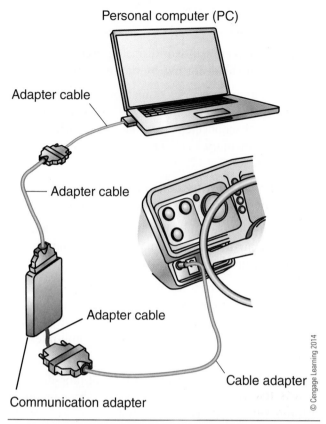

Figure 11-66 Communications adapter provides interface between the vehicle J1939 CAN bus and PC.

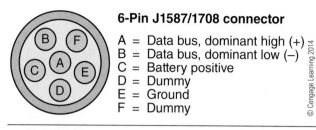

6-Pin J1587/1708 connector

A = Data bus, dominant high (+)
B = Data bus, dominant low (–)
C = Battery positive
D = Dummy
E = Ground
F = Dummy

Figure 11-67 6-Pin J1587/J1708 diagnostic connector pin assignments.

9-pin diagnostic connector may be used by some OEMs to identify 500k bps J1939 networks. This identification of the J1939 transmission rate is necessary because attempting to communicate at the wrong transmission rate will cause all CAN communications to cease.

Special software packages provided by the engine, automatic transmission, and ABS system manufacturers provide troubleshooting assistance for these complex systems. Collectively, the PC, communications adaptor and the software packages may be referred to as an **electronic service tool (EST)**. Some examples of ESTs will be introduced in later chapters.

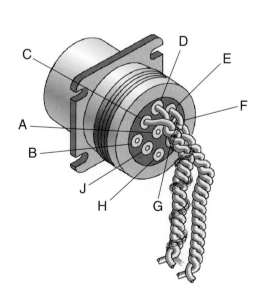

Front view
of connector

Cavity	Label
A	Battery ground
B	+12V DC
C	J1939 data link (+)
D	J1939 data link (–)
E	J1939 shield
F	J1587 data link (+)
G	J1587 data link (–)
H	Plug
J	Plug

Figure 11-68 9-Pin diagnostic connector pin assignments.

Summary

- A modern truck may have several different types of sensors used to measure temperature, pressure, speed, and position.

- A variable reluctance sensor is a small version of an AC generator. These types of sensors are often used to measure a rotational speed such as an ABS wheel speed sensor.

- The Hall effect describes the generation of voltage in the presence of a magnetic field. The voltage that is generated is proportional to the magnetic field strength. Unlike electromagnetic induction, no magnetic lines of force have to be cut to generate this voltage.

- An analog signal is a signal that can be any level within a range. A digital signal can only be one of two different values.

- Logic gates are hardware devices used to make a decision. There are several types of logic gates. A truth table describes the decision made by the logic gate to all the possible inputs.

- A microprocessor is an integrated circuit that contains thousands of logic gates. The microprocessor only accepts digital information and provides results in digital form. An analog to digital converter is used to convert real-world analog signals into the digital equivalent for use by the microprocessor.

- The two main categories of memory are volatile and non-volatile. Volatile memory is lost when electric power is disconnected from the device while non-volatile memory is retained thorough power disconnection.

- The internal circuitry in an electronic module input typically has a pull-up resistor or pull-down resistor.

- Electronic module inputs can be digital or analog types of inputs. These digital or analog inputs can be pulled-up or pulled-down.

- Multiplexing refers to methods used to combine more than one channel of information into a common signal path. Common forms of multiplexing used on modern trucks include analog multiplexing and time-division multiplexing.

- Serial data communication is a form of time-division multiplexing. An example of serial data communication is the SAE J1939 data link.

- A controller area network (CAN) is a communications network without a master control device.

Suggested Internet Searches

Try these web sites for more information on CAN, SAE J1939, sensors, and electronics:
http://www.vector.com
http://www.sae.org
http://www.ngk-detroit.com
http://www.sensata.com
http://www.wema.com
http://www.conti-online.com

Review Questions

1. Which of the following is a true statement?

   A. A variable reluctance sensor is a Hall effect device.

   B. Air gap refers to the space between the center electrode and ground electrode in a diesel engine spark plug used in the combustion chamber to ignite the diesel fuel.

   C. The frequency of the AC voltage generated by a variable reluctance sensor increases as the speed of the target passing in front of the sensor increases.

   D. The amplitude of the AC voltage generated by a variable reluctance sensor decreases as the speed of the target passing in front of the sensor increases.

2. How would a thermistor with an NTC respond to an increase in temperature?

   A. The voltage produced by the sensor would increase.

   B. The voltage produced by the sensor would decrease.

   C. The resistance of the sensor would decrease.

   D. The resistance of the sensor would increase.

3. A potentiometer would probably be used to measure which of the following?

   A. Engine coolant temperature

   B. Pre-2007 diesel engine intake throttle position

   C. Vehicle speed

   D. Accelerator position

4. Which of the following would be considered digital?

   A. Mechanical clock

   B. Sound of a single guitar string being plucked

   C. Phonograph record

   D. Wall switch used to control an electric light

5. The odometer information in an electronic engine ECM would probably be stored in which type of memory?

   A. Non-volatile memory

   B. RAM

   C. Flash memory

   D. Floppy disk memory

6. An electronic module input terminal that is internally pulled-down and is connected to a switch that is open would probably detect approximately how much voltage at that input terminal?

   A. Near battery positive voltage level

   B. About 0V

   C. Depends if switch is normally open or normally closed

   D. A negative voltage level

7. What is the difference between a digital input and an analog input?

   A. A digital input is always pulled-up and an analog input is always pulled-down.

   B. A digital input is typically connected to an on-off type of switch and an analog input is typically connected to something for which a range of voltages must be measured, such as a sensor.

   C. Digital inputs are used with microprocessors; analog inputs are used with transistors.

   D. Both types of inputs are the same.

8. Which module is in charge of a CAN used on a modern truck?

   A. Jefe module

   B. Router module

   C. Whichever module is larger

   D. No single module is typically in charge of a controller area network

9. Which best describes data communications with a J1939 specification CAN?

   A. Modules are assigned a priority. When a module with a higher priority has information to communicate, all message traffic on the data link stops in the middle of a message so that the highest-priority module can communicate.

   B. Any electronic module can communicate when the data link is not being used by another module.

   C. The engine ECM controls the network and decides which modules it wants to permit to communicate.

   D. All modules are simultaneously communicating information.

10. Analog multiplexing describes which of the following?

   A. Use of a single digital input to measure several different analog voltages

   B. The use of several analog inputs to measure several different digital devices at the same time

   C. Modulating a high-frequency signal onto a pair of wires

   D. The use of an analog input connected to several parallel-connected switches, each of which has a different value of series resistance

11. Where are the terminating resistors typically located on a J1587/J1708 serial communications network?

   A. At the extreme ends of the network

   B. In the exact center of the network

   C. In the hub or router

   D. External terminating resistors are not typically used on J1587/J1708 networks

12. Which conductor has green wire insulation in a J1939/11 data link cable?

   A. CAN neutral

   B. CAN + or CAN H

   C. CAN ground

   D. CAN – or CAN L

13. A two-input AND gate is most like which of the following?

   A. Two switches wired in parallel

   B. Two switches wired in series

   C. Two switches wired in a series-parallel configuration

   D. An SPST switch

14. A two-input OR gate is most like which of the following?

   A. Two normally open switches wired in series

   B. Two switches wired in parallel

   C. Two DPST switches

   D. Two switches wired in a series-parallel configuration

15. The output of a variable reluctance sensor is which of the following?

   A. Variable amplitude AC voltage

   B. Variable frequency AC voltage

   C. DC voltage

   D. Both A and B

12 Instrumentation

Learning Objectives

After studying this chapter, you should be able to:

- List the various types of electromechanical gauges and explain the operation of each type.
- Describe the four types of magnetic gauges.
- Explain the purpose of a bucking coil in a three-coil gauge.
- Modify the vehicle speed pulses-per-mile setting.
- Troubleshoot a problem with conventional instrumentation.
- Explain the concept of multiplexed instrumentation using J1587/J1708 or J1939 data link.
- Describe the operation of a stepper motor used in an instrument panel cluster.
- Diagnose multiplexed instrumentation using OEM information.

Key Terms

bimetallic gauge

breakout box

breakout T

bucking coil

DIP switch

instrument panel cluster (IPC)

instrument voltage regulator (IVR)

ISO symbol

liquid crystal display (LCD)

magnetic gauge

programmable parameter

pyrometer

sender

sending unit

stepper motor

INTRODUCTION

Trucks have a variety of gauges, indicator lamps, and audible alarms to keep the truck operator informed of the status of various truck systems. Collectively, these devices are referred to as instrumentation. Most OEMs place a group or cluster of these gauges and lamps into a package called an **instrument panel cluster (IPC)**. However, OEMs may utilize proprietary terms for their instrument panel cluster, as will be discussed in **Chapter 13**.

In the past, many gauges found on a truck were mechanical or electromechanical-type gauges. Most trucks now use multiplexed electronic instrumentation.

This chapter is divided into two main sections. The first section discusses conventional non-multiplexed instrumentation. The second section discusses multiplexed instrumentation and stepper-motor-driven gauges.

CONVENTIONAL INSTRUMENTATION

Conventional instrumentation describes non-multiplexed IPCs in which each electrically controlled gauge and indicator lamp is individually hardwired to a sensor or switch. These conventional IPCs may also contain mechanical gauges such as air pressure gauges that require plumbing between the gauge and the pressure source.

The gauges found in a conventional IPC are typically analog gauges (**Figure 12-1**). The indicator needle in the gauge sweeps through a circular arc. This type of gauge is described as analog because the movement of the gauge is continuous, like the movement of the hands of a clock.

The two main categories of analog gauges are mechanical and electromechanical.

Mechanical Gauges

Mechanical gauges were utilized extensively in the past and may still be found on some modern trucks. The conventional Bourdon tube-type air pressure gauge is an example of a mechanical gauge. The air pressure gauges in the instrument panel display the pressure in the primary and secondary air systems. The air pressure indicator may be contained within one gauge housing on some trucks and have two separate needles, similar to the two hands of a clock, that are typically colored green (primary) and orange (secondary). Air lines must be routed from the air tanks to the pressure gauges. Some trucks may also have an air brake application gauge that indicates the pressure being supplied to the primary brake chambers.

Speedometers in trucks of the past were typically cable-driven gauges. A gear in the transmission or transfer case causes the cable to rotate at a rate corresponding to the vehicle speed. The speedometer needle responds to the rate of the cable rotation to indicate vehicle speed.

Mechanical speedometers have several drawbacks. These devices are often inaccurate. They also require a gearing change if rear-axle ratio or tire sizes are changed. Mechanical speedometers may also make it difficult to remove and install the IPC because of the cable connection. The cable must also be lubricated periodically to minimize needle pulsation.

Because diesel engines do not have an ignition system like that found on a gasoline engine, the tachometer on older mechanical diesel engines was typically a mechanical cable-driven gauge similar to a speedometer. The tachometer cable was typically driven from the fuel injection pump.

Electromechanical Gauges

Electromechanical gauges are electrically operated gauges that cause a mechanical movement of the gauge needle or pointer. Regardless of whether the gauge is used to indicate fuel level, coolant temperature, or oil pressure, the electromechanical gauge is really an indication of the level of electric current flowing through the gauge's coils. Most electromechanical gauges use a sensing device known as a **sender** or **sending unit** to transform some physical quantity such as fuel level into a variable resistance. The variable resistance caused by the sending unit is connected in series with the corresponding gauge's coils. The resistance of the sending unit controls the current flow through the gauge's coils, causing the analog gauge needle to indicate the corresponding oil pressure, coolant temperature, fuel level, or other quantity. There are two main types of electromechanical gauges: bimetallic and magnetic.

Bimetallic Gauges. **Bimetallic gauges** make use of a bimetallic strip, like that found in circuit breakers,

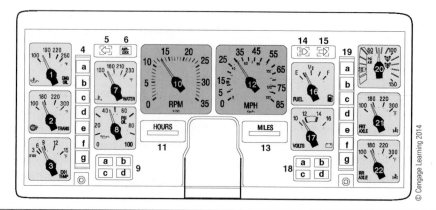

Figure 12-1 Conventional instrument panel cluster.

viscous fan drives, and other temperature-dependent devices. A bimetallic strip is made from two thin strips of dissimilar metals such as brass and steel that are bonded together (**Figure 12-2**). In this example, the two strips of metal are both the same length at 70°F (21°C). When the bimetallic strip is heated to 140°F (60°C), both the steel and the brass will expand. However, the amount of expansion that will occur for a given temperature is not the same for all metals. Brass will expand much more than steel when heated. When heated, both of the individual strips of dissimilar metal forming the bimetallic strip must become longer due to the expansion. The bimetallic strip is bent into an arc or section of a circle because of the greater length of the brass strip compared to the steel strip. The strip will bend in such a way that the brass material is on the outside of the bend to permit the brass strip to be longer than the steel strip.

Both metals also will not contract (shrink) the same amount when cooled. When the strip is cooled to 0°F (−18°C), the brass will contract more than the steel, causing the strip to deflect in the opposite direction.

The amount of curvature of the bimetallic strip that occurs is directly proportional to the temperature of the bimetallic strip. This permits the strip deflection to act as an indicator of strip temperature. To take advantage of the properties of a bimetallic strip, a coil of wire is wrapped around the bimetallic strip as shown in the fuel gauge example in **Figure 12-3**. The heat generated by the current flow through the heating coil causes the bimetallic strip or spring to bend into an arc. The more current that flows through the heating coil, the more heat that is generated in the heating coil, and the more the bimetallic strip bends. The gauge is designed to transform the bending of the bimetallic strip

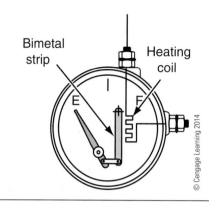

Figure 12-3 Bimetallic gauge.

into gauge needle movement. Thus, the current flow through the heating coil causes the bimetallic strip to heat, which causes the gauge needle to indicate the amount of current flowing through the heating coil.

The sending unit for bimetallic gauges is typically a variable resistance device. The example shown in **Figure 12-4** is a fuel level gauge circuit. The fuel level sending unit shown is a rheostat. A float in the fuel tank is mechanically connected to the rheostat wiper. This causes the rheostat resistance to change in accordance with fuel level (**Figure 12-5**). In this example, the resistance of the rheostat decreases as the tank is filled. This decrease in resistance causes an increase in current passing through the heating coil. The increase in current through the heating coil causes the bimetallic strip to deflect, resulting in movement of the gauge needle toward the full indication on the gauge face.

A bimetallic gauge depends on the current through the heating element surrounding the bimetallic strip to accurately display the quantity being measured. Therefore, changes in the system voltage such as engine running (14V) and engine not running (12.6V) would cause a large amount of inaccuracy in bimetallic gauges if the gauge heating coils were supplied with battery voltage. Ohm's law indicates that more current will flow through the heating coil when the engine is running than when the engine is not running because of the increase in system voltage with the engine running. This change in system voltage will result in gauge inaccuracy. To prevent fluctuating electrical system voltage from causing gauge inaccuracy, an **instrument voltage regulator (IVR)** is used to provide a constant voltage source to the gauge heating coils. In the past, IVRs used rapidly opening and closing contacts to generate a pulse width modulated (PWM) type of voltage that equated to an average voltage of about 5V DC (**Figure 12-6**). Modern IVRs are electronic devices with no moving components,

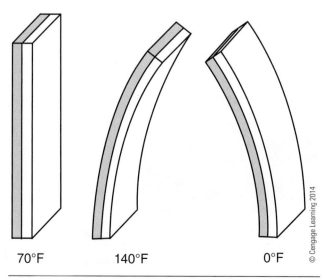

| 70°F | 140°F | 0°F |

Figure 12-2 Bimetallic strip made of brass and steel.

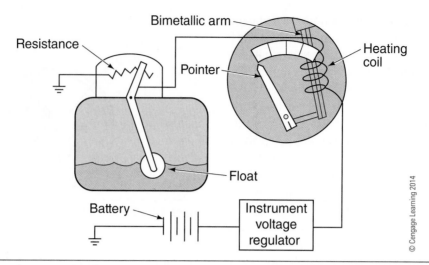

Figure 12-4 Fuel level sensor circuit.

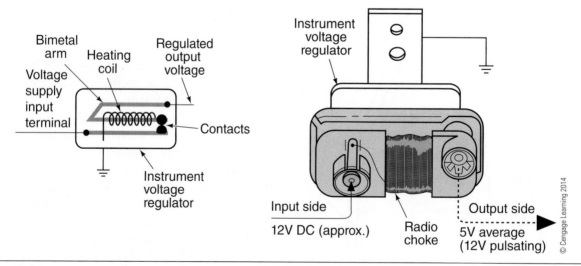

Figure 12-5 Fuel level sensor or sending unit.

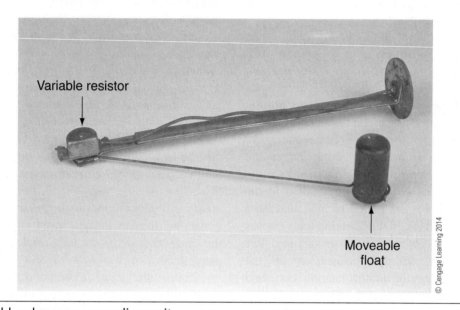

Figure 12-6 Instrument voltage regulator (IVR).

which supply a constant voltage to the gauge heating coils.

Because bimetallic gauges make use of a heating element, rapid changes in the quantity being measured—such as fuel level in the fuel tank on a bumpy road—do not cause rapid movement of the gauge needle. Bimetallic gauges may also take a minute or more when cold to accurately indicate fuel level or other quantity.

Magnetic Gauges. **Magnetic gauges** use the principles of electromagnetism to produce gauge needle movement. Conventional magnetic gauges have coils that are directly connected to the sensor or voltage source being measured. Modern magnetic gauges may use an electronic module in the IPC to control the voltage supplied to the gauge coils. The sensor acts as an input to the electronic module with these electronically controlled magnetic gauges. The module then controls the current through the gauge coils. These electronically controlled magnetic gauges will be addressed later in this section.

There are four main types of magnetic gauges: d'Arsonval, three coil, two coil, and air core.

D'Arsonval Gauges

A d'Arsonval-type gauge makes use of a permanent horseshoe magnet that surrounds a moveable electromagnet or armature attached to the gauge needle (**Figure 12-7**). Current flow through the armature winding causes the gauge needle to move due to interaction between the magnetic field caused by the permanent magnet and the magnetic field caused by the electromagnet. Increasing the current flow through the electromagnet causes the gauge needle to deflect or

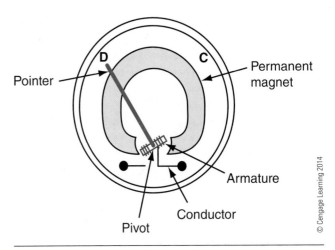

Figure 12-7 D'Arsonval gauge movement.

rotate accordingly. Reversing the direction of current flow through the conductor will cause the needle to deflect in the opposite direction. Ammeters are often d'Arsonval-type movements because current flow through the conductor in the opposite direction will cause the gauge needle to rotate in the opposite direction.

Ammeters are zero-center gauges with a D or – sign to indicate the batteries are discharging and a C or + sign to indicate the batteries are charging (**Figure 12-8**). The ammeter shunt is located in the charging system circuit such that cranking-motor current is not measured by the ammeter. When there is zero current flow into or out of the batteries, the ammeter will indicate zero or will be centered. When current is flowing from the batteries, the ammeter will indicate a negative value of current or discharging. When current is flowing into the batteries, the ammeter will indicate a positive value of current or charging.

Ammeter conditions

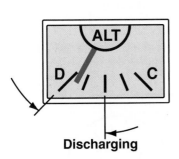

Discharging

Battery is discharged. AC generator is not charging or is not maintaining vehicle's electric needs.

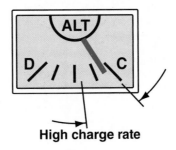

High charge rate

Battery is partially charged and AC generator is recharging it.

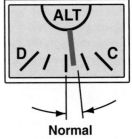

Normal

Battery is charged and AC generator is supplying the vehicle's electrical needs.

Figure 12-8 Ammeter needle movement indicating charging or discharging batteries.

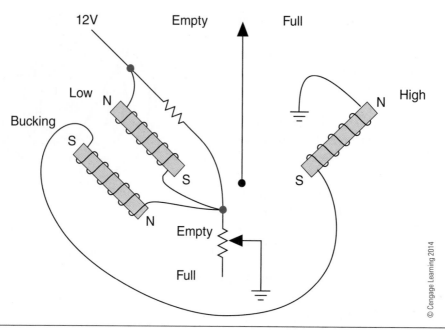

Figure 12-9 Three-coil gauge.

THREE-COIL GAUGES

Three-coil gauges make use of three coils, as the name implies. The three coils form three electromagnets (**Figure 12-9**). These three coils are called the high reading coil, the low reading coil, and a **bucking coil**. The gauge needle has a permanent or electromagnet attached to it. A three-coil gauge is designed so that the magnet on the needle is more attracted to either the high reading coil or the combination of low reading coil and bucking coil magnetic fields. The resistance of the sending unit determines which resulting magnetic field is stronger and thus the position

that the needle will maintain. Note that the low reading coil and the bucking coil are wound in opposite directions. The two coils are also placed next to each other so that current flow through the bucking coil causes a magnetic field that is opposite to or ''bucks'' the magnetic field formed by the low reading coil. This causes these two opposing magnetic fields to attempt to cancel each other out.

The three coils inside the gauge and the sending unit form a series-parallel circuit (**Figure 12-10**). Ignition voltage (+12V) is supplied to the low reading coil. The fixed resistor in parallel with the low reading

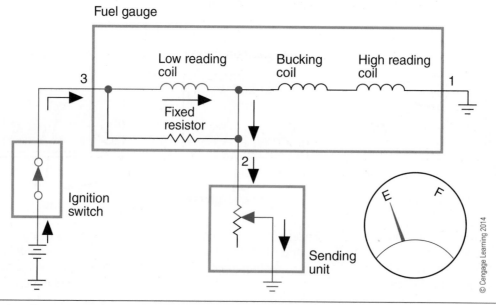

Figure 12-10 Sending unit resistance is low, indicating an empty fuel tank.

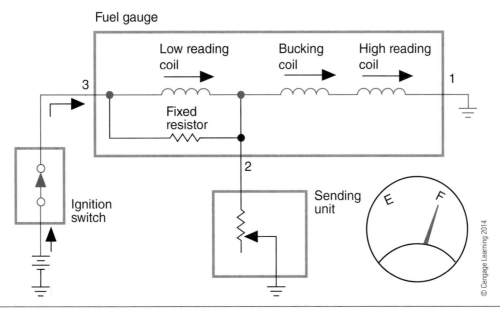

Figure 12-11 Sending unit resistance is high, indicating a full fuel tank.

coil is used to fine-tune the gauge during the manufacturing process and can be ignored. Note there are two paths for current flow after it leaves the low reading coil. Some of the current will flow through the series-connected bucking coil and high reading coil to ground and some of the current will flow through the sending unit variable resistance to ground. The amount of current that flows through the sending unit determines the needle position.

The sending unit resistance will be low in this example when the fuel level is low, as shown in **Figure 12-10**. More current will flow through the low reading coil than flows through the series-connected bucking and high reading coils because of the low resistance path to ground through the sending unit. The sending unit resistance and the series-connected bucking and high reading coils are in parallel with each other. The majority of the current that flows through the low reading coil will flow through the sending unit to ground because of the sending unit's low resistance. The rules of parallel circuits indicate that some current will also flow through the bucking coil and high reading coil. However, because more current is flowing through the low reading coil than through the bucking and high reading coils, the magnet on the gauge needle will be more attracted to the low reading coil than to the high reading coil and will indicate some value of fuel level near the empty mark on the gauge.

In this example, filling the fuel tank will cause the resistance of the fuel sending unit to increase. This causes the majority of the current that leaves the low reading coil to flow through the bucking coil and the high reading coil (**Figure 12-11**). Some current will also flow through the sending unit. Even though more current is flowing through the low reading coil than through the high reading coil, the bucking coil causes a portion of the magnetic field created by the low reading coil to be cancelled out. This is due to the opposite direction of winding on the bucking coil as compared to the low reading coil. Thus, the magnetic field created by the high reading coil causes the magnet on the needle to be attracted to the high reading coil and to indicate some value of fuel level near the full mark on the gauge.

As the fuel level decreases, the sending unit resistance will decrease. This causes the level of current flow though the bucking and high reading coils to decrease accordingly as more current is diverted through the sending unit. The gauge needle will indicate this decrease in sending unit resistance by pointing away from the full mark and toward the empty mark.

The three-coil gauge typically does not require a regulated voltage supply. Any change in the truck's electrical system voltage will cause a corresponding change in current flow through all three of the coils.

TWO-COIL GAUGES

The two-coil gauge is also known as a balancing coil gauge. There are two different designs of two-coil gauges, depending on the application of the gauge. Gauges designed for temperature indication such as coolant temperature are constructed so that ignition voltage is supplied to one side of the two-gauge coils (**Figure 12-12**). The coils in this example are

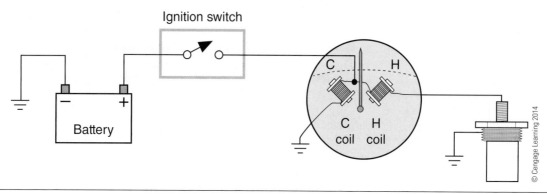

Figure 12-12 Two-coil temperature gauge.

designated C and H. The coil designated as the C coil (cold temperature) is designed to have a weaker magnetic field than the H coil (hot temperature) when both coils are supplied with the same voltage. The other side of the C coil is connected to a ground, indicating that the current flow through the C coil and the associated magnetic field will be some constant value depending on system voltage. The other end of the H coil is connected in series with a variable resistance temperature-sending unit. The temperature-sending unit is typically a thermistor or variable resistance device with a negative temperature coefficient (NTC), like that shown in **Figure 12-13**. Therefore, as the temperature increases, the resistance of the sending unit or sensor decreases. Because the sending unit is in series with the H coil, the current flow through the H coil and the associated magnetic field around the H coil are dependent on the resistance of the sending unit.

An increase in coolant temperature causes the sending unit resistance to decrease. This decrease in resistance causes an increase in the current flow through the H coil. This current flow increase results in a magnetic field around the H coil that is stronger than the constant magnetic field surrounding the C coil.

The needle then deflects toward the H coil to indicate an increased engine coolant temperature.

Another type of two-coil gauge is shown in **Figure 12-14**. Note that this application is a fuel level gauge. The coils are designated as E and F for empty and full, respectively. The E coil has fewer turns than the F coil. This type of gauge places the F coil in parallel with the sending unit. The E coil is in series with this parallel combination. When the fuel level is low, the resistance of the sending unit is also low, resulting in most of the current that is passing through the E coil to flow to ground through the sending unit. This causes the needle to deflect toward the E coil and indicate empty on the gauge face.

As fuel is added to the tank, the resistance of the sending unit will increase. This causes the current through the F coil to increase. This increase in the F coil's magnetic field causes the needle to deflect toward the F coil and provide an indication toward the full mark on the gauge face.

Neither version of the two-coil gauge requires a regulated voltage supply because changing the

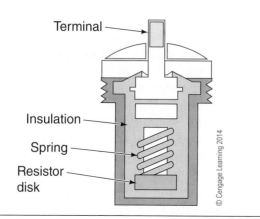

Figure 12-13 Variable resistor used to sense engine coolant temperature.

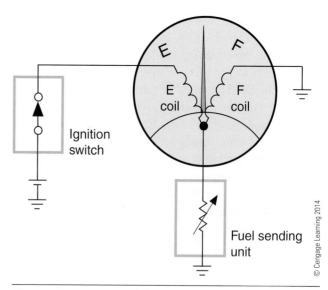

Figure 12-14 Two-coil fuel level gauge.

system voltage changes the current through both coils accordingly.

AIR CORE GAUGES

The air core gauge is like the two-coil gauge except that the needle is attached to a permanent magnet. The two windings have no core (that is, no metallic object that the coils are wrapped around). A coil that does not have a magnetic core is called an air core coil, hence the name for this type of gauge. The gauge is designed so that the permanent magnet needle is placed inside the two windings (**Figure 12-15**). The interaction of the resulting magnetic fields causes the permanent magnet needle to move in accordance with the current flow through the sending unit.

Electronically Controlled Gauges

Electronically controlled gauges are controlled by an electronic module or by electronics contained within the gauge assembly. The actual gauge is typically a magnetic type of gauge. Unlike conventional magnetic gauges discussed in previous sections, the sending unit or sensor acts as an input to the electronic module associated with the gauge instead of directly controlling the gauge's magnetic coils. The electronics then supply voltage to the gauge's magnetic coils to drive the gauge needle to the appropriate location.

Trucks built during the early development of truck electronics may only have an electronically controlled speedometer and tachometer, with a mixture of conventional gauges for the remaining instrumentation. These IPCs typically used a variable reluctance sensor mounted in the flywheel housing for the tachometer and a variable reluctance sensor in the tail shaft of the transmission for the speedometer. The teeth of the ring gear act as the tone wheel or target for the variable reluctance sensor used for the tachometer (**Figure 12-16**). A target or gear on the transmission output shaft is used to cause the variable reluctance sensor in the transmission to generate an AC voltage that is used for vehicle speed calculations (**Figure 12-17**). On some trucks, the IPC contains the speedometer, the tachometer, and the electronic module used to control these two gauges within a single package.

The actual speedometer and tachometer are typically air core magnetic gauges with the electronic control module providing the voltage to control the magnetic fields of the two gauge coils. These two coils are physically positioned at 90-degree angles relative to each other. The gauge's needle is mounted to a permanent magnet that is placed in the magnetic field of the coils (**Figure 12-18**). The design permits a continuous 360-degree rotation of the gauge's needle, like an electric motor. Typically, this rotation of the gauge needle is limited to approximately 270 degrees.

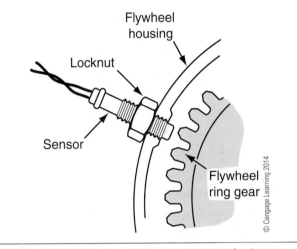

Figure 12-16 Tachometer sensor and ring gear installation.

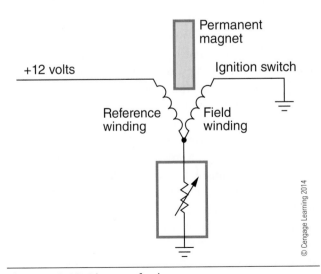

Figure 12-15 Air core fuel gauge.

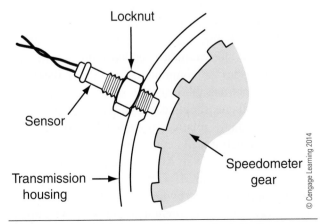

Figure 12-17 Vehicle speed sensor installation.

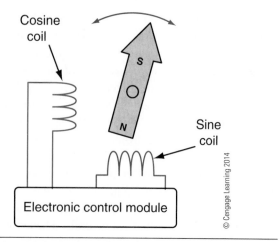

Figure 12-18 Electronically controlled magnetic gauge.

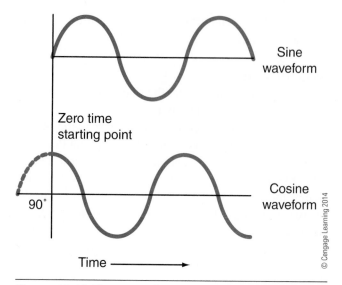

Figure 12-19 Sine waveform versus cosine waveform.

The large rotational span and fast response time of electronically controlled magnetic gauges make them ideal for tachometers and speedometers.

Some OEMs also use electronically controlled magnetic gauges for fuel level, oil pressure, water temperature, and other gauges. The conventional sensor or sending unit for these gauges is an input to the electronic module that controls the corresponding gauge. The electronic module associated with the gauge then provides the control for the gauge coils.

The electronic module that controls these magnetic gauges is connected to each of the two-gauge coils, which are positioned at a 90-degree angle relative to each other. The electronic module controls the voltage supplied to the two-gauge coils as needed to cause the needle containing a permanent magnet to be driven to the desired position.

These types of magnetic gauges may be referred to as sine-cosine gauges. The shape of a sine waveform and a cosine waveform are identical, other than having different amplitudes at the starting point of 0 degrees (**Figure 12-19**). This difference in starting points between two waveforms is referred to as a phase shift. It is as though the cosine waveform has had a head start of 90 degrees before the sine waveform. Many alternating current electric motors called induction motors have no brushes or commutator. Some of these induction motors use the principle of a phase shift between the voltages supplied to two windings placed at a 90-degree angle relative to each other to start the motor. The electronically controlled magnetic gauge is like a small version of an AC electric motor that can be stopped at any desired position by controlling the voltage supplied to the two-gauge coils.

The gauge coils in conventional magnetic gauges discussed earlier in this chapter were directly connected to the sensor or sender. It is important to note that there is no direct connection between the sensor and the gauge coils with electronic gauges. The sensors are connected to the electronic module, and the module then controls the gauge coils accordingly through four wires (**Figure 12-20**). Note in this example that the engine oil pressure sending unit is hardwired to the engine ECM. The speedometer/tachometer module controls the oil pressure gauge via four wires. The connections at the gauge are identified as S+, S−, C+, and C−. These letters stand for sine (S) and cosine (C).

Speedometer and Tachometer Programming. Variable reluctance sensors are typically used to provide the indication of engine and vehicle speed for the tachometer and speedometer. In the case of a tachometer, the AC signal provided by the variable reluctance sensor in the flywheel housing generates a complete AC cycle (sine waveform) each time a tooth and valley of the ring gear passes in front of the sensor. By measuring the frequency of this waveform, the microprocessor located in the electronic module for the tachometer can determine the number of flywheel teeth that pass in front of the sensor per second. However, a crucial piece of information is missing to convert this frequency into engine revolutions per minute. The microprocessor must "know" the number of teeth on the ring gear because not all engines have the same number of teeth on the ring gear. This programming for the number of flywheel ring gear teeth may be performed through the use of **DIP switches** (dual in-line pin) on the gauge or IPC electronic module on older trucks (**Figure 12-21**).

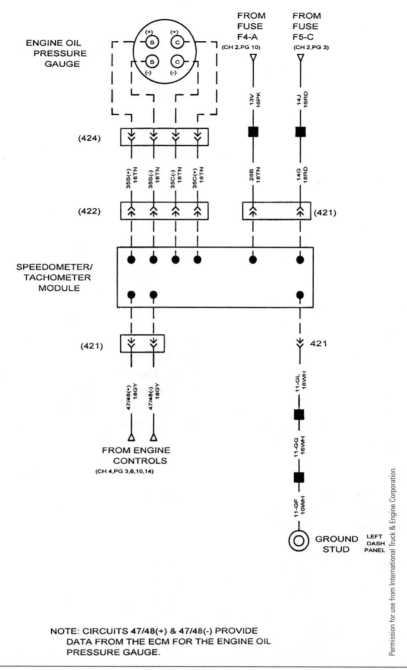

NOTE: CIRCUITS 47/48(+) & 47/48(-) PROVIDE
DATA FROM THE ECM FOR THE ENGINE OIL
PRESSURE GAUGE.

Figure 12-20 Electronically controlled magnetic engine oil pressure gauge circuit.

The DIP switches are placed in either the 1 or the 0 positions, based on the number of flywheel teeth (**Figure 12-22**). Alternatively, this information may be programmed using an electronic service tool on some IPCs. The AC signal from the tachometer variable reluctance sensor is an input to the microprocessor contained in the electronic control module that is used to drive the tachometer gauge coils. The frequency of the signal provided by the flywheel sensor and position of the DIP switches are used by the microprocessor to determine where the tachometer needle will be driven to display the calculated engine speed.

The electronically controlled speedometer on trucks with a manual transmission uses a variable reluctance sensor located in the transmission; this sensor is typically called a vehicle speed sensor (VSS). This sensor does not measure vehicle speed but instead indicates transmission output shaft speed. To convert the information supplied by the VSS into vehicle speed requires three pieces of information: the number of teeth on the target on the transmission output shaft, the rear-axle ratio, and the number of tire revolutions per mile. The standard number of transmission output-shaft tone-wheel teeth is 16. The rear-axle ratio

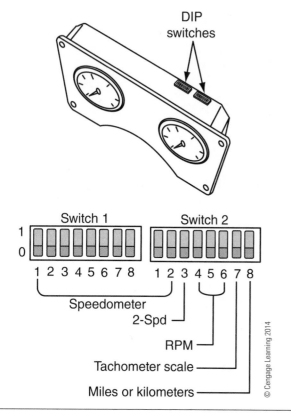

Figure 12-21 DIP switches used to program speedometer and tachometer on older trucks.

Switch Position			Teeth per Revolution	Engine or Output
4	**5**	**6**		
1	1	0	103	Cummins
0	0	1	113	Caterpillar
1	1	1	113	J1708 Data Link
0	0	0	12	12 Pulse/Rev
				NGD Electronic
0	1	1	138	Navistar PLN

Figure 12-22 DIP switch settings example for tachometer.

information is found on a tag on the rear axle, and information about the tire revolutions per mile is available from the tire manufacturer. These three numbers are then multiplied together to provide the total pulses-per-mile value. Alternatively on some vehicles, just the tire revolutions per mile and the rear-axle ratio are multiplied together to provide the drive shaft revolutions-per-mile value. This figure is then converted to a DIP switch setting using a chart

provided by the OEM. Additionally, a DIP switch typically permits programming for miles per hour or kilometers per hour speedometers.

Alternatively, the pulses-per-mile value is programmed into the IPC memory on some trucks using an electronic service tool instead of DIP switches. Modifying the tire size or rear-axle ratio requires a programming change for the speedometer to provide an accurate speedometer reading.

The electronic module in the IPC measures the frequency of the signal provided by the VSS and applies the programmed pulses-per-mile value to calculate vehicle speed. The electronic module in the IPC also supplies voltage to the appropriate gauge coils to cause the speedometer needle to indicate the calculated vehicle speed.

Pyrometer. An optional electronically controlled gauge specific to diesel engines is an exhaust temperature gauge or **pyrometer**, as it is known. The pyrometer is a magnetic gauge controlled by an electronic module. A thermocouple, typically type J iron-constantan, is installed in the exhaust manifold. The thermocouple is hardwired to the electronic module. The electronic module converts the low-level voltage produced by the thermocouple into a proportional voltage used to drive the pyrometer gauge coil.

Digital Instrument Displays

Some OEMs use digital displays instead of analog gauges to indicate the various pieces of information. There are two common methods of displaying information digitally on truck instrument panels: light-emitting diodes (LEDs) and **liquid crystal display (LCD)**. Digital displays require an electronic control module for control of the LEDs or LCD.

LED Displays. LEDs in the shape of line segments can be used to display the numbers 0 to 9 through the use of seven line segments (**Figure 12-23**). You have probably seen this type of LED display on older calculators, gas station fuel pumps, and many other places. An electronic module is used to cause certain segments to illuminate so that a digit can be displayed. These seven-segment displays are crude by modern standards and the digits displayed are somewhat square looking. Some trucks of the past made use of seven-segment LED displays to indicate vehicle speed, temperatures, and so on (**Figure 12-24**). LEDs shaped like bars can also be arranged in the shape of a circular arc and illuminated to correspond to fuel level or vehicle speed.

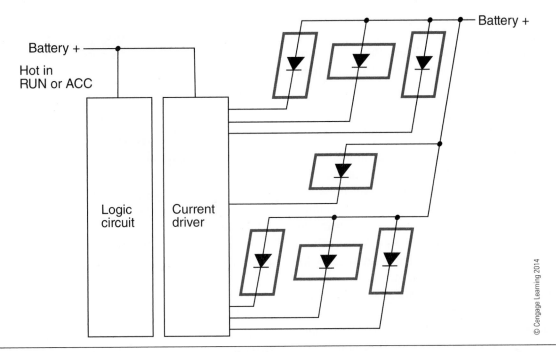

Figure 12-23 Seven-segment LED.

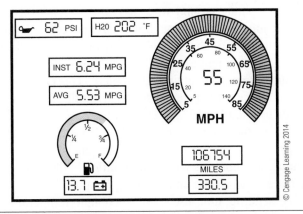

Figure 12-24 Truck instrument panel using LEDs to display information.

Another type of LED display is the dot matrix display panel. The dot matrix panel consists of a large number of LEDs arranged in rows and columns as shown in **Figure 12-25**. Letters and digits are formed on the dot matrix display panel when specific LEDs are switched on.

Liquid Crystal Display (LCD). LCDs contain a liquid crystal fluid and make use of the physics of light, including polarization. These principles are beyond the scope of this book. A character formed on an LCD display consists of segments, like the LED. However, the

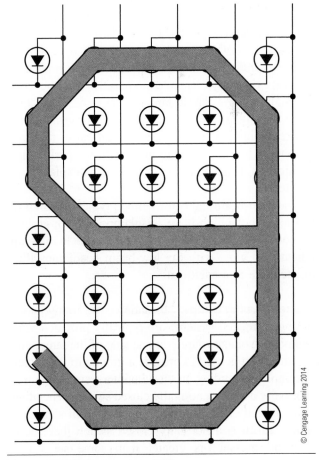

Figure 12-25 Dot matrix LED.

LCD segment does not generate light as an LED segment does. With the application of a voltage to the fluid, a segment will be either light or dark. Backlighting or sunlight provides the contrast between dark and light segments.

Warning and Indicator Lamps

Warning and indicator lamps are the various lights in the instrument panel used to alert the truck operator of certain conditions such as low engine oil pressure. These lights are known by several names, such as idiot lights or tell-tales. Many passenger cars use warning lamps instead of gauges to indicate a problem such as low oil pressure or high coolant temperature. Trucks have many of the same warning lamps found on passenger cars, along with several others specific to air brakes and truck components.

The light source for conventional instrument panel warning lamps is incandescent lamps. A colored film with a graphic is placed in front of the lamp and provides the color and the lettering such as OIL or the **ISO symbol** of an oil can. ISO is the International Organization for Standardization and provides recommendations for standardized symbols used in industry. Some newer trucks may use colored LEDs instead of incandescent lamps. The color is provided by the LED.

Lighting Indicator Lamps. Turn signal indicator lamps are typically connected in parallel with the front turn signal lamps. These lamps flash at the same rate as the turn signals. On some trucks, a poor ground of a front combination turn/park lamp can cause the corresponding turn indicator to glow with the park lamps illuminated. This is due to a backfeed, as explained in Chapter 9.

The high-beam indicator lamp is typically wired in parallel with the high-beam headlamps. Loss of ground for one of the headlamps may cause the high-beam indicator to glow with the low-beam headlamps illuminated due to a backfeed.

Engine Warning Lamps. Incandescent warning lamps for engine-related systems are typically provided with an ignition voltage source on one side of the lamp, and a sending unit is used to provide a ground to illuminate the lamp. The sending unit used with a warning lamp is a normally open or a normally closed switch, depending on the circuit. The oil pressure sending unit shown in **Figure 12-26** is a normally closed switch that opens when oil pressure is greater than some set value. The closed switch provides a ground for the oil pressure warning lamp, causing the lamp to illuminate.

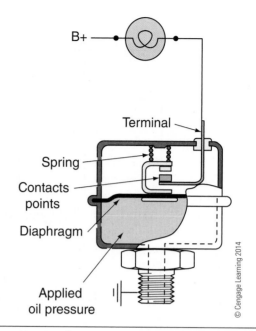

Figure 12-26 Oil pressure sending unit used with a low oil pressure warning lamp.

The ground provided by the sending unit may be through the sending unit housing, which is threaded into the grounded engine block. Some sending units may have a separate ground wire instead of relying on the connection to the engine block ground through the screw threads of the sensor. When the key switch is placed in the ignition position with the engine not running, there is no engine oil pressure. This causes the engine oil pressure lamp to illuminate until oil pressure builds to a level sufficient to open the normally closed switch in the sending unit. If the engine oil pressure drops below some value with the engine running, sending unit contacts will close resulting in illumination of the engine oil pressure warning lamp.

A sending unit used for a coolant temperature warning lamp typically contains a bimetallic strip (**Figure 12-27**). The strip deflects as temperature

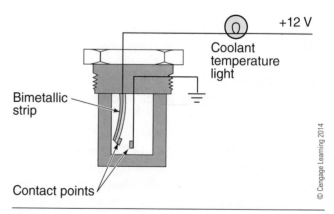

Figure 12-27 Temperature-sending unit for a coolant temperature warning lamp.

increases, causing the normally open contacts in the sending unit to close and provide a path to ground for the coolant temperature warning lamp.

Modern diesel engines may not have engine oil pressure switches or coolant temperature switches. Instead, the warning lamp may be controlled directly by the engine ECM. When the oil pressure sensor or coolant temperature sensor indicates an abnormal condition such as low oil pressure, the engine ECM provides a ground for the warning lamp using a low side driver. This causes the oil or coolant warning lamp to illuminate. Conversely, the ECM may instead source a voltage using a high side driver to a grounded lamp to illuminate the lamp, depending on the engine ECM's design.

The engine ECM may also directly control other warning or indicator lamps in the same manner. For example, the wait-to-start lamp is used to indicate that the preheating device, such as an intake air heater or glow plugs, is heating to alert the truck operator to wait until the lamp is extinguished before cranking the engine.

Brake System Warning Lamps. Some of the most important warning lamps in a truck are those related to the brake system. Trucks with air brakes are required to have a visual indication of low air pressure, in addition to the air pressure gauges. The visual indicator is typically an air pressure warning lamp that is designed to illuminate if primary or secondary air pressure is below 60 psi (414 kPa). This is typically accomplished by the use of parallel-connected normally closed pressure switches in the primary and secondary air system (**Figure 12-28**). The switches are designed to

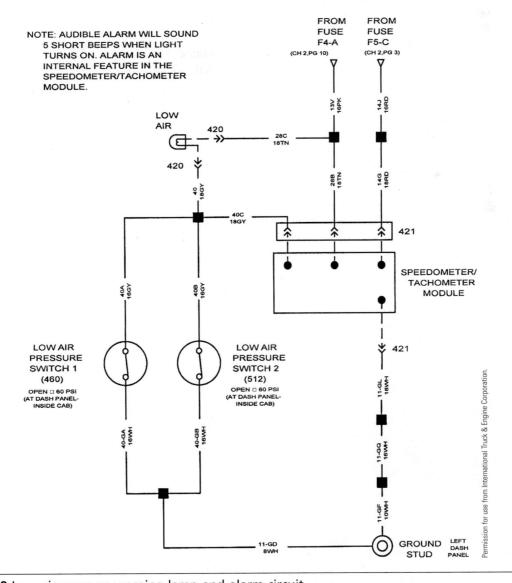

Figure 12-28 Low air pressure warning lamp and alarm circuit.

close when air pressure drops below 60 psi. The closed switch causes the low air pressure warning lamp to illuminate. Most trucks also have an audible alarm such as a buzzer or beeper that is controlled by the same switches.

Trucks with hydraulic brakes often use the Hydro-Max booster system described in **Chapter 10**. There is typically one or two warning lamps associated with the Hydro-Max system. The normally closed power steering fluid flow switch in the Hydro-Max assembly closes to provide a ground when there is no power steering fluid flow. This acts as an input to the monitoring module, which provides a ground for a brake pressure warning lamp and audible alarm (**Figure 10-33** in **Chapter 10**). The monitoring module also provides a ground for the warning lamp and alarm if the power steering flow switch does not toggle from closed to open when the engine is started. This would occur if the flow switch electrical connector were disconnected.

A differential pressure switch indicates a difference in pressure between the two halves of the split hydraulic brake system, indicating a hydraulic leak or other problem (**Figure 12-29**). This differential pressure switch provides a path to ground when a difference in pressure exists. This switch typically provides a ground for a hydraulic brake warning lamp. This hydraulic brake warning lamp may also be lamp-tested during engine crank on some trucks. A terminal on the key switch provides a ground for this lamp check during engine crank.

Most trucks have a parking brake switch that causes a parking brake warning lamp to illuminate. The parking brake switch on trucks with hydraulic brakes is mounted on the parking brake lever or pedal. Lifting the parking brake lever or depressing the parking brake pedal causes the parking brake switch to close.

Trucks with air brakes typically use a normally closed parking brake pressure switch. The switch monitors the pressure in the parking brake supply circuit downstream of the parking brake valve (yellow knob). When pressure in this circuit drops below the switch pressure setting, the parking brake warning lamp is illuminated by the path to ground provided by the switch.

Trucks sold in the United States are required to have ABS. FMVSS 121 requires an amber ABS warning lamp in plain view of the truck operator. This warning lamp must illuminate if the ABS system has detected problems that inhibit or degrade ABS performance. Additionally, the ABS warning lamp must illuminate for a lamp check when the key is first placed in the ignition position. This lamp is controlled by the ABS ECU. This is discussed in more detail in **Chapter 15**.

Modern trucks that are designed to pull a trailer equipped with air brakes must also have a trailer ABS warning lamp. The trailer ABS warning lamp located in the instrument panel is controlled by the truck ABS ECU. The truck ABS ECU typically provides a ground (low side driver) for the trailer ABS warning lamp when the truck ABS ECU senses that the trailer ABS ECU has detected a trailer ABS system fault. The trailer ABS ECU signals the truck ABS ECU through a method known as power line carrier (PLC). This communication technique is discussed in more detail in **Chapter 15**.

Charging System Warning Lamps. Some alternators have an output terminal that is used to illuminate a

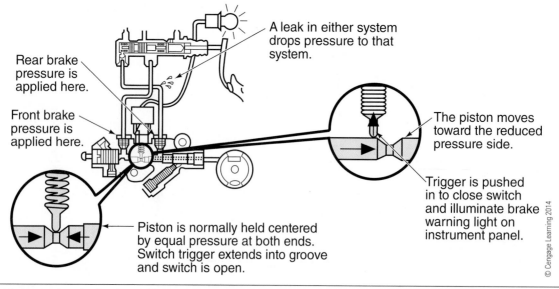

A leak in either system drops pressure to that system.

Rear brake pressure is applied here.

Front brake pressure is applied here.

The piston moves toward the reduced pressure side.

Trigger is pushed in to close switch and illuminate brake warning light on instrument panel.

Piston is normally held centered by equal pressure at both ends. Switch trigger extends into groove and switch is open.

© Cengage Learning 2014

Figure 12-29 Hydraulic brake differential pressure switch.

warning lamp if the alternator is not producing sufficient voltage, as was explained in **Chapter 7**.

Troubleshooting Conventional Instrumentation

In general, troubleshooting non-multiplexed instrumentation for a gauge indication problem typically involves installing a variable resistance device in place of the sending unit to simulate the sending unit resistance. This is useful for determining if the problem is caused by a faulty sending unit or if the problem lies between the sending unit connection and the gauge. Test tools containing variable resistance values corresponding to the sending units throughout the truck are available; these simulate the sending unit resistance to test the gauge and the associated wiring.

The use of an ohmmeter to measure resistance throughout the circuit is typically also a step in the troubleshooting process. This determines if the problem is due to an open or shorted circuit between the gauge and the sending unit. Other troubleshooting steps may include verifying the gauge power and ground circuits and IVR, if applicable. Lastly, the gauge is replaced if everything else checks out as okay.

Tech Tip: The OEM's troubleshooting information should be consulted when troubleshooting instrumentation. Indiscriminate shorts to ground of instrumentation circuits or other random test procedures can cause damage to gauges and electronics.

Troubleshooting warning lamp circuits, in cases in which a warning lamp is illuminated when it should not be, typically involves disconnecting the connector for the switch that controls the warning lamp. A warning lamp that remains illuminated with the switch disconnected may be an indication of a short to ground between the warning lamp and the switch. A warning lamp that is extinguished with the switch disconnected may be an indication of a faulty switch.

WARNING *A warning lamp that is illuminated may be due to an actual problem such as low power steering flow or low engine oil pressure. Use a pressure gauge or other test tool as necessary to verify that the pressure, temperature, and so on are within specifications. Never permanently disconnect a switch for a critical warning device as a means of extinguishing a warning lamp.*

The failure of a warning lamp to illuminate when it should be illuminated may be caused by an open circuit, a faulty switch, or a burned-out lamp.

Testing a Fuel Level Sending Unit.   A fuel level sending unit is a variable resistor. An ohmmeter can be used to bench test a fuel level sending unit. The ohmmeter measures the resistance of the sending unit throughout the sweep of the float arm. The resistance that is measured is compared to the OEM's specifications. Be aware that the float can become saturated or partially filled with fuel, causing the float to be less buoyant than it was originally designed to be resulting in a low fuel level reading.

MULTIPLEXED INSTRUMENTATION

The introduction of electronically controlled diesel engines, electronically controlled automatic transmissions, ABS, body control modules, and other systems in modern trucks has resulted in significant changes in instrumentation in recent years. The various electronic control modules in the truck communicate with each other using one or more serial communications networks or data links. These communication networks include the SAE J1587/ J1708 and the J1939 serial communication data links discussed in the previous chapter.

The data transmitted by the engine control module onto the data link includes information such as engine speed, oil pressure, coolant temperature, and vehicle speed. The IPC on modern trucks is connected to one or more of the truck's data links. The IPC contains an electronic module equipped with a microprocessor. The electronic module in the IPC uses information on the data link to determine where to drive the electric gauges and indicator lamps in the IPC.

J1587/J1708 Multiplexed Instrumentation

The first uses of multiplexed instrumentation in trucks were limited to mostly engine-associated gauges and vehicle speed. A conventional gauge cluster requires a hardwired connection to each sensor such as those for engine oil pressure, coolant temperature, and engine speed. Individual wires must also be routed from engine-mounted switches to the IPC warning lamps for low oil or high coolant warning. This means that a large number of wires must be routed from the engine to the conventional IPC. Modern electronic diesel engine ECMs require information such as coolant temperature, engine speed, and vehicle speed to provide control over the fuel system. Therefore, sensors such as the engine coolant temperature sensor and the vehicle speed sensor are inputs to the engine ECM. From these hardwired sensors, the engine ECM "knows" information such as coolant temperature, engine speed, and vehicle speed. The engine is capable of transmitting this information over the J1587/J1708 ATA data link so that other electronic modules on the truck can make use of this data.

The J1587/J1708 multiplexed IPC only requires a hardwired connection to the J1587/J1708 data link to obtain all of the information being transmitted by the engine ECM (**Figure 12-30**). This provides a significant reduction in the number of wires routed to the IPC. The electronics in the IPC are designed to extract information such as engine speed from the information present on the data link. The IPC then controls the individual gauges accordingly. The actual gauges are typically electronically controlled magnetic-type gauges. The sine-cosine method described earlier can be used to control these magnetic gauges.

Not all of the gauges in a J1587/J1708 multiplexed IPC are electronically controlled. Air pressure gauges, engine oil temperature, and other gauges may be standard mechanical or electromechanical gauges.

Warning and indicator lamps in multiplexed IPCs may be conventional hardwired lamps that use external switches, such as low coolant level. However, some warning lamp information, such as low oil pressure, may be obtained from the information broadcast over the J1587/J1708 data link instead. These warning lamps, such as that for low oil pressure, are controlled by the IPC.

A variable reluctance vehicle speed sensor on trucks with manual transmissions or a variable frequency pulse generated by the automatic transmission ECU is used as an input to the engine ECM for determining vehicle speed. The vehicle speed sensor or automatic transmission variable frequency pulse is only an indication of transmission output shaft speed. Therefore, the engine ECM must be programmed with a pulses-per-mile value. The pulses-per-mile value is based on the number of transmission output shaft pulses per revolution, the rear-axle ratio, and the tire revolutions per mile, as described earlier in this chapter. The engine ECM uses this pulses-per-mile value to convert the vehicle speed sensor output frequency into vehicle speed. The calculated value of vehicle speed is then transmitted over the data link by the engine ECM. Any changes to drive axle tire size or rear-axle ratio require modification of an engine ECM

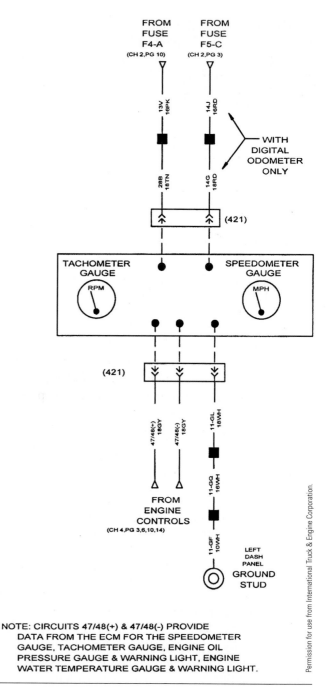

NOTE: CIRCUITS 47/48(+) & 47/48(-) PROVIDE DATA FROM THE ECM FOR THE SPEEDOMETER GAUGE, TACHOMETER GAUGE, ENGINE OIL PRESSURE GAUGE & WARNING LIGHT, ENGINE WATER TEMPERATURE GAUGE & WARNING LIGHT.

Figure 12-30 Instrument panel cluster using data link information to drive engine-related gauges and warning lamps.

programmable parameter to correct the pulses-per-mile value accordingly using an electronic service tool, as shown in **Figure 12-31**. Programmable parameters are typically EEPROM locations in an electronic module, which contain data unique to a vehicle or engine, or owner preferences. For example, road speed limiting and idle shutdown time limit are typical fleet-controlled programmable parameters. OEMs permit these programmable parameters to be

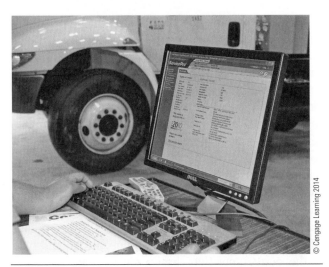

Figure 12-31 Electronic service tool used to modify ECM programmable parameters after tire size or rear-axle ratio change.

modified using an electronic service tool. In the case of fleet-owned vehicles, the programmable parameters are typically password-protected to prevent tampering.

J1939 Multiplexed Instrumentation

The expanded use of electronics in trucks has resulted in increased use of multiplexing for instrumentation. The introduction of the J1939 data link with its increased bandwidth (higher data transmission rate) permits substantially more information to be broadcast by the various electronic modules on the truck. Electronic modules throughout the truck—such as the engine ECM, automatic transmission ECU, ABS ECU, and body control modules—broadcast a variety of information onto the J1939 data link. Modern IPCs are microprocessor controlled and are connected to the J1939 data link.

The engine ECM broadcasts a variety of data onto the J1939 data link, including engine speed, oil pressure, coolant temperature, and vehicle speed. Data such as air system pressures and fuel level may be broadcast onto the J1939 data link by a body control module. Information for indicator and warning lamps—such as turn signals, parking brake, and high-beam headlamps—may also be broadcast onto the J1939 data link by the body control module. The IPC's microprocessor uses this information to control the corresponding gauges and warning lamps. Information from other electronic modules, such as the automatic transmission ECU, is used for the transmission oil temperature gauge or transmission

warning lamps. The ABS ECU may also provide indication of a problem detected with the ABS system that has reduced ABS system performance using a J1939 message. The IPC may use this information from the J1939 data link to cause the ABS warning lamp to illuminate.

Stepper-Motor-Driven Gauges. Stepper-motor-driven gauges are common in modern IPCs. A stepper-motor is an electric motor that can be driven to a precise position and held in that position. A stepper motor is an electric motor that has no brushes or commutator. All of the windings are typically located in the stator and the rotor is typically a permanent magnet-type rotor. The rotor shown in **Figure 12-32** has six poles that are shaped like pieces of a pie and are identified by N and S markings. The stator in this example has four pole pieces. This motor has two coils, each of which are separated into two halves. A portion of the coil is wound around one pole piece; the other half of the coil is wound around the pole piece on the opposite side of the stator. The other coil is wrapped around the other two pole pieces in the same direction. The wire ends of the two coils are marked as 1A, 1B and 2A, 2B. These wire ends of the two coils are connected to an electronic control module that can control the polarity of the voltage supplied to the two coils.

The electronic control module for the stepper motor supplies a DC voltage to coil 1 with a polarity as shown in **Figure 12-32** resulting in current flow

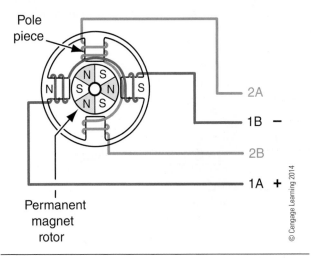

Figure 12-32 Stepper motor. The rotor is constructed of permanent magnets. The controller is shown energizing stator coil 1, causing the stator to have the magnetic polarity shown. This causes the rotor to rotate to the position shown and remain locked in this position until the stator coil is no longer energized.

through the two halves of coil 1. This causes the pole pieces around which the two halves of coil 1 are wrapped to become magnetized. The two pole pieces have opposite magnetic polarity on the side facing the rotor. This causes the permanent magnet rotor to rotate slightly so that it aligns with the pole pieces accordingly. The rotor is locked magnetically in this position with direct current flowing through the coil.

To cause the motor to rotate clockwise, the motor controller would stop supplying voltage to coil 1 and would supply voltage to coil 2 with the polarity shown in **Figure 12-33**. This would cause the rotor to be attracted to pole piece 2, resulting in clockwise rotation of the rotor. The rotor would then be locked in the position shown in **Figure 12-33** with direct current flowing through coil 2. The motor has rotated one step from the original position. Each time that the motor controller supplies a voltage pulse to one of the coils and de-energizes the other coil, the motor steps to whichever pole piece has been energized. The motor controller can change the polarity of the voltage supplied to the coils to drive the stepper motor in either direction. The motor controller is capable of rapidly pulsing the voltage on the coils to make the motor rotate rapidly while counting each step in order to maintain motor position control.

The output shaft of the stepper motor for a typical gauge is gear reduced, which causes each step that the motor is driven to move the gauge needle a small amount. A gauge with some 300 degrees of needle sweep may be designed so that more than 2000 steps

of the motor are required to rotate the gauge's needle throughout the entire sweep.

There is typically no feedback to the IPC motor control module regarding the location of the needle for a stepper-motor-driven gauge. Instead, the IPC control module will cause the gauge needle to drive in a counterclockwise direction toward the peg located beneath the lowest indicated point of the gauge when power is first supplied to the IPC. This causes the stepper motor to stall when the needle is rotated into the peg. This provides the IPC control module with an initial starting or zero step point for the gauge. From this starting point, the stepper motor is pulsed a specific number of times to cause the needle to be driven to the correct position, based on the data received. Some gauges may be driven to the maximum span point from the zero point, as part of the initialization process. The IPC control module keeps track of the present step position of each gauge, following initialization.

A stepper-motor-driven tachometer may require 2000 or more steps of the stepper motor to rotate the gauge needle from the peg to the maximum indicated position of perhaps 3000 rpm. When the gauge cluster is first initialized, the tachometer needle is driven in the counterclockwise direction to the peg located below 0 rpm. This causes the motor to stall when the needle hits the peg. From this point, the tachometer may then be driven 2000 steps in the clockwise direction so that the needle sweeps throughout the entire range of travel. This is commonly done to prove out the gauge and for appearance. From the 2000-step position, the gauge may be driven 1900 steps counterclockwise to the 0-rpm position just above the peg because the engine is not running. The information received over the data link indicates that the engine speed is zero. When the engine starts, the engine controller will broadcast engine speed on the J1939 data link as perhaps 700 rpm. The IPC control module then drives the stepper motor the required number of steps from the 0-rpm position in the clockwise direction to cause the gauge needle to indicate 700 rpm. As the information received over the J1939 data link by the IPC control module regarding engine speed changes, the tachometer will be driven the corresponding number of steps in the correct direction to cause the needle to indicate the correct speed. The change in needle position occurs rapidly and smoothly in response to a rapid engine speed change.

Other engine-related gauges such as oil pressure, coolant temperature, and speedometer are controlled

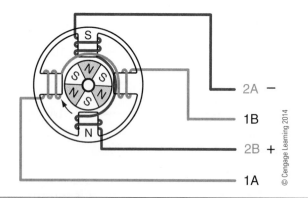

2A −

1B

2B +

1A

© Cengage Learning 2014

Figure 12-33 Stepper motor rotating clockwise. Coil 1 is no longer energized; coil 2 is now energized with voltage polarity shown, causing pole piece 2 to become magnetized as shown. Rotor moves clockwise to align with the pole piece magnet.

in the same manner as the tachometer. Data received from the engine controller is used to drive the applicable gauge to the corresponding position. Other electronic modules, such as the body control module, transmit J1939 messages indicating parameters such as primary air pressure and fuel level. The IPC controls the associated gauges in the same way as the engine-related gauges. Body control modules will be addressed in the next chapter.

Multiplexed Indicator and Warning Lamps. Warning and indicator lamp information is also made available on the J1939 data link. For example, the engine ECM may transmit a J1939 message indicating wait-to-start to permit the preheating device such as an intake air heater to heat. The IPC responds by using a low side or high side driver to illuminate the wait-to-start lamp or LED. The use of LEDs as indicator lighting has become common in modern trucks. These LED indicators should never require replacement for the life of the truck.

A body control module may be used to control headlamps, turn signals, and many other body-related systems. Information such as left turn signal flashing or high-beam headlamps on is transmitted by the body control module on the J1939 data link serial communications network. The IPC control module uses this information to cause the turn signal indicators or high-beam headlamp indicator to illuminate using a high side or low side driver to directly control the indicator LED.

Troubleshooting Multiplexed Instrumentation

Troubleshooting multiplexed instrumentation is typically not as difficult as it might seem. Fault tree-type diagnostic information and electronic service tools are used to troubleshoot instrumentation problems.

When troubleshooting a multiplexed instrumentation problem, it is useful to know where the source of information is obtained for each gauge and warning/indicator lamp. The multiplexed IPC merely displays the information transmitted by the various electronic modules throughout the vehicle. For example, the engine ECM typically transmits the information for all engine-related gauges and warning lamps, as well as vehicle speed. The body control module may transmit information for primary and secondary air pressure, fuel level, and turn signal indicators. Other modules such as the ABS ECU may

transmit information such as ABS warning lamp status, trailer ABS warning lamp status, and traction control warning lamp information.

....

Tech Tip: By knowing the source of information for each multiplexed gauge and warning lamp, you may be able to narrow the instrumentation problem to a specific module or a specific electrical connector based on which gauges or warning lamps are functional and which are not functional.

....

For example, a truck with a J1939 IPC with an inoperative tachometer, speedometer, oil pressure, and coolant temperature gauges but all other gauges functional would be an indication of a J1939 communications problem between the IPC and the engine ECM. The problem might not be at the engine ECM, though. Knowing the layout or network topology of the J1939 data link on the truck you are working with is another key factor for troubleshooting J1939 data link communication problems. In addition, knowing how the other modules on the truck react when a loss of communications occurs with other electronic controllers, such as the engine ECM, can assist in troubleshooting. This will be addressed in more detail in later chapters.

Troubleshooting sensor circuits that are inputs to the various electronic modules throughout the truck may require a **breakout T** (**Figure 12-34**) or **breakout box** (**Figure 12-35**) to be used along with a DMM. These tools reduce connector and terminal damage and prevent accidental shorts to ground and other potential damage to expensive truck electronic modules. These tools are used in conjunction with the OEM's EST and service information.

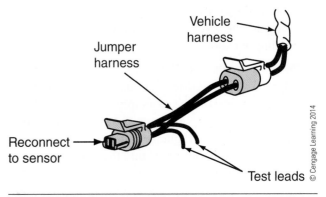

Figure 12-34 Sensor breakout T used to measure sensor voltage with a DMM voltmeter.

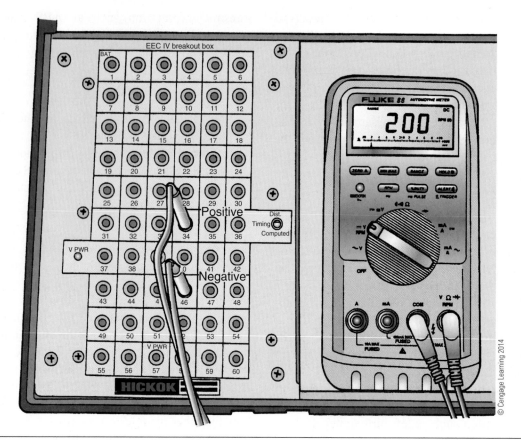

Figure 12-35 Breakout box used with a DMM voltmeter.

Summary

- Instrumentation refers to the gauges, indicator lamps, and audible alarms used to inform the truck operator of the status of the truck's various systems.

- A sensing device called a gauge sender or sending unit is used to transform pressure, temperature, level, or other physical value into a corresponding value of resistance. The varying resistance causes the current flow through a coil or heating element in a conventional gauge to increase or decrease, resulting in the gauge needle moving to the corresponding location.

- The two main categories of electromechanical gauges are bimetallic and magnetic. Bimetallic gauges operate on the principle that current flow through a heating element causes a bimetallic strip to deflect. Magnetic gauges contain coils (inductors) and operate on the principles of electromagnetism.

- An instrument voltage regulator is often used with bimetallic gauges to maintain a constant voltage supply for the gauge heating element.

- The four main types of magnetic gauges are d'Arsonual, three coil, two coil, and air core.

- Electronically controlled gauges use an electronic module to control the current flow through the magnetic-type gauge coils. The sending unit or sensor is an input to the electronic module. The electronic module controls the current supplied to the gauge accordingly.

- Electronically controlled speedometers and tachometers typically must be programmed to correspond to the vehicle if changes such as rear-axle ratio are made. This programming is performed on some trucks through a series of DIP switches in the IPC or through the modification of a programmable parameter in the engine ECM or other control module.

- Trucks may have a variety of warning and indicator lamps. Each lamp is controlled by a switching device that typically provides a path to ground to illuminate the indicator lamp.

■ Multiplexed instrumentation describes instrumentation that receives information from a serial communications network, commonly known as a data link. The various electronic modules on the vehicle, such as the engine ECM, broadcast information such as engine coolant temperature onto the data link. The IPC "listens" for this information on the data link and drives the appropriate gauge or warning lamp accordingly.

■ The term *stepper motor* describes a type of motor that can be driven to a specific position and held in that position. Many modern truck IPCs make use of stepper-motor-driven gauges.

Review Questions

1. Which of the following is a true statement?

   A. A variable reluctance sensor would typically be used as a fuel level sending unit.

   B. The speedometer cable that is used on most modern trucks requires frequent maintenance.

   C. An air application gauge informs the truck operator of how much air pressure remains in the air application tank.

   D. Both primary and secondary air pressure may be displayed in a single gauge with two needles.

2. A bimetallic gauge takes advantage of what property of metals?

   A. Metal produces a voltage when heated.

   B. The expansion rates of different types of metal when heated are not the same.

   C. A magnetic field is formed when a coil of wire is heated.

   D. Metal can be magnetized by heating.

3. A bimetallic gauge would typically use which type of sensor or sending unit?

   A. Variable capacitance

   B. Variable reluctance

   C. Thermocouple

   D. Variable resistance or rheostat

4. A truck operator indicates that the ammeter shows a slight discharge with the engine idling at night in rainy weather. Of the following choices, which is the most likely cause?

   A. The truck's electrical loads exceed the current supplied by the alternator, causing the truck's batteries to supply the remainder of the current.

   B. Ammeters are always inaccurate and should be ignored.

   C. The IVR is not operating correctly.

   D. The ammeter shunt is not communicating on the J1939 data link.

5. The rear-axle ratio is being changed in a truck with a variable reluctance speed sensor, located in the transmission, which is wired directly to the IPC. After the truck is driven, the speedometer is found to be inaccurate. How would this speedometer inaccuracy problem most likely be resolved?

   A. VSS air gap increase or decrease

   B. IVR adjustment

   C. DIP switch change of the speedometer or IPC electronic control module, based on OEM information

   D. Tire size change would be necessary to compensate for rear-axle ratio change

6. An exhaust pyrometer typically uses what type of sensor?

   A. Variable capacitance   C. PTC thermistor

   B. Thermocouple   D. Thermonuclear

7. An oil pressure switch used to control a low oil pressure warning lamp is typically what type of switch?

   A. Bimetallic   C. Normally open

   B. Normally closed   D. Pull-up resistor

8. The parking brake warning lamp remains illuminated on a truck with air brakes, even though the parking brake is released. The parking brake switch on this truck is a normally closed pressure switch located in the parking brake supply line. This line is pressurized when the parking brake is released. Which of the following is the most likely cause?

   A. The parking brake switch is disconnected.

   B. There is too much air pressure in the parking brake supply line.

   C. The yellow diamond-shaped parking brake control knob is not fully pulled out.

   D. The circuit between the warning lamp and the parking brake switch is shorted to ground.

9. The oil pressure gauge does not operate and the engine warning lamp is illuminated on a truck with a J1587/J1708 data link controlled IPC. The truck operator indicated that the engine warning lamp came on at the same time that the oil pressure gauge quit working. All other engine-related gauges are working correctly. Which of the following is the most likely cause?

   A. J1587/J1708 data link is open between the engine ECM and the IPC.

   B. The engine ECM has detected a faulted engine oil pressure sensor circuit.

   C. The engine oil pressure switch is defective.

   D. The engine oil pressure gauge is defective.

10. A truck with a multiplexed IPC that is using information from the J1939 data link for all gauges and warning lamp information has several gauges that do not operate. These inoperative gauges include the speedometer, tachometer, engine oil pressure, and engine coolant temperature gauges. The air pressure, voltmeter, and fuel gauges are working correctly. Which of the following is the most likely cause?

    A. The J1939 CAN + and CAN – circuits are shorted together.

    B. The J1939 CAN + circuit is shorted to ground.

    C. The IPC is not receiving any J1939 data link information.

    D. The J1939 data link is open between the engine ECM and the IPC.

11. Of the choices listed, which is the best method of troubleshooting an instrumentation problem?

    A. Pierce the sending unit wires with a test light to determine if voltage is present.

    B. Disconnect the sending unit connector and place a paper clip between the connector terminals.

    C. Replace components until the instrumentation is working correctly.

    D. Use OEM-provided information and procedures to troubleshoot the cause of the instrumentation problem.

12. A stepper motor used in instrumentation is typically controlled by which of the following devices?

    A. Commutator

    B. Electronic control module

    C. Magnetic switch

    D. All of the above

13. A truck with hydraulic brakes has a brake pressure warning light that remains illuminated at all times. Disconnecting the brake differential pressure switch causes the light to be extinguished. What should the technician do to resolve this problem?

    A. Leave the differential pressure switch disconnected.

    B. Remove the brake pressure warning lamp bulb.

    C. Replace the differential pressure switch.

    D. Determine why the differential pressure switch is providing an indication of a problem.

14. A truck runs out of fuel when the fuel gauge still indicates that 1/4 tank of fuel remains. What could be the cause of this problem?

    A. An incorrect fuel level sending unit has been installed in the tank.

    B. The float on the sending unit has developed a hole.

    C. The fuel level cable is not adjusted correctly.

    D. The fuel level sensor air gap is not adjusted to the correct value.

15. The charging system warning lamp remains illuminated at all times with the engine running. Which of the following is the most likely cause?

    A. An open circuit in the warning lamp circuit.

    B. The CAN + and CAN – wires are shorted together.

    C. The alternator drive belt is broken.

    D. Defective alternator variable reluctance sensor.

CHAPTER

13 Body Control Modules

Learning Objectives

After studying this chapter, you should be able to:

- Explain the concept of virtual fusing.
- List the types of inputs and outputs used by a typical body control module.
- Discuss the reasons that most electronic modules used in modern trucks make use of a reference ground system.
- Explain the differences between a conventional switch and a diagnosable switch.
- Describe how a body control module can act as a turn-signal flasher.
- Retrieve DTCs stored in memory related to the body controller.
- List the main components of the International Diamond Logic® and the Freightliner® SmartPlex™ multiplexed electrical systems.
- Describe how a Freightliner Smart Switch operates.
- Explain the concept of ghost voltage and describe how it can lead to confusion when diagnosing an electrical problem.

Key Terms

accessory air valve assembly (AAVA)

air management unit (AMU)

body controller

bulkhead module (BHM)

chassis module (CHM)

diagnosable switch

diagnostic trouble code (DTC)

electrical system controller (ESC)

electromagnetic interference (EMI)

electronic gauge cluster (EGC)

failure mode indicator (FMI)

ghost voltage

hardwire

instrument cluster unit (ICU display)

interlocks

leakage current

microswitches

out of range high (ORH)

out of range low (ORL)

reference ground

resistive ladder network

smart switch

virtual fusing

INTRODUCTION

Body control modules are electronic modules used to control various truck electrical features. These features may include headlamps, turn signals, horn, and windshield wipers, to name a few of the components managed by a body control module. These tasks may be performed by one module on some trucks, or the tasks may be split up between two or more modules on other trucks.

Modern trucks with body control modules make use of the SAE J1939 data link as a means to communicate with other electronic modules on the truck, such as the engine ECM and electronic instrument panel cluster (IPC). Information for many of the gauges and warning lamps in the IPC may be provided by the body control module via the J1939 data link.

Many of the concepts introduced in this chapter are common industry practices for automotive electronics and do not apply to only a body control module. Electronic modules on a modern truck such as the engine ECM use many of these same techniques.

This chapter will highlight Freightliner® Business Class M2™ and SD models and International® High Performance Vehicle models.

INTERNATIONAL MULTIPLEXED ELECTRICAL SYSTEM

International refers to their multiplexed electrical system as the Diamond Logic® electrical system. The body control module used in 2001–2006 International High Performance Vehicles is called an **electrical system controller (ESC)**. The ESC is located beneath the left kick panel inside the cab (**Figure 13-1**). The ESC has two electrical connections inside the cab. An opening in the left lower dash panel permits electrical connection to three electrical connectors located outside of the cab as well as to the ESC's main battery supply cable(s). For 2007, the name of the International body control module was changed to the **body controller**, which is fully contained within the cab beneath the left kick panel (**Figure 13-2**). Additional inputs, outputs, and enhanced functionality were added to the body controller, but the base functionality of the ESC and the body controller are identical. The body controller is used in International's Star series trucks, including ProStar® class 8 trucks and DuraStar® medium-duty models. Unless indicated otherwise, the term body controller will be used in this chapter to describe both the ESC and the body controller.

Overview of International Body Controller

Like other electronic control modules found on a modern truck, the International body controller contains a microprocessor. The microprocessor controls the body controller's outputs based on the status of its inputs. The microprocessor evaluates the information received from its inputs and makes decisions based on its programming instructions. These decisions are then

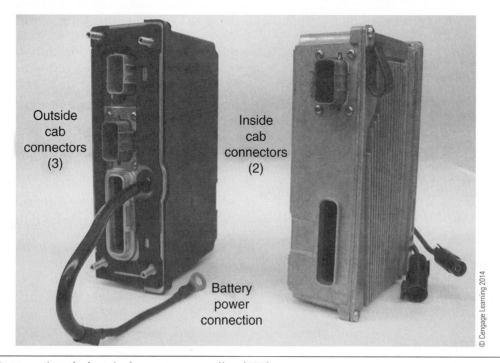

Outside cab connectors (3)

Inside cab connectors (2)

Battery power connection

© Cengage Learning 2014

Figure 13-1 International electrical system controller (ESC).

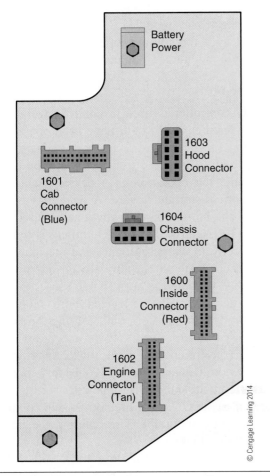

Figure 13-2 International body controller.

used to control the various microprocessor outputs. Note that much of the information in this section describes general automotive electronics industry practices which may apply to other electronic modules, such as the engine ECM.

The body controller has an assortment of inputs, including switches and sensors (**Figure 13-3**). Other inputs to the body controller include data from the J1939 data link, which is also connected to the other electronic modules on the truck such as the engine ECM, transmission ECU, and ABS ECU. The body controller is able to obtain a great deal of information from the J1939 data link. The body controller also has a proprietary (private) switch data link meeting the standards of SAE J1708/J1587 to receive information from instrument panel mounted switch modules known as switch packs. The body controller is not connected to the truck's standard J1708/J1587 ATA data link.

Outputs of the body controller include high side drivers and low side drivers. High side drivers source current (supply a positive voltage), while low side drivers sink a path for current to ground as discussed in **Chapter 6**. Some of these high side drivers can source up to 20A of current similar to the output terminal of a relay. These high side drivers are used to source high-current loads such as headlamps and turn signals. The body controller's low side drivers are used to control

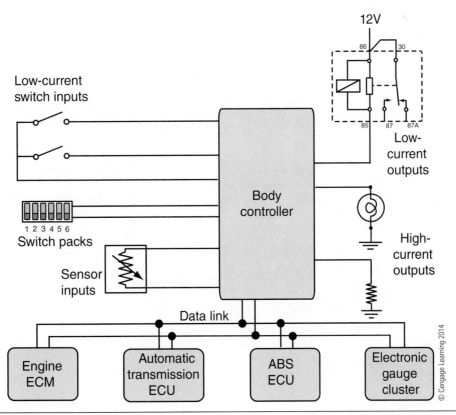

Figure 13-3 Overview of body controller.

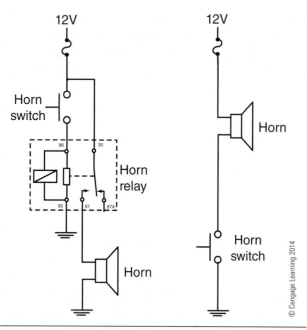

Figure 13-4 Conventional electrical system where horn switch controls the horn relay or directly supplies ground for the horn.

low-current loads that draw less than 1A of current, such as relay coils and small air solenoids.

Other outputs of the body controller include the messages that are broadcast onto the J1939 data link and received by other electronic modules on the truck. For example, the body controller is used to transmit messages via the J1939 data link to the engine ECM requesting that the cruise control be engaged or that the cruise set speed be increased.

Tech Tip: Think of the J1939 data link as being both an input and an output device for

most electronic modules. The messages received by a module are inputs; the messages sent by a module are outputs.

The body controller is used to control several truck electrical features that would have been operated by conventional switches and relays in previous models. In a conventionally wired truck without a body control module, the operation of a device such as the electric horn is straightforward. A relay may be used to control the horn with a conventionally wired truck (**Figure 13-4**). One of the horn's terminals is typically wired directly to chassis ground, and the other horn terminal is connected to the normally open terminal of a horn relay. Closing the horn switch causes the horn relay to energize. The energized horn relay then supplies voltage to the horn coil, causing the horn to sound.

Some conventionally wired trucks may instead supply the horn with a constant +12V battery source on one horn terminal and connect the other horn terminal to the horn switch. Closing the horn switch completes the ground circuit for the horn, causing the horn to sound (**Figure 13-4**).

The electric horn circuit in a truck with a body controller is much different from a conventionally wired truck's electric horn circuit. The electric horn is controlled by the microprocessor in the body controller. In the example shown in **Figure 13-5**, the horn switch acts as an input device to the body controller. Closing the horn switch provides a path to a reference ground. For now, consider this reference ground to be about the same as chassis ground. Reference ground will be explained in more detail later in this chapter.

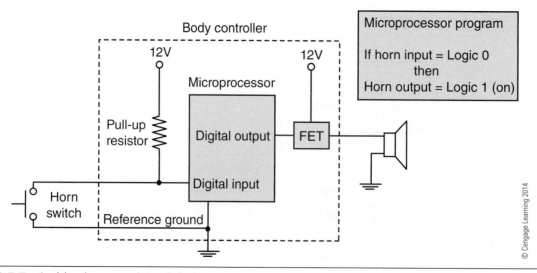

Figure 13-5 Typical body control module horn circuit.

The microprocessor detects the closed horn switch because the voltage at the body controller's pulled-up digital input assigned to the horn switch drops from approximately 12V with the horn switch *not* depressed to 0V with the horn switch depressed. The microprocessor's programming instructions cause the microprocessor to switch on a high side driver in response to the state of the horn switch input circuit. The body controller's electric horn high side driver then sources current to the horn, causing the horn to sound.

In the conventional electric horn circuit, the horn switch directly controlled the horn via the horn relay or the horn switch supplied a direct ground for the horn. In the body controller version of the horn circuit, the horn switch is just an input to the microprocessor. The microprocessor evaluates the horn switch input to determine whether the horn switch is being depressed by the truck operator. The microprocessor then makes the decision based on its programming instructions to determine if the high side driver used to supply voltage to the horn should be switched on or switched off. This all occurs so fast that you would not be able to determine whether the horn was being controlled by a microprocessor or was being directly controlled by the horn switch.

This concept of a switch acting as an input to a microprocessor and not directly controlling an electrical device is very important to understand when troubleshooting a truck with a body controller. There is no continuity or direct electrical connection between the horn switch and the horn in the body controller version of a horn system.

Body Controller Outputs

The body controller has three main types of outputs:

- High side drivers
- Low side drivers
- Messages transmitted on the J1939 data link

High Side Drivers. High side drivers, which are capable of sourcing a high level of current, are an important part of body control modules. The high side drivers used in the body controller are power MOSFETs (FETs). Some of these FETs are capable of sourcing up to 20A continuously and many times that amount of current for brief periods of time, such as with the surge current of a pair of headlamp lamps. Cold-filament headlamp lamp surge current may exceed 60A, as explained in **Chapter 9**.

Virtual Fusing. The FETs used as high side drivers in the body controller are integrated circuits and contain other components besides a transistor acting as the electronic equivalent of a relay or switch. These other components permit the amount of current that the FET is conducting to be measured. Such FETs are often referred to as smart FETs.

The FET shown in **Figure 13-6** has four terminals in addition to a ground terminal (not shown). One of the FET's terminals acts as the power supply and is connected to a +12V source inside the body controller. This power supply circuit will supply all of the load current (10A in this example) to the device being controlled by the FET, such as the low-beam headlamps. Another of the FET's terminals is the output terminal, which is connected to low-beam headlamps or similar load. The output terminal will conduct all of the current being sourced to the load (10A in this example), since the FET is in series with the supply voltage and the load. The FET's control terminal acts in the same manner as voltage that is supplied to a relay coil to energize a relay. A digital control voltage supplied to this control terminal (gate) by the microprocessor causes the FET to "switch on" similar to an energized relay. The fourth FET terminal is a current feedback terminal, which provides a voltage output that is proportional to the load current being sourced by the FET. As the current being sourced by the FET increases, the current feedback voltage also increases proportionally.

The FET's current feedback terminal is connected to a microprocessor analog input. The microprocessor converts this voltage into the appropriate current value. Thus, the microprocessor "knows" how much current is flowing through the circuit being sourced by the FET.

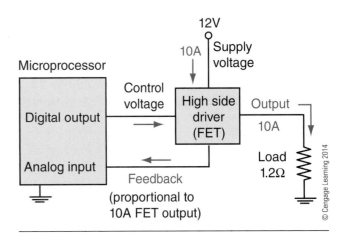

Figure 13-6 The high side driver measures the current being conducted and feeds back this information to the microprocessor as a proportional voltage.

The microprocessor can be programmed to cause the FET to be switched off if the circuit is drawing more current than expected. For example, a short to ground of the wire between the load, such as a horn, and the FET output causes a large amount of current to flow through the FET when the FET is switched on. The FET provides indication of the current flow to the microprocessor, which can be programmed to shut the FET off if current exceeds the programmed amount. Therefore, there is no need to use a conventional circuit protection device, such as a fuse, between the body controller horn output terminal and the horn, to protect the wiring. The current feedback information provided by the FET is supplied to the microprocessor, which causes the FET to act in a manner similar to a circuit breaker. The microprocessor can even be programmed to retry the circuit again to determine if the short to ground was intermittent, similar to the functionality of an automatic reset circuit breaker. This **virtual fusing** simplifies the electrical system by eliminating many of the conventional circuit protection devices. Virtual fusing is reset when the ignition is cycled.

The current feedback feature of the FET can also be used for diagnostics. This type of self-diagnosis performed by an electronic module is sometimes referred to as on-board diagnostics (OBD). The term on-board diagnostics or OBD has been used in the past by some OEMs to describe general self-diagnostics. However, the term OBD is now reserved for exhaust emissions related systems, as will be discussed in **Chapter 14**.

An example of self-diagnostics related to virtual fusing will be examined. Suppose the truck operator presses the electric horn switch on a truck with a body controller that has a short to ground of the circuit between the body controller and the horn. The microprocessor in the body controller determines that the horn circuit is drawing too much current, based on the current feedback from the FET, and shuts off the FET that is sourcing the horn circuit. This information can also be used to cause a specific **diagnostic trouble code (DTC)** or fault code to be stored into non-volatile

memory indicating that too much current flow through the horn circuit had been detected. This DTC can be retrieved by the technician when the truck is brought in for service. The DTC description would inform the technician that the reason the horn does not operate is that too much current is being drawn. Diagnostic trouble codes typically do not indicate specifically where the problem is located. The technician still has to find the cause of the excessive current draw in the circuit. A DTC can be useful in situations in which the circuit is only failing intermittently, such as in the case of a wire with insulation that has worn through on a sharp edge that is only occasionally making contact with chassis ground.

A disconnected horn connector or a wiring harness terminal that has become badly corroded would cause zero current or less than normal current to flow in the horn circuit when the horn FET is switched on. The FET measures this lower than expected current when the horn FET is switched on and sends this information back to the microprocessor. The microprocessor could cause a specific DTC to be stored into its non-volatile memory, indicating that there is less current flow than expected through the horn circuit. The technician would now be looking for an open circuit, not a shorted-to-ground circuit, for the cause of the inoperative horn.

Low Side Drivers. Low side drivers with current ratings of 1A or less are common in body control modules. A low side driver sinks a path to ground instead of sourcing current like a high side driver. These smaller electronic switching devices are typically used to energize relay coils or small solenoids, such as air-switching solenoids for compressed air-controlled features like power divider lock or sliding fifth wheel lock.

Interlocks. Using the body controller to control an air solenoid for features such as a power divider lock or sliding fifth wheel lock permits **interlocks** to be easily added to the control of these features. Interlocks are additional requirements that are necessary to permit some action to occur besides just depressing a switch or turning a key. For example, the interlock designed into your passenger car's electrical system prevents the engine from cranking unless the automatic transmission is in park or neutral. Interlocks are used to improve safety or to prevent damage from occurring from inadvertent activation of a feature.

The interlocks for many different features controlled by the body controller are written into the body controller's programming instructions (software). For example, the software can be written to not permit the tractor's sliding fifth wheel lock to be released if vehicle speed is over some value such as 5 mph (8 km/h), even if the truck operator depresses the switch for the fifth wheel lock. The body controller obtains the vehicle speed information from the J1939 data link, and uses this information in the decision-making process. Air solenoids for other truck features, such as locking differential, air suspension dump, and power take off (PTO), can be controlled by low side drivers located in the body controller. Information such as vehicle speed, engine speed, current automatic transmission range, clutch pedal state, and so on is included in the data that is transmitted on the J1939 data link by other electronic modules on the truck or by sensors or switches that may be wired directly to the body controller. This information can be used in the body controller's decision-making process.

J1939 Data Link Outputs. The body controller is connected to the J1939 data link. The body controller transmits various messages onto this data link, such as primary air pressure and cruise control switch state. Recipients of these messages include the electronic instrument panel cluster (IPC), the engine ECM, and the air ABS ECU.

Body Controller Inputs

There are three main types of inputs to the body controller:

- **Hardwired** inputs from switches and sensors
- Messages received from a proprietary (private) switch data link
- Messages received from the J1939 data link

Hardwired Inputs. The simplest type of body controller input is a hardwired switch. *Hardwire* describes the use of conventional electrical wiring techniques to

connect electrical devices. One of the important details about hardwired inputs is the method of grounding devices that act as inputs to an electronic module such as the body controller.

Reference Ground. The body controller has several pulled-up digital inputs and pulled-up analog inputs. A pulled-up input is an input that is connected to a pull-up resistor inside of an electronic module, as explained in **Chapter 11**. The switch that is connected to a pulled-up digital input terminal of the body controller provides a path to ground when closed. This causes the voltage measured at the body controller's pulled-up input terminal to decrease to near 0V. A common practice in automotive electronics is to make use of a reference grounding scheme instead of permitting any switch or sensor to provide a direct path to chassis ground for an electronic module input circuit. The electronic module provides this **reference ground** terminal, which is shared by all sensors and switches that act as inputs to the module. This reference ground is actually connected to chassis ground, but at only one location inside the electronic module. For example, the body controller has a ground terminal that is hardwired to chassis ground. Inside the body controller, the microprocessor, A/D converters, and all other devices are connected to a common ground point inside the body controller. This common ground point is then connected to one chassis ground location (**Figure 13-7**). International calls this reference ground the zero-volt reference (ZVR) for the body controller related circuits. Other OEMs may refer to their reference ground by other terminology.

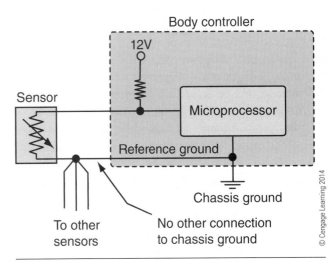

Figure 13-7 The common reference ground point inside of body controller is connected to chassis ground at only one location.

It may seem that it would be easier to connect a wire to chassis ground near the switch or sensor to use as a ground instead of supplying a reference ground for each sensor and switch that completes a ground circuit for a body controller input. There is a twofold reason for this standard industry practice of using a reference ground for all switches and sensors that act as inputs to a microprocessor-controlled device such as the body controller.

The first reason for routing a reference ground wire to each switch or sensor is that the analog to digital (A/D) converter inside the electronic module is like a DMM voltmeter. Any voltage that the microprocessor or the A/D converter measures is referenced or measured with respect to the reference ground, which is ultimately connected to chassis ground. Chassis ground is not at the exact same voltage level everywhere throughout the truck because of circuit resistance and because of current flow through the chassis ground resistance. The difference in the voltage level between different parts of the chassis may only be a few tenths of a volt, but this difference can cause a substantial inaccuracy in the voltage measured by the microprocessor at one of its inputs. For example, a 200mV inaccuracy in measuring air conditioning system high side pressure translates to a measurement error of approximately 20 psi (138 kPa).

The other reason for using a reference grounding scheme instead of connecting each sensor and switch to chassis ground involves susceptibility or vulnerability of the electronic module to high-frequency electromagnetic fields. Electromagnetic fields make wireless communications possible such as radio and cell phones. These invisible electromagnetic fields can cause major problems for truck electronic systems. An example of this was seen in the early attempts of truck air ABS in the mid-1970s. These ABS systems were very susceptible to electromagnetic fields and were ultimately removed from trucks because of this susceptibility, among other reasons. Disturbances caused by electromagnetic fields are referred to as **electromagnetic interference (EMI)**.

Vast improvements to automotive electronics and wiring techniques have occurred over the years to make vehicles much less susceptible to EMI. Some of these changes involve reference grounding techniques for sensors and switches that act as inputs to modules.

The susceptibility of the electrical system to EMI on modern trucks is significantly reduced if there is only one connection to chassis ground for the electronic module. This includes the ground circuit completed by each sensor and each switch that acts as an input to the electronic module. One reason for this is

that an open switch that provides a path to ground acts as a capacitor. The switch contacts are the two plates of the capacitor and air is the dielectric. A capacitor is an open circuit to direct current; but a capacitor is a like a short circuit to high-frequency alternating current, such as that produced by high-frequency electromagnetic fields. The chassis and wiring harnesses can act as an antenna for electromagnetic fields. Thus, a switch connected to chassis ground that acts as an input to an electronic control module can provide a path for high-frequency AC signals to enter the electronic module and cause a variety of intermittent problems. The unfortunate truck technician who has to troubleshoot these intermittent problems may never find the root cause of the complaint.

Each electronic module on the truck—such as the engine ECM, transmission ECU, and body control module—typically has its own independent reference ground networks for the device's hardwired input circuits.

> **CAUTION** *It is vital that any electronic system that uses a reference ground only be grounded in accordance with OEM recommendations. Even though the reference ground may appear to be the same as chassis ground, connecting the reference ground to chassis ground outside of the electronic module may result in the truck having intermittent EMI induced problems that are nearly impossible to duplicate or troubleshoot.*

The reference ground for each electronic module must also be kept isolated from other electronic modules' reference ground and from chassis ground. Otherwise, this second ground path increases the susceptibility of the body controller and any other electronic module to EMI. Also, do not use the reference as a ground for any device such as a light or solenoid. As current flow through the reference ground increases, the measurement error of any sensor input that uses that reference ground also increases.

Even though many electronic modules may have a metal case and may be bolted to cab sheet metal or the frame rail, the module's case is typically not internally directly connected to the module's reference ground. Instead, the metal case is typically connected to the internal reference ground through a capacitor inside the device to shed high-frequency electrical noise. However, other electronic modules such as some engine ECMs may be mounted in rubber isolators to keep the module case from contacting the chassis and from becoming grounded. It is important not to permit electronic module cases designed to be isolated from ground to contact a grounded component such as a dipstick tube or metal air compressor discharge line. Such contact can result in increased susceptibility of the module to EMI.

Electronic components can also emit or radiate high-frequency electromagnetic fields due to improper grounding techniques, thus causing EMI for other devices. These can cause audible noise in AM and FM radio reception or disturb two-way radio communications such as CB radios.

Diagnosable Switches.   A common electrical switch only has two states, open and closed. A switch is typically connected to the digital input of an electronic module in a microprocessor-controlled system. A digital input can only resolve two ranges of voltage levels, high and low. Voltages above some value are considered as logic high (1), and voltages below some value are considered as logic low (0). Based on the voltage measured at the digital input, the microprocessor can determine if the switch is opened or closed (**Figure 13-8**). A closed switch should have near 0Ω of resistance, while an open switch has a very high (infinite) resistance.

A standard stop lamp (brake) switch could be connected to the digital input of a body control module and the microprocessor could be programmed so that it would be able to detect that the brake pedal has been depressed based on the voltage level at the corresponding digital input. The body control module could use the state of the standard stop switch to control the truck's stop lamps and to disable cruise control.

However, an open circuit between the stop switch and a digital input cannot be detected by a microprocessor because an open switch with brake pedal not depressed and an open circuit due to a cut wire or disconnected switch "look" the same to the

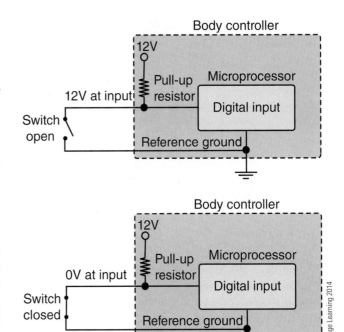

Figure 13-8 Open switch (upper) and closed switch (lower) with digital input.

microprocessor (**Figure 13-9**). This example shows a pulled-up digital input that is connected to a stop switch. The voltage at the module's digital input is 12V. Pressing on the brake pedal closes the stop lamp switch, but the voltage at the module's stop lamp digital input remains at 12V because of the open circuit. The stop lamps do not illuminate nor does the cruise control disengage if cruise was active when the brake pedal was depressed. The engine ECM does not "know" that the brake pedal has been depressed due to the open circuit in the stop switch circuit and does not disengage the cruise control when the brake pedal is depressed.

An open circuit in the stop lamp circuit would be considered a safety problem if it is not detected. One

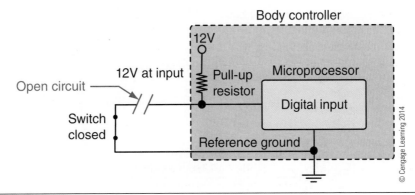

Figure 13-9 Open circuit prevents digital input from detecting that switch is closed; input voltage should be 0V with switch closed, but is actually +12V.

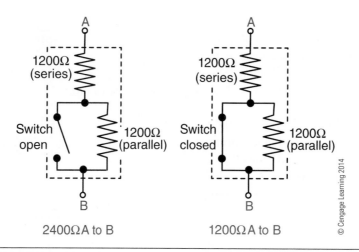

Figure 13-10 Diagnosable switch with contacts open and closed.

of the enhancements from using a body control module over a conventional electrical system is the ability to use a **diagnosable switch** for important circuits such as the stop switch circuit.

A diagnosable switch typically consists of two resistors and a switch contact contained inside the switch assembly. One resistor is in parallel with the switch contacts, and the other resistor is in series with the parallel-connected switch contacts and resistor (**Figure 13-10**). This may seem like an odd switch. After all, a switch is supposed to have near 0Ω of resistance with the switch closed and near infinite resistance with the switch open. However, a diagnosable switch is never used to directly control a load such as a light bulb. A diagnosable switch is only used as an input to an electronic module.

A conventional on-off switch, such as the electric horn switch, would typically be connected to a digital input terminal of an electronic module. However, a diagnosable switch is always connected to an analog input terminal of an electronic module. If the diagnosable switch contact is open, the two resistors inside

the diagnosable switch are connected in series. If this diagnosable switch were connected to a pulled-up analog input, the pull-up resistor inside the electronic module and the two resistors inside the diagnosable switch would all be in series. The resistors form a series voltage divider circuit when the pull-up voltage supplied by the module is applied to the circuit. The resistance of this particular diagnosable switch with the switch contacts open is 2400Ω. The voltage measured at the A/D converter or microprocessor at the input terminal of the electronic module would be the same as the voltage that is dropped across the combination of the two 1200Ω resistors inside the diagnosable switch, as shown in **Figure 13-11**. In this example, the voltage measured at the analog input is about 8V with the diagnosable switch contacts open. The microprocessor is programmed to "know" that a value of 8V present at this analog input means that the switch contact is open.

Closing the switch contact inside the diagnosable switch causes the resistor that is in parallel with the switch to be shorted by the switch. The equivalent

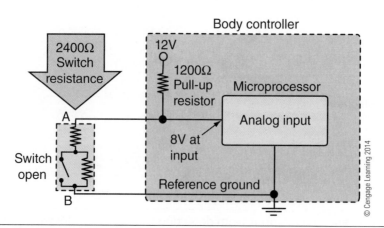

Figure 13-11 Diagnosable switch open; 8V at input terminal.

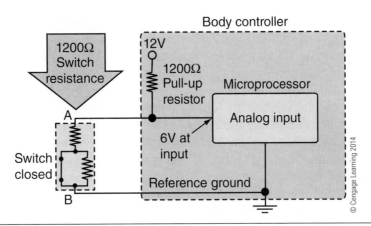

Figure 13-12 Diagnosable switch closed; 6V at input terminal.

resistance of the 0Ω switch and 1200Ω parallel resistor is 0Ω. Therefore, the resistance measured between the terminals of this particular diagnosable switch decreases to 1200Ω, as shown in **Figure 13-12**. The decrease in resistance for the diagnosable switch will result in a decrease in the voltage dropped across the terminals of the diagnosable switch. This will also cause a decrease in the voltage measured by the microprocessor or A/D converter at the analog input terminal connected to the diagnosable switch. In this example, the voltage decreases from 8V with the switch contacts open to 6V with the switch contacts closed. The microprocessor is programmed to "know" that 6V measured at this diagnosable switch analog input means that the diagnosable switch contacts are closed.

The reason for using a diagnosable switch is to verify that the switch circuit is intact even with the switch contacts open. It would not be possible to use a

standard switch as a diagnosable switch because it would not be possible for the electronic module to determine the difference between an open switch and an open circuit in the wiring harness between the switch and the electronic module.

Disconnecting the connector on a diagnosable switch or opening the circuit somewhere between the switch and body controller causes the voltage at the analog input to increase to 12V (**Figure 13-13**). In this example, the 1200Ω pull-up resistor in the body controller is supplied with battery voltage. The microprocessor could be programmed to "know" that a measurement of battery voltage (+12V) at this analog input indicates that the switch circuit is open. The microprocessor can then set a DTC for this open circuit and take other action as necessary. If the diagnosable switch is used as the stop lamp switch, then the body control module can request that the engine ECM inhibit cruise control operation, via a J1939 data link

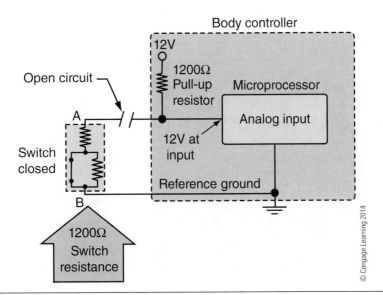

Figure 13-13 Diagnosable switch open circuit; 12V at input terminal.

message, when the stop switch circuit is detected as being faulted. Otherwise, depressing the brake pedal with an open stop lamp switch circuit would not cause cruise control to disengage.

The other failure that a diagnosable switch can detect is a shorted-to-ground circuit. Shorting the diagnosable switch circuit to ground causes the voltage at the body control module analog input to drop to 0V. The microprocessor can be programmed to "know" that a measurement near 0V at this analog input indicates that the stop switch circuit is shorted to ground somewhere between the body control module stop switch input terminal and the switch. The microprocessor can then set a different DTC for this shorted circuit and take action, such as inhibiting cruise control, because it is not possible to determine if the stop switch is closed or open with the stop switch circuit shorted to ground.

Switch Packs. International trucks with body controllers may have switch packs that communicate with the body controller using a J1708/J1587 specification data link. This proprietary data link is not the same as the J1708/J1587 data link that is connected to the powertrain control modules, such as the engine ECM. International refers to this proprietary data link as the switch data link.

An International truck with a body controller may have two or more 6-switch packs and one 12-switch pack of switches. The switches control optional features such as fog lamps, mirror heat, air suspension dump, and other electric or air-controlled devices. The status of all of these switches is communicated to the body controller via the switch data link. This is a form of time-division multiplexing, as explained in **Chapter 11**.

The rocker switch in these switch packs is not an electrical switch at all, but rather a switch actuator. The actuator snaps into place in the switch pack and has graphics that indicate what feature the actuator controls, such as mirror heat. The switch actuators may be two- or three-position latching-type switches or momentary-type switches, depending on the feature that the switch controls. For example, the actuator used to control sliding fifth wheel lock is a two-position momentary-type switch, while the fog lamp actuator is a two-position latching-type switch. In all cases, protrusions on the back side of the actuator cause one of two different **microswitches** mounted on a circuit board to close (**Figure 13-14**). Microswitches are small switches that are contained in a sealed package. One microswitch is beneath the top switch actuator protrusion; the other switch is located beneath the lower switch actuator protrusion. Placing a switch actuator, such as the fog lamp switch actuator, in the ON position causes the lower microswitch beneath the switch actuator to be closed, while placing a switch in the OFF position causes the upper microswitch beneath the switch actuator to be closed. Three-position switch actuators cause neither microswitch to be closed with the switch actuator in the center position.

The switch pack communicates with the body controller via the switch data link. The switch pack transmits digital messages to the body controller indicating the status of each microswitch on the switch pack circuit board. The body controller microprocessor is specifically programmed for each truck to "know" what each switch in each switch pack is designed to control. For example, the first switch location in a six-switch pack may be designated as mirror heat. The body controller microprocessor is programmed to "know" that if a message is received over the switch data link indicating that the first

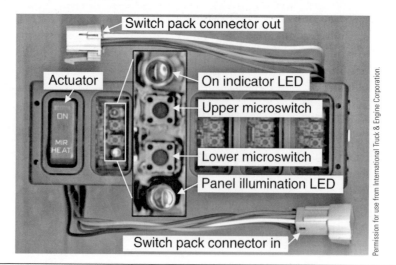

Figure 13-14 Switch pack with switch actuators removed.

switch actuator is in the ON position, the truck operator is requesting mirror heat. The body controller microprocessor responds by energizing the mirror heat high side driver provided all other conditions in body controller programming for mirror heat activation are also met.

Problems such as missing switches or extra switches in any of the switch packs cause the body controller to set a corresponding DTC. Also, placing a three-position switch in a switch pack position that has been programmed as a two-position switch will also cause the body controller to set a DTC with the switch in the center position. It is important to note that the body controller cannot determine the information label on the switch such as fog lamps or mirror heat. Swapping the fog lamp and mirror heat switch positions in a switch pack will cause the fog lamps to illuminate when the mirror heat switch is placed in the "ON" position and vice-versa. The body controller programming can be modified using International's Diamond Logic® Builder electronic service tool to accommodate the relocation of switches within switch packs or to add or remove electrical system features.

J1939 Data Link Inputs. The body controller is connected to the SAE J1939 data link. This data link is connected to other electronic modules, including the electronic instrument panel cluster, engine ECM, air ABS ECU, and automatic or automated mechanical transmission ECU. The body controller receives a great deal of information from these modules. This information acts as an input to the body controller.

INTERNATIONAL DIAMOND LOGIC ELECTRICAL SYSTEM DETAILS

This section provides detailed information about many of the electrical features controlled by the body controller.

Exterior and Interior Lighting

The body controller manages most of the vehicle lighting through the use of high side drivers. These high side drivers are power MOSFETs (FETs) capable of supplying 10A or 20A of current.

Tech Tip: Think of high side drivers as being like conventional relays. A small signal provided by the microprocessor is used to control a large amount of current.

Headlamps and Running Lamps. The body controller uses one high side driver to supply the pair of low-beam headlamps and another high side driver to supply the pair of high-beam headlamps (**Figure 13-15**).

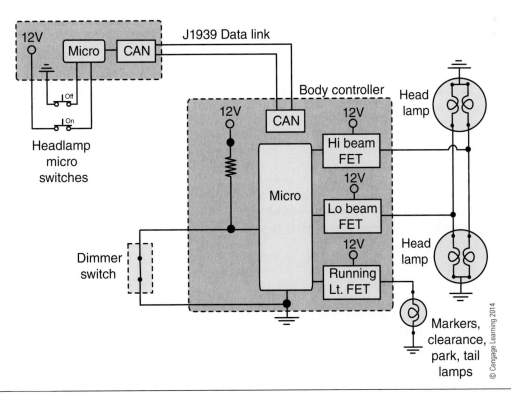

Figure 13-15 Body controller headlamp circuit. EGC contains headlamp switch.

The multiplexed instrument panel cluster, referred to as an **electronic gauge cluster (EGC)** by International, contains the headlamp switch actuator. The small three-switch circuit board on the bottom left side of the gauge cluster contains two microswitches for each switch actuator and is similar to the six-switch packs described earlier.

The EGC communicates with the body controller via the J1939 data link. Placing the headlamp switch in the park lamp position causes the EGC to transmit a message onto the J1939 data link requesting the body controller to switch on the running lamp FET. The body controller receives this J1939 message and switches on the running lamp FET. This FET sources +12V (battery voltage) to all of the parallel-connected park and running lamps on the truck. Each lamp is grounded to chassis ground in the same manner as any other truck.

It is important to note that the body controller is only providing the high side (+12V) for the loads such as headlamps and running lamps. A large electrical cable provides the battery positive voltage supply to the body controller, yet the body controller ground wire is just a 12 AWG wire. The current that leaves the body controller via the high side drivers to the various loads, like headlamps, does not return to the body controller but rather to the battery negative terminal through the chassis ground.

Placing the headlamp switch in the headlamp position causes the EGC to transmit a message onto the J1939 data link requesting the body controller to switch on the low-beam headlamp FET and the running lamp FET. This causes these two FETs in the body controller to supply +12V to the low-beam headlamps and the running lamps.

The headlamps can be switched to high beams by use of the dimmer switch. The dimmer switch is part of the multifunction turn-signal switch similar to those found in many passenger cars. This dimmer switch acts as a digital input to the body controller. Toggling the dimmer switch causes the body controller to switch on the high-beam FET and switch off the low-beam FET. The body controller also transmits a J1939 message indicating that the high-beam headlamps are illuminated. The EGC detects this message and causes the high-beam indicator LED in the instrument cluster to illuminate for as long as the headlamps are in the high-beam position. Toggling the dimmer switch again causes the headlamps to return to low beams. The body controller also transmits a message on the J1939 data link indicating that the high-beam headlamps are off, which causes the EGC to switch off the high-beam indicator LED.

Daytime running lights (DRL) describe reduced-intensity low-beam headlamps on some trucks. DRL on International trucks are obtained by the body controller switching the low-beam headlamp FET off and on at a rapid rate so that the level of headlamp illumination is decreased. This is often referred to as pulse width modulation (PWM). The same concept is used for controlling the current flow in the alternator rotor for voltage regulation and for instrument panel illumination. The body controller can be programmed so that the DRLs are illuminated when the park brake is released. The park brake switch is a digital input to the body controller.

An optional marker interrupt switch is located in a switch pack. Truckers use the marker interrupt switch to flash their running lamps off and on as a means of signaling a "thank you" when completing a pass of another truck and the other truck has flashed its high-beam headlamps to indicate that it is clear to change lanes. Placing the momentary contact marker interrupt switch in the ON position causes the switch pack to generate a corresponding digital message that is transmitted over the switch data link. The body controller interprets this digital message on the switch data link and causes the running lamp high side driver to be switched off for as long as the marker interrupt switch is held in the interrupt position. In addition, depressing the marker interrupt switch with the running lamps off causes the running lamps to illuminate for as long as the marker interrupt switch is held in the interrupt position.

Turn-Signal and Hazard Lamps. In a conventionally wired truck turn-signal system, the turn-signal switch is used to direct current flow to the left or right side turn signals. The turn-signal flasher opens and closes a set of contacts to make and break the current through the circuit to cause the turn-signal lamps to flash. The stop lamps and the hazard flasher may also be integrated into this conventional system, as described in **Chapter 9**.

The turn-signal system used in International trucks with a body controller is much different than a conventionally wired truck turn-signal system (**Figure 13-16**). The turn-signal switch contains two switches that act as two separate inputs to the body controller. One switch contact is for right turn, and the other is for left turn. The two switches are wired to two body controller pulled-up digital inputs. The body controller's turn-signal outputs are four independent 10A high side drivers, one for each corner of the truck.

Placing the turn-signal switch in the left turn position completes a path to reference ground for the corresponding pulled-up body controller input terminal. The closed left turn switch pulls down the voltage at the

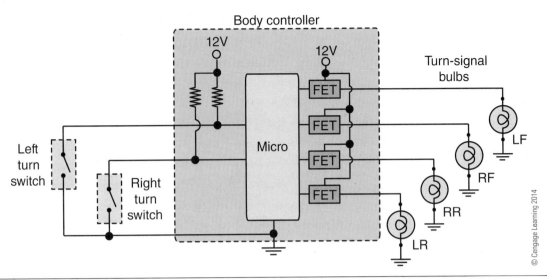

Figure 13-16 Turn-signal system on International trucks with a body controller. Two switches are inputs; four high side drivers are outputs.

body controller left turn input terminal to near 0V. The microprocessor in the body controller detects 0V at the left turn signal digital input. This causes the microprocessor to switch on and off the left front and left rear turn signal high side drivers to directly supply current to the left turn signal lamps. There is no flasher in this system because the microprocessor is programmed to pulse the two left or two right side turn signal high side drivers on and off at a specific rate of one-third second on and one-third second off (90 flashes per minute).

The right turn signal operates in the same manner as the left turn signal except that the right turn signal digital input and the right side turn signal FETs are used.

The hazard switch (four-ways) causes both the left and right turn signal switches located in the multifunction switch assembly to close. This causes the voltage measured at the pulled-up body controller right and left turn signal switch inputs to drop to near 0V. This causes the body controller to flash the turn signal lamps on all four corners of the truck by cycling the four turn signal high side drivers off and on in the same manner as during a turn.

The left and right turn signal FETs for the front turn signals can also be used to energize the high side of the trailer turn signal lamp relay coils. Highway trailers typically have separate stop/turn lamps. The front turn signal FETs are used to control the trailer's turn signal relays because the truck's front turn signals continue to flash when the brake pedal is depressed, unlike the combination stop/turn lamps at the rear of the truck or tractor.

Stop Lamps. The diagnosable stop lamp switch described earlier in this chapter is an input to the body controller. Combination stop/turn lamps at the rear of the truck use the rear turn signal filaments as the stop lamps, as discussed in **Chapter 9**. Because four separate high side drivers or FETs are used to control each of the truck's turn signals independently, activating the stop lamps only requires the body controller to switch on the two rear turn signal FETs. Turn signal and stop lamps together are obtained by the body controller flashing one of the rear turn signals (cycle the FET off and on) and illuminating the other turn signal steadily to act as a stop lamp.

Stop lamps for trailers with separate stop/turn lamps, such as those on most highway trailers, are provided by a body controller-controlled relay. The high side of the relay coil is connected to battery-positive voltage; the low side of the relay is controlled by a body controller low side driver. The body controller energizes this low side driver anytime the brake pedal is detected as being depressed. The output of the trailer stop lamp relay is supplied to the trailer stop lamp circuit. This permits the truck or tractor to have combination rear stop/turn lamps and the trailer to have separate stop/turn lamps.

Tech Tip: The body controller only powers the applicable trailer lighting relay control circuit. The body controller does not directly supply the current to the trailer lighting. A conventional CPD such as a fuse or circuit breaker in the PDC is used to protect the trailer wiring, not virtual fusing.

Fog Lamps. Optional fog lamps are controlled by a switch in one of the switch packs. Closing the fog lamp switch causes the corresponding switch pack to generate a digital message on the switch data link. The body controller responds to the request for fog lamps from the switch pack and evaluates whether the low-beam headlamps are illuminated. If the low-beam headlamps are illuminated, the body controller will switch on a high side driver to illuminate the fog lamps. The body controller will also transmit a message on the switch data link to command the switch pack to illuminate the green indicator LED in the switch pack beneath a clear window in the fog lamp switch actuator, indicating to the truck operator that fog lamps are illuminated. If the high-beam headlamps are then switched on, the fog lamp FET is switched off and the body controller commands the microprocessor in the switch pack via the switch data link to switch off the fog lamp green indicator in the switch pack.

Dome Lamp. The body controller also controls the interior dome lamp and optional courtesy lamps. The parallel-connected door switches drive a body controller pulled-up digital input down to near 0V when a door is opened. The body controller responds by switching on a high side driver that supplies voltage to the dome lamp and courtesy lamps. The interior illumination is also dimmed theater-style when the door is closed through PWM of the dome lamp high side driver. This causes the dome lamps to decrease gradually in intensity when the door is closed. The body controller can also be programmed to switch off the dome lamp after a period of time if the door is left open with the key switch in the OFF position.

Air Conditioning

The air conditioning (A/C) system on 2001–2006 International High Performance Vehicles with an ESC is a cycling clutch orifice tube (CCOT) type system. The ESC uses a high side driver to supply voltage to the A/C clutch coil. This system is capable of diagnosing several A/C system problems and warns the truck operator of A/C system problems, such as low refrigerant level, through a Check A/C indicator lamp in the EGC. The A/C system on 2007 and later models with a body controller is a thermal expansion valve system but still makes use of much of what is described in this section.

There are several inputs to the ESC that are used by the ESC's microprocessor to determine when to cycle the A/C clutch high side driver off and on to provide the optimal cooling performance (**Figure 13-17**).

The HVAC control head assembly provides a ground to an ESC pulled-up digital input when the controls are in a position where A/C is being requested. The ESC detects the A/C request based on the voltage level at this digital input.

Two thermistors are inserted in the refrigerant flow in the low side of the A/C system. One thermistor is located downstream of the orifice tube at the evaporator inlet and the other is located at the evaporator outlet. The two thermistors are hardwired to two of the ESC's analog inputs. These thermistors are used to measure the temperature of the refrigerant at the evaporator inlet and evaporator outlet.

The information that the ESC receives from the evaporator inlet thermistor is used to cause the ESC to cycle the A/C compressor clutch off and on based on evaporator temperature. This is similar to the function of the cycling switch used in conventional CCOT air

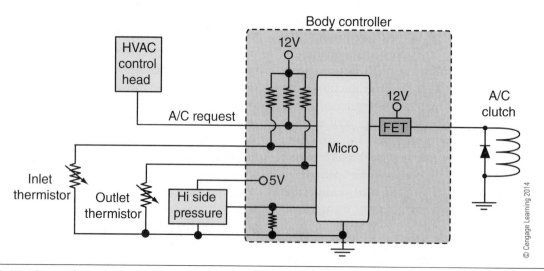

Figure 13-17 Air conditioning system inputs; output is A/C clutch high side driver.

conditioning systems. The A/C clutch on any CCOT system must be cycled off during A/C operation when evaporator core temperature drops near 32°F (0°C) to prevent moisture on the exterior of the evaporator from freezing and blocking the flow of air through the evaporator.

The difference in temperature measured by the evaporator inlet and evaporator outlet thermistors is used to determine the refrigerant charge level in the system. A low refrigerant charge level causes the evaporator inlet port to maintain a much lower temperature value than the evaporator outlet port. This temperature difference between evaporator inlet and outlet temperatures becomes much greater as the refrigerant charge level in the system decreases. If the ESC determines that the refrigerant charge level is low, the ESC will transmit a message on the J1939 data link requesting that the EGC illuminate the Check A/C indicator lamp. The ESC will also inhibit A/C compressor operation if refrigerant charge drops to a very low level to protect the compressor.

A pressure sensor in the high side of the A/C system is hardwired to an ESC analog input. This pressure sensor is used for three purposes. First, the A/C pressure sensor is used to determine whether there is sufficient refrigerant in the system for safe A/C operation. An A/C system that has a very low level of refrigerant charge should not be operated. The pressure of R134A refrigerant is also very low when ambient temperatures are low. Thus, the high side pressure is also an indication of ambient temperature if the compressor has not been recently operated. The ESC inhibits A/C operation in low ambient temperatures based on the combination of pressure sensor and thermistor inputs.

Second, the high side pressure sensor is used as a fan switch if the truck has an on-off type of engine cooling fan drive. When high side pressure is greater than approximately 285 psi (1.9 MPa), the ESC transmits a message on the J1939 data link requesting that the engine ECM switch on the fan drive until the high side pressure decreases.

Third, the high side pressure sensor is used to protect the A/C system from excessive high side pressures. This is similar to the function of a high-pressure cut-out switch used in conventional A/C systems. Problems with the A/C system, such as restricted airflow through the condenser, will cause the high side pressure to rise to very high levels. The ESC will switch off the A/C clutch if high side pressure exceeds the high-pressure limit of 420 psi (2.9 MPa).

Cruise Control

The body controller is used in the cruise control system. Four switches in the steering wheel—on, off, set, and resume—are reduced to a single body controller analog input through analog multiplexing (**Figure 13-18**), as described in **Chapter 11**. A clockspring in the steering column is used to route the circuits between the body controller and the steering wheel switches. This clockspring contains a tape with four circuits: cruise control, electric horn, air horn, and reference ground. ProStar models may also have additional circuits for headlamp control and engine brake.

Depressing any of the cruise switches in the steering wheel causes the voltage at the body controller cruise control switch analog input to decrease as the pull-up resistor and the resistor in series with the corresponding cruise control switch form a series voltage divider circuit. This causes the body controller to transmit a message on the J1939 data link to the

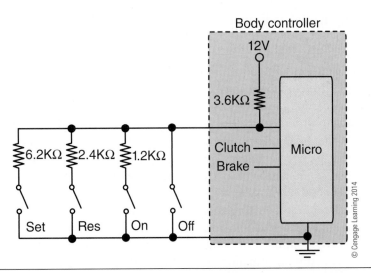

Figure 13-18 Cruise control related inputs; output is J1939 message to engine ECM.

engine ECM corresponding to the cruise control switch that was depressed. The engine ECM has full authority over cruise control and causes the appropriate action if all conditions for cruise control are met. The body controller also uses a diagnosable clutch switch and the diagnosable stop switch as inputs. The status of these switches is also transmitted over the J1939 data link. Depressing the brake pedal or the clutch pedal causes the body controller to transmit a J1939 message indicating that the brake or clutch is depressed. The engine then takes the appropriate action, such as deactivating cruise control.

The diagnosable stop and clutch switches are important parts of this system. A wiring problem such as an open circuit in the stop switch circuit causes the body controller to transmit the status of the stop switch as faulted. The body controller then inhibits cruise control operation until the problem is repaired. Otherwise, a faulted stop switch circuit would prevent the engine ECM from detecting that the brake pedal has been depressed and it would not disengage cruise control when the brake is depressed.

The cruise control switches are also used as fast idle or PTO speed control switches. This permits the engine speed to be increased and decreased. The J1939 messages sent by the body controller are identical to the messages sent for cruise control, except that the engine ECM uses the information to regulate engine speed instead of vehicle speed.

Instrumentation and the Body Controller

The body controller and the EGC communicate with each other over the J1939 data link. The body controller is used to control several warning and indicator lamps in the EGC. The body controller transmits messages over the J1939 data link indicating that the left or right turn signal is active, high-beam headlamps are on, parking brake is set, and so on. The EGC responds by illuminating or flashing the appropriate indicator LED. The EGC also contains a speaker that can be used as an audible alarm.

The body controller is also used to provide information for several gauges in the EGC. Air pressure sensors are located in the primary and secondary air system. These sensors are analog inputs to the body controller (**Figure 13-19**). The body controller supplies a regulated +5V and reference ground for these sensors. The pressure in the primary and secondary air system is transmitted by the body controller over the J1939 data link. The EGC uses this information to drive the stepper-motor-type primary and secondary

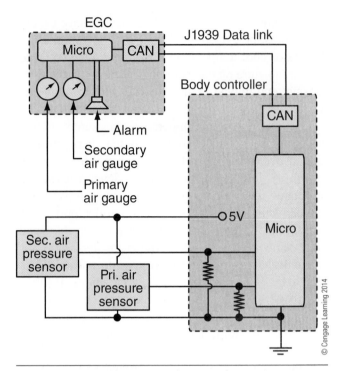

Figure 13-19 Primary and secondary air pressure measurement system.

air pressure gauge needles to indicate the corresponding system pressures. The EGC will also illuminate a warning LED and sound the audible alarm if either the primary or secondary air pressure drops below about 72 psi (496 kPa).

If either of the body controller's air pressure sensor circuits is detected as an open circuit or shorted to ground by the body controller, the body controller will command the EGC to drive the corresponding gauge's needle to the six o'clock position. This straight down position is well outside the normal limits of the gauge and indicates that the corresponding circuit is faulted. Most other gauges respond in the same manner when the associated hardwired circuit is faulted. This provides an indication to the truck operator that the information for the gauge is not available.

Windshield Wipers

The body controller is used to control the windshield wipers. A high side driver (20A FET) is used to provide power for the windshield wiper motor. This high side driver is switched on anytime the ignition is active. Two low side drivers are also used to control two wiper relays identified as the wiper power relay and the high-low speed relay (**Figure 13-20**).

The wiper switch shown in **Figure 13-21** acts as three different digital inputs to the body controller. These three switches are identified as switch 0,

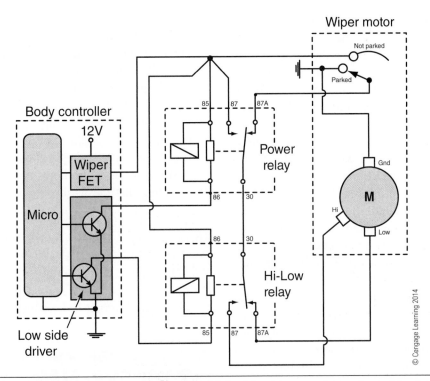

Figure 13-20 Windshield wiper system outputs; one high side driver and two low side drivers.

switch 1, and switch 2. There are eight wiper switch positions: off, five intermittent wiper levels, low speed, and high speed. Placing the wiper switch in each of these eight positions causes a different combination of the switches to be closed or opened, as shown in **Figure 13-21**. You may notice that this figure has a truth table for the various wiper switch positions. The body controller digital inputs detect which of the three wiper switches are closed and which switches are opened. The body controller microprocessor is programmed to "know" what each switch combination means and controls the wipers accordingly. For example, the combination where switch 2 and switch 1 are both open and switch 0 is closed indicates that the wiper switch is in the low-speed position.

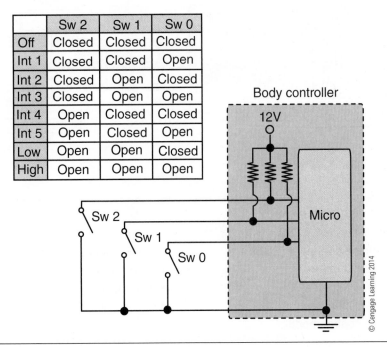

	Sw 2	Sw 1	Sw 0
Off	Closed	Closed	Closed
Int 1	Closed	Closed	Open
Int 2	Closed	Open	Closed
Int 3	Closed	Open	Open
Int 4	Open	Closed	Closed
Int 5	Open	Closed	Open
Low	Open	Open	Closed
High	Open	Open	Open

Figure 13-21 Wiper switch inputs and truth table.

If the body controller detects that the wiper switch is in the low-speed position, the body controller switches on the low side driver for the wiper power relay shown in **Figure 13-20**. This causes the wiper power relay to close and voltage to be provided to the wiper motor through the normally closed contacts of the wiper high-low speed relay.

If the wiper switch is placed in the intermittent 1 position, switch 2 and switch 1 would both be closed and switch 0 would be open. This switch combination would cause the body controller to determine that the wiper switch is in the intermittent 1 position. The difference between the five intermittent wiper positions is the time delay or dwell between wiper pulses. In the intermittent 1 position, the body controller would pulse the wiper power relay on for about one-half second to cause the wiper motor to move off the park switch. The wipers then make one sweep and park through the normal wiper park circuit. The body controller would not provide another half-second pulse until 15 seconds later. Changing to a different intermittent wiper position causes the body controller to adjust the dwell time between pulses accordingly.

The windshield washer switch is also monitored by the body controller. Depressing the windshield washer switch with the wipers off causes the body controller to switch on the low-speed wiper relay for several seconds to clean the windshield.

Tech Tip: An open windshield wiper switch connection on an International truck with a body controller causes the wipers to operate at high speeds any time the key switch is in the ignition position. Additionally, a short to ground of the windshield washer pump control circuit will cause the windshield wipers to operate at low speed any time the key switch is in the ignition position because the body controller is falsely detecting that the washer switch is depressed.

There are many other features of the International Diamond Logic electrical system not covered in this chapter. Additional information on the International multiplexed electrical system can be found by searching the Internet links at the end of this chapter.

FREIGHTLINER MULTIPLEXED ELECTRICAL SYSTEM

Freightliner refers to their multiplexed electrical system as the SmartPlex™ electrical system. This system is found on Freightliner Business Class M2 models, 108SD, and 114SD severe duty models. These trucks have at least two standard electronic modules to control body and chassis electrical system features (**Figure 13-22**) as well

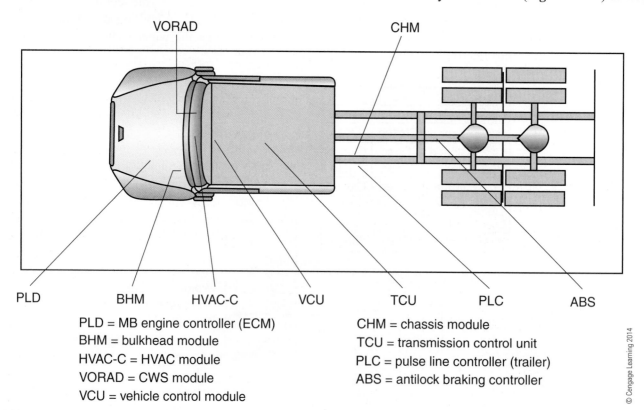

PLD = MB engine controller (ECM)
BHM = bulkhead module
HVAC-C = HVAC module
VORAD = CWS module
VCU = vehicle control module

CHM = chassis module
TCU = transmission control unit
PLC = pulse line controller (trailer)
ABS = antilock braking controller

© Cengage Learning 2014

Figure 13-22 Freightliner SmartPlex electronic module locations.

Figure 13-23 Freightliner bulkhead module (BHM).

as several optional electronic modules. The **bulkhead module (BHM)** acts as the primary command module for the body and chassis electrical system. The bulkhead module is mounted to the left side of the front bulkhead (front wall) and is connected to the external truck wiring harness through four sealed electrical connections, as shown in **Figure 13-23**. An opening in the front bulkhead behind where the BHM is mounted permits the BHM to be connected to the interior cab wiring harness through three sealed electrical connectors. The **chassis module (CHM)** shown in **Figure 13-24** is the other standard electronic module and is typically mounted to the chassis behind or beneath the cab. The CHM is a slave or dependent module that receives its commands from the BHM via the J1939 data link.

Bulkhead Module (BHM)

The BHM is the main electronic controller for the Freightliner SmartPlex electrical system. The BHM is

Figure 13-24 Freightliner chassis module (CHM).

connected to the J1939 data link and transmits and receives information from the other electronic modules on the truck, including the engine ECM, multiplexed **instrument cluster unit (ICU display)**, and CHM. The BHM is hardwired to several inputs in the cab. These inputs include a horn switch, hazard switch, passenger and truck operator door switches, back-up lamp switch, headlamp switch, A/C request, and input from optional **smart switches**, as discussed later in this section. Other inputs to the BHM are messages received on the J1939 data link transmitted by the other electronic modules on the truck.

Direct hardwired high side driver outputs of the BHM include horn, dome lamp, left side low-beam and high-beam headlamps, tail lamps, A/C clutch, windshield washer pump, and windshield wipers. Other outputs of the BHM are the messages transmitted by the BHM on the J1939 data link.

Smart Switches. Smart switches, as described in the discussion of the Freightliner electrical system, are special switches used to control a specific electrical feature such as fog lamps or engine brake. Each smart switch is hardwired to three BHM analog inputs and one BHM low side driver. The BHM also provides the power and ground for smart switch panel illumination backlighting. Up to five smart switches can be directly connected to the BHM.

The smart switch contains two identification resistors with values between 1020Ω and $10k\Omega$, as shown in **Figure 13-25**. These two resistors shown as ID 1 and ID 2 are used by the BHM to identify the specific function of the smart switch, such as mirror heat. A mirror heat switch has a value of 1020Ω for ID 1 resistor and 1620Ω for ID 2 resistor. A marker interrupt switch has a value of 1020Ω for both ID resistors. Other switches have unique values for the combination of ID 1 and ID 2. The smart switch identification resistors are connected to switch terminals 7 and 8 as shown in **Figure 13-25**. For a smart switch that is connected to the BHM, smart switch terminals 7 and 8 are hardwired to two BHM analog inputs. The BHM can determine the specific function of the smart switch based on the voltage measured at these two analog inputs.

A smart switch provides three different values of resistance when the switch is placed in each of the three different positions (up, center, down). Note that some smart switches are two-position on-off switches and do not latch in the center position. The smart switch position output terminal, shown as terminal 2 in **Figure 13-25**, is hardwired to a third BHM analog input. The BHM determines what state the switch is in

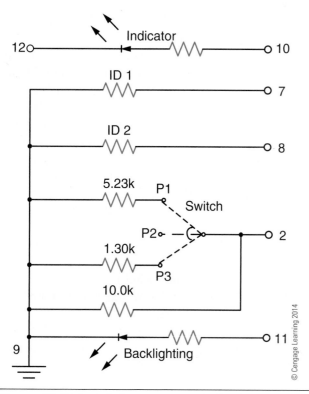

Figure 13-25 Freightliner smart switch schematic.

based on the voltage measured at this analog input. For a three-position smart switch, in position P1 the switch provides a resistance of 3.43kΩ (5.23kΩ in parallel with 10.0kΩ), in position P2 a value of 10.0kΩ, and in position P3 a value of 1.15kΩ (1.30kΩ in parallel with 10.0kΩ). The resistance of the smart switch acts as a voltage divider with the pull-up resistor within the BHM. The BHM then causes the requested switch action to occur, such as mirror heat on or off, if conditions for the output action are correct. The output action of the BHM may include commanding the CHM via the J1939 data link to energize one of its hardwired outputs if the device controlled by the smart switch is hardwired to the CHM.

An indicator LED in the smart switch is controlled by a BHM low side driver. The BHM energizes the driver to illuminate the indicator LED in the smart switch when the indicated switch function is active. If an error condition exists, the BHM can cause the indicator LED in the smart switch to flash.

An additional six smart switches can be connected to optional switch expansion modules (SEM). The SmartPlex electrical system can accommodate up to four switch expansion modules. Electrically, the functionality is the same if the smart switch is hardwired to a SEM instead of the BHM. The SEM is a dependent module and transmits smart switch identification and switch position information onto the

J1939 data link. The BHM contains the logic (program instructions) to determine the appropriate action based on the information transmitted by the SEM. The BHM also commands the SEM via the J1939 data link to switch on or switch off the switch indicator LEDs on smart switches that are hardwired to the SEM.

The BHM programming instructions are unique to each truck. Therefore, the BHM can detect problems with smart switches such as: duplicate smart switches, extra smart switches, or missing smart switches. The BHM can set DTCs indicating the specific problem and disable the applicable feature controlled by the smart switch. The BHM can also be reprogrammed using Freightliner ServiceLink® electronic service tool to accommodate the installation of additional electrical system features controlled by smart switches.

Chassis Module (CHM)

The chassis module (**Figure 13-24**) is mounted to the frame rail at the rear of the cab or other chassis location such as beneath the cab. The CHM contains high side driver FETs that source current to several chassis electrical features. The CHM is a dependent module of the BHM and receives its commands from the BHM via the J1939 data link. The CHM also has some hardwired inputs including low air pressure, service stop switch, and park brake switch. The CHM transmits the status of its hardwired inputs onto the J1939 data link.

High side driver outputs of the CHM include park and marker lamps, back-up lamps, turn signal lamps, right side high- and low-beam headlamps, and fog lamps. Other hardwired outputs of the CHM are utilized to control up to four air solenoids in the **air management unit (AMU)** discussed later in this section.

The CHM also acts as a pass-through device for some circuits. For example, the tail lamps are supplied current by a BHM high side driver. The BHM output for the tail lamps is connected to a CHM pass-through terminal. This tail lamp circuit passes through the CHM to two CHM tail lamp output terminals and a CHM trailer tail lamp relay control output terminal.

An optional expansion module (EXM) is found on some trucks to control additional electrical features. The EXM has the same electrical pin-out and connectors as the CHM.

Air Management Unit (AMU)

The optional AMU shown in **Figure 13-26** contains air system switches and air solenoids which are used to control various air-powered features on the truck.

Figure 13-26 Air management unit (AMU).

Solenoids in the AMU are controlled by the CHM via hardwire. The AMU also contains some air switches such as the low air pressure and park brake switches. The switches are hardwired inputs to the CHM, which transmits the status of the switches onto the J1939 data link.

Accessory Air Valve Assembly (AAVA)

The **Accessory Air Valve Assembly (AAVA)** has replaced the AMU on model year 2010 and later trucks. The AAVA contains only modular air solenoids.

Air pressure switches that were previously located in the AMU have been relocated inside the cab as individual pressure switches. The air solenoids in the AAVA may be controlled by the CHM or some other electronic module such as the engine ECM.

Instrument Cluster Unit (ICU display)

The instrument cluster unit (ICU display) found on the Freightliner Business Class M2 and SD models is a multiplexed instrument cluster that transmits and receives messages on the J1939 data link. The ICU has an LCD display for displaying odometer and other information (**Figure 13-27**).

The ICU also has three analog inputs that are hardwired to a multifunction switch. The multifunction switch is the combination turn signal, wiper/washer, and headlamp dimmer switch. Additionally, a hazard switch is hardwired to the BHM. The multifunction switch has three **resistive ladder networks**, also known as ladder switches, which are used for turn signal, wiper speed or intermittent dwell selection, and headlamp dimmer state. The three resistive ladders are a form of analog multiplexing, as described in

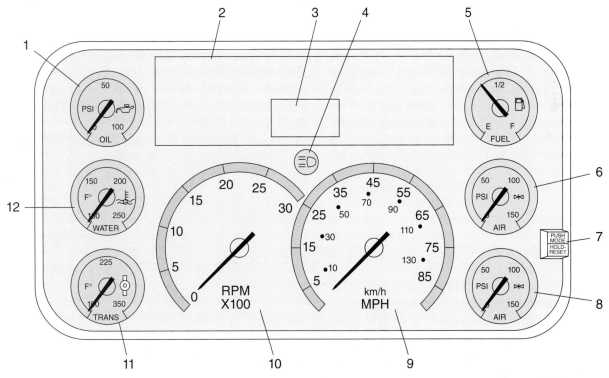

1. Engine oil pressure gauge
2. Dash message center
3. Dash driver display screen
4. Headlight high-beam indicator
5. Fuel level gauge
6. Primary air pressure gauge
7. Mode/reset switch (optional)
8. Secondary air pressure gauge
9. Speedometer (U.S. version)
10. Tachometer (optional)
11. Transmission temperature gauge (optional)
12. Coolant temperature gauge

Figure 13-27 Freightliner Business Class M2 instrument cluster unit (ICU display).

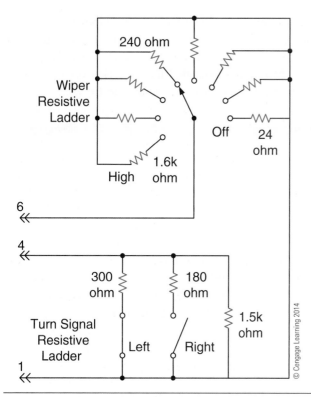

Figure 13-28 Freightliner multifunction switch.

The turn signal resistive ladder is a diagnosable switch. With the multifunction switch in the center (not turning) position, the resistance between terminals 1 and 4 is 1.5kΩ. In the left turn position, illustrated by the closed left turn switch in **Figure 13-28**, the combined 300Ω and 1.5kΩ result in a resistance of approximately 250Ω. In the right turn position, the resistance between terminals 1 and 4 is approximately 161Ω. The ICU display determines the turn signal switch state based on this resistance.

The ICU transmits the status of the three multifunction switch analog inputs onto the J1939 data link for use by the BHM.

Example of SmartPlex Multiplexing

The control of the headlamps on a Freightliner SmartPlex system involves the use of three multiplexed electronic modules, as shown in **Figure 13-29**. The headlamp switch is a hardwired BHM input. The high/low beam switch contained in the multifunction switch is an ICU hardwired input. When the BHM detects that the headlamp switch is on, the BHM switches on the low-beam headlamp FET. This FET sources current to the left-side low-beam headlamp. The BHM also transmits a J1939 message commanding the CHM to switch on its low-beam headlamp FET. The CHM switches on its low-beam headlamp FET, which sources current to the right-side low-beam headlamp. The CHM also transmits a J1939 message indicating that it has switched on the low-beam FET, as a confirmation of the completion of the task.

If the ICU later detects that the high beams have been requested based on the status of the hardwired multifunction switch, the ICU transmits a J1939 message indicating high-beam headlamps are requested. The BHM then switches off the low-beam FET and

Chapter 11. **Figure 13-28** illustrates the wiper and turn signal resistive ladder switches. The headlamp dimmer switch with windshield washer is not shown, but it is similar in function to the turn signal switch. In **Figure 13-28**, the wiper resistive ladder switch is shown in the intermittent wipers #4 position. This causes 240Ω between multifunction switch terminals 1 and 6. Moving the wiper switch to the high-speed position would cause this resistance to increase to 1.6kΩ. The ICU display analog input terminal for the wiper switch determines the wiper switch position based on this resistance.

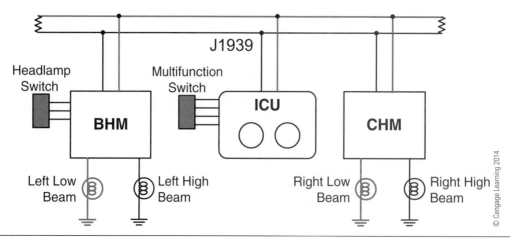

Figure 13-29 Freightliner multiplexed headlamp control.

switches on the high-beam FET. The BHM also transmits a J1939 message to the CHM commanding the CHM to switch off the low-beam FET and to switch on the high-beam FET. The ICU also illuminates the high-beam indicator.

TROUBLESHOOTING THE MULTIPLEXED TRUCK

Troubleshooting a multiplexed truck with a body control module can be very different from troubleshooting a truck with a conventional electrical system. This section will address troubleshooting an International model that uses a body controller to control various body and chassis electrical features.

Because trucks with body controllers can be very different from conventional electrical systems, it is vital that OEM service information be consulted prior to performing any troubleshooting or repairs.

PC-Based Tool

Troubleshooting the electrical system controlled by the body controller is best performed using a PC equipped with OEM diagnostic software, referred to as an electronic service tool (EST). The software for diagnosing the Freightliner SmartPlex electrical system is called ServiceLink®. The software designed for troubleshooting International trucks with body

controllers is called Diamond Logic® Builder. This software permits viewing diagnostic trouble codes generated by the body controller and EGC, along with fault descriptions. The software also permits the status of body controller inputs and outputs to be viewed graphically (**Figure 13-30**). For example, the voltage levels measured by the body controller at each analog input can be viewed without having to use a voltmeter and breakout box. Display screens in the software indicate the voltage measured at each analog input. The status of digital inputs is indicated by a check box. A check in the box indicates that the digital input is considered as on or active. The status of outputs is also indicated by check boxes. A check indicates that the body controller has switched on the corresponding output. Diamond Logic Builder can also be used to temporarily force on or force off some body controller outputs and inputs for troubleshooting.

The instrument cluster can also be diagnosed using Diamond Logic Builder. A virtual instrument cluster screen in the software permits graphically viewing the gauge and warning lamp information being transmitted on J1939 (**Figure 13-31**). This tool can also be used to drive a gauge needle to a specific location for testing.

Self-Diagnostics

Troubleshooting a body controller-related circuit can also be performed using standard diagnostic tools

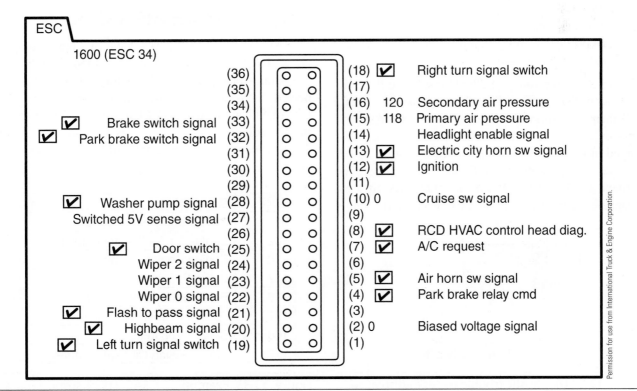

Figure 13-30 Diamond Logic Builder connector view showing status of inputs and outputs.

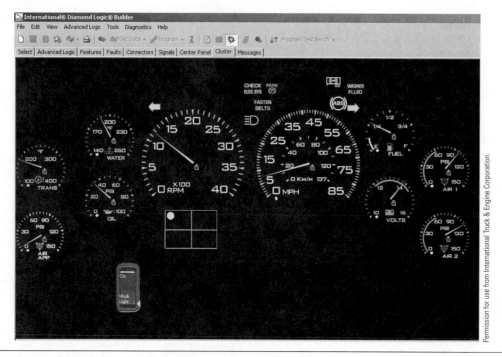

Figure 13-31 Diamond Logic Builder instrument panel cluster view.

such as a DMM and a breakout box. The LCD display in the instrument cluster can be used to indicate DTCs that are currently active or that are stored in memory. The International body controller and EGC are placed in the diagnostic mode by performing the following steps:

1. Place key switch in the ignition or accessory position.
2. Set park brake.
3. Depress the cruise control ON and RESUME switches at the same time.

This sequence causes the LCD display in the EGC to indicate the number of DTCs that exist. The DTCs will then be displayed one after the other for several seconds each. The format of the DTC messages displayed in the EGC LCD is as shown in **Figure 13-32**.

The example shown in **Figure 13-32** would be defined as DTC 611-14. International uses an "A" designator to indicate that the DTC is currently active;

that is, the conditions for setting this DTC have been recently detected. A previously active DTC, such as a DTC caused by an intermittent problem, would be identified by a "P". This DTC could be looked up in an OEM-provided listing of diagnostic trouble codes to determine what the DTC number combination means and how to troubleshoot the problem.

SAE J1939 Diagnostic Trouble Codes

SAE J1939 defined DTCs have two main components. The suspect parameter number (SPN) indicates the parameter or measurable factor that is not as expected. SPNs were introduced in **Chapter 11**. Examples of SPNs are shown in **Figure 13-33**. The **failure mode indicator (FMI)** indicates the type of failure that has been detected. Examples of FMIs are shown in

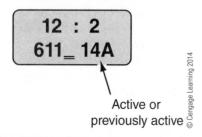

Active or
previously active © Cengage Learning 2014

Figure 13-32 Diagnostic trouble codes as displayed in instrument panel cluster odometer display.

SPN	Description
91	Accelerator Pedal Position
102	Engine Intake Manifold Pressure
110	Engine Coolant Temperature
117	Brake Primary Pressure
118	Brake Secondary Pressure
157	Fuel Rail Pressure
168	Battery Voltage
177	Transmission Oil Temperature

© Cengage Learning 2014

Figure 13-33 Examples of SAE J1939 suspect parameter numbers (SPN).

FMI	Description
00	Data valid but above normal operating range - most severe level
01	Data valid but below normal operating range - most severe level
02	Data erratic, intermittent, or incorrect
03	Voltage above normal or shorted to high source
04	Voltage below normal or shorted to low source
05	Current below normal or open circuit
06	Current above normal or grounded circuit
07	Mechanical system not responding or out of adjustment

© Cengage Learning 2014

Figure 13-34 Examples of SAE J1939 failure mode indicators (FMI).

Figure 13-34. There are currently 32 different J1939 FMIs.

In addition to the SPN and FMI, the source address (SA) of the device that has detected the failure is typically included in the DTC along with the occurrence count (OC). The OC indicates the number of times that the DTC has transitioned between active and previously active, or the number of times the ignition has been switched off for an active DTC.

Figure 13-35 illustrates FMIs related to general temperature measurement. FMIs indicating low or high temperatures are shown on the horizontal axis in **Figure 13-35**. These FMIs indicate temperatures within the range that the sensor is capable of measuring. For example, an FMI of 00 indicates that the temperature being measured by the applicable sensor is indicating an extremely high value. For engine coolant temperature, a temperature of 230°F (110°C) would be an example of an extremely high value that could cause an active DTC 110-00. The SPN of 110

indicates engine coolant temperature, and the FMI of 00 indicates extremely high.

There are also two FMIs, 03 and 04, shown on the vertical axis of **Figure 13-35**. The vertical axis indicates the voltage measured at the input terminal for the applicable sensor at the electronic module. FMIs 03 and 04 indicate that the voltage is outside the possible or plausible range of the sensor connected to that input terminal. For example, if the voltage at the input terminal of a module is 5V and the maximum voltage expected at the input terminal with a properly functioning sensor is 4.8V, then an active DTC with an FMI of 03 would be set. This DTC indicates that the voltage measured at the input terminal of the electronic module is **out of range high (ORH)**. An open circuit for an electronic module input that has an internal pull-up resistor would result in a DTC with an FMI of 03, as shown in **Figure 13-36**.

If the voltage at the input terminal of a module is 0V and the minimum voltage expected at the input terminal with a properly functioning sensor is 0.2V, then an active DTC with an FMI of 04 would be set. This DTC indicates that the voltage measured at the input terminal of the electronic module is **out of range low (ORL)**. If a pulled-up analog input circuit were shorted to ground, the voltage at the input would drop to 0V (**Figure 13-37**). This would result in a DTC with an FMI of 04.

An out-of-range low condition would also occur when a pulled-down analog input circuit is open (**Figure 13-38**). Therefore, an open circuit causes an FMI of 04 for a pulled-down input. However, as shown in **Figure 13-36**, an open circuit causes an FMI of 03 for a pulled-up input. Therefore, an open circuit can cause either an FMI of 03 or an FMI of 04, depending on the input being pulled-up or pulled-down inside the electronic module. This difference is important to understand when troubleshooting out-of-range DTCs.

A short to ground of a pulled-down analog input circuit looks the same to the microprocessor as an open

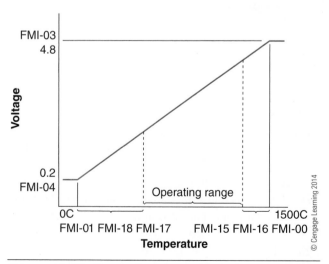

Figure 13-35 FMIs related to general temperature measurement. Horizontal FMIs are in range; vertical FMIs are out of range.

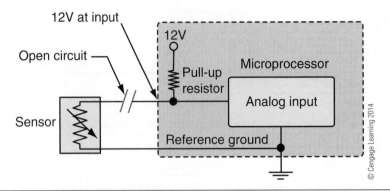

Figure 13-36 ORH condition at a pulled-up input with an open circuit.

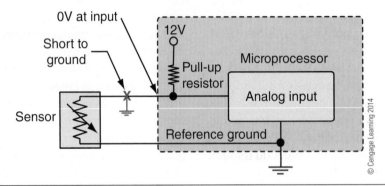

Figure 13-37 ORL condition at a pulled-up input with a grounded circuit.

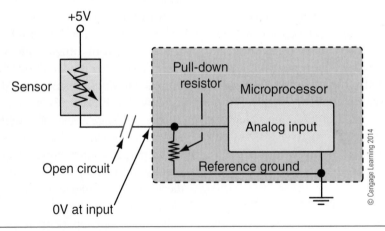

Figure 13-38 ORL condition at a pulled-down input with an open circuit.

circuit (0V at the input). Either an open circuit or a short to ground would set an out-of-range low DTC with an FMI of 04 for a pulled-down analog input.

Other DTCs set by the body controller and other electronic modules on a truck indicate the loss of specific J1939 messages that are expected but are not being received. For example, an open J1939 data link circuit between the engine ECM and the body controller will result in several features on the truck, such as cruise control, not operating. The body controller sets a DTC for each expected J1939

message that is missing. A general method of troubleshooting J1939 data link problems will be discussed in **Chapter 15**.

Electronic modules like the body controller set DTCs based on many factors. The simplest DTCs are those that indicate that an output circuit is shorted to ground or open, based on the current supplied by a high side driver. The DTC only indicates that the current in the circuit is not what is expected. The technician must still determine where the actual problem in the circuit is located.

Most OEMs also have a proprietary DTC numbering system. For example, Cummins uses fault number 151 within their service information to indicate engine coolant temperature is too high. However, the engine ECM also transmits the SAE J1939 SPN/FMI combination of 110-00 for this condition. Most OEMs have DTC cross-reference information available online.

SAE J1587/J1708 DTCs

Although not used for body controller diagnostics, SAE J1587/J1708 DTCs utilize a combination of parameter identifier (PID) and FMI to describe a DTC. Additionally, a subsystem identification number (SID) may also be used in instances where a PID is not available to accurately describe the system. The message identification (MID) is also a part of the DTC and indicates the electronic module or system that has detected the failure. PID and MID were discussed in **Chapter 11**. There are 16 different SAE J1587/J1708 FMIs, these being nearly the same as FMIs 00 through 15 used for J1939 DTCs.

Troubleshooting Using a DMM

Measuring the voltage using a voltmeter at the input and output terminals of an electronic module like the body controller requires some precautions to prevent damage to the electronics. A breakout box as shown in previous chapters is sometimes used when troubleshooting truck electronic modules. The breakout box is connected in series with the electronic module and the truck wiring harness. The connectors on the breakout box are designed to be the same as those on the truck wiring harness and the electronic module. The breakout box has terminals that are used to act as voltage test points for the voltmeter. This permits the technician to measure the voltage present at inputs and outputs without damaging the wiring harness and to make closed circuit voltage measurements. The cavity numbers of some connectors may be very difficult to see. The corresponding terminals on a breakout box are clearly marked to reduce the chances of misdiagnosing a problem due to measuring the voltage at the wrong connector terminal.

Ghost Voltage. The use of a test light can damage some circuits that are designed for low current only. Most OEMs specify that a DMM should be used for all electrical troubleshooting. However, the high internal resistance of a DMM voltmeter can lead to some confusion when troubleshooting. This section may

make more sense if you first study the Extra for Experts section in **Chapter 2**.

The internal resistance of many DMM voltmeters is 10 MΩ. This means that a 10 MΩ resistor is inside the DMM between the two voltage measurement ports of the meter as shown in **Figure 13-39**. The DMM internal circuitry measures the voltage that is dropped across this internal resistor. The high internal resistance of the DMM prevents loading down electronic circuits during troubleshooting because a voltmeter is connected in parallel with whatever voltage is being measured. However, this high internal resistance in the DMM can result in misdiagnosis when working with power MOSFETs like those found in the body controller. Most FETs have some small amount of **leakage current** between the source and the drain terminal. This means that when the FET is switched off, a small amount of current leaks between the voltage supply and output terminals of the FET through the internal resistance of the FET. This internal resistance of a typical FET is approximately 1 MΩ. This is a very high level of resistance. However, this 1 MΩ of resistance is only one-tenth the internal resistance of a common DMM voltmeter.

To illustrate the problem that FET leakage current can cause, if the connectors for both truck headlamps were disconnected with the headlamps switched off on a truck with a body controller and a DMM were used to measure the voltage between the wiring harness connector terminal for the headlamps to ground, the DMM would indicate the presence of approximately 90 percent of battery voltage. Even though the low-beam headlamp FET is switched off, approximately 11V is being measured by the DMM between the disconnected headlamp harness connector terminal and ground. This occurs because of the rules of

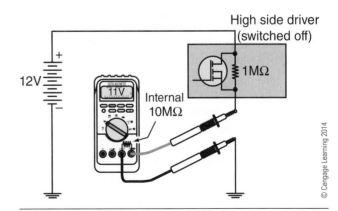

Figure 13-39 Leakage current and DMM measuring open circuit voltage from FET output to ground. Ghost voltage would incorrectly indicate that FET is switched ON.

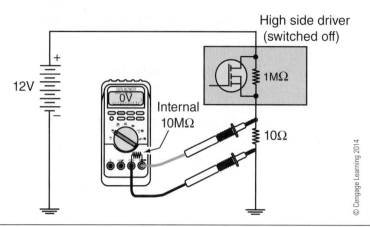

Figure 13-40 Closed circuit DMM measurement indicates the actual voltage available at load (0V).

series circuits. The resistance of a switched-off FET is similar to a floating 1 MΩ resistor that is connected to battery positive on one end. The 10 MΩ internal resistance of the DMM is connected in series between the other end of this floating 1 MΩ resistor and ground, as shown in **Figure 13-39**. This is just a simple series voltage divider circuit. You could apply Ohm's law to this circuit and find that approximately 11V will be dropped across the internal resistance of the DMM. The switched-off FET's leakage current causes the DMM to indicate that there is 11V present between the open headlamp circuit and ground. This would lead you to believe that when this circuit is connected to the headlamps, the headlamps will illuminate because 11V should be enough to cause the headlamps to illuminate. The problem is that when the circuit is connected to the headlamps, the low resistance of the headlamps will cause nearly all of the voltage to now be dropped across the FET's 1 MΩ of leakage resistance (**Figure 13-40**). The headlamps will not illuminate, even though it would seem that they should, given the 11V of open circuit voltage measured at the headlamp connector.

It might seem that this 11V of open circuit voltage measured at the headlamps would cause the batteries to discharge. A very small amount of parasitic load current is always flowing through the headlamps with the headlamps switched off, but the amount of leakage current is so small that it would require several months to discharge the batteries due to this leakage current alone.

This type of problem, where a DMM appears to indicate falsely the presence of a voltage, causes a great deal of misdiagnosis and confusion for technicians. The voltage measured by a high internal resistance DMM that only seems to be present with the circuit open is known as **ghost voltage** or some similar term. High resistance caused by terminal corrosion or

even broken wires, with corrosion bridging the gap of the broken wire, will also cause these ghost voltages even in a circuit that is otherwise energized but the load cannot operate because of the high resistance. Ghost voltage is merely the rules of series circuits related to voltage dividing proportional to resistance. The high internal resistance of the DMM forms a series circuit with the overall resistance of the circuit. The DMM merely displays the amount of voltage that is dropped across its 10 MΩ of internal resistance. Just a few ohms of resistance in a circuit due to corrosion can prevent a device from operating. The difference between open-circuit DMM voltage readings obtained from a circuit with 1000Ω of series resistance due to corrosion and the voltage readings taken in a circuit with near 0Ω of series resistance is insignificant. Both voltage readings would indicate nearly the same voltage, even though the 1000Ω of series resistance will keep most devices from operating or may cause measurement error in the case of sensor circuits.

Tech Tip: When open circuit voltage readings obtained using a DMM indicate that there is voltage present but the known-good device still does not operate, suspect that the high input impedance of the DMM is much higher than the series resistance in the circuit that you are measuring, resulting in a ghost voltage.

One of the ways to avoid ghost voltage is to use a breakout box or breakout T to measure the loaded voltage instead of the open circuit voltage, as shown in **Figure 13-40**. However, there are times that you must obtain open circuit voltage measurements when troubleshooting due to connector design or other limitations. If breakout boxes or breakout Ts are not

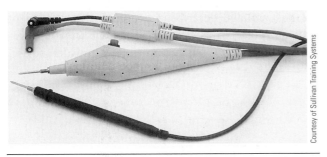

Courtesy of Sullivan Training Systems

Figure 13-41 LOADpro® dynamic voltmeter leads.

available, some technicians may back-probe the terminals of a connector that is still connected to the device or the mating harness connector to avoid making open circuit voltage measurements. Some OEMs indicate that terminals should not be back-probed because doing so may damage wire seals and terminals. However, special curved probes, known as diagnostic spoons, designed for back-probing terminals

should always be used when back-probing a terminal. This will minimize damage to the wire seals and terminals compared to paper clips and similar non-approved methods, such as piercing wire insulation with a sharp probe.

A special set of enhanced DMM test leads called LOADpro® dynamic voltmeter leads have been designed for automotive technicians that causes a 510Ω parallel resistance to be switched on across the DMM's internal resistance when a pushbutton on the lead is depressed (**Figure 13-41**). If you observe a sizable decrease in the open circuit voltage indicated by the DMM when the button is depressed (i.e., greater than 0.5V decrease), there is a high amount of series resistance present somewhere in the circuit, or the open circuit voltage that you are measuring is caused by leakage current through an electronic component, such as an FET, that is not switched on.

Summary

- The body control module on International High Performance Vehicles is called the electrical system controller (ESC) on model year 2001–2006 trucks and the body controller on 2007 and later year trucks. The body controller or ESC contains a microprocessor. The body controller or ESC uses information obtained from the input sources to control the outputs.

- Outputs of the body controller include high side drivers, low side drivers, and messages on the J1939 data link.

- Input sources for the body controller include messages from the J1939 data link, messages from the switch data link, and hardwired inputs such as switches and sensors.

- A reference ground scheme is commonly used in automotive electronics to minimize the effects of electromagnetic interference and to improve measurement accuracy for sensor circuits.

- A diagnosable switch is a special switch that provides specific values of resistance, unlike a conventional switch, which is either an open circuit or near 0Ω. A diagnosable switch is connected to an analog input. This permits circuit failures such as an open circuit or a shorted-to-ground circuit to be diagnosed.

- The body controller controls several electrical system features, including headlamps, turn signals, stop

lamps, and windshield wipers. The various switches act as inputs to the body controller through either hardwiring or multiplexing. The body controller microprocessor makes decisions based on its programming and provides an output in the form of energizing a high side driver, energizing a low side driver, or transmitting a data link message.

- Self-diagnostics assist in troubleshooting the body controller. Diagnostic trouble codes may be logged to indicate a circuit that is out of range high or out of range low.

- The Freightliner SmartPlex system uses two or more separate modules to control body electrical features. The bulkhead module (BHM) controls the chassis module (CHM) via the J1939 data link.

- The Freightliner SmartPlex uses optional smart switches to control electrical system features. Each smart switch contains two identification resistors that uniquely define the switch function.

- The instrument cluster in the Freightliner SmartPlex system is called the ICU. The ICU acts as the input device for the stalk-mounted multifunction switch. The ICU transmits the status of the multifunction switch inputs onto the J1939 data link.

- Diagnostic trouble codes (DTCs) are used to indicate that an electronic module has detected an abnormal condition. DTCs may indicate a sensor in-range operating condition, such as engine coolant temperature is too high. DTCs may also indicate sensor out of range conditions, such as open circuits or shorts to ground. A J1939 DTC consists of an SPN and an FMI.

- Ghost voltage describes open-circuit measurement of a voltage by a DMM in a circuit with high resistance. Ghost voltage can result in misdiagnosis.

Suggested Internet Searches

Try the following web sites for more information:
http://www.internationaltrucks.com

http://www.freightliner.com
http://www.brighterideas.com

Review Questions

1. Which of the following is a true statement concerning an International truck with an electrical system controller (body controller)?

   A. The headlamp ground circuit is completed through the body controller ground.

   B. A low side driver in the body controller is used to control the horn relay coil low side.

   C. An electric horn circuit that is shorted to ground will result in a blown body controller fuse.

   D. A high side driver in the body controller supplies +12V directly to the electric horns when the body controller detects that the horn input is active.

2. The body controller reference ground should not be connected to chassis ground or used as a chassis ground for what reasons?

   A. To prevent eddy currents and to prevent interfering with the CB radio reception.

   B. To reduce electromagnetic susceptibility and reduce voltage measurement inaccuracies at inputs.

   C. To prevent back EMF from damaging body controller internal components and to keep digital inputs from becoming logic 0.

   D. The body controller reference ground is the same as chassis ground and can be used as a ground for low-current devices (under 5A).

3. A diagnosable switch would typically be connected to which type of input?

   A. J1939 data link input

   B. Pulled-up digital input

   C. Pulled-down digital input

   D. Analog input

4. The resistance across a diagnosable switch is found to be approximately 1200Ω with the switch closed and approximately 2400Ω with the switch open. What does this indicate?

   A. The switch is defective because open switch resistance should be near 0Ω and closed switch resistance should be near infinite ohms.

   B. The switch is defective because closed switch resistance should be near 0Ω and open switch resistance should be near infinite ohms.

   C. The switch has probably been exposed to moisture and has corrosion on the contacts.

   D. The switch may be working as designed.

5. The mirror heat does not work on an International truck with a body controller. The mirror heat switch is located in a switch pack. The switch pack communicates with the body controller through the switch data link. Which of the following would *not* be a cause of the inoperative mirror heat?

   A. There is an open circuit between the body controller high side driver and the heated mirrors.

   B. There is an open circuit between the mirror heat switch and the body controller mirror heat digital input.

   C. The switch pack is not communicating with the body controller.

   D. The mirror heat circuit is shorted to ground.

6. The left side turn signals, both front and rear, do not work on an International truck with a body controller. The left side turn signals also do not work with the hazard switch in the ON position. The truck has combination stop/turn rear lamps and the left rear stop lamp illuminates when the brake is depressed. What could be a possible cause?

   A. A defective left turn-signal flasher.

   B. The left turn-signal message is not being sent by the body controller over J1939.

   C. An open circuit between the body controller left turn-signal input and the turn-signal switch.

   D. The left rear turn-signal circuit is shorted to ground.

7. The windshield wipers on an International truck with a body controller operate at high speed anytime the key is in the ignition position, regardless of wiper switch position. What is a possible cause?

   A. A defective wiper park switch.

   B. A disconnected wiper switch.

   C. The body controller wiper high-speed input circuit is shorted to ground.

   D. The wiper-off command is not being received by the body controller over the J1939 data link.

8. The air pressure gauges, fuel level gauge, and voltmeter do not work on an International truck with a body controller. All other gauges, such as tachometer and engine oil pressure, are operating correctly. Which of the following is the most likely cause of the inoperative gauges?

   A. The EGC is not communicating over the J1939 data link.

   B. The engine ECM is not communicating with the EGC.

   C. The J1939 CAN + circuit is shorted to ground.

   D. The body controller is not communicating with the EGC over the J1939 data link.

9. The cruise control does not work on an International truck with a body controller. Which of the following is *not* a likely cause?

   A. A disconnected stop switch connector

   B. A broken clockspring

   C. No J1939 communication between the body controller and the engine ECM

   D. All of the above are possible causes

10. Two technicians are discussing the HVAC system on a 2001 International truck with an ESC. Technician A says that one purpose of the high side pressure sensor is to act as a high-pressure cut-out switch. Technician B says that one purpose of the high side pressure sensor is to cause the ESC to switch on a viscous-type fan clutch to increase high side pressure in cold weather. Who is correct?

    A. A only

    B. B only

    C. Both A and B

    D. Neither A nor B

11. The multifunction switch on the Freightliner Business Class M2 is hardwired to which device?

    A. AMU   C. BHM

    B. CHM   D. ICU

12. Smart switches used on the Freightliner Business Class M2 are identified by the BHM through which method?

    A. Binary on-off combination of   B. A unique J1939 source address is transmitted by each smart
       two switches inside the smart   switch.
       switch that are connected to
       two BHM digital inputs.   C. Two specific values of resistance inside the smart switch that are
                                          connected to two BHM analog inputs.

                                       D. Time-division multiplexing.

13. An out-of-range low DTC might be set for which condition?

    A. A digital input circuit is   C. An analog input circuit is open.
       shorted to ground.
                                       D. Both B and C could cause this DTC to be set.
    B. An analog input circuit is
       shorted to ground.

14. Which best describes ghost voltage?

    A. The low internal resistance of   C. The rules of a series voltage divider circuit as it relates to the
       a DMM ammeter causing a   high internal resistance of a DMM voltmeter
       false voltage to be displayed
       on the DMM   D. High DMM internal resistance in parallel with a low value of
                                           circuit resistance
    B. The voltage supplied by the
       DMM ohmmeter interacting
       with other voltage sources on
       the truck

15. Two technicians are discussing DTC 110-03. The service information indicates that this DTC is set for an out-of-range high condition of the engine coolant temperature sensor circuit. Technician A says this DTC is set due to abnormally high coolant temperature, which could be caused by poor airflow through the radiator. Technician B says that this DTC can be set due to abnormally high coolant temperature caused by a stuck thermostat. Who is correct?

    A. A only   C. Both A and B

    B. B only   D. Neither A nor B

14 Diesel Engine Electronics

Learning Objectives

After studying this chapter, you should be able to:

- Discuss the various types of electronically controlled diesel fuel systems.
- Describe why the current flow through an inductor cannot instantly change.
- Explain the fundamentals of a closed loop control system.
- Discuss how an in-range failure might not cause a DTC to be set.
- Explain the difference between in-cylinder emissions reduction and exhaust aftertreatment.
- Discuss how a variable geometry turbocharger can be used to regulate EGR flow rate.
- Explain how particulate matter and NO_x exhaust emissions are controlled.
- Describe the requirements of HD-OBD and discuss some of the benefits of this regulation for technicians.

Key Terms

aftertreatment

closed loop control

confirmed DTC

derate

desired output

error

exhaust gas recirculation (EGR)

freeze frame

H-bridge

heavy-duty on-board diagnostics (HD-OBD)

hydraulically actuated unit injector (HEUI)

in-range failure

injection control pressure (ICP)

injection pressure regulator (IPR)

learn cycle

lookup tables

malfunction indicator lamp (MIL)

measured output

measurement error

open loop control

oxides of nitrogen (NO_x)

particulate matter (PM)

pending DTC

permanent DTC

previous MIL-on DTC

proportional solenoid

readiness

selective catalytic reduction (SCR)

smart actuator

smart sensor

variable geometry turbocharger (VGT)

virtual sensor

INTRODUCTION

A diesel engine wiring harness from a 1970s truck may have consisted of one wire for the fuel cut-off solenoid, not including the wiring for the starter motor and alternator. Compare this single wire to the Volvo VE-D16 engine wiring harness shown in **Figure 14-1**. One of the primary reasons for the addition of electronic controls to diesel engines was the reduction of exhaust emissions, as required by agencies such as the U.S. Environmental Protection Agency (EPA). Diesel engines have always had a reputation for being reliable and efficient. However, diesel engines of the past were also smelly, noisy, and polluting. **Figure 14-2** illustrates the reduction in EPA-mandated on-road diesel engine exhaust emissions since 1970. The most visibly evident change is **particulate matter (PM)** reduction, which includes all types of exhaust smoke. It is now rare to observe an on-highway truck belch out black smoke (soot), although this was a common sight just a few years ago. The reduction of **oxides of nitrogen (NO$_x$)**, one of the pollutants responsible for the formation of smog, has also required substantial changes in diesel engines. It is estimated that one model year 1988 HD truck produced the same amount of pollutants as 65 model year 2010 trucks.

From an electrical perspective, there is not much new material in the final two chapters. Instead, the application of some of the fundamental concepts of electricity and electronics will be explored as they apply to a modern diesel engine and truck.

THE ELECTRONICALLY CONTROLLED DIESEL ENGINE

The central component in modern diesel engines is the engine electronic control module (ECM). Some OEMs refer to the ECM as an engine control unit (ECU) or as an engine electronic control unit (engine ECU). These terms all mean the same thing in the context of this chapter, an electronic component that contains a microprocessor. For simplification, the acronym ECM will be used throughout this chapter to refer to this device or, in some cases, multiple devices.

The microprocessor in the ECM is programmed to make logical "decisions" on how it will control its *outputs* based on the information it receives from its *inputs*. This concept was introduced in **Chapter 13** for body control modules. The ECM inputs include a variety of sensors, such as those that would connect to the engine wiring harness shown in **Figure 14-1**. However, the primary ECM input is the crankshaft position and speed, which is typically obtained from the combination of camshaft and crankshaft position sensors, as explained in **Chapter 11**. The truck operator's torque or speed request, obtained from the accelerator position sensor (APS), is also an important ECM input.

Wiring Diagram

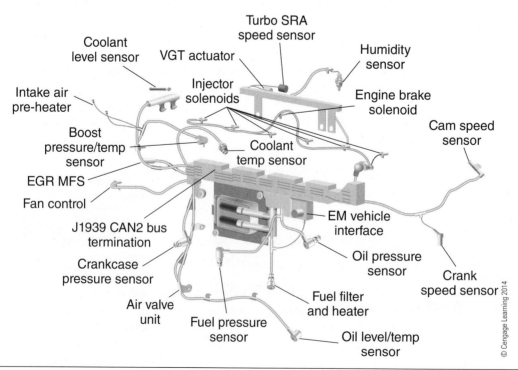

Figure 14-1 Typical 2010 engine wiring harness.

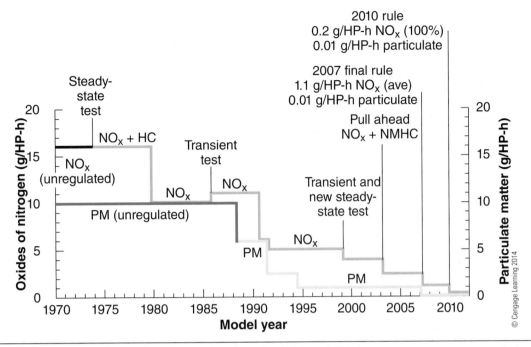

Figure 14-2 EPA heavy-duty diesel exhaust emissions for PM and NO$_x$.

The primary ECM outputs are related to the control of the fuel system and the exhaust emission control system.

FUEL SYSTEMS

There are three basic tasks of any diesel fuel system:

1. Provide injection metering (quantity of fuel).
2. Atomize the fuel into a fine mist that is capable of penetrating the densely packed air in the combustion chamber.
3. Provide injection timing control.

Diesel fuel systems of the past were strictly hydrome-chanical systems and performed these three tasks reasonably well. These mechanical systems provided good engine performance and reliability, but poor exhaust emissions control. To reduce exhaust emissions as well as provide improved performance and fuel economy, the fuel system must also be capable of performing these basic tasks with extreme precision and at much higher injection pressures. In addition, the fuel system must:

1. Control fuel delivery independent of engine speed and camshaft geometry (rate shaping).
2. Perform multiple injection events per cycle.

To perform these tasks with the necessary precision, electronic control of the fuel system is necessary.

Figure 14-3 shows the phasing of diesel delivery, injection, and combustion events. These events are applicable to most types of diesel fuel systems with some minor differences. During the delivery phase, the pressure within the injector or nozzle increases after the port is closed. The pressure increases within the injector or nozzle until the nozzle opening pressure (NOP) is reached and fuel is injected into the dense, compressed air in the combustion chamber. Combustion then occurs after an ignition delay or lag. The injector continues to inject fuel under pressure until the nozzle closure. In a strictly hydromechanical fuel system, all of these events are controlled without any electrical components. In an electronically controlled fuel system, these events are controlled by a combination of hydromechanical and electrical means.

The earliest electronic diesel engine controls were an adaptation of the existing system; that is, electronic control was added to the existing fuel system. As the exhaust emissions standards became increasingly more stringent, entirely new electronically controlled diesel fuel systems have been introduced as outdated designs became obsolete. Examples of this progression will be reviewed.

Electronically Controlled Pump-Line-Nozzle Systems

Pump-line-nozzle (P-L-N) fuel systems include port-helix and rotary distributor pump systems. These hydromechanical fuel systems were enhanced with electronic controls permitting these systems to meet EPA standards for a few more years of production. P-L-N systems were not used much in on-highway engines much after 1997 because they could not deliver the fuel economy and emissions reduction of more advanced fuel systems. Although these fuel systems are simple by today's standards, they permitted advances such as

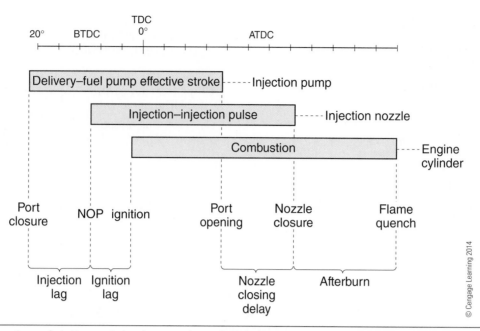

Figure 14-3 Phasing of delivery, injection, and combustion events.

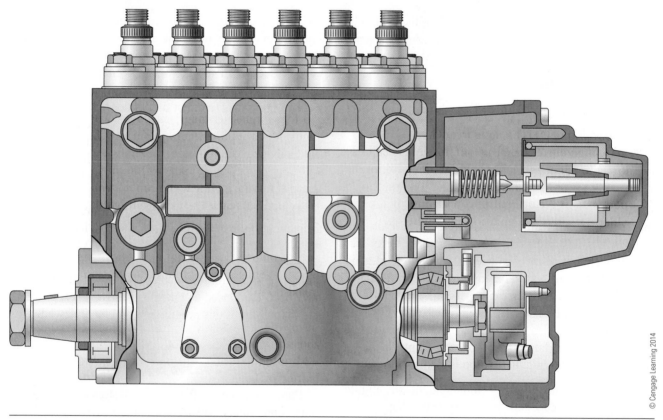

Figure 14-4 Bosch port-helix injection pump with RE30 rack actuator.

cruise control and road speed limiting. Fuel temperature and air temperature compensation were also made possible by these electronic enhancements.

Port-Helix Injection Pumps. Port-helix injection pumps utilize a control rack to regulate the effective stroke of each pump plunger. In a conventional port-helix system, a hydromechanical governor controls the rack position, thus controlling fuel metering. In the electronically controlled version of the port-helix system, the hydromechanical governor is replaced by an electronic governor (rack actuator) as shown in **Figure 14-4**.

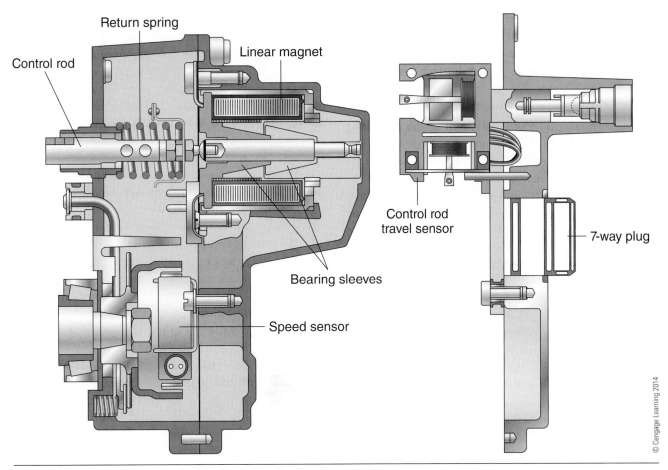

Figure 14-5 Bosch RE30 electronically controlled rack actuator.

Details of the rack actuator are shown in **Figure 14-5**. The control rod is attached to the pump control rack. The control rod travel sensor provides an indication of the control rod position to the ECM. The ECM causes the control rod to move to the desired position by controlling the current flow through a **proportional solenoid** or linear magnet as shown in **Figure 14-5**. A conventional solenoid like that used in a starter motor is an on-off (two-position) digital-type device, like a switch. Unlike a conventional solenoid, a proportional solenoid is an analog device that can be moved and held at any position throughout its range of travel based on the amplitude of current flowing through the solenoid's electric windings. The control rod is spring applied in the zero fuel position. Therefore, an interruption of electric current flow through the proportional solenoid's windings results in engine shutdown. The rack actuator also contains a variable reluctance speed sensor and timing event sensor, which act as inputs to the ECM.

Rotary Distributor Injection Pumps. Rotary distributor pumps were also adopted for electronic control, such as the Bosch VE rotary distributor injection pump shown in **Figure 14-6**. The hydromechanical governor is replaced by a rotary solenoid actuator in the

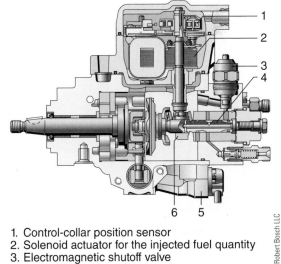

1. Control-collar position sensor
2. Solenoid actuator for the injected fuel quantity
3. Electromagnetic shutoff valve
4. Delivery plunger
5. Solenoid valve for start-of-injection timing
6. Control collar

Figure 14-6 Bosch electronically controlled VE distributor injection pump used on smaller diesel engines.

electronically controlled version of a Bosch VE pump. The ECM controls the position of the control collar by regulating the position of the rotary electrical solenoid actuator. A potentiometer is used as a control collar position sensor and is an ECM input. The control collar is used to regulate the opening and closing of the spill/cutoff ports to control injected fuel quantity. A start of injection solenoid valve in the pump is controlled by the ECM to regulate injection timing.

The pintle-type nozzles may also contain a nozzle valve motion sensor, shown in **Figure 14-7**, which is an input to the ECM. The nozzle valve motion sensor is used to determine precisely when the start of injection has occurred. A variable reluctance engine speed sensor also acts as an ECM input. **Figure 14-8** illustrates the waveforms generated by the needle valve motion sensor, identified as NBF (waveform 1), and the engine speed sensor (waveform 3). The ECM converts or conditions these raw analog signals into their respective digital representations in waveforms (2) and (4) shown in **Figure 14-8**. The raw analog signals produced by these sensors are not usable by the ECM's microprocessor until they are converted to a digital signal.

The ECM can be programmed to respond to specific failures, such as the loss of information from a

Sensor signals

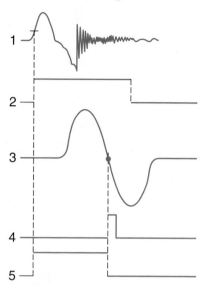

1. Untreated signal from the needle-motion sensor (NBF)
2. Signal derived from the NBF signal
3. Untreated signal from the engine-speed signal
4. Signal derived from untreated engine-speed signal
5. Evaluated start-of-injection signal

Courtesy of Robert Bosch LLC

Figure 14-8 NVMS and engine speed sensor waveforms.

sensor, by applying a **derate** which is a reduction of maximum allowable engine torque or speed or initiating a limp-home (default) strategy. An interruption of current flow through the fuel metering solenoid results in engine shutdown.

Electronic Unit Injector Systems

Unit injector systems describe diesel fuel systems where the pump and the injection nozzle are combined into one unit. The pump can be actuated by the camshaft, as shown in **Figure 14-9**, or high-pressure engine oil can be used to generate injection pressures, as described in the next section.

Cam-actuated electronic unit injectors (EUIs) were introduced in the North American market in 1987. These single actuator EUIs with two electrical terminals (one solenoid) were widely utilized in North America until 2007. The electronically controlled actuator replaces the mechanical rack system used to control metering in previous mechanical unit injector (MUI) systems. To increase injection pressures and to more precisely control timing and metering, dual actuator EUIs with four electrical terminals (two solenoids) were introduced in 2007. Both of these systems will be described.

Nozzle-and-holder assembly with needle-motion sensor (NBF)

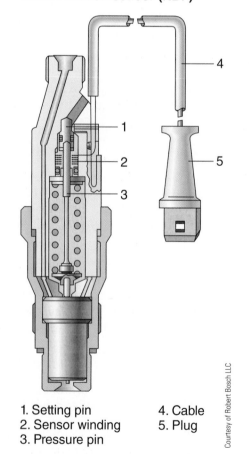

1. Setting pin   4. Cable
2. Sensor winding   5. Plug
3. Pressure pin

Courtesy of Robert Bosch LLC

Figure 14-7 Nozzle valve motion sensor (NVMS).

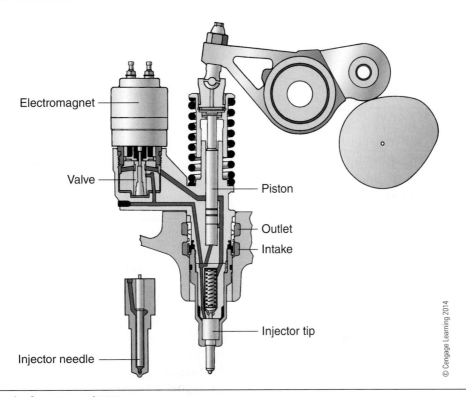

Electromagnet —

Valve —

Piston

Outlet

Intake

Injector tip

Injector needle —

© Cengage Learning 2014

Figure 14-9 Camshaft actuated EUI.

Single Actuator EUI. Single actuator EUIs, like that shown in **Figure 14-10**, utilize a normally open spill valve (4) which is closed when an electric current flows through the solenoid valve assembly (2). The ECM injector drivers (FETs) supply the control current for the solenoid valve assembly. Alternately, some OEMs utilize a separate injector drive module to supply this current.

The ECM determines when to energize the solenoid valve based on its inputs and programming instructions. Energizing the solenoid valve causes the spill valve to close. Once the spill valve is closed, the pressure within the pumping element increases rapidly until the nozzle opening pressure (NOP) is reached, at which time fuel is injected into the combustion chamber. The NOP is determined by mechanical means in the same manner as a mechanical unit injector system. Typical NOP is 5000 psi (34 MPa), but the injection pressures can rise to as high as 30,000 psi (207 MPa).

Once the ECM has determined that the desired fuel quantity has been metered, the ECM shuts off the current flow through the solenoid valve assembly, which causes the normally open spill valve to open. This causes the pressure within the pumping element to rapidly decrease. The nozzle will then close once the pressure decreases below the nozzle closing pressure resulting in the end of injection. Looking back at **Figure 14-3**, the ECM is only controlling the port closure and the port opening to control the pump

effective stroke. This would have been controlled mechanically on a typical unit injector system of the past.

It is important to note that there is no injection pressure sensor within the unit injector and there is no direct measurement of the quantity of fuel being injected. In addition, there is no direct measurement of when the nozzle actually opens and closes. However, all of this information is vital for the ECM to be able to provide precise fuel metering and injection timing. Therefore, the ECM is programmed to "know" precisely when current should be switched on to source the solenoid valve assembly and when it must be switched off to meter a precise amount of fuel at the exact time. The OEM has determined all of this information through extensive testing in a laboratory environment during the fuel system development.

Figure 14-11 illustrates the relationship between the voltage sourced by the ECM, the current that flows through the injector solenoid, the spill valve movement, and the fuel injection rate (all with respect to time on the horizontal axis). Notice in **Figure 14-11** that although the voltage supplied by the ECM instantly rises from 0V to some level (often 50V or more), the current flow through the control valve solenoid gradually ramps up from 0A to some value. Inductors were introduced in **Chapter 3**. Recall that when current flows through an inductor, a magnetic field is generated. This magnetic field passes through the windings of the inductor (coil), which cause a

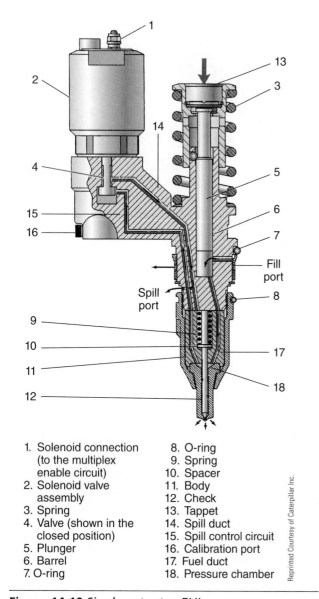

1. Solenoid connection (to the multiplex enable circuit)
2. Solenoid valve assembly
3. Spring
4. Valve (shown in the closed position)
5. Plunger
6. Barrel
7. O-ring
8. O-ring
9. Spring
10. Spacer
11. Body
12. Check
13. Tappet
14. Spill duct
15. Spill control circuit
16. Calibration port
17. Fuel duct
18. Pressure chamber

Reprinted Courtesy of Caterpillar Inc.

Figure 14-10 Single actuator EUI.

voltage to be induced. The polarity of this induced voltage is such that it opposes the voltage that caused the original current flow through the inductor. This was defined as self-inductance or CEMF in **Chapter 3**. To simplify, inductors resist a change in current flow through the inductor.

This leads to an important fact:

Important Fact: The current flow through an inductor cannot instantly change from one level to another. However, the voltage measured across an inductor can instantly change from one level to another.

To illustrate this concept, **Figure 14-12** shows a simple electrical circuit with a battery, a switch, and a

6Ω resistor. The graph below the circuit in **Figure 14-12** shows the circuit current on the vertical axis and time on the horizontal axis, which is the same as would be displayed on an oscilloscope using a clamp-on current probe. When the switch closes, the current flow through the circuit instantly increases from 0A to 2A, as predicted by Ohm's law. In **Figure 14-13**, the resistor is replaced with an inductor that has 6Ω of wire resistance. Both the inductor and the resistor would each indicate 6Ω if an ohmmeter were used to measure their resistance. Notice in the graph shown in **Figure 14-13** that the current does not instantly increase from 0A to 2A when the switch is closed but rather the current ramps up to 2A over time. However, the voltage across the inductor does instantly change from 0V to 12V when the switch is closed. **Figure 14-14** illustrates the opposition of the induced voltage generated by the expanding magnetic field cutting through the inductor to the applied voltage. The result is that until the expanding magnetic field becomes stationary, the current flow through the inductor will be reduced. The time that it takes for the current to rise to the peak value is dependent upon the value of inductance (henries).

Looking again at the injector voltage and current traces shown in **Figure 14-11**, the current through the solenoid does not instantly rise to the peak value (I_P) but instead ramps up to this peak value over time. The ECM must take this current rise time into account, as well as the time for the spill valve to actually close, the time for the pressure to rise to the NOP, and other mechanical time constants. The OEM has factored all of these electrical and mechanical lag times into the ECM programming instructions so that the nozzle opening pressure is reached at precisely the optimal time for injection to begin.

The injector solenoid current rise time is also important for failure detection. The injector solenoid resistance as measured by an ohmmeter may be less than 1Ω on some injectors and is very dependent on temperature because copper has a positive temperature coefficient. If the insulation coating on the injector solenoid windings becomes damaged by heat and vibration and causes windings to be shorted together, the resistance of the solenoid as measured with an ohmmeter may not decrease much at all, but the inductance of the solenoid will be decreased substantially. The inductance of a coil cannot be easily measured without specialized test equipment. The decreased inductance due to the shorted windings can cause the strength of the magnetic field to be reduced such that the solenoid cannot cause the spill valve to close and the cylinder will misfire. However, the ECM may be programmed to monitor the time that it takes for the injector current

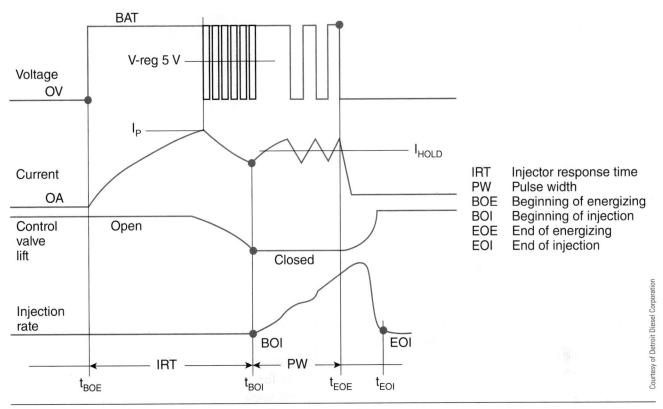

Figure 14-11 EUI voltage and current waveforms.

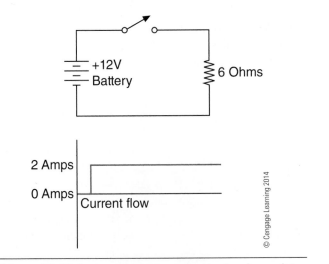

Figure 14-12 Instantaneous current rise in resistor circuit.

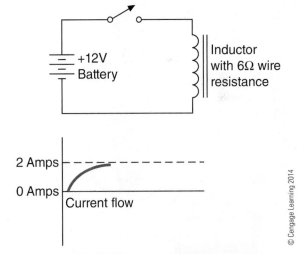

Figure 14-13 Ramped current rise in inductor circuit.

to rise to the peak value. If the injector solenoid has failed such that the windings have shorted together, the peak current rise time will be reduced to resemble that of a resistor. The ECM detects this reduced peak current rise time and may set a DTC indicating that the injector is shorted. For example, a J1939 DTC with SPN 652 and FMI 06 might be set if injector #2 is detected as being internally shorted.

One other item of interest in **Figure 14-11** is the PWM of the voltage sourced by the ECM once the

peak current has been reached (I_P). Recall from **Chapter 8** the discussions on solenoid pull-in and hold-in current. The peak injector current serves as the solenoid pull-in current. Once the solenoid armature has moved to the closed position, the current through the solenoid can be reduced to minimize self-heating of the injector solenoid and to reduce the stress on the FETs in the ECM. A reduced average hold-in current (I_{HOLD}) sustained by PWM of the supply voltage is all that is necessary to hold the spill valve closed.

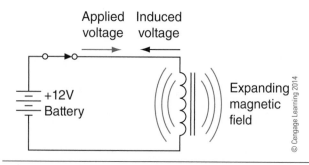

Figure 14-14 Induced voltage (CEMF) opposes the applied voltage.

Dual Actuator EUIs. One disadvantage of single ac-tuator EUIs is that the NOP is still mechanically controlled. Dual actuator EUIs have two solenoids and two control valves; thus, these injectors typically have four electrical terminals. Dual actuator EUIs were in-troduced in North America in 2007. **Figure 14-15** illustrates a Delphi E3 dual actuator EUI. The elec-trically controlled spill valve (SV) works in the same way as described in the previous section for a single actuator EUI. The needle control valve (NCV) is

controlled by a second electrical coil. The NCV is designed to keep the nozzle closed at pressures above the hydromechanical NOP. Energizing the NCV causes the nozzle control pressure to vent, which allows the nozzle to open and inject as shown in **Figure 14-16**. To end injection, both the NCV and SV are de-energized by the ECM.

The main advantage of the dual actuator EUI is that the ECM has control over the NOP. This permits much higher injection pressures, which reduces the fuel droplet size (better fuel atomization). The dual actua-tor EUI can also be used for multiple injection events (multipulse injection) by cycling the NCV off and on during the injection event. One type of multipulse in-jection is pilot injection. Pilot injection reduces the familiar diesel engine knock heard at low engine speeds and reduces white smoke at cold starts by re-ducing ignition delay. A small quantity of pilot fuel is injected into the combustion chamber a few hundred microseconds before the main injection pulse. The pilot fuel is ignited after a brief ignition delay. The main fuel charge is then injected into the ignited pilot fuel. Pilot injection has been described as injecting fuel into fire. The result is smooth combustion resulting in reduced noise and emissions.

Figure 14-17 illustrates a variation of the EUI called the electronic unit pump (EUP). The EUP is a unit injector that is separated into the pumping com-ponent and the injector nozzle component. A tube

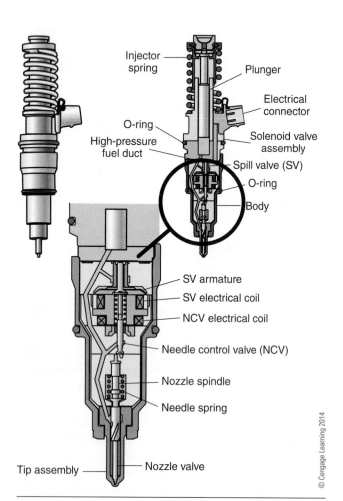

Figure 14-15 Delphi E3 dual actuator EUI.

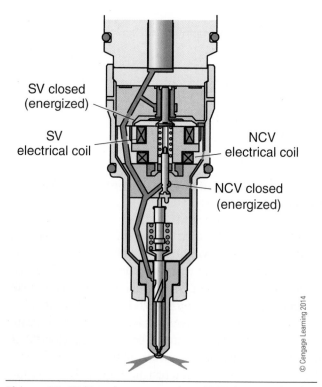

Figure 14-16 Nozzle opens when NCV is energized.

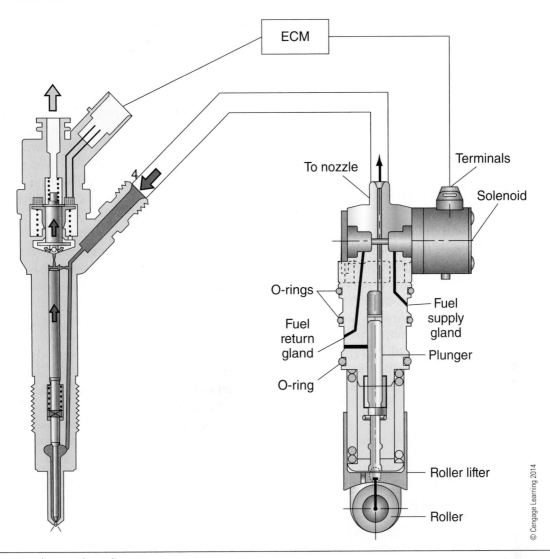

Figure 14-17 Electronic unit pump.

connects the nozzle to the unit pump element. The ECM controls the valves in the nozzle and the pump in a manner similar to EUIs.

Hydraulically Actuated Unit Injector System

Hydraulically actuated unit injectors (HEUI) describe unit injectors that utilize engine lubrication oil under pressure to actuate the pumping element similar to the action of a hydraulic jack. The internal HEUI injector components cause fuel to be pressurized up to seven times greater than the pressure applied to the lubrication oil, as will be explained later in this section. The HEUI system is still in use by Navistar as of 2012. The main advantages of the HEUI system over camshaft actuated unit injectors is the injection cycle is not limited to camshaft lobe profile and injection pressures can be controlled independent of

engine speed permitting high injection pressures and injection rate shaping. Details of a HEUI injector can be found in most diesel fuel system textbooks.

The main components of the HEUI system shown in **Figure 14-18** include:

1. High-pressure oil pump, which develops the high-pressure engine oil called the **injection control pressure (ICP)**.
2. An electric **injection pressure regulator (IPR)**, which is controlled by the ECM.
3. An ICP sensor, which acts as an input to the ECM.
4. The electronic unit injectors, which are controlled by the ECM.
5. A high-pressure oil manifold, which supplies engine oil under high pressure to each injector.

The component locations on a typical Navistar DT engine are shown in **Figure 14-19**.

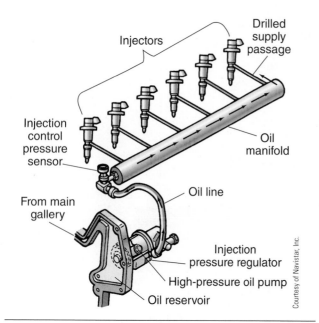

Figure 14-18 Navistar HEUI injection pressure control system components.

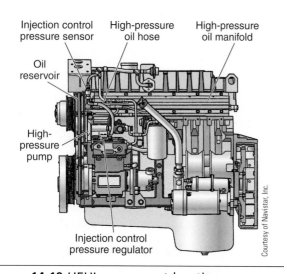

Figure 14-19 HEUI component locations.

The engine ECM inputs and outputs for a pre-2004 HEUI DT engine are shown in **Figure 14-20**. Navistar has introduced several generations of the HEUI control system over the years. These have been known informally in part by the number of electronic control modules used. The first generation was a three-box system and had a separate ECM, an injector drive module (IDM), and a vehicle personality module (VPM). The next generation was a two-box system with an ECM and an IDM. The third-generation consolidated engine controller (CEC) is a single module system. The separation of tasks into individual electronic modules was common practice throughout the industry during the early years of electronic controls. Most OEMs now consolidate engine control into one module, but a separate electronic module may control some specific tasks, such as exhaust aftertreatment.

Since injection control pressure (ICP) has a direct correlation to fuel injection pressure, the ECM is programmed to drive the ICP to a specific pressure for the engine's current operating conditions. The injection pressure regulator (IPR) shown in **Figure 14-21** is an on-off type solenoid, which is controlled by the ECM. The ECM provides a 400 Hz PWM signal to the IPR and varies the duty cycle (percentage of on-time) to control the ICP. When the IPR is not energized, the ICP decreases; when the IPR is energized, the ICP increases. Therefore, a disconnected IPR results in a no-start condition because without sufficient ICP, fuel injection pressure cannot be developed by the unit injectors. Conversely, if the IPR becomes physically stuck in the energized position, the ICP will increase rapidly until a mechanical pressure relief valve in the high-pressure oil manifold opens at 4000 psi (28 MPa) to dump the high-pressure lubrication oil back into the sump. Either of these conditions will result in a corresponding DTC being set by the ECM.

Closed Loop Control. The ECM is programmed to be able to determine the desired ICP based on a variety of inputs and operating conditions. The ECM determines the measured ICP in the high-pressure oil manifold based on the voltage that the ICP sensor supplies to the ECM. The ECM converts this voltage into a pressure based on its programming instructions, as explained in **Chapter 11**. Therefore, the ECM knows the measured ICP and the desired ICP. If the measured ICP is different from the desired ICP, the ECM is able to increase or decrease the ICP by increasing or decreasing the duty cycle of the PWM voltage that the ECM is supplying to the IPR. This is referred to as **closed loop control** or feedback control.

An example of a simple closed loop control system is shown in **Figure 14-22**. In this system, a clear plastic tank of water is being used for some industrial process. The water exits the tank at a varying rate through a pipe at the bottom of the tank. A desired level of water in the tank has been established. To maintain the level of water in the tank at the desired level, the amount of water entering the tank must be regulated. The person shown in **Figure 14-22** has been given a task of maintaining the tank at the desired level shown by opening or closing the valve in the overhead fill pipe. If the level of water in the tank drops below the desired level, the valve must be opened to add water to the tank. Conversely, the valve must be closed if the level of water in the tank is above the desired level. A closed-loop control system consists of three

DT 466E & International @ 530E Electronic Engine Controls

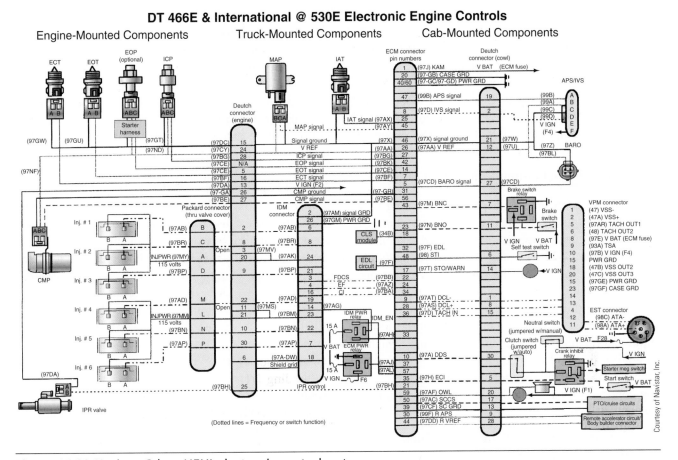

Figure 14-20 Navistar 3-box HEUI electronic control system.

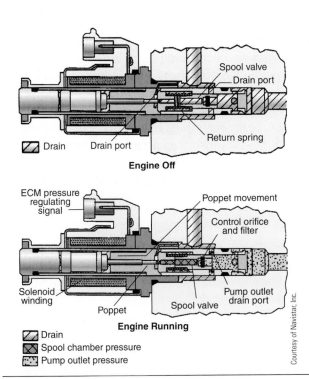

Figure 14-21 Injection pressure regulator (IPR) valve.

main components: a controller, a plant or process referred to as a system, and a means of measuring the actual system output. The actual system output as determined by the measurement device is referred to as the **measured output**. In this case, the person's brain is the controller, the system is a tank with a valve in a water inlet pipe, and the means of output measurement are the eyes of the person. The other components of a closed loop control system are a **desired output**, also known as setpoint, and a means for the controller to change the system, such as an actuator. In the example shown in **Figure 14-22**, the desired output is the

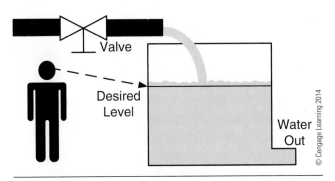

Figure 14-22 Closed loop control system.

established desired tank level, the measured output is the actual tank level observed by the person, and the means for the controller to change the system is a combination of the person's muscles and the valve.

Figure 14-23 shows a block diagram for a typical closed loop control system. In a closed loop control system, the difference between the desired output and the measured output is referred to as the **error**. In other words:

error = desired output − measured output

The error can be positive or negative, depending upon the desired output being greater or less than the measured output. For example, if the measured level in the tank in **Figure 14-22** were above the desired level, the error would be negative. The controller uses the amplitude and sign of the error input to determine what it needs to do with its controller output to cause the error to go to zero. An error of zero indicates that the desired output and the measured output are the same, which is the goal of a closed loop control system.

In the case of the injection control pressure closed loop control system, the following specific definitions apply:

- System = Regulation of the injection control pressure by the IPR
- System Output = Injection control pressure
- Measurement = ICP sensor
- Measured Output = Measured pressure as indicated by the ICP sensor
- Desired Output = Desired ICP determined by the ECM
- Error = Difference (+/−) between the desired and measured ICP
- Controller = ECM
- Controller Output = PWM duty cycle of the IPR control signal

The ECM attempts to make the error go to zero, which will occur when the measured ICP is the same as the desired ICP. The ECM will increase the duty cycle of the IPR signal if the measured ICP is less than the desired ICP. Conversely, the ECM will decrease the duty cycle of the IPR signal if the measured ICP is greater than the desired ICP.

The block diagram shown in **Figure 14-23** can be applied to many other types of closed loop control systems in modern trucks. Other examples include cruise control, automatic HVAC cab temperature control, and exhaust aftertreatment temperature control.

Open Loop Control. If the ECM detected some problem with the ICP sensor, such as an out-of-range low condition caused by a disconnected ICP sensor connector, the ECM would not have any knowledge of the ICP. Without the ICP sensor information, there can be no closed loop control of the injection control pressure. In this case, the ECM could still attempt to drive the IPR with some set duty cycle sufficient to still permit the engine to run. This is referred to as **open loop control**. The ECM could be programmed to drive the IPR at a varying duty cycle dependent upon engine speed and load. The disconnected ICP sensor would likely result in a DTC and an engine speed or torque derate, but the engine should still provide sufficient power to permit the truck to limp-in to the next repair facility.

Open loop control can be described as the use of **lookup tables** within the ECM software. The mileage chart at the bottom of a roadmap is an example of a lookup table. If you wanted to find the distance from one city on a map to another, you could use the lookup table. Through extensive testing, the OEM populates various lookup tables and programs these into the ECM software. For example, at an engine speed of 1200 rpm and at 50 percent load, driving the IPR at a fixed duty cycle of 30 percent should provide sufficient control of ICP.

Open loop control is often used on other truck systems besides instances where a failure has been detected. For example, open loop control of a system is common during transient (changing) engine load

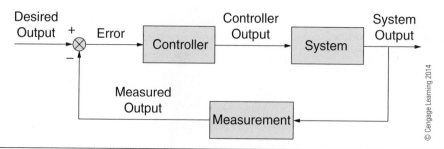

Figure 14-23 Closed loop control block diagram. When measured output and desired output are equal, error is zero.

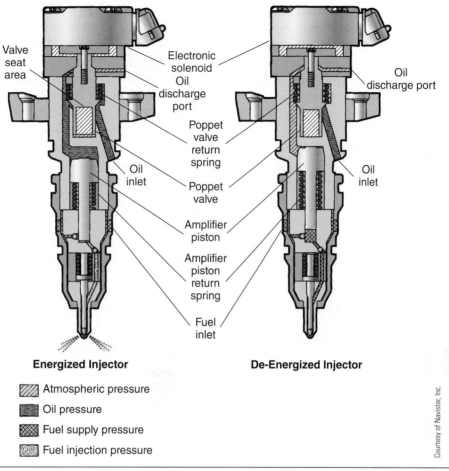

Figure 14-24 Navistar single coil HEUI injector internal components.

conditions and before the engine has reached operating temperature.

HEUI Injectors. There have been two major versions of HEUI injectors used by Navistar. **Figure 14-24** shows a single-coil version of the injector. The ECM (or IDM) supplies 115V DC to the solenoid to open a port to permit high-pressure oil to act on an amplifier piston. This piston has a larger diameter on the ICP engine lubrication oil end (top) than the fuel end (bottom). The piston downward movement causes the fuel to be pressurized seven times greater than the ICP. For example, if the ICP is 3000 psi (21 MPa), the fuel injection pressure will be 21,000 psi (145 MPa).

Figure 14-25 shows the second-generation Navistar HEUI injector produced by Siemens, which has two electrical coils. The two coils are used to control the position of a spool valve. The spool valve shuttles horizontally to permit high-pressure engine oil to act on the intensifier piston in one direction and to dump this oil back to sump in the other direction. There is no return spring on the injector spool. The ECM alternately energizes one solenoid while de-energizing

the other to cause the spool to move one way or the other.

The voltage supply is reduced to 48V DC for these second-generation injectors. **Figure 14-26** shows the high side and low side drivers (FETs) inside the ECM, which are used to control the injector solenoids. The 12V supplied by the truck's batteries is stepped up to 48V or 115V by a DC-to-DC converter within the ECM or IDM. DC-to-DC converters were introduced in **Chapter 7**.

WARNING *Follow the OEM's specific safety recommendations when troubleshooting any electronically controlled diesel fuel system to prevent fatal electric shock.*

Common Rail Fuel Systems

If you are familiar with sequential port fuel injection systems found on many current gasoline engines, then you already understand some of the fundamentals of a diesel common rail fuel system. In both systems, each injector is supplied by a common fuel rail.

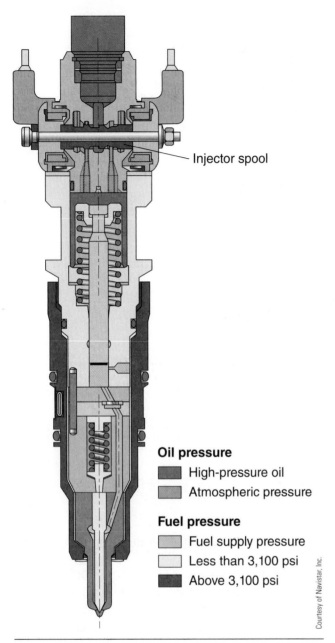

Injector spool

Oil pressure

■ High-pressure oil

■ Atmospheric pressure

Fuel pressure

■ Fuel supply pressure

□ Less than 3,100 psi

■ Above 3,100 psi

Courtesy of Navistar, Inc.

Figure 14-25 Navistar second-generation two-coil HEUI injector.

The injectors are electrically actuated in both systems at a specific time resulting in an injection event. However, the biggest difference between gasoline sequential port fuel injection and common rail diesel injection is the fuel pressure. In a gasoline port injection system, the fuel pressure is typically not higher than 100 psi (690 kPa) while in a diesel common rail system, the pressures can exceed 35,000 psi (241 MPa). In addition, the solenoid actuator in a diesel injector does not directly lift the needle to start injection as it does in a gasoline port fuel injector. Details on the internal workings of an electro-hydraulic injector are provided in most diesel fuel system textbooks.

A typical common rail fuel system is shown in **Figure 14-27**. The ECM controls the high-pressure pump solenoid(s) (**Figure 14-28**) and the electric solenoids within each injector nozzle (**Figure 14-29**). A fuel rail pressure sensor is an input to the ECM. Closed loop control by the ECM, as described in the previous section, is used to maintain the desired rail pressure. The ECM programming makes use of all of the ECM sensor inputs to determine the optimal injection timing and fuel metering quantity.

WARNING *Never crack common rail fuel lines open with the engine running for any reason, including checking for a misfiring cylinder. Doing so could cause fuel under extreme pressure to penetrate your skin resulting in death or serious injury. Follow OEM's instructions for rail pressure bleed down before loosening any component in the high-pressure fuel system. Never install a standard pressure gauge in a common rail system. The extreme pressures will cause the gauge to explode.*

Measurement Error in Closed Loop Control Systems
Because of the extreme fuel pressures in a common rail system, a pressure relief valve (PRV) is typically located in the fuel rail. This mechanical relief valve opens to dump fuel back through the return if rail pressure exceeds a predetermined value. The mechanical relief valve should never open during normal operation, but problems such as an **in-range failure** of the fuel rail pressure sensor can cause the ECM to not be able to determine the actual fuel rail pressure correctly. An in-range failure describes a condition where the voltage measured at the input of the controller is not out-of-range high or out-of-range low (defined in **Chapter 13**) but is within the acceptable normal operating range for the sensor. An in-range failure is a voltage that is within the acceptable range, but it does not accurately represent the actual pressure, temperature, or flow rate that an accurate gauge or other measurement device would indicate. In other words, an in-range failure is a **measurement error**. For example, excessive resistance on the terminals of the rail pressure sensor and the wiring harness terminals caused by fretting corrosion can cause the rail pressure sensor to incorrectly indicate that the pressure within the rail is much lower than the actual rail pressure. The ECM only "knows" the measured rail pressure based on voltage supplied by the sensor at the applicable ECM input terminal. If an electrical problem causes this voltage to be incorrect but still

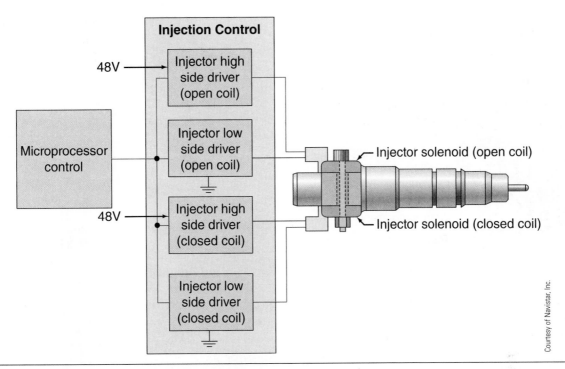

Figure 14-26 High and low side drivers in ECM control injector coils.

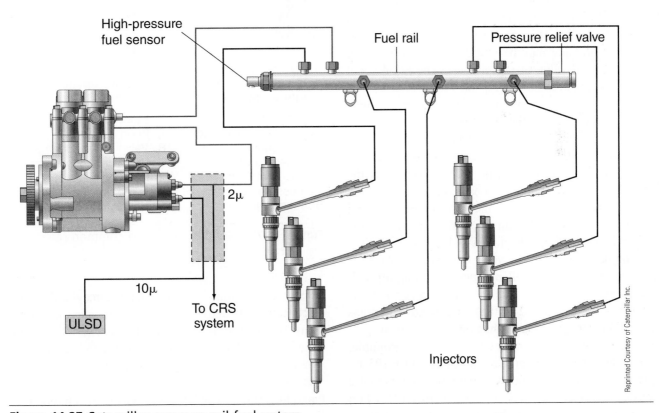

Figure 14-27 Caterpillar common rail fuel system.

within the normal operating range for the sensor, no DTC will likely be set indicating a problem with that particular sensor circuit. However, DTCs indicating a potential problem with some other system may actually set due to the in-range failure. These in-range failures can be some of the more difficult electrical problems to resolve because a measurement error of temperature, pressure, or flow rate can cause a variety of unexpected problems in any closed loop control system.

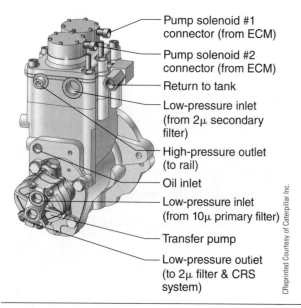

Pump solenoid #1 connector (from ECM)

Pump solenoid #2 connector (from ECM)

Return to tank

Low-pressure inlet (from 2μ secondary filter)

High-pressure outlet (to rail)

Oil inlet

Low-pressure inlet (from 10μ primary filter)

Transfer pump

Low-pressure outlet (to 2μ filter & CRS system)

©Reprinted Courtesy of Caterpillar Inc.

Figure 14-28 Common rail high-pressure pump with control solenoids.

A phrase borrowed from computer science, which describes in-range failures, is "garbage in, garbage out." If fretting corrosion causes high resistance in the rail pressure sensor circuit, the actual value of rail pressure is unknown by the ECM. Therefore, the measured rail pressure may be much less or much greater than the actual value of rail pressure. The ECM could control the high-pressure pump such that the actual rail pressure increases to the point that the mechanical pressure relief valve in the rail opens to prevent damage to the fuel system. In this case, the incorrectly measured rail pressure due to the in-range failure could be considered as *garbage in* and the ECM's control of the high-pressure pump could be considered as *garbage out*.

In many cases, observing the pressure, temperature, or other parameter using an electronic service tool (as described later in this chapter) with the key on, engine off and determining if the value displayed is plausible (believable) can be excellent clues indicating a possible in-range failure. For example, if the rail pressure is indicated as 3000 psi (21 MPa) on an engine that has not been running for quite some time, you may suspect a rail pressure measurement error. In another example, if engine coolant temperature indicates 200°F (93°C) and manifold air temperature and all other temperature sensors on the engine indicate close to 70°F (21°C) on a cold engine, you may suspect a measurement error of coolant temperature (provided no electric block heater is currently being used).

Some OEMs may perform such plausibility checks at engine start up and while the engine is running to cause a DTC to set indicating a possible sensor measurement error.

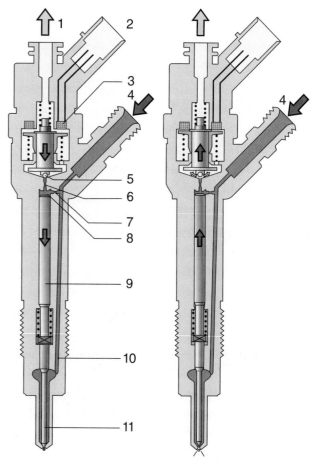

a. Injector closed (at-rest status)

1. Fuel return
2. Electrical connection
3. Triggering element (solenoid valve)
4. Fuel inlet (high pressure) from the rail
5. Valve ball

b. Injector opened (injection)

6. Bleed orifice
7. Feed orifice
8. Valve control chamber
9. Valve control plunger
10. Feed passage to the nozzle
11. Nozzle needle

Courtesy of Robert Bosch LLC

Figure 14-29 Solenoid-controlled common rail injector nozzle.

Some common rail fuel systems may utilize an electrically actuated PRV in the fuel rail. The ECM controls the PRV to cause rail pressure to rapidly decrease when zero fuel is being commanded, such as when the truck is coasting down a hill.

Piezo Injectors. The latest generation of common rail injectors contain a piezo actuator instead of an electromagnetic solenoid actuator. These injectors are often called piezo injectors. Piezoelectric pressure sensors were introduced in **Chapter 11**. Piezo injectors contain a stack of crystal wafers as shown in **Figure 14-30**. Recall form **Chapter 11** that if a crystal is squeezed, it will produce a proportional voltage. This process is also

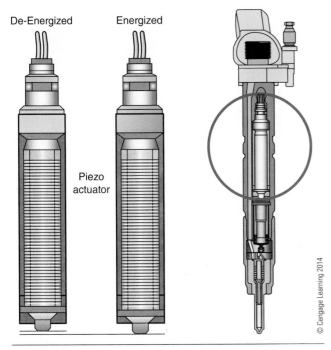

De-Energized   Energized

Piezo
actuator

© Cengage Learning 2014

Figure 14-30 Piezo-actuated common rail injector.

reversible. If a voltage is applied to a crystal, the crystal will expand. Piezo injectors take advantage of this principle. Although the stack of crystal wafers may only expand by perhaps 0.004 in. (0.16 mm) when 100V is applied by the ECM, this movement can be translated through an internal hydraulic system within the injector to cause the injector to actuate and inject.

The advantage of piezo actuators over solenoid actuators is the reduced response time. Recall in **Figure 14-11** the time that was necessary for the current level through the solenoid to increase to a value sufficient for the solenoid actuator to respond. Since piezo injectors contain no inductor, there is no associated current rise time permitting very fast actuation and multiple injection events per cycle. One diesel engine manufacturer indicates that up to seven separate injection events per combustion cycle can be performed using piezo injectors. Piezo injectors also require less electric power to operate than solenoid actuated injectors.

Piezo injectors are currently used on many automotive diesel engines and some smaller truck engines. However, piezo injectors will likely come into more widespread use in on-highway truck engines as the technology matures.

Injector Replacement. Most EUIs or common rail injectors have a calibration or trim code. This calibration code indicates specific flow characteristics for the component and compensates for manufacturing tolerances. The calibration code for each injector or

EUI is stored in the ECM. The ECM decrypts this calibration code and modifies the pulse width accordingly for each injector to reduce fuel-metering errors caused by manufacturing tolerances. If injectors or EUIs that have calibration codes are replaced, the new calibration information must be written to the ECM using the appropriate electronic service tool.

DIESEL EXHAUST EMISSIONS CONTROL

Looking back at **Figure 14-2**, the EPA mandated reduction in diesel exhaust emissions has led to an influx of technology into on-highway diesel engines. The advances in fuel delivery systems discussed in the previous section are responsible for a large percentage of the overall exhaust emissions reduction. The ability of modern fuel systems to meter a precise amount of atomized fuel at just the right time permits engine manufacturers to balance performance, fuel economy, and exhaust emissions. However, the fuel system is just one aspect of exhaust emissions reduction as will be discussed in this section. Exhaust emissions reduction can be divided into two major areas: in-cylinder and aftertreatment.

In-Cylinder Exhaust Emissions Reduction

In-cylinder exhaust emissions reduction refers to reduction of emissions in the combustion chamber. The two main regulated diesel engine pollutants are PM and NO_x. It would be relatively easy to reduce one or the other of these pollutants within the combustion chamber. However, almost anything that is done in-cylinder that causes a reduction in soot (PM) results in an increase in NO_x and vice versa. For example, advancing the injection timing increases peak combustion temperatures resulting in reduced soot, but causes higher NO_x creation. Conversely, retarding injection timing reduces peak combustion temperatures, which reduces NO_x creation but causes higher levels of soot to be generated. This dilemma is known as the soot-NO_x tradeoff.

Exhaust Gas Recirculation (EGR). Exhaust gas recirculation (EGR), as the name implies, is recirculation of exhaust gasses back into the combustion chamber. The exhaust gas is somewhat inert and displaces combustion oxygen, resulting in a decrease in peak combustion temperature and a resulting reduction in NO_x. Exhaust gas also contains water vapor, which has a very high specific heat. Therefore, the water

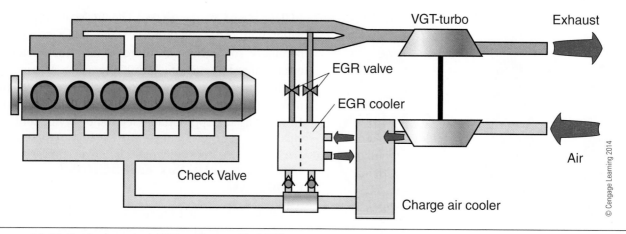

Figure 14-31 Typical cooled EGR system with VGT.

vapor in the recirculated exhaust gas acts as a heat sink to further reduce peak combustion temperatures. The exhaust gas is taken from the exhaust manifold and directed through plumbing to the intake manifold, as shown in **Figure 14-31**. An EGR valve, which is controlled by the ECM, is used to only permit EGR to flow when specified by the ECM. The exhaust gas passes through an EGR cooler to reduce the gas temperature and increase the gas density.

The intake manifold pressure on a turbocharged diesel engine may be nearly the same as the exhaust manifold pressure. Therefore, it is necessary to cause the exhaust manifold pressure to become greater than the intake manifold pressure for sufficient EGR to flow. This can be accomplished using a **variable geometry turbocharger (VGT)**. The term variable turbine geometry (VTG) is also used to describe the technology. The operation of a VGT turbocharger is described in detail in most diesel engine textbooks. VGTs are also used to control the air-to-fuel ratio, especially at low engine speeds. In the case of using a VGT for EGR control, driving the VGT vanes closed increases the exhaust manifold pressure (backpressure). This also increases the intake manifold pressure, but the increase in the exhaust manifold pressure is more than the increase in intake manifold pressure due to the inherent inefficiencies of the system.

The ECM controls the VGT vane position using an actuator, like that shown in **Figure 14-32**. This example uses a control valve that contains an electric solenoid, which regulates the flow of engine lubrication oil under pressure to control the VGT vane position. A VGT vane position sensor is an input to the ECM. The ECM may use a PWM output to control the electric solenoid. Therefore, this is a closed loop control system. The ECM determines a desired VGT position, the vane position sensor indicates the measured VGT position, and the ECM regulates the duty

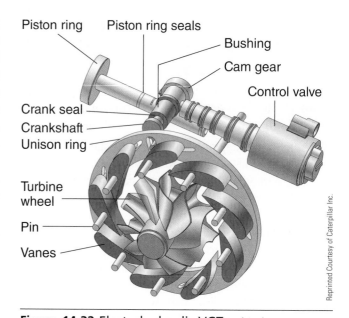

Figure 14-32 Electrohydraulic VGT actuator.

cycle of the PWM output supplied to the solenoid to attain the desired VGT vane position.

An electric VGT actuator is shown in **Figure 14-33**. These actuators typically contain a stepper motor and the control circuitry for the stepper motor. The earliest versions of these actuators received the desired VGT position from the ECM by way of an ECM provided PWM signal or with some other simple digital signal. This was one-way communication; the actuator only received a desired VGT position from the ECM but could not communicate any information back to the ECM. The latest versions of VGT actuators are capable of communicating with the ECM via a J1939 data link. This permits the actuator to detect problems such as stuck vanes, a common problem on variable geometry turbochargers, and provide this information back to the ECM. The ECM may then set a DTC indicating that the desired VGT position could not

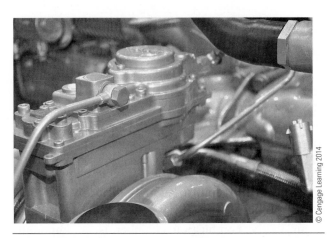

Figure 14-33 Electronic VGT smart actuator.

be attained. This type of actuator may be known as a **smart actuator**. A smart actuator is an actuator that contains electronic circuitry such as a microprocessor and a communications interface. A smart actuator is typically managed by another controller, such as the engine ECM. A J1939-based smart actuator would typically only require four wires: +12V, ground, and the two J1939 circuits.

Most VGTs have a speed sensor. This may be a variable reluctance or Hall effect type sensor. The sensor measures the speed of the shaft connecting the turbine and compressor wheels. A flat section milled on the shaft acts as the target for the turbocharger speed sensor. The turbocharger speed is an input to the ECM and is used to prevent the turbocharger from exceeding the maximum design speed. The ECM monitors the turbocharger speed and will adjust the vane position or derate to reduce the turbocharger speed to prevent damage to the turbocharger.

The turbocharger speed, along with the intake air density, can also be used by the ECM to estimate the intake manifold absolute pressure (MAP). Although most 2004 and later diesel engines with EGR have a MAP sensor, the estimated manifold pressure can be compared to the value indicated by the sensor. If the ECM determines a large discrepancy between the estimated and measured values for intake manifold pressure, a DTC can be set indicating such. Possible causes include large boost leaks or sensor measurement error caused by a wiring harness problem or an in-range MAP sensor failure. The estimation of MAP by the ECM using turbocharger speed and intake air density is known as a **virtual sensor**. A virtual sensor is an estimated value obtained from mathematical models of the system. These mathematical models are developed using physical sensor readings installed on engines during development to calculate the estimated value for a virtual sensor. Virtual sensors are used for some diesel exhaust emission estimations. The use of virtual sensors throughout a truck electrical system will likely expand in the future as the processing power of microprocessors continues to increase.

Smart actuators, such as a VGT actuator, typically perform a **learn cycle** at initialization (power-up). A learn cycle can be described as a zero and span calibration; the actuator is driven to the minimum position (zero) and then driven to the maximum position. In the case of a stepper motor controlled actuator, the number of steps necessary to travel from the minimum to the maximum position is the span or operating range. For actuators directly controlled by the ECM, the ECM may perform an initial learn cycle of applicable actuators when the engine is new and store the actuator's position sensor minimum and maximum voltage information in non-volatile memory. The ECM then performs a learn cycle of the actuator at each initialization (or at engine shutdown for some actuators) and compares these values to the values stored in memory. If the most recent learn cycle values deviate by more than some allowable tolerance, the ECM may set a DTC indicating an invalid learn cycle of that component. This invalid learn cycle may be caused by a mechanical problem such as stuck vanes on a VGT or by heavy carbon deposits on an EGR valve that prevent the valve from fully closing. The invalid learn cycle may also be due to an electrical problem, such as wiring harness failures like shorted and open circuits.

Performing a learn cycle on an actuator permits the ECM to compensate for manufacturing tolerances of the actuator and to verify that the actuator is capable of moving throughout its full range of travel. Replacement of some engine electrical components that have stored initial learn cycle values may require the use of an electronic service tool to inform the ECM that the component has been replaced so that the ECM can perform a new initial learn cycle calibration of the component. Consult the specific OEM's service literature for more information.

An EGR valve is typically an electric motor operated device with an integral valve position sensor which acts as an input to the ECM. The EGR valve is typically a normally closed valve and may be a poppet or a butterfly-type valve. The ECM controls the EGR valve position by supplying current to the motor in the valve. For the circuit shown in **Figure 14-34**, the ECM is able to drive the EGR valve to the open or the closed position by reversing the current flow through the motor. This type of circuit is known as an **H-bridge** because the layout looks like the letter ''H''. The ECM switches on only high side driver A and low

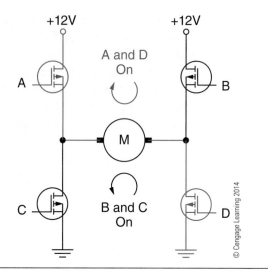

Figure 14-34 H-bridge used to control EGR valve.

side driver D as shown in **Figure 14-34** to cause conventional current to flow through the motor from left to right to cause the motor to turn clockwise, which opens the EGR valve. To reverse the direction of the motor and close the EGR valve, the ECM switches on only high side driver B and low side driver C. This same H bridge type control may be applied to other devices, such as an intake air throttle.

On many diesel engines, the ECM controls the amount of EGR flowing into the intake manifold by opening the EGR valve to a specific position and regulating the VGT position to increase or decrease the exhaust manifold pressure. Some engines have an exhaust manifold pressure sensor, which is an input to the ECM. This sensor along with the intake MAP sensor can be used by the ECM to determine EGR flow rate. Several 2007 and later diesel engines make use of an EGR flow rate sensor. The EGR flow sensor measures the differential pressure (delta pressure) across a venturi in the EGR plumbing. The ECM then calculates the EGR flow rate based on this differential pressure. This is similar to the calculation of electric current flow by measuring the voltage drop across a known resistance. Some engines may use three temperature sensors to determine EGR flow rate. The temperature sensors measure the charge air cooler outlet air temperature (temperature of the air that is entering the intake manifold), the post EGR cooler exhaust gas temperature, and the manifold air temperature. The ECM can calculate EGR flow rate using these three temperatures. This is analogous to a water faucet with separate controls for the hot and cold water that flow out of one faucet. If you know the temperature of the hot water (exhaust gas temperature) and the cold water (charge air cooler outlet temperature) and the temperature of the resulting warm water

flowing out of the faucet (manifold air temperature), it is not difficult to calculate the percentage of hot water making up the warm water flowing from the faucet.

EGR flow rate is typically under closed loop control by the ECM. The ECM determines the desired EGR flow rate based on its program instructions. The measured EGR flow rate is determined by the ECM based on the applicable sensor input values. On engines so equipped, the ECM then controls the EGR valve position and VGT position to attain the desired EGR flow rate for the current operating conditions.

The percentage of EGR compared to the percentage of air in the combustion chamber is an important measurement for a modern diesel engine. Too much EGR results in reduced air-to-fuel ratio, which causes poor performance, black exhaust smoke (soot), and high exhaust manifold temperatures. Conversely, too little EGR causes the engine to produce excessive NO_x, resulting in non-compliance with exhaust emissions standards. An in-range failure caused by a defective sensor or a wiring harness problem can cause the ECM to incorrectly calculate the percentage of EGR.

Diesel Exhaust Aftertreatment

Since it is not possible to sufficiently reduce both soot and NO_x with in-cylinder techniques alone, the 2007 EPA emissions standards resulted in the addition of exhaust **aftertreatment** devices to most on-highway diesel engines. As the name implies, aftertreatment is modification to the exhaust gasses outside of the combustion chamber.

2007 EPA Emissions. For the 2007 EPA requirements, most North American OEMs chose to control NO_x in-cylinder and handle the PM (mostly carbon-based soot) reduction using aftertreatment. A diesel particulate filter (DPF) is an aftertreatment device used to trap the soot particles in a porous ceramic filter material. The soot in the DPF is periodically removed through a process referred to as regeneration. In a DPF regeneration, high temperatures of up to 1165°F (630°C) are used to cause the soot to oxidize. This oxidation process is similar to what happens when barbequing with charcoal briquettes. Fire is initially used to heat the charcoal briquettes. Once the fire burns out, the briquettes continue to glow orange for a prolonged period of time as the charcoal (carbon material similar to soot) combines with oxygen in the air (oxidizes) to form CO_2 gas and heat. Eventually, all of the carbon is oxidized and nothing remains of the charcoal briquettes except for a small amount of ash.

There are two primary methods of creating the heat in the diesel exhaust system necessary to cause DPF regeneration. Most North American OEMs use a diesel oxidation catalyst (DOC) upstream of the DPF to generate heat. A DOC is similar to a catalytic converter found in automobiles. A catalyst is a substance which increases the rate of a chemical reaction. The catalyst material in a DOC is typically a platinum metal. The DOC converts hydrocarbons (HC) and exhaust oxygen (O) into CO_2, H_2O, and heat energy. Note that there is no flame or ignition involved. Since diesel fuel is a hydrocarbon compound, a system of injecting diesel fuel into the exhaust stream is used to create the heat necessary to regenerate the DPF. This hydrocarbon injection system is controlled by the ECM or by a separate aftertreatment control module in some systems. Temperature sensors at the DOC inlet, the DOC outlet (or DPF inlet), and the DPF outlet are typically used as inputs to the ECM. These temperature sensors are thermocouples, thermistors, or RTDs (PRTs), as described in **Chapter 11**. Thermocouple type sensors are typically hardwired to a remote aftertreatment temperature module instead of being wired directly to the ECM. This module communicates temperature information to the ECM using a CAN network or some other digital network system such as a local interconnect network (LIN). A typical DOC/DPF system is shown in **Figure 14-35**.

The ECM monitors the soot loading of the DPF to determine when DPF regeneration is necessary. A differential (delta) pressure sensor across the DPF is typically an input to the ECM to assist in this calculation of DPF soot loading. As the soot loading increases, the differential pressure increases. However, most ECMs also calculate the soot in the DPF based on fuel consumption and many other factors. When the calculated soot level increases above a threshold, the ECM will initiate the DPF regeneration. One fundamental of chemistry is that the rate of a chemical reaction increases as the temperature increases. Therefore, prior to diesel fuel being injected into the

exhaust upstream of the DOC, the DOC temperature must be greater than approximately 575°F (300°C) to make sure that most of the fuel will be converted into heat in the DOC. This is known as the DOC light-off temperature. If the engine is sufficiently loaded, the DOC temperature may already be above the light-off temperature. Otherwise, many 2007 and later diesel engines have an intake air throttle at the intake manifold inlet. The ECM has control over this normally open butterfly-type throttle valve. Reducing the air-to-fuel ratio of a diesel engine by supplying less air to the engine results in an increase in exhaust temperatures. Some engines may use an exhaust throttle instead of an intake air throttle for the same purpose.

Once the DOC light-off temperature is reached, the ECM can start the injection of diesel fuel into the exhaust. This is referred to as fuel dosing. Many 2007 and later engines utilize electric solenoids, which are controlled by the ECM to meter diesel fuel under moderate pressure, typically less than 250 psi (1.7 MPa), through a dosing injector into the exhaust upstream of the DOC. Here is another closed loop control system—the desired DOC outlet temperature and the measured DOC outlet temperature.

A fuel dosing system supplied by Bosch called the Departronic® system is used by many OEMs. This system has the fuel dosing control components all contained in a single block, as shown in **Figure 14-36**. This assembly contains an inlet and outlet pressure sensor, a fuel temperature sensor, and an inlet and outlet control solenoid. The pressure sensors can be utilized for fuel dosing system diagnostics permitting the ECM to set a DTC if a problem is detected. The dosing injector used with this system is a

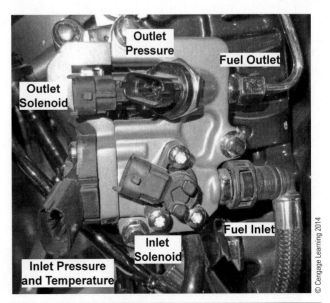

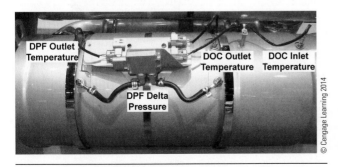

Figure 14-35 DOC/DPF system.

Figure 14-36 Bosch Departronic fuel dosing system.

hydromechanical injector, which opens and closes up to 1000 times per second to atomize the fuel. Other systems may use an electric solenoid-actuated dosing injector, which is controlled by the ECM.

The ECM will steadily increase the DOC outlet temperature to approximately 1100°F (600°C) using closed loop controlled metering of the dosing fuel. At this temperature, the soot in the DPF is typically oxidized within 20 minutes without flame, similar to glowing barbeque charcoal briquettes.

Some OEMs utilize in-cylinder dosing as the source of exhaust hydrocarbons instead of using a separate dosing injector in the exhaust. In-cylinder dosing refers to a late charge of fuel injected into the combustion chamber after combustion is complete. This charge of unburned fuel is pushed out by the piston during the exhaust stroke and is oxidized in the DOC.

Caterpillar used a different approach for 2007–2009 engines compared to most other North American OEMs called the Caterpillar Regeneration System (CRS) as shown in **Figure 14-37**. Instead of using a DOC to generate heat necessary for DPF regeneration, CRS uses a spark plug near the fuel injector, shown as the central component in **Figure 14-38**, to ignite the diesel fuel to produce a flame in a burner unit within the exhaust upstream of the DPF. Compressed air is also introduced into the exhaust to aid combustion. The spark plug and ignition coil are similar to those found in gasoline engines. An ignition coil is a type of transformer, as discussed in **Chapter 3**.

The preceding described active DPF regeneration. Another form of DPF regeneration is passive regeneration. Passive regeneration is described as natural regeneration. Soot in the DPF will oxidize at a slow rate at temperatures as low as 500°F (260°C) in an environment containing moderate levels of nitrogen dioxide (NO_2). One additional function of the DOC is to convert the nitric oxide (NO) that makes up the exhaust NO_x into nitrogen dioxide to promote passive regeneration, thus reducing the need for active DPF regeneration. Ideally, an engine that is operated under load will naturally produce sufficient exhaust temperatures in the DPF to permit all of the soot that is being created by the engine to be passively oxidized. Additional information on passive DPF regeneration can be found in the Internet links at the end of this chapter.

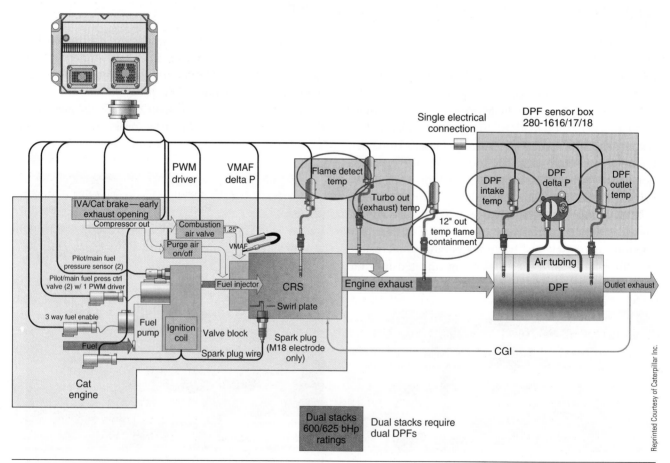

Figure 14-37 Caterpillar CRS system. Compressed air is used to purge the injector and to aid in combustion.

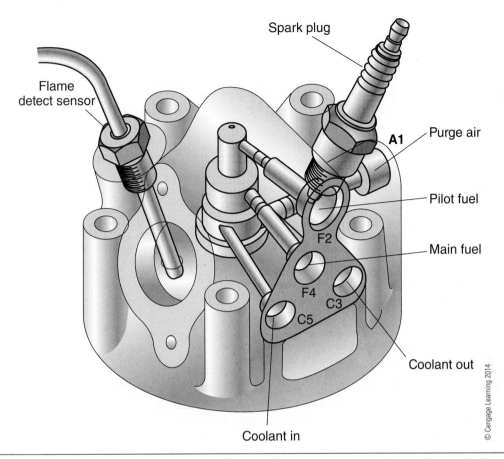

Figure 14-38 Caterpillar CRS injector and spark plug details. Engine coolant is used to cool the injector.

2010 EPA Emissions. The 2010 EPA emissions regulations required a further reduction in NO_x emissions. To meet this requirement, most North American on-road diesel engine manufacturers opted to use **selective catalytic reduction (SCR)** aftertreatment systems. As of 2013, all North American truck manufacturers will use SCR. The term *selective* means that only the oxygen in the exhaust bonded with nitrogen is targeted. SCR uses ammonia (NH_3) passing through a catalyst in the exhaust system to convert most of the NO_x into harmless nitrogen (N) and water (H_2O). The ammonia is created by the injection of a urea-water compound into the exhaust called diesel exhaust fluid (DEF) in North America and AdBlue® in other parts of the world. DEF may also be referred to as a reductant, which means an NO_x reducing agent. A typical SCR system diagram is shown in **Figure 14-39**. Note that there are three components shown within the catalyst assembly. The SCR catalyst and hydrolysis catalyst are utilized for the NO_x and DEF conversion processes. The oxidation catalyst located at the exit of the catalyst assembly converts any excess ammonia that remains or *slips* through the SCR catalysts into nitrogen (N), nitrous oxide (N_2O), and nitric oxide (NO). This prevents ammonia from being exhausted

by the system. However, the nitrous oxide (N_2O) and nitric oxide (NO) are regulated exhaust emissions so this is not a desirable condition.

Most 2010 systems still have a DOC and DPF for PM emissions. However, most 2010 and later diesel engines have substantially less soot emissions because of the soot-NO_x tradeoff discussed earlier in this chapter. SCR has permitted OEMs to mostly control soot in-cylinder and control NO_x in the aftertreatment system. Model year 2010 and later engines using SCR are said to rarely require DPF regeneration and achieve improved fuel economy compared to 2007 engines.

The DEF that is injected into the exhaust dissociates or breaks down into ammonia in the hot exhaust stream upstream of the SCR catalyst. A mixer swirl plate is typically placed in the exhaust downstream of the DEF dosing injector to assist in breaking down the DEF into ammonia. Temperature sensors at the SCR catalyst inlet and outlet act as inputs to the ECM to make sure the temperature is sufficient for decomposition of the DEF into ammonia. The SCR temperature sensors may be thermocouples, thermistors, or RTDs. Two NO_x sensors are also typically used in an SCR system. A NO_x sensor in the exhaust

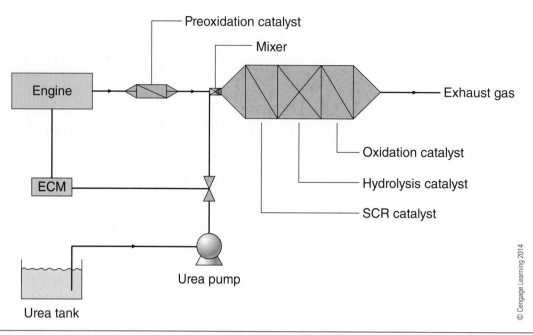

Figure 14-39 SCR system components.

upstream of the SCR catalyst inlet and another NO_x sensor at the SCR catalyst outlet are inputs to the ECM. The ECM or a separate aftertreatment control module has control of the DEF dosing system. The inlet NO_x sensor is typically used by the ECM to determine the quantity of DEF to be dosed. The outlet NO_x sensor may be utilized for diagnostics, such as the detection of poor quality or diluted DEF. Alternatively, some OEMs may use a virtual sensor for the inlet NO_x sensor, meaning that there is no physical inlet NO_x sensor and that the NO_x present at the SCR catalyst inlet is estimated by the ECM using mathematical models.

NO_x sensors are typically **smart sensors**. A smart sensor is a sensor that contains signal processing and a communication interface. The communication interface is often J1939, permitting the smart sensor to communicate directly with the ECM over a J1939 data link. Thermocouple type aftertreatment temperature sensors are also typically smart sensors. The output of a smart sensor is a message indicating NO_x information, aftertreatment temperature, DEF quality, or other parameter. Some OEMs may have a proprietary (private) J1939 data link used for smart sensors and other engine-based components, such as a VGT actuator, instead of connecting these components to the vehicle J1939 data link. Other networks such as LIN may also be used for smart sensors to communicate with the ECM. A J1939-based smart sensor typically requires four wires: 12V power, ground, and the J1939 circuits.

NO_x sensors operate on principles similar to wideband oxygen sensors. For more information on

NO_x sensors, see the Internet links at the end of this chapter.

To meter the DEF, many OEMs use the Bosch Denoxtronic® DEF injection system shown in **Figure 14-40**. The DEF injection module for this system contains an electric motor driven DEF pump, electric heating element, temperature sensor, and a reversing valve. Because DEF is mostly purified water, the freezing temperature of DEF is about 12°F (−11°C). Therefore, it is necessary to evacuate DEF from the lines and the pump to prevent expansion damage caused by frozen DEF. The reversing valve in the DEF injection module is controlled by the ECM. At engine shutdown, the reversing valve is actuated causing DEF to be evacuated from the system and returned to the DEF tank.

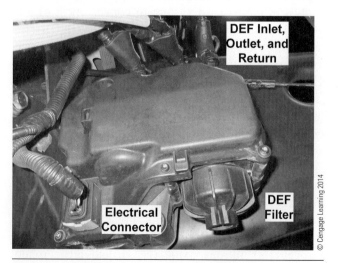

Figure 14-40 Bosch Denoxtronic system.

Electrically heated DEF supply and injection lines to thaw a frozen DEF system are common and most DEF tanks are heated by an engine coolant heater (**Figure 14-41**). An electric control valve in the tank header is controlled by the ECM. A temperature sensor in the DEF tank is an ECM input. The ECM only permits coolant to flow through the DEF heater if the tank temperature sensor indicates the DEF might be frozen. DEF is not damaged if it is frozen and it can typically be thawed by the system within a few minutes of operation. However, DEF rapidly deteriorates if its temperature exceeds 140°F (60°C) so it is important that the DEF not be overheated.

Typical problems with SCR systems include poor quality DEF caused by dilution with water. The water used in DEF is purified and deionized. The use of tap water to dilute DEF can cause the SCR catalyst to be permanently contaminated by the minerals in the tap water. DEF dilution is a form of emissions tampering.

Sulfur contamination is another SCR system problem. The use of ultra-low sulfur diesel fuel is critical for aftertreatment devices. The sulfur in the fuel contaminates both the DOC and the SCR catalyst rendering them ineffective. In some cases, the sulfur can be removed with heat by performing a DPF regeneration. The heat during the regeneration may also dissolve DEF deposits in the SCR mixer, another common problem with SCR systems.

DEF quality sensors are used on some trucks outside of North America and their use will likely become mandated in the United States by 2014 to reduce the likelihood of intentional DEF dilution. These sensors determine if the DEF has been diluted with water or other contaminates by measuring the viscosity, specific gravity, or conductivity of the DEF. Information on DEF quality sensors can be found in the Internet links at the end of this chapter.

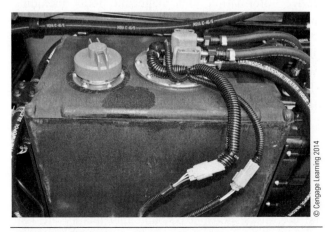

Figure 14-41 DEF tank header with engine coolant heater.

Ammonia sensors will also be used in SCR systems in the near future. Ammonia sensors are similar to NO_x sensors. The ammonia sensor is located within the SCR catalyst and is used by the ECM for closed loop control of the DEF injection quantity. The ammonia sensor will also reduce the possibility of excess ammonia being generated and passing through the SCR catalyst. Excess ammonia is falsely detected as NO_x by the outlet NO_x sensor.

Soot sensors are also likely in the future. A cracked or melted DPF can cause excessive PM emissions. To detect this condition, a soot sensor will be placed in the exhaust downstream of the DPF outlet. A soot sensor contains two closely spaced electrical conductors. As soot (carbon) particles accumulate in the sensor, the electrically conductive carbon causes a change in the resistance between the electrical conductors. This resistance indicates the amount of soot in the exhaust. To remove the accumulated soot from the sensor, an electric heater in the sensor is switched on periodically to burn off the soot particles.

HD On-Board Diagnostics

Heavy-duty on-board diagnostics (HD-OBD) describes the ability for a diesel engine to monitor if its exhaust emissions exceed EPA thresholds and then to alert the vehicle operator by means of a **malfunction indicator lamp (MIL)**. A DTC consisting of the SPN and FMI must also be stored in the ECM memory indicating the detected malfunction. The MIL graphic is defined by the EPA as an amber-colored ISO engine symbol. You are likely familiar with the term OBD II, which refers to passenger car exhaust emissions. The term OBD has been utilized in the past by some OEMs to define non-emissions related system self-diagnostics such as a body controller setting a DTC for headlamp current being too low. Since HD-OBD requirements for on-road diesel engines started coming into effect in 2010, most OEMs now reserve the term OBD for EPA mandated exhaust emissions on board diagnostics. By 2014, all EPA on-highway diesel engines will be subject to some form of HD-OBD regulations.

HD-OBD may also include cylinder misfire detection because a misfire can cause large amounts of unburned fuel (hydrocarbons) to be present in the exhaust. Currently, this detection is limited to lower engine speeds only, but will likely expand to all speeds and loads in the future. The hydrocarbons that are present in the exhaust due to a misfire can cause damage to the DOC and DPF. The hydrocarbons due to a misfire are oxidized in the DOC and generate heat, as discussed earlier in this chapter. Hydrocarbons (HC) are also a regulated exhaust emission so any excess hydrocarbons that pass through or slip through the DOC and DPF cause the engine to be non-complaint. Cylinder misfire is typically detected by changes in the crankshaft speed as indicated by the crankshaft position sensor.

In addition to exhaust emissions, HD-OBD also addresses engine crankcase emissions (blow-by). Crankcase filters and closed crankcase systems are used to control crankcase emissions. Cooling system thermostat performance is also monitored in HD-OBD systems.

HD-OBD DTC Types. There are several types of HD-OBD DTCs. When the ECM first detects most types of emissions related malfunctions, a **pending DTC** is stored in the ECM memory, but this will not illuminate the MIL lamp. If the same malfunction is not detected during a second EPA defined drive cycle, the ECM erases the pending DTC. However, if the same malfunction is detected on a second consecutive drive cycle, the MIL lamp must be illuminated and the pending DTC will become a **confirmed DTC**, which is also referred to as a MIL-on DTC. A confirmed DTC also results in the ECM storing a **permanent DTC** in non-volatile memory. A permanent DTC cannot be cleared with an electronic service tool or by disconnecting the vehicle batteries. The MIL lamp will remain illuminated until the problem is no longer detected by the ECM for three consecutive drive cycles, at which time the confirmed DTC and permanent DTC will be erased by the ECM and become a **previous MIL-on DTC**. Previous MIL-on DTCs are erased by the ECM after 40 engine warm-up cycles, provided the same problem has not been detected again.

Although some locales have existing truck emissions inspection and maintenance (I/M) programs such as smoke opacity tests, as of 2012 emissions inspections for on-road trucks are not required in most of the United States. However, these I/M inspections will likely become mandated in the future. The inspection on HD-OBD vehicles would likely consist of verifying with an electronic service tool that the MIL lamp is not being commanded on by the ECM and that no pending, confirmed, or permanent DTCs are stored. Prior to an emissions inspection, the engine must have been operated long enough and under the required conditions for the ECM to have performed all internal emissions systems tests. This is referred to as **readiness** status or I/M readiness. The OBD readiness status indictor in the ECM must indicate "ready" before an emissions test can be performed.

The readiness status indicator is also useful for technicians. After performing an emissions related repair, confirmed DTCs can be cleared using an electronic service tool. However, since two drive cycles are required to cause the MIL lamp to illuminate, just clearing the confirmed DTCs, starting the engine, and checking for an illuminated MIL lamp is typically not sufficient to verify the repair has fixed the problem. The readiness indicator and pending DTCs can be used to validate the repair. A road test of 20 minutes or more may be necessary to cause the readiness indicator to indicate "ready." If a pending DTC is indicated, the repair may have not fixed the problem even though the MIL lamp is not yet illuminated.

HD-OBD also requires the ECM to store a snap shot of specific engine operating conditions in non-volatile memory when the MIL lamp has been illuminated. This information, known as a **freeze frame** data, can be retrieved by an electronic service tool with the intent of assisting a technician in the diagnosis of the problem. Freeze frame data includes engine parameters that were recorded when the failure was detected such as engine speed, load, and coolant temperature among others.

Additional HD-OBD Provisions. Unlike the two consecutive trip MIL illumination process described earlier, some failures of the SCR system may result in the ECM imposing severe derate penalties and illuminating the MIL upon the first detection of the failure. This includes an empty DEF tank and disconnected SCR system electrical connectors. The derate penalties are necessary because an empty DEF tank or many other types of emissions system failures may not otherwise result in any operator detectable engine performance issues.

HD-OBD also has provisions for access to OEM's service information and diagnostic tools including electronic service tools. This permits non-OEM independent repair facilities to purchase the OEM's service information and tools necessary to diagnose and perform repairs of the HD-OBD system. Much of HD-OBD as it pertains to service information is still under discussion as of 2012.

DIESEL ENGINE DIAGNOSIS

As complicated as a modern diesel engine may appear to be, checking the fundamentals such as fuel and air filters, boost leaks, and fuel quality are still essential for proper engine performance. Battery state of charge and charging system voltage are other basic items that cannot be overlooked. Once the fundamentals are ruled out, more advanced diagnostic techniques and tools may be necessary to find the cause of the problem.

Electronic Service Tools

Electronic service tools (ESTs) refer to devices that read information from the various electronic modules connected to the vehicle data link. Early ESTs were often handheld display devices known as scan tools, such as the popular Pro-Link 9000 shown in **Figure 14-42**. This tool is still widely used today to diagnose legacy electronic diesel engines as well as some vehicle systems such as ABS.

Most OEMs now use a PC-based EST, as shown in **Figure 14-43**. These tools can display ECM parameters, such as the measured intake manifold air temperature or the measured intake manifold pressure. This can be very useful in diagnosing in-range measurement errors, which may not be causing a DTC to be set. For example, an engine that has not been operated for several hours should indicate similar temperatures for all temperature parameters such as coolant temperature, manifold air temperature, or other similar parameter with ignition on and engine off. A substantial difference in one reading may indicate a measurement error due to a wiring harness problem or a defective sensor.

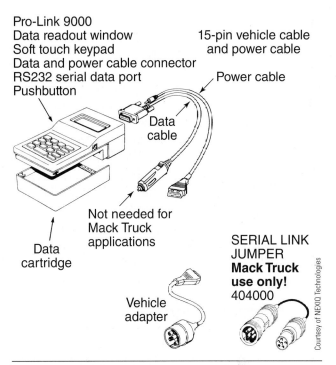

Figure 14-42 Pro-Link 9000 scan tool.

A variety of tests can be performed with most ESTs, such as a cylinder cutout test. This test causes fueling to each cylinder to be disabled resulting in that cylinder not firing. When a cylinder that is contributing is disabled, a change in the way the engine sounds will be evident along with increased vibration.

Figure 14-43 PC-based electronic service tool.

However, disabling a cylinder that is misfiring due to an injector problem or some mechanical issue will not cause a noticeable change in the engine sound or vibration. Engine parameters such as injection pulse width may also be displayed on the EST during the cutout test. The example shown in **Figure 14-44** indicates that when the number 2 cylinder was cut out, the injection pulse width for the other five cylinders did not increase as would be expected. If a contributing cylinder is cut out, the load on the engine increases and the ECM responds by supplying more fuel to all of the cylinders to compensate. If a dead or minimal contributing cylinder is cut out, the load on the engine does not change so there is minimal change in the quantity of fuel that is supplied to the other cylinders.

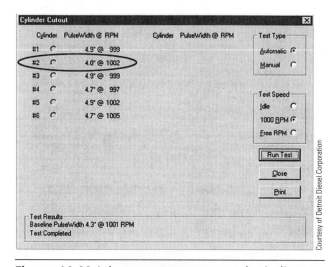

Figure 14-44 Injector cutout test results indicate a problem with the #2 cylinder.

Therefore, **Figure 14-44** indicates a problem with the number 2 cylinder.

Most PC-based ESTs can command an active DPF parked regeneration on applicable engines, which is useful for diagnosing DPF regeneration problems. The graphical display of a DPF regeneration shown in **Figure 14-45** is an example of the type of information that can be displayed by a PC-based EST.

Oscilloscopes

Portable hand-held oscilloscopes have become an essential tool for truck technicians. Oscilloscopes permit a rapidly changing electrical signal to be evaluated. Modern oscilloscopes are sometimes referred to as digital storage oscilloscopes (DSO) because these are digital devices with the capability of storing waveforms in memory for later reference. Oscilloscopes display voltage amplitude on the vertical axis and time on the horizontal axis. Current can also be displayed using a clamp-on ammeter. The volts per amp setting of the clamp-on ammeter can be entered into the oscilloscope settings so that properly scaled current is displayed on the oscilloscope screen.

Multiple-channel (multiple input) oscilloscopes permit two or more signals to be evaluated relative to each other at the same time as shown in **Figure 14-46**. For example, the engine crankshaft position sensor and camshaft position sensor signals could both be displayed on an oscilloscope at the same time to detect problems such as a slipped or loose target. When combined with a clamp-on ammeter, voltage and current can both be displayed at the same time, such as injector supply

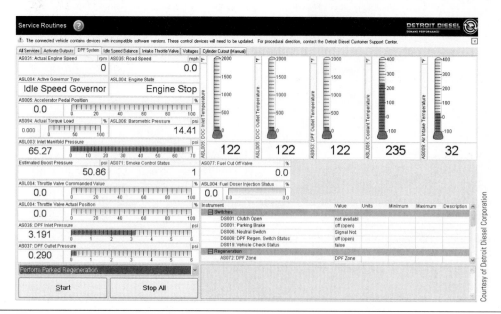

Figure 14-45 DPF parked regeneration parameters monitored with electronic service tool.

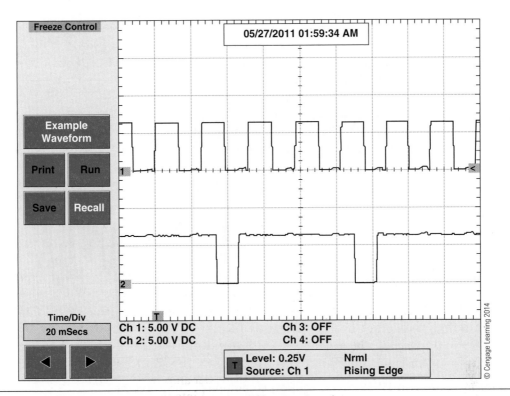

Figure 14-46 Two-channel oscilloscope displays two different signals.

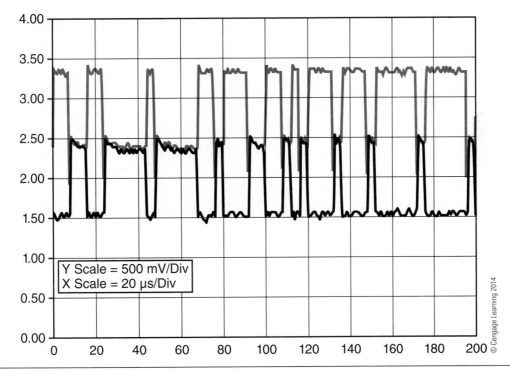

Figure 14-47 Typical J1939 voltage levels using two-channel oscilloscope referenced to ground.

voltage and injector current draw. Comparing the waveforms from a properly running engine with those observed on an engine that is exhibiting a problem can be very useful information.

Another use of an oscilloscope is to diagnose problems on a communication network, such as the J1939 data link. Intermittent open or shorted circuits cause a change in the signal pattern, as does electrical noise. The oscilloscope can be monitored for change while wiggling the wiring harness containing the data link circuits and while switching on and off potential electrical noise sources. **Figure 14-47**

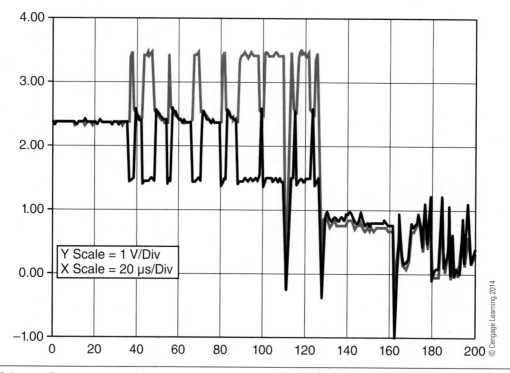

Figure 14-48 Intermittent open J1939 circuit causing communication problems.

illustrates the typical voltage levels observed with a two-channel oscilloscope monitoring J1939 data link activity with no problems present. One channel is connected between CAN+ and ground and the other channel is connected between CAN– and ground. In **Figure 14-48**, an intermittent open circuit has been captured.

Hand-held battery-powered oscilloscopes are reasonably priced, but another option is a PC-based oscilloscope, such as the PicoScope® shown in **Figure 14-49**. This oscilloscope uses the PC as the display screen via a USB connection. Waveforms can be stored on the PC and viewed later.

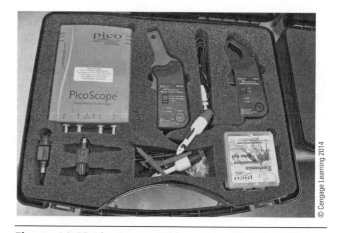

Figure 14-49 PicoScope PC-based oscilloscope kit.

Summary

- Vehicles with diesel engines sold in the United States are subject to EPA exhaust emissions regulations. These regulations have resulted in substantial decrease of particulate matter (PM) and oxides of nitrogen (NO_x) emissions. To attain these requirements, electronic engine controls were necessary.

- Electronic unit injector (EUI) systems with single actuators control the spill valve opening and closing. Dual-actuator EUIs control both the spill valve and the nozzle opening.

- Self-inductance prevents the current flow through an inductor from changing instantly.

- Hydraulically actuated electronic unit injectors amplify engine lubrication oil under pressure to generate fuel injection pressures.

- A closed loop control system compares a desired output to the measured output with the difference being referred to as the error. A controller drives the output such that the error becomes zero, indicating the desired output and the measured output are the same. Closed loop control systems are used in a variety of truck systems.

- An in-range failure is a condition where the input voltage at an electronic control module is within the normal operating voltage range, but does not correctly represent what is being measured.

- Exhaust emissions can be controlled by a combination of in-cylinder techniques and by exhaust aftertreatment systems.

- A variable geometry turbocharger (VGT) can be used to regulate EGR flow.

- An H-bridge is an electronic circuit that causes the polarity of two wires to reverse to permit directional control of an electric motor.

- Soot is captured in the diesel particulate filter (DPF). A periodic regeneration of the filter to oxidize the soot is sometimes necessary. Heat for the regeneration is produced either by diesel fuel being injected over a diesel oxidation catalyst (DOC) or by igniting the fuel in a burner in the exhaust.

- Selective catalytic reduction (SCR) uses ammonia to reduce NOx. The ammonia is produced when a urea water solution called diesel exhaust fluid (DEF) is injected into the exhaust stream.

- HD-OBD describes the ability of a diesel engine to determine that its exhaust emissions have exceeded a threshold and to illuminate a malfunction indicator lamp (MIL) to alert the operator.

Suggested Internet Searches

Try the following web sites for more information:
http://cumminsemissionsolution.com
http://www.bosch.com
http://factsaboutscr.com
http://www.conti-online.com
http://www.epa.gov
http://www.picotech.com

Review Questions

1. Technician A says that cam-actuated unit injectors use a common high-pressure pump to develop injection pressures. Technician B says that a single-actuator EUI typically has four electrical terminals. Who is correct?

   A. A only

   B. B only

   C. Both A and B

   D. Neither A nor B

2. The ICP of an HEUI fuel system is controlled by which device?

   A. HEUI regulator

   B. Open loop control valve

   C. Camshaft control regulator

   D. IPR

3. Which of these is a true statement regarding a closed loop control system?

   A. The desired output and the measured output are *never* the same value.

   B. When the error is zero, the desired output and the measured output must *both* also be zero.

   C. If the desired output and the measured output are the same value, the error is zero.

   D. The desired output is always greater than the measured output.

4. Technician A says that the current flow through an inductor can instantly change. Technician B says that an in-range failure always causes a DTC to be set. Who is correct?

   A. A only

   B. B only

   C. Both A and B

   D. Neither A nor B

5. A high-pressure common rail injector is being replaced. Which of the following may be necessary?

   A. Switch the ignition on with engine off for 5 minutes to permit the ECM to perform a learn cycle of the new injector.

   B. Use an EST to enter the specific calibration data for the new injector.

   C. Adjust injector preload and rack timing.

   D. No specific action is ever required when replacing any electronic injector.

6. An active DPF regeneration is occurring on an EPA 2007 engine equipped truck that has a DOC. Which of the following is likely *not* present in this system:

   A. DPF

   B. Spark plug and ignition coil

   C. DOC inlet temperature sensor

   D. DPF differential (delta) pressure sensor

7. Which of these is most likely to be a smart sensor?

   A. Coolant temperature sensor with two electrical terminals

   B. Manifold temperature sensor with two electrical terminals

   C. Virtual exhaust manifold temperature sensor

   D. NO_x sensor with four electrical terminals

8. HD-OBD includes which of the following?

   A. Permanent DTC

   B. Pending DTC

   C. Confirmed DTC

   D. All of the above

9. Which of the following would be the most likely device to be controlled by an H-bridge?

   A. Smart sensor

   B. HEUI injector

   C. DEF injector

   D. EGR valve

10. Technician A says that a rail pressure sensor wiring problem can cause the pressure relief valve in a high-pressure common rail system to open. Technician B says that DEF that has been diluted with water may result in a DTC. Who is correct?

   A. A only

   B. B only

   C. Both A and B

   D. Neither A nor B

15 Modern Truck Electrical System

Learning Objectives

After studying this chapter, you should be able to:

■ Explain the basic functionality of an electronically controlled automatic transmission.

■ Discuss the functionality of an automated manual transmission.

■ Define the terms PRNDL, NSBU, VIM, chuff test, differential braking, and ABS modulator.

■ Explain how an air ABS system and traction control operates.

■ Test an ABS wheel-speed sensor circuit.

■ Describe how a fault detected by a trailer ABS ECU causes the trailer ABS warning lamp to illuminate in the truck instrument panel.

■ Discuss the operation of an Eaton electric hybrid system.

■ Describe the difference between a series hybrid and a parallel hybrid.

■ List the steps to troubleshoot a wiring problem with the J1939 data link.

Key Terms

anti-spin regulation (ASR)

automated manual
    transmission (AMT)

automatic traction control (ATC)

chuff test

data mining

differential braking

drive-axle ABS event

dual-mode hybrid system

electronic on-board recorder
    (EOBR)

linear Hall effect sensor

modulators

parallel hybrid system

power line carrier (PLC)

series hybrid system

telematics

torque-speed control (TSC)

vehicle onboard radar (VORAD)

INTRODUCTION

This chapter discusses details of major systems on a modern truck not already covered in previous chapters and how these systems interact. Emerging technologies, such as hybrid electric trucks, are also discussed along with general troubleshooting tips for modern truck electrical and electronic systems.

TRANSMISSIONS

Historically, most truck transmissions were manually-shifted standard transmissions with the only electrical components being a reverse lamp switch and perhaps a vehicle speed sensor. However, electronically controlled automated manual transmissions and electronic automatic transmissions can now be found in trucks of all weight classes.

Electronic Automatic Transmissions

Electronically controlled automatic transmissions are common in vocational trucks and school buses. Although Caterpillar has entered the electronic automatic transmission market in recent years, most automatic transmissions found in North American built trucks are manufactured by Allison Transmission®. Details on the most common models of Allison transmissions will be discussed, but these principles can be applied to most other electronic automatic transmissions.

Allison World Transmission. The Allison World Transmission (WT) 3000/4000 series is widely used in medium-duty vocational trucks, but may also be found in some heavy-duty applications. A larger 5000/6000 series is used in heavy-duty off-highway trucks. This transmission can be described as a full-authority electronically controlled automatic transmission because the transmission will only shift into a range when commanded by the transmission ECU. Solenoid valves in the transmission valve body are controlled by the transmission ECU to provide the various transmission ranges. The electronic control system is known as the World Transmission Electronic Controls (WTEC). A simplified WTEC block diagram is shown in **Figure 15-1**. The WTEC ECU has three 32-pin electrical connectors that are color-coded gray, black, and blue for identification. Three generations of WTEC systems have been used; the last being the WTEC III.

Typical electrical layout for a World Transmission is shown in **Figure 15-2**. The +12V power supply and ground for the transmission ECU is typically made directly at the batteries in the same manner as in most electronic engine ECMs. Some OEMs refer to this direct battery power and ground connection for electronic modules as "clean power" or similar terminology. Many trucks use the Allison vehicle interface module (VIM) shown in **Figure 15-2** as a means of isolating the transmission ECU from the truck wiring. The VIM contains relays, fuses, and jumper wires.

The World Transmission uses an electronic shifter to permit the truck operator to select the desired range. This shifter may be a pushbutton-type shifter or a lever-type shifter (**Figure 15-3**). There is no mechanical connection, such as a shift cable, between the shifter and the WT automatic transmission. The electronic shifter is hardwired to the transmission ECU. The range selected by the truck operator is communicated from the electronic shifter to the transmission ECU via four hardwired circuits. The four circuits are connected to four transmission ECU digital inputs. These four digital inputs make up a 4-bit binary number corresponding to the selected range. The shifter causes the voltage present on the four ECU inputs to be high or low, corresponding to the selected range.

The transmission ECU may also use information from the J1708/J1587 or J1939 data links to obtain engine load information. This engine load information includes accelerator-position information and is used by the transmission ECU to determine shift points. Alternatively, an accelerator-position sensor may be hardwired directly to the transmission ECU in some applications.

The transmission has three variable reluctance–type speed sensors, which are used to measure engine speed,

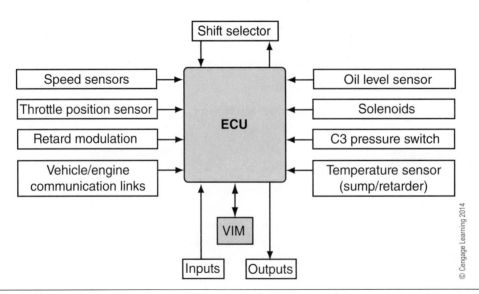

Figure 15-1 Allison WTEC block diagram.

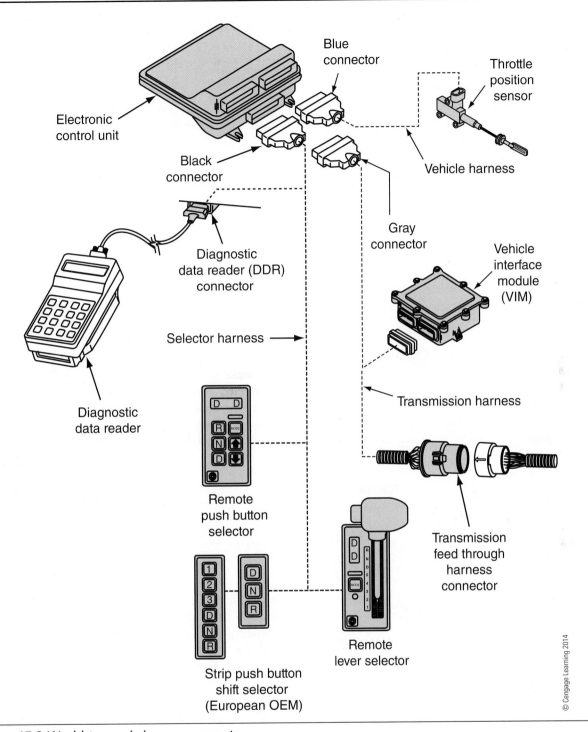

Figure 15-2 World transmission components.

turbine speed, and output shaft speed (**Figure 15-4**). The transmission ECU determines shift points based on all of its hardwired information, and information obtained from the data links indicating accelerator position and engine load.

The main output of the transmission ECU is the control of several electric shift solenoids within the electrohydraulic control module located inside the transmission (**Figure 15-5**). These solenoids control the transmission fluid flow to the clutch packs within the transmission. There are two types of solenoids: normally closed (**Figure 15-6**) and normally open (**Figure 15-7**). The transmission ECU supplies a pulse width modulation (PWM) control voltage, like that shown in **Figure 15-8**, to the applicable combination of solenoids to obtain the desired range. For example, **Figure 15-9** illustrates the elements in use to obtain third range. Solenoids B, C, F, and G are energized by

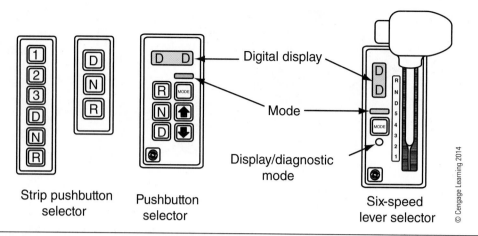

Figure 15-3 Allison WT electronic shift selectors.

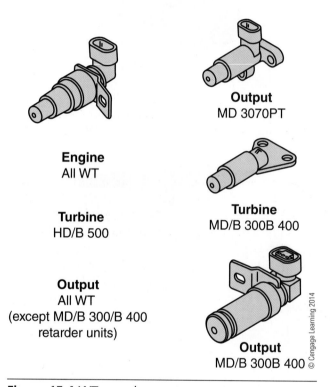

Figure 15-4 WT speed sensors.

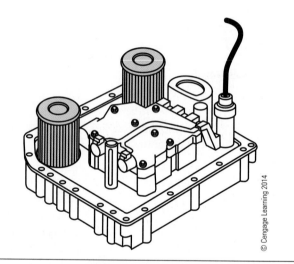

Figure 15-5 WT electrohydraulic control module.

the transmission ECU to apply clutches C1 and C3, and the lockup torque converter clutch.

J1939 message outputs of the transmission ECU include transmission output shaft speed, selected range, current attained range, and transmission oil temperature. There are also typically two or more transmission indicator lamps that the transmission ECU can cause to illuminate if a failure or other condition is detected. Much of this information is transmitted on the J1939 data link by the transmission ECU for use by other electronic modules.

The transmission ECU also provides a hardwired indication of transmission output shaft speed that is often used by the engine ECM or electronic speedometer to determine vehicle speed. Transmission output shaft speed is provided by the transmission ECU as a pulsing 12V signal (square waveform). The frequency of the square waveform produced by the ECU is proportional to the vehicle speed. It is important to note that this output shaft speed signal is produced by the transmission ECU, not the transmission's output speed sensor. The transmission output speed sensor is a variable reluctance sensor, which produces an AC sine wave. This AC signal is an input to the transmission ECU. The transmission ECU then produces a variable frequency output shaft speed signal based on the information the transmission ECU receives from its output shaft speed sensor. This square waveform transmission output shaft speed signal produced by the transmission ECU is typically hardwired to the engine ECM or electronic speedometer. The engine ECM or electronic speedometer then converts the frequency of this square wave into vehicle speed. The engine ECM

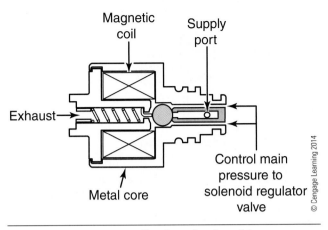

Figure 15-6 WT normally closed solenoid.

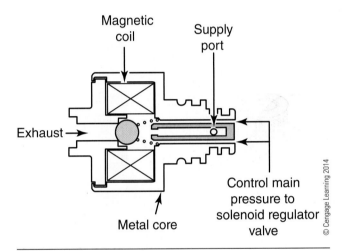

Figure 15-7 WT normally open solenoid.

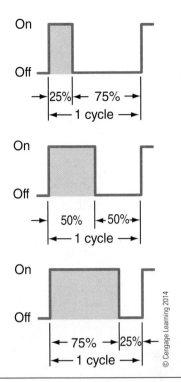

Figure 15-8 Pulse width modulation (PWM) voltage supplied by transmission ECU controls WT solenoids.

or electronic speedometer is able to convert transmission output shaft pulse frequency into vehicle speed through a pulses-per-mile value that has been programmed into the engine ECM or electronic speedometer. The engine ECM can then transmit this vehicle speed information onto the J1939 and J1708/J1587 data links for use by all other electronic modules on the truck, including the instrument panel cluster (IPC).

The transmission ECU also provides an indication of neutral by supplying +12V at a specific output terminal if the transmission is in neutral. FMVSS requirements indicate that the cranking motor in a truck with an automatic transmission can only be operated if the transmission is in park or neutral. This ECU neutral indication terminal is used in the engine crank inhibit control circuit on most trucks with automatic transmissions.

Allison 2000 Series. A lighter-duty Allison 2000 series automatic transmission is also used in some

medium-duty trucks. This transmission uses a cable-type shifter that controls a manual selector valve in the transmission valve body. Unlike the World Transmission series, a complete loss of electrical control of the automatic transmission still permits a forward range (third range) and reverse to permit the truck to "limp home." The 2000 series transmission ECU has two 32-pin electrical connectors that are color-coded red and gray for identification. The ECU controls electric solenoids located in the valve body in a manner similar to the WTEC series.

A multifunction switch assembly located on the side of the transmission known as the NSBU (neutral-start back-up) switch provides an indication of requested gear to the transmission ECU. The NSBU contains four switches that are hardwired to the transmission ECU to provide an indication of shifter position. The NSBU also contains two additional switches that are typically connected to the truck's chassis wiring harness. One switch provides an indication of park or neutral position for engine crank inhibit. The other switch is used to control the back-up lights. The NSBU does not contain any electronic components.

The 2000 series transmission ECU is connected to the J1939 data link. Information is exchanged with other electronic modules on the truck, such as the engine ECM via the J1939 data link.

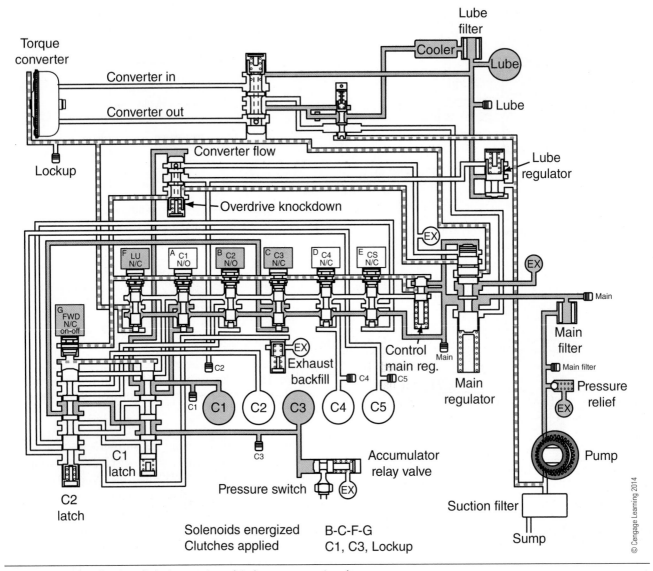

Figure 15-9 WT hydraulic schematic—third range attained.

Allison 4th Generation Controls. In 2006, Allison electronic automatic transmission systems were updated and are referred to as 4th Generation Controls. All 2000, 3000, and 4000 series 4th Generation transmissions use a common transmission ECU. The calibration (programming) in the ECU varies for each model. The 4th Generation transmission ECU has a single 80-pin electrical connector.

The electronic shifter on 3000 and 4000 series 4th Generation transmissions is connected to the truck's J1939 data link and communicates with the transmission ECU via the J1939 data link. A hardwired circuit between the transmission ECU and shifter also provides a confirmation of requested range using a PWM signal.

A cable shifter is used on 2000 series 4th Generation transmissions to control a manual selector valve in the transmission. However, the external NSBU switch

mounted on the external shift linkage in prior 2000 series transmission has been eliminated on these models. Switches inside the transmission are hard-wired to the transmission ECU to provide an indication of shifter position based on the manual selector valve position.

Allison Transmission Diagnostics. Allison Transmission's Diagnostic Optimized Connection (DOC™) PC-based electronic service tool (EST) is the best method of troubleshooting an Allison automatic transmission. The interface device is connected to the truck's diagnostic connector and uses J1708/J1587 (World Transmission) or J1939 (2000 series and all 4th Generation Controls) to communicate with the transmission ECU. The DOC electronic service tool permits diagnostic trouble-code retrieval and observation of the status of the transmission's inputs and outputs as well as providing some transmission tests and calibration procedures.

The Pro-Link 9000 and other hand-held EST can also be used to troubleshoot the World Transmission series. Additionally, the World Transmission series and 4th Generation Controls models with electronic shifters permit the retrieval of diagnostic trouble codes stored in the transmission ECU on the shifter's digital display when a combination of buttons on the shifter is depressed.

Automated Manual Transmissions

An electronically controlled **automated manual transmission (AMT)** can be thought of as a standard manual transmission without a shift lever. The shift lever has been replaced by two electric motors (**Figure 15-10**) that are controlled by an electronic transmission ECU.

AMTs can be divided into two main types: two-pedal and three-pedal systems. This refers to the presence of a clutch pedal in the cab of the truck. In a two-pedal system, the clutch is controlled by the transmission ECU. A three-pedal system has a standard clutch pedal and is controlled by the truck operator, although the clutch is only used for start-up and stopping the vehicle.

Several manufacturers now produce AMTs, but the most common are the Eaton Fuller Roadranger family of AMTs. Details on some models of these transmissions will be discussed.

Eaton AMT Inputs. An electronic truck operator interface shift control, like those shown in **Figure 15-11**, communicates with the transmission ECU via a private J1939 communication network called the EPL. The primary hardwired inputs to the transmission ECU are

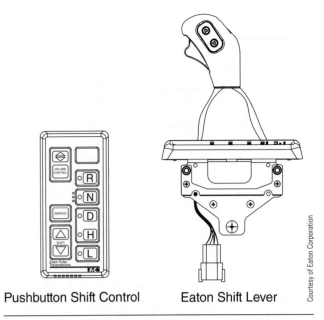

Pushbutton Shift Control   Eaton Shift Lever

Figure 15-11 Eaton driver interface.

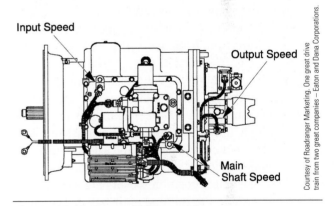

Figure 15-12 Eaton AMT speed sensors.

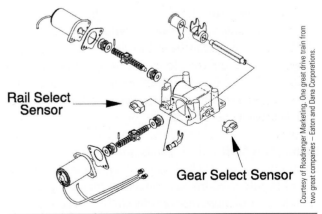

Figure 15-13 Gear and rail select position sensors.

obtained from three transmission-mounted speed sensors (**Figure 15-12**) and the rail select and gear select position sensors (**Figure 15-13**). These position sensors may be potentiometers on early Generation 1

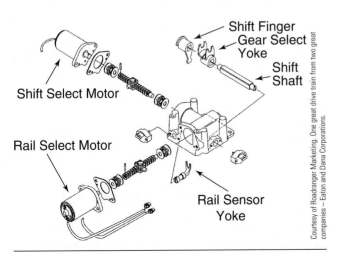

Figure 15-10 Eaton AMT electric shift control assembly.

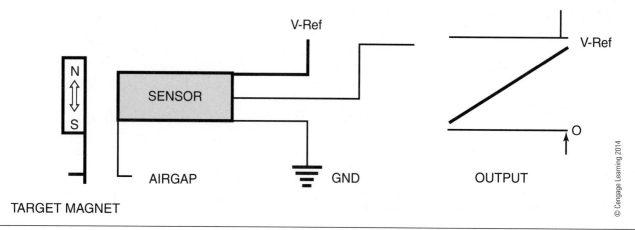

© Cengage Learning 2014

Figure 15-14 Linear Hall effect position sensor principle.

AMTs, or **linear Hall effect sensors** in later generations. A linear Hall effect sensor produces a linear output voltage relative to the location of a target magnet, as shown in **Figure 15-14**. Additionally, engine load information is also obtained by the transmission ECU through a J1939 network connection, as well as information from the ABS ECU and body control module, if applicable.

The primary outputs of the transmission ECU are the control of the rail select motor and the shift select motor, shown in **Figure 15-10**. The motors drive ball screws, which move the internal transmission shift yokes in the same way as a standard transmission shift lever operated by the truck operator. The rail select and gear select position sensors are used by the ECU to determine the rail and shift ball screw positions to control the two motors to obtain the desired range.

The transmission ECU calibrates the shift control system each time the ignition is switched off. A series of clicking noises can be heard as the shift and rail select motors are driven to each end of their available travel. This calibration information is stored in the transmission ECU non-volatile memory for use at the next key cycle.

Eaton AutoShift. The Eaton AutoShift is a three-pedal version of AMT. The truck operator controlled clutch is only used for start-up and stopping. Once the vehicle is in motion, the transmission ECU commands the engine ECM to reduce torque via a J1939 message to shift, by matching engine speed to the transmission speed. The truck operator can leave his foot on the accelerator throughout the shift. The J1939 message controls engine speed much in the same way that a skilled truck operator feathers the accelerator to shift once the truck is underway without depressing the

clutch (although this practice is not recommended). The J1939 message used to reduce the engine torque or speed is called **torque-speed control (TSC)**.

Eaton UltraShift. The Eaton UltraShift is a two-pedal version of AMT. The transmission ECU controls the clutch. There have been several types of clutches throughout the various generations of UltraShift including centrifugal and wet clutches. However, the latest (2012) version of UltraShift Plus uses a conventional disc-type clutch, which is operated by an electronic clutch actuator (ECA). An ECA is shown in the section on Hybrid Electric Trucks (**Figure 15-39**). The ECA is controlled by the transmission ECU via a private communication network. The ECA contains a brushless DC electric motor and planetary gear set to produce the torque necessary to release and engage the clutch. The transmission ECU can release the clutch during each upshift and downshift, as well as during initial start-up and when stopping the truck.

Eaton AMT Diagnostics. Eaton ServiceRanger™ is the PC-based EST used to diagnose an Eaton AMT. ServiceRanger permits AMT DTCs to be viewed and cleared. Various transmission parameters, such as shift rail position and shaft speeds, can be viewed in real time and tests and calibrations can also be performed using ServiceRanger.

ANTILOCK BRAKING SYSTEMS

Antilock braking systems (ABS) have been required in the United States for many years for both air and hydraulic truck brake systems and for trailers with air brakes.

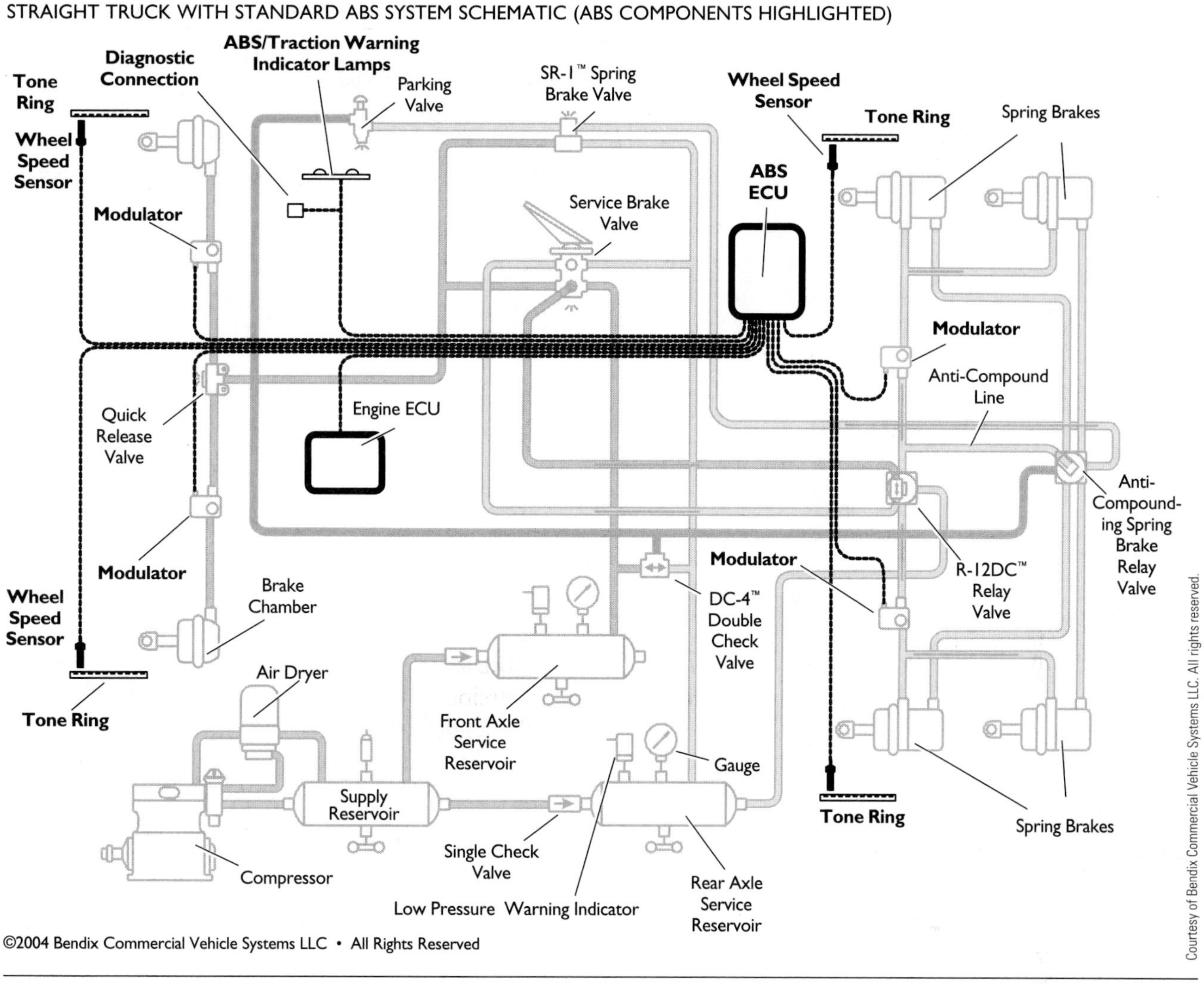

STRAIGHT TRUCK WITH STANDARD ABS SYSTEM SCHEMATIC (ABS COMPONENTS HIGHLIGHTED)

Figure 15-15 ABS schematic for straight truck.

Tractor and Truck Air ABS

An ABS electronic control unit (ECU) is the electronic module that controls the air ABS system. Variable reluctance–type speed sensors are used to measure individual wheel speeds at both front wheels and at least two of the rear wheels on tandem-axle trucks. Electrically controlled ABS valves are used to control the air supplied to the service brake chambers to prevent wheel lockup. The electrical and pneumatic layout for a straight truck ABS application is shown in **Figure 15-15**.

Wheel-Speed Sensors. Most ABS wheel-speed sensors are variable reluctance sensors containing a permanent magnet with a coil of wire wrapped around the magnet. A variable reluctance sensor is a two-wire sensor that generates an AC voltage signal when a low-reluctance target passes in front of the sensor. The wheel-speed sensors are inputs to the ABS ECU. A typical ABS

wheel-speed sensor with an integral wiring harness and an electrical connector is shown in **Figure 15-16**.

The wheel-speed sensor target for drum brakes is typically constructed of stamped metal and is installed on the wheel hub, as shown in **Figure 15-17**. The target may be known by several names, including tooth wheel, tone wheel, tone ring, reluctor wheel, and exciter ring. A typical truck ABS target has 100 teeth. The wheel-speed sensor is installed in a mounting block on the axle so that the target will pass in close proximity to the tip of the wheel-speed sensor when the hub is installed. A sensor clamping sleeve is typically used to retain the sensor in the mounting block.

The target passing in front of the tip of the wheel-speed sensor causes an AC voltage to be produced by the variable reluctance wheel-speed sensor, as described in **Chapter 11**. The frequency and amplitude of the voltage produced by the wheel-speed sensor is directly proportional to wheel speed. The ABS ECU measures the frequency of the AC waveform produced

Figure 15-16 Typical ABS wheel-speed sensor with integrated wiring harness and connector.

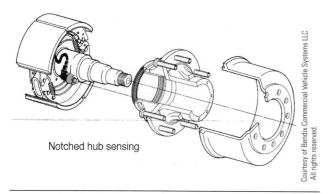

Notched hub sensing

Figure 15-17 ABS tone wheel and wheel-speed sensor for a drum brake system.

by the sensor to determine the speed of each wheel that is equipped with a wheel-speed sensor.

The electrical resistance of an ABS wheel-speed sensor is typically between 1000 and 2000 ohms. This is the resistance of the coil of very small gauge wire inside the sensor. The wheel-speed sensor is connected to the ABS ECU through two wires that are twisted together to reduce electrical interference caused by changing magnetic fields near the wiring. Neither of these wires is directly connected to ground or to battery positive inside the ABS ECU or inside the sensor.

The ABS ECU continuously monitors the integrity of the individual wheel-speed sensor circuits for opened or shorted circuits. If the ABS ECU detects a faulted wheel-speed sensor circuit, the ABS ECU sets a diagnostic trouble code (DTC) indicating which sensor circuit is faulted. The ABS ECU then disables a portion of the ABS system due to the faulted sensor circuit and causes the ABS warning lamp in the instrument panel cluster (IPC) to be illuminated.

Wheel-speed sensors are typically installed in each of the two front (steer) axle wheel ends. Trucks with tandem rear axles often only have wheel-speed sensors in one of the rear axles. The two rear sensors are installed at the wheel ends of the forward-rear axle or rear-rear axle, depending on the truck's rear suspension type. This four-sensor four-ABS valve system shown in **Figure 15-18** is known as a four-channel ABS system. Other trucks with tandem rear axles may have sensors in each of the rear-axle wheel ends for a total of six sensors and six ABS valves; this is known as a six-channel ABS system.

Modulators. ABS **modulators**, also known as ABS valves, are used to control the air in the service brake

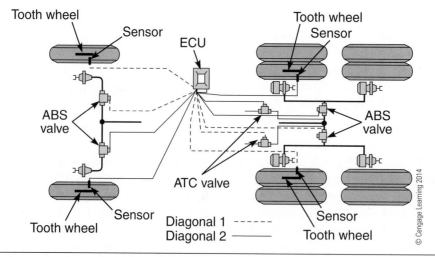

Figure 15-18 Four-channel ABS with ATC.

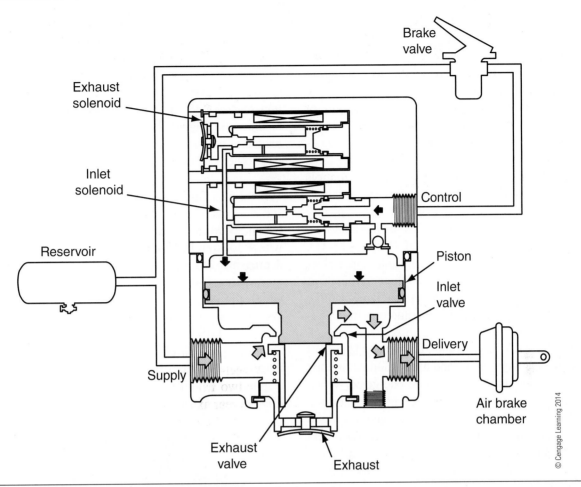

Figure 15-19 Air ABS brake modulator with integral relay valve.

chambers to prevent wheel lockup. The modulators contain two solenoids (coils) referred to as the inlet solenoid and the exhaust solenoid. The two solenoids may also be known as the hold and dump solenoids, respectively. The two solenoids control the inlet and the exhaust valves inside the modulator as shown in **Figure 15-19**. The modulator depicted is a combination modulator-relay valve.

Modulators on modern trucks are typically plumbed in series with each of the truck's service brake chamber supply lines. These modulators have an air supply port, air delivery port, an air exhaust port, and an electrical connector. Two modulators that are plumbed to a quick-release valve for a front-axle installation are shown in **Figure 15-20**. The air supply ports for these modulators are plumbed to the quick-release valve shown in the center of the assembly. The modulator air delivery ports are the openings at the opposite ends of this assembly and are plumbed to the service brake chambers. The air exhaust ports are at the bottom of the modulator. Some modulators are also designed to act as quick-release valves, permitting the elimination of standard quick-release valves from the air brake system.

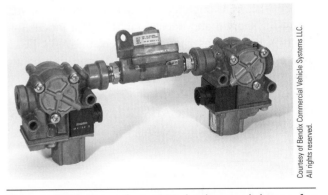

Figure 15-20 Pair of air ABS brake modulators for front axle plumbed to a quick-release valve.

Air brake modulators are designed so that if the ABS system is not functional, normal braking is not impacted. The modulator inlet valve is normally open, and the modulator exhaust valve is normally closed. When neither modulator solenoid is energized, the truck's brake system is like a traditional non-ABS system due to the normal state of the modulator solenoid valves. Air passes directly through the modulator to the respective service brake chamber with neither modulator solenoid energized.

Energizing the modulator's normally open inlet solenoid causes the modulator to block the flow of air into the respective service brake chamber. Energizing the normally closed exhaust solenoid then causes some of the compressed air in the respective service brake chamber to be vented to atmosphere through the modulator's exhaust port. This action reduces the brake force on the wheel that is locking up. The ABS ECU controls these solenoids as necessary to prevent wheel lockup.

The two coils in each modulator are typically connected together on one end inside the modulator so that they share a common terminal (**Figure 15-21**). Therefore, the typical modulator only has three electrical terminals. This is similar to the three terminals of a combination high-beam low-beam headlamp. Unlike a headlamp, though, none of the modulator terminals is connected directly to ground. The three wires of each modulator are hardwired to the ABS ECU. The ABS ECU supplies power and ground to the modulator coils only when a modulator coil is to be energized. When neither modulator solenoid is being energized by the ABS ECU, neither battery positive voltage nor ground is being supplied to any of the three modulator terminals. Because both the power and the ground for the modulator solenoids are controlled, a shorted-to-ground or shorted-to-battery-positive circuit between the modulator and the ABS ECU will not interfere with normal braking.

The ABS ECU continuously monitors the modulator wiring for shorted and opened circuits. If a modulator circuit is detected as shorted or open, the ABS ECU disables that modulator and causes illumination of an ABS warning lamp in the instrument panel.

ABS ECU. The air brake ABS ECU controls the ABS system. The air brake ABS ECU may be located outside or inside the cab. If the ABS ECU is located outside of the cab, the ECU is a sealed-type module that uses sealed connectors. ABS ECUs mounted inside the cab may be non-sealed modules. An example of a sealed ABS ECU is shown in **Figure 15-22**. A non-sealed ABS ECU designed for interior mounting is shown in **Figure 15-23**. Most ABS ECUs are connected to the J1939 data link to permit communication with other electronic modules on the truck.

Typical ABS Operation. The ABS ECU monitors the speed of each wheel that has a sensor to determine if wheel speed is decreasing too rapidly, indicating that wheel lockup is about to occur. If wheel speed is decreasing too rapidly, the ABS ECU controls the modulators to prevent wheel lockup. The action in which the ABS ECU is controlling the modulators is known as an antilock event or ABS event.

If the ABS ECU determines that wheel lockup is about to occur based on the frequency generated by a wheel-speed sensor, the ABS ECU energizes the corresponding modulator inlet valve solenoid to block the flow of air into the service brake chamber. The ABS

Figure 15-22 Bendix EC30/EC17 ABS electronic control unit connector view. This unit is sealed and can be mounted outside of the cab.

Figure 15-23 Bendix EC60 ABS electronic control unit with advanced features designed for interior mounting.

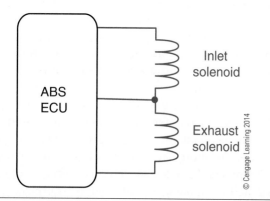

Figure 15-21 Each brake modulator has three electrical terminals. The common circuit may be positive or negative, depending on the ABS system.

ECU may also simultaneously energize the corresponding modulator exhaust valve solenoid. This causes the air in the corresponding service brake chamber to be partially vented to atmosphere, reducing braking force on the wheel that is nearing lockup.

The reduction in braking force on the wheel that was nearing lockup as air in the service chamber is vented to atmosphere through the modulator exhaust port causes the speed of the wheel to increase. The ABS ECU then de-energizes the modulator exhaust and inlet valve solenoids for that wheel. This permits compressed air to flow back into the service chamber to resume normal braking. The process repeats as necessary until wheel speed stabilizes. This series of actions of holding and exhausting the air by the solenoids occurs very rapidly and can be heard as a series of air venting sounds during an antilock event.

The ABS ECU continuously monitors all of the wheel-speed sensors and is capable of independently controlling each of the modulator solenoids as necessary to prevent wheel lockup. However, the brakes on the front (steering) axle are usually controlled differently than brakes on the rear axle. A front wheel that begins to lock up causes most ABS ECUs to actuate both front modulators at the same time, even though only one wheel may be locking up. This is done so as not to interfere with steering the truck, which might occur if only one front brake were applied. However, some ABS systems with advanced stability features may control each front brake independently as will be explained later in this section.

Chuff Test. When the key switch is first cycled on, the ABS ECU performs a **chuff test** of all of the modulators. A chuff test is a brief actuation of each modulator coil by the ABS ECU to provide a functional test of the system. The order in which the modulators are chuffed or energized follows a specific pattern, such as clockwise starting at the left front modulator.

....

Tech Tip: The chuff test can provide you with a great deal of information. Depressing the brake pedal during the chuff test causes air to be exhausted from the modulators when the modulator inlet and exhaust solenoids are tested. This permits you to hear the chuff test during initialization. With experience, you will learn what a good chuff test sounds like versus a chuff test where there is a wiring problem or a defective modulator.

....

Drive-Axle ABS Event. When the ABS ECU detects that the wheels on the drive axle are decreasing in speed too rapidly, the ABS ECU will signal that a **drive-axle ABS event** is occurring. The signal may be in the form of energizing a low side driver in the ABS ECU that controls an external ABS event relay. The output of this relay acts as an input to the engine ECM and transmission ECU, if applicable. The ABS ECU on newer trucks may instead transmit a J1939 message indicating that a drive-axle ABS event is occurring.

The drive-axle ABS event information is important for the engine ECM and the automatic transmission ECU. The automatic transmission ECU unlocks the torque converter clutch and disables the transmission retarder, if applicable, when the ABS ECU signals that a drive-axle ABS event is occurring. This reduces the load on the torque converter clutch and prevents the engine from stalling. Two-pedal AMT transmissions may also release the clutch during a drive-axle ABS event.

The engine ECM uses the ABS event information to disable the engine brake if a drive-axle ABS event is occurring. The reason for disabling the engine brake and transmission retarder during a drive-axle ABS event is that the cause of the drive-axle ABS event could be the braking force provided by the engine brake or the retarder, and not due to service brake application. This is especially true when the truck is operated on slippery or icy roads. The engine brake or transmission retarder will be engaged once again by the engine ECM or transmission ECU after the drive-axle ABS event has ceased.

Automatic traction control (ATC) or **anti-spin regulation (ASR)** is optional with most air ABS systems. Traction control is used to limit wheel slip under acceleration. The wheel-speed sensors are used to determine if the rear wheels are rotating faster than the front wheels or if one rear wheel is rotating faster than the rear wheel on the other side of the same drive axle. This would occur if one drive-axle wheel end is in mud or ice and the other side is on dry pavement.

Traction control can consist of two types of control, depending on whether one rear drive wheel or both rear drive wheels are spinning. When both rear wheels are spinning, such as when the truck is being operated on snow or ice-covered roads, a request for torque reduction is transmitted by the ABS ECU to the engine ECM over the J1939 or J1922 data link. The engine ECM responds to the request for torque reduction by reducing fueling, regardless of accelerator position. This operation is sometimes known as engine torque limiting (ETL) or torque-speed control (TSC).

If only one rear wheel is spinning, such as when one wheel is on dry pavement and the other is in mud, the ABS ECU can use **differential braking**. Differential braking describes the action of the ABS ECU applying the brake on the wheel that is spinning to cause the differential in the axle to drive the non-spinning wheel.

During differential braking, an electric solenoid contained in the traction control valve is energized to supply air to the primary brake system relay valve (rear brakes) as though the brake pedal had been depressed. The normally closed traction control valve is supplied with primary air pressure. The traction control solenoid is hardwired to the ABS ECU. When the ABS ECU energizes the traction control solenoid, air is supplied to the rear brake modulators as though the foot valve (brake pedal) had been depressed. The ABS ECU then blocks air to the wheel that is not spinning by energizing the modulator inlet solenoid associated with that wheel. However, the ABS ECU does not energize either modulator solenoid of the wheel that is spinning. The normal state of the modulator causes air to be supplied to the brake chamber of the wheel that is spinning. The result is that the rear brake is only applied to the wheel that is spinning. This causes the differential in the axle to drive the non-spinning wheel.

An indicator lamp in the IPC illuminates steadily or flashes when the ABS ECU detects that a traction control event is occurring. This indicator lamp in the instrument panel may be hardwired to the ABS ECU or controlled by a J1939 message transmitted by the ABS ECU.

Tech Tip: The engine brake or retarder may be disabled if the ABS system detects a fault. When diagnosing an engine brake or retarder issue, check for active ABS DTCs. Additionally, a flashing traction control indicator light does not necessarily mean that there is an ABS malfunction. Operators often misunderstand this feature and may report as a malfunction.

A means of temporarily defeating the traction control is usually provided. This mode is necessary for dynamometer testing of a truck and when the truck is driven in some low-traction situations in which wheel slip is desired. A traction control disable switch in the instrument panel is typically hardwired to the ABS ECU.

On International truck models with a body control module, the traction control disable switch is located in one of the switch packs. The body control module

detects that the traction control switch is in the disable position from a switch data link message and transmits a J1939 message to the ABS ECU to disable traction control. The traction control warning indicator in the instrument panel is then illuminated via a J1939 message transmitted by the ABS ECU when the traction control system is disabled.

Some trucks may also have a switch for a reduced level of traction control that may be called a mud-snow mode, off-road mode, or other similar name. This switch does not defeat traction control but provides a less aggressive reduction in engine torque, permitting some wheel slip to occur.

> **WARNING** *Differential braking can cause a truck to drive off of a jack stand if only one drive-axle wheel end is lifted off the ground and the driveline is operated. Traction control must also be disabled if the truck is operated on a dynamometer.*

Advanced Air ABS Features. ABS systems may also have advanced features to improve vehicle stability. These systems are classified as roll stability control (RSC) or electronic stability control (ESC), with ESC being the more advanced system. These advanced features require additional ABS system sensors, including:

- Steering wheel angle sensor
- Air application pressure sensors
- Air suspension pressure sensor for vehicle load-shift determination
- Yaw rate sensor that detects vehicle rotation (spin) around a vertical axis at the center of gravity as viewed from overhead
- Lateral acceleration sensors (accelerometers)

Some of these sensors are shown in **Figure 15-24**. Using these sensors, the advanced ABS ECU is able to

Figure 15-24 Advanced air ABS sensors.

anticipate that a vehicle rollover, jack-knife, or other instability is about to occur and command the engine ECM to reduce torque and apply the engine brake or retarder via a J1939 message to decrease vehicle speed. If necessary, the ABS ECU and advanced trailer ABS ECU can also automatically apply or release individual brakes on the vehicle to attempt to correct the instability (**Figure 15-25**).

Advanced ABS features also make use of radar. A first generation radar system called **vehicle onboard radar (VORAD)** was originally produced by Eaton, but the technology has since been purchased by Bendix and has been integrated into the ABS system. VORAD is a collision warning system. The system components are shown in **Figure 15-26**. The antenna assembly is mounted to the front of the truck and emits radar microwave pulses that bounce off metal objects in front of the vehicle. These pulses are used to determine the relative speed and distance of objects in the path of the vehicle. The truck operator display unit provides visual and audible indication of objects being detected in the path of the vehicle. A blind spot sensor is a motion detector mounted to the side of the truck. Motion detected by the sensor causes the blind spot display to provide visual indication to the truck operator.

The VORAD system can also be integrated into the vehicle cruise control system to reduce engine torque and apply the engine brake via a J1939 message to maintain a safe following distance. This is referred to as adaptive cruise control. The latest generation

Bendix Wingman® ACB system is also capable of applying the foundation brakes to increase following distance or to prevent an impending collision.

Future ABS systems will include video systems to alert the truck operator of lane departure and automatically apply the brakes if necessary. The video system can also read traffic signs, such as speed limit signs, and alert the truck operator or even automatically reduce vehicle speed via a J1939 TSC message transmitted to the engine ECM if the speed limit is being exceeded. Nonmetal objects such as animals or pedestrians can also be recognized by the system and alert the truck operator of the hazard.

ABS Warning Lamp. FMVSS 121 requires that an amber-colored ABS warning lamp must be in clear view of the truck operator and that a lamp check must be performed when the key is first placed in the ignition position. The ABS warning lamp must also illuminate if the ABS system detects a problem that limits the ABS system performance. This also means that the ABS warning lamp must illuminate as a truck operator warning even when the ABS ECU is not powered due to a disconnected ABS power supply connection.

There are three main methods of controlling the ABS warning lamp. One method of controlling the ABS warning lamp is a directly wired warning lamp as shown in **Figure 15-27**. One side of the ABS warning lamp filament is connected to an ignition source. The other side of the warning lamp filament is

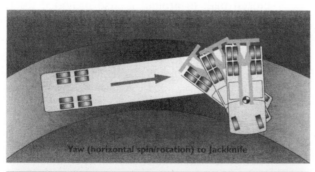

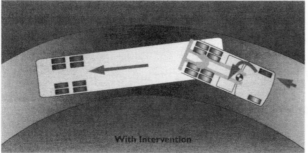

Driving Scenario:

Driving speed exceeds the threshold, and the resulting lateral force causes the vehicle to slide or jackknife on lower-friction surfaces.

Action by Bendix Stability Solutions:

System applies the appropriate brakes to reduce speed and properly align the vehicle, thereby reducing the tendency to slide or jackknife.

Figure 15-25 Stability control.

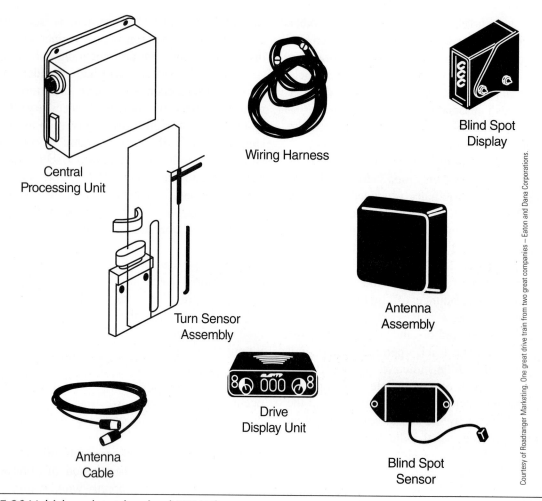

Figure 15-26 Vehicle onboard radar (VORAD).

Central
Processing Unit

Wiring Harness

Blind Spot
Display

Turn Sensor
Assembly

Antenna
Assembly

Antenna
Cable

Drive
Display Unit

Blind Spot
Sensor

Courtesy of Roadranger Marketing. One great drive train from two great companies – Eaton and Dana Corporations.

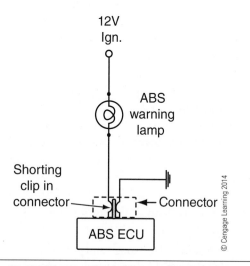

Figure 15-27 Direct-wired ABS warning lamp.

12V
Ign.

ABS
warning
lamp

Shorting
clip in
connector

Connector

ABS ECU

© Cengage Learning 2014

connected to the ABS ECU. The ABS ECU provides a normally closed path to ground inside the module for the warning lamp if a condition that would cause reduced ABS performance is detected. This normally closed path to ground is also provided for the initial ABS lamp test at key-on. If the ABS ECU is not powered due to a blown fuse or other reason, the normally closed path to ground is still provided inside the ABS ECU, causing the ABS warning lamp to illuminate continuously. This type of system typically has a shorting clip located in the connector containing the ABS power supply and ground. The shorting clip is designed to be held open by a plastic spacer on the ABS ECU when the connector is connected to the ABS ECU. The shorting clip completes a path to ground for the ABS warning lamp if the ABS ECU connector has been disconnected.

The second method of controlling the ABS warning lamp is through the use of a fail-safe relay, as shown in **Figure 15-28**. The ABS warning lamp is connected to an ignition source on one side of the filament and the normally closed contacts of the ABS fail-safe relay on the other side of the lamp filament. The normally closed contacts of the ABS fail-safe relay provide a

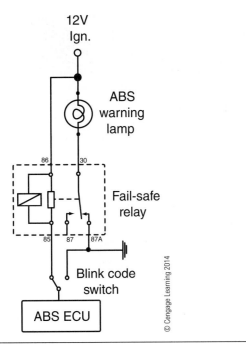

12V
Ign.

ABS
warning
lamp

86   30

Fail-safe
relay

85   87   87A

Blink code
switch

ABS ECU

© Cengage Learning 2014

Figure 15-28 Fail-safe relay used to control ABS warning lamp.

path to ground. The control coil of the fail-safe relay is wired to ignition and the ABS ECU. The ABS ECU must energize the fail-safe relay coil for the ABS warning lamp to be extinguished. The ABS ECU de-energizes the fail-safe relay coil to illuminate the ABS warning lamp for initial ABS warning lamp check at key-on and when a condition that causes reduced ABS system performance is detected. Disconnecting the ABS ECU connector also causes the fail-safe relay to be de-energized, which results in illumination of the ABS warning lamp. A normally closed momentary contact blink code switch is often placed in series between the ABS ECU and the fail-safe relay coil. The blink code switch provides a ground to the ABS ECU warning lamp output terminal when held in the check-faults position. The ABS ECU then causes the ABS warning lamp to flash out any diagnostic trouble codes in memory when the blink code switch is returned to the normal position. This blink code switch may also be used to temporarily disable traction control with some ABS systems by toggling the switch a specific number of times.

The third method of controlling the ABS warning lamp is the use of the J1939 data link with a J1939 multiplexed IPC. The ABS ECU communicates the appropriate J1939 message, indicating whether the ABS warning lamp should be illuminated or extinguished. The multiplexed IPC detects this message and causes the ABS warning lamp in the IPC to

illuminate. This type of system is also designed to cause the ABS warning lamp to illuminate when J1939 communications between the ABS ECU and multiplexed IPC are interrupted or when the ABS ECU is not communicating on J1939. A "heartbeat" message from the ABS ECU is transmitted over the J1939 data link several times per second indicating that the ABS ECU is "alive" and capable of communicating a message to illuminate the ABS warning lamp, if necessary. The multiplexed IPC is designed to cause the ABS warning lamp to illuminate if the heartbeat message from the ABS ECU is missing for a period of time.

WARNING *It is important to note the activation of the ABS warning lamp when the key switch is first placed in the ignition position. If the ABS warning lamp fails to illuminate for the lamp check, the operational status of the ABS system cannot be determined. The cause of the ABS warning lamp check not illuminating for the lamp check or remaining illuminated after the lamp check should be evaluated immediately.*

Diagnosing Air ABS. The ABS ECU is capable of detecting several types of problems that would impact ABS performance. The modulator circuits and wheel-speed sensor circuits are continuously monitored for open and shorted circuits. Other problems, such as a loose wheel bearing or partially pushed-out wheel-speed sensors, can also be detected by the ABS ECU. If the ABS ECU detects a problem, a DTC can be set by the ABS ECU to aid in troubleshooting. The ABS ECU typically causes the ABS warning lamp to be illuminated if a DTC is set.

Legacy Bendix EC15, EC17, and EC30 ABS ECUs may have a series of LEDs on the ECU that is used to indicate a problem that causes a DTC to be set with a sensor or modulator circuit without the use of ESTs (**Figure 15-29**). A magnet placed above the reset switch located inside these ECUs can be used to clear faults and to cause the ECU to reconfigure itself, if new features are added.

Most Meritor-Wabco ABS ECUs use an ABS warning lamp blink code switch to permit a DTC to be retrieved. The ABS warning lamp flashes a specific number of times to indicate the DTC number when the blink code switch is closed and released, as shown in **Figure 15-30**. A blink code switch may also be used on some trucks with Bendix ABS systems.

Figure 15-29 Bendix ABS diagnostic LEDs.

Most Bendix ABS ECUs also support a small diagnostic tool called a remote diagnostic unit (RDU) that plugs into the truck's 9-pin diagnostic connector. The diagnostic tool causes LEDs in the tool to illuminate to indicate a DTC, such as an open left front wheel-speed sensor circuit (**Figure 15-31**). A magnet can also be held near the RDU to cause DTCs to be cleared, similar to legacy Bendix ABS ECUs.

For more in-depth troubleshooting, Bendix ACom™ PC-based diagnostic software can be used to troubleshoot a Bendix air ABS system. Meritor-Wabco

TOOLBOX™ is the PC-based diagnostic software used to troubleshoot a Meritor-Wabco air ABS system. These PC-based ABS ESTs permit viewing ABS sensor-indicated wheel speed and other ABS-related inputs. Individual modulators can also be activated to assist in diagnosing wiring problems. The ESTs also permit the retrieval of DTCs that are inactive but are stored in memory. This diagnostic trouble-code history is useful in troubleshooting intermittent ABS complaints. The Pro-Link 9000 and other hand-held ESTs can also be used to diagnose many legacy air ABS systems.

Diagnosing Wheel-Speed Sensors. The most common ABS DTCs are those related to wheel-speed sensors. The wheel-speed sensor is typically installed so that the tip of the sensor is contacting the target. The distance or air gap between the target and the sensor has a large impact on the amplitude of the AC voltage produced by the sensor.

The wheel-speed sensor is typically retained in its mounting hole by a spring clip for a friction fit. Loose wheel bearings, brake lining replacement and other factors can cause the sensor to push away from the target. Increasing the distance between the sensor and the low-reluctance target causes the amplitude of the

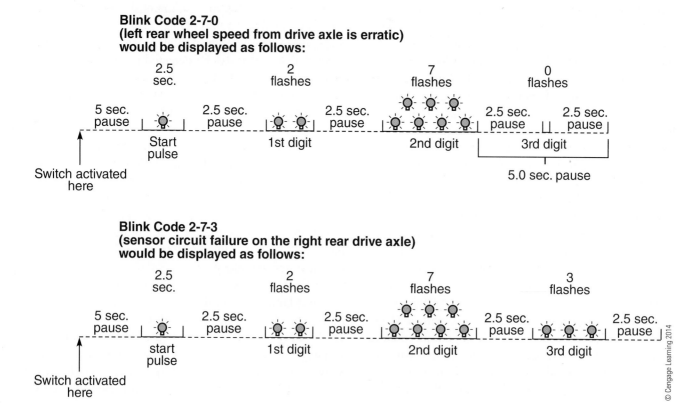

Figure 15-30 Blink codes displayed using ABS warning lamp flash sequence.

Figure 15-31 Bendix remote diagnostic unit (RDU).

voltage produced by the sensor to be too low to be detected accurately by the ABS ECU. The low signal amplitude causes the ABS ECU to set a DTC and illuminate the ABS warning lamp.

Variable reluctance type wheel-speed sensors can be tested on the truck as shown in Photo Sequence 6.

Trailer Air ABS

Trailers built for the U.S. market after March 1, 1998 with air brakes are required to have ABS. The trailer ABS ECU operates in a manner similar to the truck ABS system. Trailers have two or four wheel-speed sensors and modulators. The power supply for the trailer ABS ECU is provided through the tractor's or truck's SAE J560 7-pin trailer socket. Tractors and trucks that are designed to accommodate a trailer with air ABS have a green color-coded trailer socket and use a green color-coded trailer cord identified as ABS. The green color of the trailer socket indicates that the center pin of the trailer socket is always powered with the key in the ignition position. The green seven-way trailer cord along with an SAE J2394 designation indicates that the ground wire, center or auxiliary wire, and stop lamp wire in the trailer cord are of larger gauge than standard trailer cords. The larger wire gauge reduces the voltage drop on the wiring, providing a higher-level voltage supply at the trailer ABS ECU.

The trailer ABS ECU receives its 12V power supply through the center or auxiliary pin of the seven-way

socket. The trailer ABS ECU can also receive its power supply from the trailer stop lamp circuit in the event that the primary power supply via the center pin is interrupted. The seven-way socket ground pin provides the ground for the trailer ABS ECU.

Trucks and trailers with air brakes manufactured after March 1, 2001 for use in the United States are required to have a means to illuminate a trailer ABS warning lamp in the IPC of the tractor or truck if the trailer ABS ECU detects a condition that causes reduced trailer ABS system performance. The method of controlling the trailer ABS warning lamp in the IPC is provided by the trailer ABS ECU injecting a high-frequency signal onto its 12V power supply. The trailer ABS ECU 12V power supply is obtained from the tractor via the seven-way trailer socket. This high-frequency signal originating from the trailer ABS ECU is placed onto the tractor's 12V power supply by the trailer ABS ECU.

The truck ABS ECU is designed to detect this high-frequency signal generated by the trailer ABS ECU on its 12V power supply. Through the high-frequency signal, the truck ABS ECU can determine if the trailer ABS ECU has detected a condition that causes reduced trailer ABS system performance. This communication method is called **power line carrier (PLC)**. An example of the high-frequency signal that a trailer ABS ECU has transmitted (placed) onto its own 12V power supply is shown in **Figure 15-32**. The frequency and phase of the signal is constantly changing. This signal rides on top of the 12V DC level and typically has peak amplitude of 200mV when measured at the voltage supply of the truck ABS ECU. This communication technique is called spread spectrum and is used to ensure that other power line noise present in the electrical system does not interfere with the intended message.

The PLC signal is transmitted as a burst of information every half-second by the trailer ABS ECU. The signal indicates either that the trailer ABS ECU has detected a condition that is causing reduced ABS system performance or that it is not detecting any problems. The phase of each section of the PLC signal (starting positive or starting negative) is used to represent logic 1 or logic 0. The combination of logic 1s

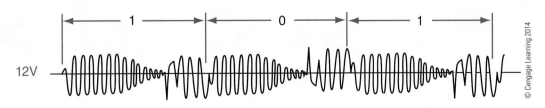

Figure 15-32 Power line carrier (PLC) signal generated by trailer ABS ECU.

PHOTO SEQUENCE 6

Diagnosing a Variable Reluctance ABS Wheel-Speed Sensor

PS6-1 Block the vehicle wheels.

PS6-2 Raise one side of the axle with a floor jack and lower the axle onto a safety stand.

PS6-3 Disconnect the wheel-speed sensor connector.

PS6-4 Connect a pair of AC voltmeter leads to the wheel-speed sensor terminals.

PS6-5 Rotate the wheel at 30 rpm (1 revolution per 2 seconds) and observe the reading on the AC voltmeter.

PS6-6 Compare the measured AC voltage to the OEM specifications. Verify the sensor air gap is correct, if applicable, and retest. Otherwise, replace the sensor and retest.

PS6-7 Reconnect the wheel-speed sensor connector.

PS6-8 Raise the axle with a floor jack and remove the safety stand.

PS6-9 Lower the wheel onto the shop floor and remove the floor jack. Clear the stored DTC.

and 0s is used to signal the status of the trailer ABS system.

The truck ABS ECU detects the trailer ABS signal (PLC) in its 12V ignition supply. When the PLC signal indicates that a trailer ABS problem exists, the truck ABS ECU causes the trailer ABS warning lamp in the IPC to illuminate. Additionally, some trailers may also have a small amber warning lamp at the left rear side of the trailer that illuminates if there is an active trailer ABS DTC. This lamp is visible in the truck operator's mirror. A loss of signal from the trailer ABS ECU also causes most truck ABS ECUs to illuminate the trailer ABS warning lamp for several seconds to warn the truck operator that the trailer ABS ECU is no longer communicating with the truck ABS ECU. This also occurs when a trailer with PLC is disconnected from the truck at the end of a trip.

The trailer ABS ECU is designed to transmit the PLC signal indicating that a trailer ABS DTC is active when power is initially provided to the trailer, such as at key-on or when the seven-way connector is first plugged into the trailer. This provides a lamp check of the trailer ABS warning lamp in the truck or tractor.

The trailer ABS warning lamp in the IPC is hardwired directly to the truck ABS ECU on some trucks. The truck ABS ECU causes the trailer ABS warning lamp in the IPC to illuminate via hardwire if the appropriate PLC signal is detected. The truck ABS ECU on International trucks with a body controller transmits the status of the trailer ABS ECU over the J1939 data link. If a trailer ABS fault is detected via PLC, the truck ABS ECU transmits a J1939 message indicating a trailer ABS fault. This causes the electronic gauge cluster to illuminate the trailer ABS warning lamp.

ABS ECUs designed for straight truck applications that are not designed to pull an air brake trailer may not have PLC capability. If it is ever necessary to replace an ABS ECU, the replacement ABS ECU must be the correct type for the truck or tractor. It is important to replace the ABS ECU in a truck or tractor built after March 1, 2001, that is designed to pull an air brake trailer with an ABS ECU equipped for PLC. Otherwise, the trailer ABS warning lamp in the truck or tractor will not illuminate if a trailer ABS problem exists. The PLC trailer ABS capability is clearly marked on applicable ABS ECUs.

Some OEMs may utilize a PLC electric filter module which is connected in series with the trailer ABS power supply circuit. The output of this filter is part of the ABS ECU power supply circuit. The PLC electric filter increases the amplitude of the PLC signal at the ABS ECU.

....

Tech Tip: A problem with the trailer ABS power supply circuit may cause the trailer ABS warning lamp to illuminate when the service or trailer brakes are applied.

....

Hydraulic ABS

Many medium-duty trucks are equipped with hydraulic brakes. ABS has also been required for trucks with hydraulic brakes built after March 1, 1999, for use in the United States. FMVSS 105 defines the requirements for hydraulic brake systems. The components of a typical hydraulic ABS brake system include wheel-speed sensors, a hydraulic brake ABS ECU, and a hydraulic control assembly.

The basic function of hydraulic ABS is similar to that of air ABS except that hydraulic brake fluid is returned to a reservoir during an ABS event. The valves in the hydraulic control assembly work similar to the modulators described in the air ABS section. The electric solenoids and hydraulic valves in the hydraulic control assembly are typically installed in a single assembly that is bolted directly to the ABS ECU. An example of a hydraulic ABS system is shown in **Figure 15-33**. Note that the normally open inlet valves and normally closed outlet valves in the hydraulic assembly permit normal non-ABS braking capability, should the electric portion of the hydraulic ABS system fail.

Four-wheel disc brakes are typically used in modern truck hydraulic brake systems. Wheel-speed sensors detect wheel-speed changes in the same way as air ABS systems, with the target teeth typically being cast into the back side of the rotor, as shown in **Figure 15-34**.

Many legacy truck hydraulic ABS systems are not capable of J1939 communications, although current generations of these systems are SAE J1939 compatible. Therefore, hardwired ABS warning lamp circuits are required for many older trucks with hydraulic ABS. Drive-axle ABS event indication from the ABS ECU may also be hardwired to the engine ECM and transmission ECU.

Some International trucks with body controllers and hydraulic ABS use the body controller for the ABS warning lamp and drive-axle ABS event status. The drive-axle ABS event and the ABS warning lamp output from the ABS ECU are hardwired to two body

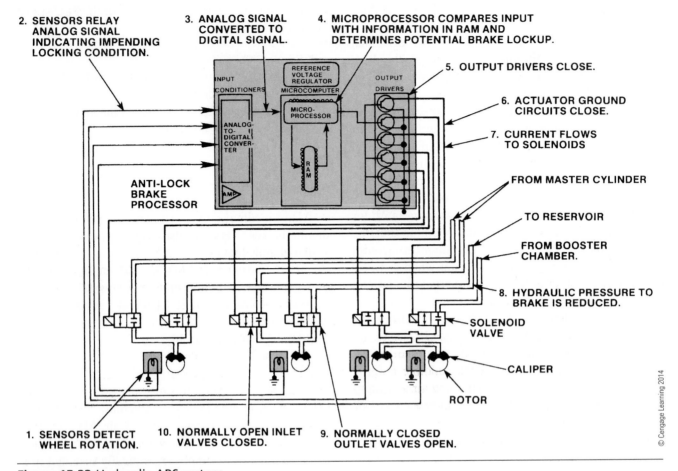

2. SENSORS RELAY ANALOG SIGNAL INDICATING IMPENDING LOCKING CONDITION.

3. ANALOG SIGNAL CONVERTED TO DIGITAL SIGNAL.

4. MICROPROCESSOR COMPARES INPUT WITH INFORMATION IN RAM AND DETERMINES POTENTIAL BRAKE LOCKUP.

5. OUTPUT DRIVERS CLOSE.

6. ACTUATOR GROUND CIRCUITS CLOSE.

7. CURRENT FLOWS TO SOLENOIDS

FROM MASTER CYLINDER

TO RESERVOIR

FROM BOOSTER CHAMBER.

8. HYDRAULIC PRESSURE TO BRAKE IS REDUCED.

SOLENOID VALVE

CALIPER

ROTOR

1. SENSORS DETECT WHEEL ROTATION.

10. NORMALLY OPEN INLET VALVES CLOSED.

9. NORMALLY CLOSED OUTLET VALVES OPEN.

REFERENCE VOLTAGE REGULATOR

INPUT CONDITIONERS

MICROCOMPUTER

OUTPUT DRIVERS

MICRO-PROCESSOR

ANALOG-TO-DIGITAL CONVERTER

RAM

AMP

ANTI-LOCK BRAKE PROCESSOR

© Cengage Learning 2014

Figure 15-33 Hydraulic ABS system.

© Cengage Learning 2014

Figure 15-34 Hydraulic ABS disc brake rotor with cast tone wheel.

controller inputs. The body controller detects the ABS warning lamp signal and the ABS event signal and transmits the status of the ABS warning lamp and drive-axle ABS event over the J1939 data link. The electronic gauge cluster then illuminates the ABS warning lamp accordingly, based on this J1939 message. The engine

ECM and automatic transmission ECU act accordingly, based on the drive-axle ABS event J1939 message transmitted by the body controller, causing the engine brake to be disengaged and the automatic transmission torque converter to unlock in the same manner as in air ABS systems.

ABS diagnostic trouble codes can be retrieved on some Meritor-Wabco hydraulic ABS systems by depressing a blink code switch. This causes the ABS warning lamp to flash in a sequence that indicates a two-digit diagnostic trouble code. Meritor-Wabco TOOLBOX™ is the PC-based EST used to diagnose problems with the Wabco hydraulic ABS system.

Hydraulic Power Brake. The Meritor-Wabco hydraulic power brake (HPB) system introduced in 2004 is significantly different from conventional hydraulic brake systems. The HPB system uses two electric pump motors to charge two gas-filled accumulators with brake fluid under pressure. The two hydraulic systems are isolated from each other; one hydraulic system controls the front brakes, and the other system controls the rear brakes. The accumulators and pump motors are components of the hydraulic compact unit (HCU).

The HCU also consists of hydraulic valves, a large brake fluid reservoir, and an HPB ECU. A dual master cylinder is also a system component. The HPB system is connected to the J1939 data link.

When the brake pedal is depressed to apply the service brakes, a control hydraulic pressure supplied by the master cylinder causes hydraulic relay valves in the HCU to provide proportionally intensified hydraulic pressure to the brake calipers. The stored energy of the brake fluid under pressure in the accumulators is similar to the stored energy in the tanks of compressed air in an air brake system.

There is a pressure sensor in each accumulator circuit. The HPB ECU monitors the hydraulic system pressure in each accumulator and controls the operation of the pump motors accordingly to maintain both system pressures at approximately 2320 psi (16 MPa). If either system pressure drops below 1523 psi (10.5 MPa), the HPB ECU transmits a J1939 message indicating that brake pressure is low. The multiplexed instrument panel cluster responds by illuminating the brake warning lamp and sounding an audible alarm. This is similar to the operation of the low air alarm in air brake systems. Unlike most other hydraulic brake systems, the force exerted by the operator's leg muscle to depress the brake pedal cannot directly apply the brake calipers. The master cylinder output is only a control pressure. Therefore, if there is not sufficient hydraulic pressure stored in the accumulators, the service brakes cannot be applied.

WARNING *Never attempt to drive a vehicle with HPB that has the brake warning lamp illuminated. You may not be able to stop the vehicle.*

With the ignition switch in the off position and the vehicle not in motion and brake pedal not depressed, the system will go into a low power consumption sleep mode and the HPB ECU will cease maintaining the brake system hydraulic pressure. The accumulator pressures will leak down after a period of time, similar to air leaking out of storage tanks in air brake systems. When the ignition is switched on or the brake pedal is depressed, the HPB ECU will awaken and control the pump motors as needed to maintain brake system hydraulic pressure.

The ABS portion of the system is similar to conventional hydraulic ABS and includes ATC and electronic brake force distribution (EBD). EBD adjusts the drive-axle brake apply pressure for more balanced braking. The ABS control valves (modulators) are located in the HCU. The ABS warning lamp is controlled by a J1939 message for use by a multiplexed IPC.

The HPB system also has optional parking brake control, as shown in **Figure 15-35**. The thicker lines in **Figure 15-35** indicate hydraulic circuits. The parking brake is typically a drum-type driveline brake, mounted to the front of the rear axle. The parking brake is cable-actuated. A spring-applied hydraulically released (SAHR) mechanical actuator supplies the energy to apply the parking brake.

Outputs of the HPB ECU for the parking brake are two high-side drivers to control two hydraulic valves identified in **Figure 15-35** as the supply valve and the cut-off valve. HPB ECU analog inputs include a diagnosable parking brake control switch. The center-stable three-position parking brake control switch mounted on the instrument panel has a yellow diamond-shaped knob similar to an air brake system parking brake. A parking brake status switch, which indicates if the parking brake is applied or released, is also an HPB digital input.

The normal state of the supply valve and cutoff valve are such that the parking brake is applied when neither valve coil is energized, as shown in **Figure 15-35**. To release the parking brake, the center-stable three-position parking brake control switch, shown in the center position, is depressed by the operator, similar to depressing the yellow parking brake knob with air brake systems. This causes a specific value of parking brake control switch resistance (560Ω for the example shown in **Figure 15-35**), which results in a voltage level at the HPB ECU parking brake control switch analog input corresponding to a parking brake release request. The HPB ECU then switches on the high side driver for the supply valve, which causes brake fluid under pressure stored in the accumulator to enter the SAHR actuator. This causes the piston in the SAHR actuator, shown in **Figure 15-35**, to move to the left. This relaxes the tension on the parking brake cable, causing the parking brake to release. A Hall effect type parking brake status switch is used by the HPB ECU to determine the status of the parking brake. A magnet on the SAHR piston acts as a target for this switch. With the parking brake applied, the magnet is directly below the parking brake status switch. When the piston moves to the left to release the parking brake, the HPB ECU detects that the parking brake is released and switches on the high side driver for the cut-off valve. This causes the brake fluid under pressure to be trapped in the SAHR actuator to maintain the parking brake in the released position. The HPB ECU will then switch off the high side driver for the supply valve.

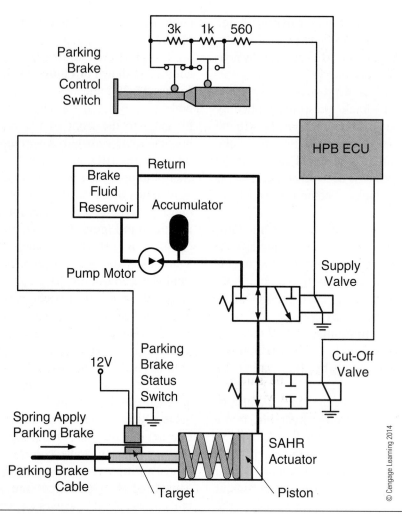

Figure 15-35 Wabco HPB spring-apply hydraulic-release (SAHR) park brake.

The HPB ECU can also use the parking brake status switch to determine that the parking brake cable requires adjustment. If the parking brake cable is too loose, the target will pass by the parking brake status switch as the parking brake is applied, causing a brief transition in the parking brake status switch as the parking brake is applied. The loose parking brake cable causes the target to rest outside the detection region of the Hall effect switch; that is, the target is not directly below the parking brake status switch with the parking brake applied.

To apply the parking brake, the parking brake control switch knob is pulled outward from the center-stable position, which changes the value of the parking brake control switch resistance to 4.56kΩ. The voltage at the HPB ECU analog input causes the ECU to detect that parking brake apply has been requested by the operator. The HPB ECU then switches off the high side driver for the cut-off solenoid, which causes the brake fluid under pressure in the SAHR actuator to return to the reservoir. The spring in the SAHR causes

the piston shown in Figure 15-35 to move to the right and apply tension to the parking brake cable.

Meritor-Wabco TOOLBOX™ is the EST used to diagnose the HPB system.

Repairing ABS Wiring

The wiring that connects wheel-speed sensors and modulators to the ABS ECU is typically twisted together, as shown in **Figure 15-36**. The twisting provides immunity to magnetic field interference that could otherwise induce a voltage difference on sensor wires that are spaced far apart. With the wires twisted together, both conductors are located about the same distance from any magnetic field source. The same amount of voltage should be induced in both sensor wires by the magnetic field if the wires are twisted. The ABS ECU measures the difference in the voltage between the two wires of the wheel-speed sensor, and not the voltage present on either wire with respect to ground. If the same voltage produced by a magnetic

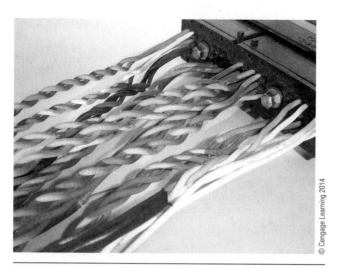

Figure 15-36 Twisted wiring for ABS wheel-speed sensors and modulators.

field is present on both wires, the difference in the voltage between the two wires due to the magnetic field is 0V.

Some ABS sensor wiring may also be shielded. Shielding is the foil and bare copper drain wire that is wrapped around wiring to provide immunity to electric fields, as described in the section on J1939 data link wiring in **Chapter 11**.

The location of ABS wheel-speed sensors makes them vulnerable to wiring harness damage, especially on front wheels. The ABS wiring must be routed to permit suspension travel and front-axle lock-to-lock steering wheel rotation. Front ABS sensor wiring can be damaged by contact with the wheel or tire, resulting in cut wiring and an ABS DTC. Even a small cut in ABS wheel-speed sensor wiring can cause a DTC even if the circuit remains intact. A small cut in the speed sensor wiring or a poorly sealed connector can provide a leakage path for current to chassis ground when exposed to water. The path to ground through the water can be sufficient to cause a DTC indicating a sensor circuit that is shorted to ground. The exposure to water also causes the wiring to rapidly corrode.

EMERGING TECHNOLOGIES

This section discusses recent advances in truck electronics. It is likely that future trucks will become even more complex than current models as rising fuel costs and increased government regulations require the trucking industry to become even more efficient.

Electronic On-Board Recorders

An **electronic on-board recorder (EOBR)** is an electronic device installed in a truck to record the

hours of service (HOS) for each truck operator. An EOBR can be considered as an electronic logbook. Although similar devices known as digital tachographs have been required for many years in EU countries, EOBRs are not required by the Federal Motor Carrier Safety Alliance (FMCSA) as of 2012. However, proposed legislation will likely require the devices for new trucks starting in 2014. Some fleets have already adopted EOBRs as a means of protecting their compliance, safety, and accountability (CSA) score.

An example of an EOBR is shown in **Figure 15-37**. The VDO RoadLog™ EOBR produced by Continental VDO has a touch screen and integrated printer, which produces a printout similar to a paper logbook page. This printout can be presented to a roadside inspector in the same way as a paper logbook. This system utilizes a unique USB key, which identifies each specific truck operator. The key is inserted in a USB port on the side of the device when a truck operator change occurs to identify the operator. The EOBR then monitors the amount of time that the truck is in motion and logs that time for the current truck operator. The HOS information from the truck operator key can also be downloaded by the trucking company for record keeping. A fleet key can also be inserted in the EOBR to retrieve information for all truck operators.

The EOBR requires a connection to the vehicle J1939 or J1708 network to acquire vehicle speed information. Alternatively, an analog input on the device can be hardwired to a vehicle speed sensor on older trucks. The EOBR also contains a GPS receiver, which is used by the device to validate the accuracy of the vehicle speed signal. The simultaneous input of the speed sensor signal and the GPS signal ensures that the EOBR is tamper resistant.

Future versions of EOBRs will permit HOS information from the device to be retrieved by fleet management via a cellular connection.

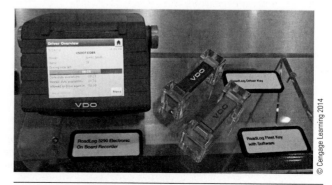

Figure 15-37 Electronic on-board recorder (EOBR).

Telematics

Telematics refers to the integration of telecommunications and information technology (informatics). General Motors OnStar® system found in passenger cars is an example of telematics. In the trucking industry, systems developed by Qualcomm® have been used for several years to track truck and trailer locations. These systems utilize a communications satellite in a geosynchronous orbit to relay information from the vehicle to the fleet office, as shown in **Figure 15-38**. Telematics hardware installed in the vehicle typically includes the following: antenna, communications modem, J1939 interface module, operator display, and associated cabling. The communications modem and J1939 interface module are often combined into an electronic module called a telematics gateway. Placement of these components varies greatly from system to system and truck to truck.

In recent years, several truck manufacturers have begun offering telematics systems that are integrated into the truck electrical system and most offer pre-wired options for popular telematics systems. The communications link for most of these systems is the cellular telephone network. Providers of these systems include PeopleNet®, Teletrac®, and DriverTech®. A telematics gateway module connected to one or more of the truck data links (J1939 and J1708) monitors DTCs and information such as vehicle speed, fuel consumption, vehicle mileage, and engine hours. This information can be transmitted back to a central location to schedule service should a DTC become active or to schedule preventative maintenance.

Most of these telematics systems include GPS capabilities. The location of hazardous cargo can be monitored by the fleet office and the system can indicate an alert to fleet management if the vehicle is located outside of a predefined geo-fence area. An emergency alert switch in the truck can also be activated by the truck operator.

The application of telematics in trucks is rapidly expanding. For technicians, telematics provides the ability to remotely diagnose problems. Remote diagnostics capability is predicted to greatly expand in the future as telematics matures.

Another use of telematics is **data mining**. Data mining refers to the use of telematics data by OEMs and fleets for the purpose of prognostics and for condition-based maintenance. Prognostic is a term borrowed from the medical field. Prognostics can be thought of as predictive diagnostics, or predicting a failure before the failure occurs using the available information. You may be tasked someday to replace a component on a truck that has no truck operator complaints and no DTCs, but the telematics prognostic system has determined that an impending failure of the component will occur. The impending failure is predicted based on trend analysis (pattern spotting). For example, fleet history may indicate that a turbocharger will likely fail shortly after a compressor surge or stall event has been detected a number of times. Using telematics data, the turbocharger might be replaced and the cause of the surge or stall corrected before a catastrophic failure of the turbocharger occurs.

Condition-based maintenance (CBM) can refer to the use of telematics data to schedule preventative maintenance (PM) based off actual vehicle usage, not just miles or hours. Information such as truck operator habits, primary region of operation, and ambient conditions can be used to either extend or reduce PM intervals as compared to hours or miles alone.

Hybrid Electric Trucks

Hybrid electric trucks have been introduced in recent years. One of the most widely used hybrid electric

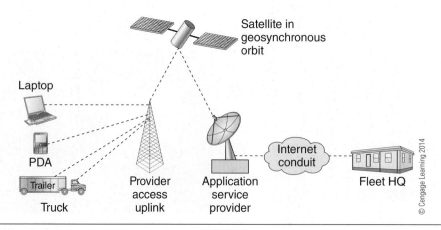

Figure 15-38 Satellite link between truck/trailer and fleet management.

truck system is the Eaton electric hybrid. The Eaton hybrid is classified as a **parallel hybrid system**. In a parallel hybrid system, the power produced by the engine is supplemented by the electric motor under certain operating conditions. The electric motor is not typically used alone to propel the vehicle in a parallel hybrid system, although this is possible with some parallel hybrid systems. By comparison, in a **series hybrid system** the vehicle is propelled only by an electric motor. The internal combustion engine in a series hybrid is used to generate voltage to power the electric motor. A diesel electric locomotive is similar to a series hybrid; many hybrid automobiles are parallel hybrids.

The Eaton hybrid system makes use of an Eaton UltraShift AMT with electronic clutch actuator (ECA), as shown in **Figure 15-39**. A three-phase AC motor/generator is sandwiched between the clutch and transmission. The permanent magnet rotor of the motor/generator is splined to the transmission input shaft. The three-phase stator surrounds the rotor, similar to the construction of an alternator as discussed in **Chapter 7**.

Figure 15-40 shows a block diagram of the Eaton hybrid system. A conventional diesel engine is the main source of power in the system. An UltraShift AMT with ECA clutch and integral motor/generator is attached to the diesel engine in the same way as a conventional transmission. The power electronics carrier (PEC) contains two banks of lithium-ion batteries connected in series to produce approximately 340V DC. The inverter is a power electronics component that converts the 340V DC provided by the batteries into a variable-frequency three-phase AC voltage of up to 500V, which is connected to the stator windings of the motor/generator.

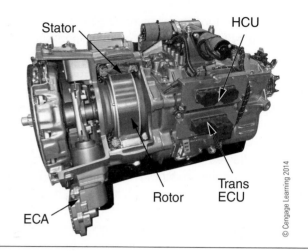

Figure 15-39 Eaton electric hybrid AMT cutaway showing motor/generator stator and rotor.

The hybrid control unit (HCU) is the central ECU for the hybrid electric system and controls the inverter and the ECA, as well as many other aspects of the system. The transmission ECU is the same found on a conventional UltraShift AMT, but with modified programming.

As indicated, the Eaton hybrid system is a parallel hybrid. With the clutch engaged, the rotor of the motor/generator and the engine crankshaft rotate at the same speed. During acceleration, the torque produced by the motor/generator acting as a motor and the diesel engine work together to propel the truck. During braking, the motor/generator acts as a generator to recharge the batteries (if necessary).

The inverter is responsible for converting DC voltage supplied by the lithium-ion batteries into a variable frequency three-phase AC voltage when the motor/generator is acting as a motor. The inverter is

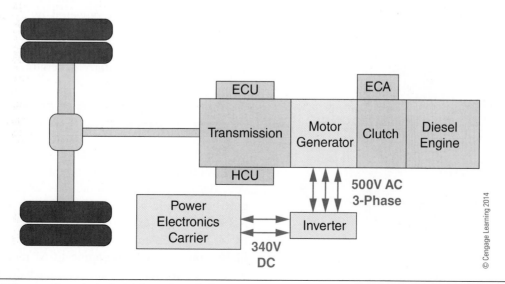

Figure 15-40 Eaton electric hybrid functional diagram.

also responsible for converting three-phase AC voltage produced by the motor/generator into approximately 340V DC to recharge the lithium-ion batteries. The inverter is a flux vector drive, meaning it is capable of producing a variable frequency three-phase AC voltage, like that shown in **Figure 15-41**. One of the principles of a three-phase AC motor is that the rotational speed of the motor is dependent upon the frequency of the AC voltage supplied to the motor. The three-phase voltage produces a rotating magnetic field around the stator windings, which causes the permanent magnet rotor to rotate at nearly the same speed as the rotating magnetic field. As the frequency of the AC voltage supplied to the stator increases, the speed of the rotating magnetic field also increases, as does the speed of the rotor.

WARNING *Lethal voltage levels are used in hybrid electric systems. You should not attempt to work on the high-voltage components of an HEV until you have had proper training and have all the proper protective equipment including insulated gloves. High-voltage wiring is identified by orange-colored insulation. Never splice into this high-voltage wiring.*

The electric motor/generator is capable of producing 59 hp (44 kW) of peak power intermittently and 35 hp

(26 kW) continuously. This is not sufficient to power the truck down the road at highway speeds, but does provide substantial assist during acceleration resulting in fuel savings. Trucks used for inner-city delivery can benefit the most from this hybrid technology.

A variation of the Eaton electric hybrid used in stationary work truck applications (i.e., bucket trucks) uses the electric motor to drive the power take off (PTO). The system shifts the transmission into neutral and releases the clutch to drive the PTO without the diesel engine running resulting in fuel savings.

The motor/generator can also act as the starter motor for the system, although a conventional starter motor is present in the event the hybrid system is not functional. The truck can also be operated using the diesel engine alone in the event of some hybrid system failures.

Future hybrid-electric truck technology includes the ArvinMeritor hybrid system for class 8 line haul trucks. At low speeds, the truck is propelled only by a three-phase electric motor that operates at 700V AC. Voltage for the motor is supplied by a three-phase inverter powered by two series-connected 350V lithium-ion battery packs producing 700V DC. The diesel engine is also used to drive a generator to supply voltage for the system, like a series hybrid system. At speeds above 48 mph (77 km/h), the drivetrain transitions to a conventional diesel-powered system with the electric motor providing power to assist the diesel engine as necessary, like a parallel hybrid system. This type of hybrid is called a **dual-mode hybrid system** because it acts as both a series and a parallel hybrid system. This system consists of two electric motor/generators and a transmission, which mechanically shifts the system between series and parallel modes. The combined system is estimated to produce 480 hp (360 kW). The inverter powered by the lithium-ion batteries also provides power for sleeper cab HVAC systems and cab electrification during overnight rest stops, similar to an auxiliary generator.

TROUBLESHOOTING A MODERN TRUCK ELECTRICAL SYSTEM

The interactions between the various electronic modules on a modern truck should now be apparent to you. Because of these interactions, an inoperative engine brake could be caused by a false indication of an ABS drive-axle event being detected by the engine ECM. This problem could be due to a wiring issue or a faulty component. The engine ECM inhibits or shuts off the engine brake by design during a drive-axle ABS event, as explained in the section on ABS systems.

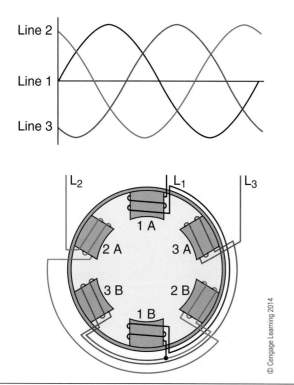

Figure 15-41 Three-phase AC voltage supplied to motor windings.

However, there may be no indication to the truck operator of a problem other than the inoperative engine brake. Such a problem would be very difficult to diagnose without the knowledge of how the ABS ECU interacts with the engine ECM to control the engine brake.

A different truck that is equipped with an automatic transmission but is not equipped with an engine brake might have different symptoms due to a false indication of an ABS drive-axle event occurring. The truck operator might have a complaint of decreased fuel economy and a higher than normal engine speed at highway speeds. The truck operator may also indicate that it feels like the truck coasts too fast during deceleration, similar to how a truck with a manual transmission feels when the clutch is depressed during deceleration. Again, the cause of all of these complaints could be a false indication of an ABS drive-axle event due to a wiring or component problem. When test-driving the truck, a technician would probably realize that the lockup torque converter clutch is not engaging and is causing all of the truck operator's complaints. The transmission ECU is programmed to unlock the torque converter clutch during an ABS event, as explained earlier. Without this knowledge about the interaction between the ABS ECU and the transmission ECU, a technician might remove and tear down the automatic transmission, looking for a problem with the lockup torque converter clutch when no such problem exists.

Both of these examples illustrate the types of problems that technicians may encounter when working on modern trucks. One of the most important tools for an electrical troubleshooter to use to solve these types of problems is a general knowledge of the operation of the entire truck. Every major truck system now has some electrical aspects. The interaction among the various systems—such as ABS, engine, transmission, and body control module—is increasing as trucks become more complex. Becoming a good electrical troubleshooter means studying and learning the details of every system on the truck. This may require a considerable amount of self-study on your own time.

Manufacturer Service Information and Tools

The days of paper service manuals sitting on a shelf in the office with mostly clean pages have ended. Electronic service information has become one of the most important tools for diagnosing and repairing modern trucks. Every major truck and engine OEM now has an online service information system (SIS). Examples include:

- Caterpillar Service Information System
- Cummins QuickServe Online
- Mack Trucks Electrical Information System
- Navistar Service Information
- Paccar DAVIE and ServiceNet
- Volvo IMPACT

In general, having no information is typically better than having the wrong information. For example, using an electrical wiring schematic that does not accurately represent the wiring harness on the truck you are repairing would likely lead you to the wrong conclusion. Most of these systems provide the correct specific service information for the truck being repaired based on the engine or vehicle serial numbers. The manufacturer can also update online information daily as relevant failures or errors in diagnostic procedures are discovered, something that was not possible with paper copies of service information. For example, if a manufacturer determines from recent warranty history that the most likely cause of DTC 123.06 is a short to ground at a specific location in the wiring harness, the inspection for this failure might become the first step in the troubleshooting procedure for this DTC. This is known as case-based reasoning and was introduced in **Chapter 4**.

Most manufacturers reference the use of proprietary PC-based ESTs in their service information. Prior to HD-OBD requirements related to service information availability, some manufacturers would not provide their EST or service information outside their dealer network. It remains to be seen how HD-OBD will affect the independent truck repair facilities since the ESTs and service information are a considerable expense for a shop that only occasionally services a particular brand of truck or engine. Additionally, HD-OBD only addresses service information related to emission-related repairs, so ESTs and information necessary to service a proprietary system such as a body control module may not be available to independent repair facilities.

For major components or systems used by several truck manufacturers, such as a Cummins engine, an Allison transmission, or a Bendix ABS system, ESTs and repair information can typically be purchased directly from the component manufacturer.

The manufacturer-provided information is the preferred place to start when troubleshooting. However, too many possible problems can occur on a modern truck to list every possible troubleshooting step. In addition to following manufacturer-supplied troubleshooting information, some suggestions for troubleshooting modern truck electrical systems are as follows:

- Determine if there is anything else that is not working correctly, not just what the truck operator has written up as the problem. Talk to the truck operator directly, if possible. Details tend to get lost and the information changes as more people are involved in relaying information.
- Use an EST to look for DTCs in all electronic modules, not just in the module controlling the system that is not working correctly. For example, the root cause of what appears to be an engine problem may be due to a problem with another electronic module or system on the truck. The fault history or historical DTCs can also provide some clues for intermittent problems that cannot be duplicated.
- Look for abnormalities in the data indicated by each electronic module using an EST. For the ABS drive-axle event example, the automatic transmission ECU and the engine ECM diagnostic tools may provide an indication of an ABS drive-axle event being detected when no ABS drive-axle event is actually occurring. Most ESTs permit the technician to view the status of various inputs and outputs for a particular electronic module, such as the engine ECM.
- Try comparing data from a truck that has a problem with a similar truck that does not have a problem. A mental database of what appears as normal and what is abnormal data takes considerable time to amass. Monitoring data from a truck without a problem can assist you in determining what abnormal data looks like. Get in the habit of looking at data from the various electronic modules on trucks that are functioning correctly.
- Most ESTs are capable of recording selected parameters such as ECM sensor input voltages. This permits a vehicle to be driven while recording data to recreate the conditions that cause an intermittent problem. The recorded data can then be evaluated for abnormalities after the test drive.

WARNING *Never attempt to monitor EST data while driving a vehicle. Ask a coworker or the vehicle operator to drive if EST data must be monitored while the vehicle is operated.*

Troubleshooting the J1939 Data Link

This section assumes that the J1939 data link uses 120Ω terminating resistors at either end of the backbone, as described in **Chapter 11**. Some J1939 networks may use electronic terminating bias circuits instead of terminating resistors. A terminating bias circuit is an electronic device that serves the same purpose as a terminating resistor in a J1939 network. However, unlike a terminating resistor the terminating bias circuit cannot be used to aid in troubleshooting the J1939 data link, as will be described in this section. If the truck on which you are working has terminating bias circuits instead of terminating resistors, consult the specific OEM information for troubleshooting procedures.

The J1939 data link is really just two copper wires covered with insulation and twisted together. The wires must pass through connectors using terminals like any other wiring used in the truck electrical system. This means that the data link wiring can fail like any other circuit, such as a short to ground, a wire-to-wire short, or an open circuit.

Troubleshooting wiring problems with the J1939 data link is usually not too difficult. SAE J1939 uses a bus-type topology. A bus topology means that there is just a single pair of twisted wires routed past each electronic module. This pair of twisted wires is called the backbone. Think of the backbone being like Main Street of a small town. At either end of the backbone, a 120Ω terminating resistor is connected between the two conductors (**Figure 15-42**).

Each of the electronic modules splices into each of the wires in the backbone through short sections of J1939 cable called stubs. Stubs may also be known as node branch circuits. Think of the stubs as being like driveways along Main Street. The houses are like the electronic modules. One of the driveways or stubs also leads to the 9-pin diagnostic connector.

For troubleshooting purposes, the J1939 data link can be considered as being just a pair of wires with a 120Ω resistor connected across either end. Several modules may be connected to the J1939 data link as well, but they can be ignored for now if the vehicle batteries are disconnected. Therefore, the J1939 data link could just be reduced to a pair of wires with a 120Ω resistor across the two ends of the pair of wires, as shown in **Figure 15-43**. The 9-pin diagnostic connector provides easy access to this pair of wires.

If the truck's batteries were disconnected, none of the modules would be able to communicate on

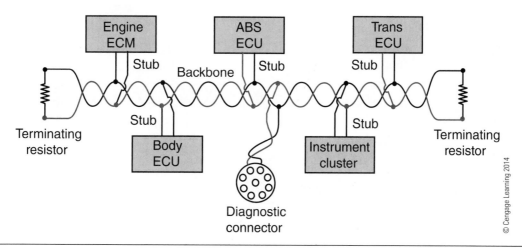

Figure 15-42 J1939 bus-type network.

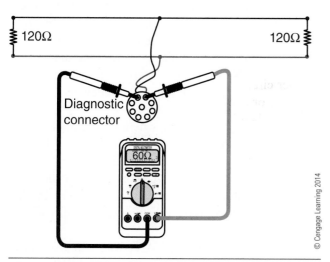

Figure 15-43 J1939 data link modeled as a pair of wires with two 120Ω resistors in parallel.

the J1939 data link. With the truck's batteries disconnected, a DMM ohmmeter connected across the diagnostic connector CAN+ and CAN– terminals (typically pins C and D of the 9-pin diagnostic connector) should show the value of the two parallel-connected 120Ω terminating resistors. If the entire data link is intact from terminating resistor to terminating resistor with the diagnostic connector located somewhere in between, the resistance measured across the diagnostic connector should be approximately 60Ω. Two identical-value resistors connected in parallel results in a total resistance that is equal to one-half the value of one of the resistors, as explained in **Chapter 2**. Therefore, the two 120Ω resistors in parallel as measured at the diagnostic connector should have approximately 60Ω of resistance.

If you measure 120Ω across the J1939 data link at the diagnostic connector instead of 60Ω, then either an

open circuit exists somewhere in the J1939 data link or one of the terminating resistors is missing. The vehicle circuit diagram is necessary at this point because the topology or layout of the J1939 data link is important to know when troubleshooting a J1939 communications problem.

One of the unique things about the CAN protocol, which includes J1939, is that there is no single module in charge of the J1939 data link. There is no network router or hub in a CAN network like that in local area networks (LANs) used to connect many PCs. If the J1939 data link were cut into two pieces so that the engine ECM, transmission ECU, and ABS ECU were still connected, and the body control module and IPC were still connected, the two pieces of the J1939 data link would each attempt to act like independent CAN networks (**Figure 15-44**).

Cutting the J1939 data link into two pieces would typically still permit the engine ECM, transmission ECU, and ABS ECU to communicate with each other, as in the example shown in **Figure 15-44**. The body control module and the instrument cluster would also still be able to communicate with each other in this example. However, the engine ECM would not be able to communicate with the body control module or the instrument cluster because of the open circuit in the J1939 data link. Additionally, an EST connected to the diagnostic connector would only be able to communicate with the instrument cluster and the body control module. The diagnostic tool that is connected to the diagnostic connector would not be able to communicate with the engine ECM, the transmission ECU, or the ABS ECU because of the location of the open circuit in this example.

Because the engine ECM and the automatic transmission ECU can still communicate with each other in

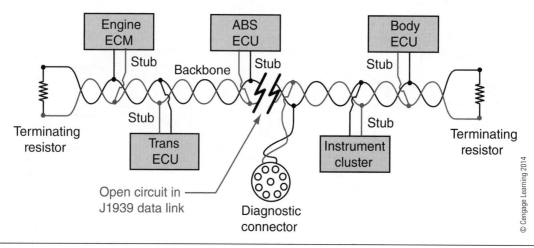

Figure 15-44 Damaged J1939 data link will act as two separate networks.

this example, the open circuit in the J1939 data link wiring will probably not cause an automatic transmission shifting problem such as delayed up-shifts. The engine ECM cannot communicate with the multiplexed instrument cluster, however, because of the open circuit, so any instrument cluster gauge that receives its data from the engine ECM via J1939 will not operate. However, any gauge that receives data from the body control module would still operate because the body control module and the instrument cluster are still able to communicate with each other. The J1939 data link has basically been transformed into two separate communication networks due to the break in the wiring. The single terminating resistor on each half of these two networks typically permits sufficient communications to occur within each half of the network.

The circuit diagram and a DMM ohmmeter are very useful in troubleshooting where the break in the J1939 data link is located. It may be necessary to disconnect connectors that the J1939 data link is passing through to find the source of the problem and use the ohmmeter to determine if the meter can "see" the resistance of the terminating resistors at either end of the data link. Make sure that the truck's batteries are disconnected first, though, because any electronic module that is communicating on the J1939 data link will cause voltage pulses to be present across the data link CAN+ and CAN– wires and will affect the ohmmeter readings.

Besides an open circuit, other failures can occur with the J1939 data link that can cause a loss of communications between the electronic modules. A wire-to-wire short circuit between the CAN+ and CAN– wires will cause all communications to cease on the entire network. This may be caused by the data

link wiring being crushed somewhere and the copper wiring piercing through the insulation of both conductors, resulting in a short circuit between CAN+ and CAN–. An ohmmeter connected across the diagnostic connector CAN+ and CAN– terminals with the two data link wires shorted together somewhere would indicate near 0Ω.

A defective module somewhere on the truck could also cause a short between CAN+ and CAN–. It may be necessary to disconnect each module one by one to find the defective module. A short between the data link CAN+ and CAN– conductors anywhere in the system will cause a loss of communications throughout the entire J1939 data link.

A short to ground of the CAN+ wire will also cause all communications to cease on the entire J1939 network. A short to ground of the CAN+ wire would not cause any fuse to blow because the current flow in the CAN+ circuit would be very small. Such a short circuit would probably not cause any permanent damage to any module, either. However, a short to ground of the CAN– circuit may only cause intermittent J1939 communication problems.

A short to positive battery voltage of the CAN– circuit will also cause all J1939 communications to cease. However, a short to battery positive of the CAN+ circuit will probably only cause intermittent communications problems, or will only cause some of the electronic modules to not communicate. Note that either of these types of short can occur within any of the electronic modules that are connected to the bus, requiring replacement of the module.

The voltage measured across the CAN+ and CAN– terminals of the diagnostic connector—with the batteries connected and the key switch in the ignition position so the various modules are communicating—can also tell you a lot about the condition of the J1939 data link. Voltage measurements referenced to ground from the CAN+ and CAN− with modules communicating are also useful for troubleshooting. The exact voltages measured will depend on your meter and the truck, but most DMMs in the DC volts setting will read these approximate values with the key switch in the ignition position:

CAN+ to CAN−	3V
CAN+ to ground	2.6V
CAN− to ground	2.4V

Voltage readings that vary greatly from these values could be an indication of a J1939 data link problem, such as a short to ground, short to +12V, or CAN+ to CAN− short circuit.

An electronic module with a defective CAN transceiver chip or a module that is continuously causing errors to be transmitted onto the J1939 data link can cause communications problems for the entire CAN network. Each electronic module that communicates on the J1939 data link can be removed one by one (cycle ignition off first) to determine which module is causing the communications problem. A wiring harness problem that causes CAN+ and CAN− to be reversed at a module connector can also cause this problem.

Tips for Troubleshooters

Troubleshooting a modern truck electrical problem can be challenging at times. This section provides some general tips for electrical troubleshooting that you may find useful someday.

Most electronic modules require a battery-positive (B+) supply, ground, and ignition signals. Series resistance or intermittent connection can cause electronic modules to respond in ways that seem to make no sense, possibly with no DTCs being set by the module. The importance of checking the electronic module's power supply cannot be overstated. Open circuit measurements of B+ power supply, ground, and ignition at a module connector using a DMM voltmeter can fool even the most experienced troubleshooter. Ghost voltage can make it appear that the B+ power supply, ground, and ignition are acceptable when significant series resistance, commonly known as green wire, may actually be present in the circuit. Obtain closed circuit measurements using a breakout box or breakout T whenever possible to minimize misdiagnosis due to ghost voltage.

Internal alternator problems, such as failed rectifier diodes, can result in high amplitude ripple voltage, as discussed in **Chapter 7**. Excessive charging system ripple voltage can cause a variety of electrical system problems. An oscilloscope or DMM voltmeter set to measure AC voltage can be used to measure charging system ripple voltage.

Intermittent open circuits caused by spread terminals or broken wires can sometimes be found using the peak hold mode on a high-quality DMM voltmeter. Gently pulling on the wiring harnesses associated with the module in question while monitoring the voltage using the DMM can cause the intermittent open circuit to be detected by the DMM's peak hold mode.

Check the simple and obvious things first, even if it does not seem like there is any possibility of these being the cause of the problem. Think of it this way: If you cannot find the cause of a problem with a truck, another technician will have to diagnose the problem. If the cause of the problem ends up being something simple that another technician finds in five minutes, your reputation as a troubleshooter will suffer.

At some point, you may have to troubleshoot a problem that another technician could not solve. In these situations, it is best to start the troubleshooting process from the beginning, as though no one else had worked on the truck. Check the fundamental things yourself, such as B+ power supply, ground, and ignition, even if you have been told that these have already been checked. Keep in mind that the previous attempts to diagnose the problem may have caused DTCs to be set that are not related to the original problem, so treat all stored DTCs as suspect.

Technical information is vital for troubleshooting most problems. However, documentation such as circuit diagrams or diagnostic procedures may contain errors or may not be exactly like the truck you are working on due to changes to wiring that have not been updated in the circuit diagrams. If something you observe in the service information does not make sense, it may be because the information does not match the truck you are working on or because of an error in the documentation. Read the information a second time or ask another technician to read the information. If applicable, check online for updated information.

Diagnostic procedures and other manufacturer-provided troubleshooting procedures cannot include every possible cause of a problem. If they could, then skilled electrical system troubleshooters would not be necessary because anyone who can read and follow directions could repair any complex electrical system problem. It is often necessary to use the diagnostic procedures to get an idea about what might be wrong and use your own troubleshooting skills to expand from there when the procedures do not get you to the source of the problem.

DTCs indicate that something is not as expected, not that some component named in the trouble-code description necessarily requires replacement. Many DTCs are caused by intermittent wiring problems. Treat a DTC as a general indication of a problem. Many manufacturers indicate what specific conditions must occur to set a specific trouble code. This information can be very useful in troubleshooting. Many electronic modules record a fault history that may be retained in memory for a long period of time. This fault history can be analyzed to aid in troubleshooting an intermittent problem. Additional information is required for HD-OBD systems, including the specific conditions required for a DTC to set. Freeze-frame information may also be valuable for troubleshooting.

Occasionally, the fault detection that the manufacturer designed may be too sensitive for all operating conditions, such as very cold ambient temperatures. A warning lamp or DTC that occurs intermittently may not be an indication that anything is actually wrong. Manufacturers may publish field information to inform technicians of these known problems with fault detection and may offer controller software updates or special procedures to correct these issues. Such manufacturer information can prevent wasted time troubleshooting a problem that you cannot solve alone.

Sloppy troubleshooting techniques have the potential to inflict serious damage on the truck's electrical system that may take considerable time and expense for the truck's owner to correct. The idea of troubleshooting is to solve the problem, not create more problems. One of the worst things that a troubleshooter can do is to pierce the wire insulation to obtain readings. When the wire is exposed to road splash, the wire will corrode in a very brief time and cause a failure later. This can lead to very expensive repairs when some other technician has to find the source of the problem that you created.

Another poor troubleshooting practice is to over-probe female terminal sockets. Connector terminals that are probed using too large a test device, such as a DMM test lead, can be spread apart, causing

intermittent open circuits that are very difficult to locate. Use breakout boxes and breakout Ts whenever possible to avoid damaging connector terminals. Many OEMs have male terminal pin tools which can be used to drag-test female terminal sockets.

One of the most difficult wiring harness problems to diagnose is multiple-terminal connectors which have one or more terminal and wire that is not inserted in the correct connector cavity (mispinned). This typically only occurs on new vehicles, but can also occur if a new wiring harness is installed or after wiring harness repairs have been performed. In some cases, two terminals are inadvertently swapped when they are inserted in the connector. The types of problems that this error can cause are nearly unlimited. Typically, the best way to diagnose two or more swapped terminals in a connector is to use a DMM to check continuity of each circuit in the suspect wiring harness using the OEM circuit diagrams as a reference.

Do not be afraid to ask for assistance when you are stumped. The days of one technician knowing it all without having to use service manuals, online information, and circuit diagrams are long gone. Some of the most useful information for newer trucks are technical service bulletins (TSBs) and other similar OEM information. These provide information on specific problems. Networking with other technicians via the Internet may also be very helpful. One such network is the International Automotive Technicians Network (iATN), which has a dedicated HD/Fleet forum. Information on this organization can be found in the Internet links at the end of this chapter.

Be careful when using "known good parts" from another truck to troubleshoot. A wiring problem on the truck that caused damage to an electronic module such as a shorted to +12V reference ground circuit may damage the known good module as well. Subtle changes, including programming differences between seemingly identical electronic modules from another truck, can also lead you to a wrong diagnosis of the real problem.

Lastly, a $400 DMM and a $2000 oscilloscope are excellent diagnostic tools, but there is no tool more valuable for electrical and electronic system troubleshooting than a thorough knowledge of the fundamental principles of electricity and electronics. These fundamentals include Ohm's law, Kirchhoff's laws, and the rules of series, parallel, and series-parallel circuits.

Summary

- Allison electronic automatic transmissions use three variable reluctance speed sensors to measure engine speed, turbine speed, and output speed. The transmission ECU controls a series of normally open and normally closed solenoids using PWM. These solenoids control the flow of transmission fluid to apply clutches used for specific range.

- Eaton automated manual transmissions (AMTs) use two electric motors to actuate the shift mechanism. AMTs are divided into two-pedal and three-pedal systems, indicating if the clutch is controlled by the truck operator or the transmission ECU. Inputs to the ECU include three speed sensors and rail and gear select positions.

- Air ABS systems use wheel-speed sensors to determine if wheel lockup is occurring or is about to occur. The ABS ECU then controls modulators to block air flowing to the brake chambers and exhaust the air out of the brake chambers to prevent wheel lockup.

- Testing wheel-speed sensors and the associated circuits may involve raising each wheel end and rotating the wheel by hand while watching for the amplitude of the AC voltage that is generated by the variable reluctance sensor.

- Trailer air ABS ECUs may use the power line carrier technique to signal the truck or tractor ABS ECU that a trailer ABS fault is active. A variable frequency signal is placed onto the 12V power supply by the trailer ABS ECU. This signal is detected by the truck ABS ECU in its 12V power supply, causing the truck ABS ECU to energize the trailer ABS warning lamp on the truck's instrument panel.

- Electronic on-board recorders (EOBR) are electronic truck operator log books designed to record the hours of service (HOS) for each truck operator.

- In a parallel hybrid system, a diesel engine and electric motor/generator are coupled together. The electric motor is used to supplement the diesel engine. In a series hybrid system, the electric motor alone is used to propel the vehicle and the engine is used to drive a generator for the electric motor. A dual-mode hybrid system acts as a series hybrid system during acceleration and as a parallel hybrid system at higher speeds.

Suggested Internet Searches

Try the following web sites for more information:
http://www.allisontransmission.com
http://www.eaton.com
http://www.bendix.com
http://www.meritorwabco.com

http://www.conti-online.com
http://www.qualcomm.com
http://www.peoplenetonline.com
http://www.iatn.net

Review Questions

1. Which of the following is a true statement for an Allison electronic automatic transmission?

   A. Electric motors are used to control the clutch packs in an Allison transmission.

   B. It is okay for the engine to crank on a truck with an automatic transmission with the shifter in positions other than park or neutral.

   C. The Allison transmission output speed sensor is typically directly hardwired to the engine ECM so that the engine ECM can determine vehicle speed.

   D. The Allison 3000/4000 series 4th Generation Controls electronic shifter (truck operator interface) communicates with the transmission ECU via the J1939 data link.

2. Which of the following is a true statement for an Eaton automated manual transmission?

   A. On two-pedal versions of AMTs there is no clutch between the engine and the transmission.

   B. On three-pedal versions of AMTs the truck operator must control the clutch for start-up and when coming to a complete stop.

   C. The ECA is controlled by the truck operator using the electronic clutch pedal.

   D. With no battery power supplied to the transmission ECU, the truck operator can still use the shifter to drive the truck.

3. Which of the following may occur when the ABS ECU detects a drive-axle ABS event?

   A. Retarder devices are disabled.

   B. Automatic transmission lockup torque converter clutch is disengaged.

   C. ABS ECU broadcasts a J1939 message indicating a drive-axle ABS event is occurring.

   D. All of the above could occur.

4. Technician A says that automatic traction control can involve the ABS ECU detecting drive wheel spin and requesting the engine to reduce fueling. Technician B says that automatic traction control (differential braking) can involve the ABS ECU, causing one of the rear service brakes to be applied. Who is correct?

   A. A only

   B. B only

   C. Both A and B

   D. Neither A nor B

5. The ABS warning lamp is illuminated on a truck with air ABS. Which of the following could be the cause of the warning lamp?

   A. ABS controller power supply fuse is blown.

   B. ABS controller connector has been disconnected.

   C. ABS controller has detected a fault condition.

   D. All of the above could cause the ABS warning lamp to illuminate.

6. The trailer ABS controller on a trailer with air brakes built after March 1, 2001, controls the trailer ABS warning lamp in the truck instrument panel through what method?

   A. A J1939 message from the trailer ABS ECU to the truck ABS ECU. The truck ABS ECU then causes the trailer ABS warning lamp to illuminate.

   B. The trailer ABS ECU causes a high-frequency signal to be modulated onto its power supply circuit. The truck ABS ECU detects this signal through its own power supply and causes the trailer ABS warning lamp to illuminate.

   C. Fiber-optic cable between the trailer ABS ECU and trailer ABS warning lamp in the IPC.

   D. GFCI-type ground modulation between ABS ECU ground and trailer ABS ECU ground.

7. The expected resistance between the J1939 data link CAN+ and CAN– measured at the diagnostic connector with the batteries disconnected on a system using terminating resistors should be approximately what value of resistance?

   A. Infinite resistance; there should be no path between CAN+ and CAN–

   B. Near 0Ω of resistance

   C. 120Ω of resistance

   D. 60Ω of resistance

8. An ABS wheel-speed sensor has been pushed partially out of its mounting hole, causing a large air gap between the sensor and the target. How would the ABS ECU detect that the sensor air gap is too large?

    A. The frequency of the signal generated by the sensor would be much greater than the other sensors.

    B. The resistance of the sensor increases.

    C. The Hall effect device located in the sensor would signal the ABS ECU of the reduction in magnetic field strength.

    D. The amplitude of the signal generated by the sensor would be decreased.

9. A tractor built after March 1, 2001, is being used to pull a trailer with trailer ABS that is also built after March 1, 2001. Why should a seven-way trailer electrical interconnection cable identified as SAE 2394 be used with this combination vehicle?

    A. The SAE 2394 trailer cable contains special electronics that are necessary for the PLC signal.

    B. The power and ground wires are twisted together in the SAE 2394 cable for immunity to electromagnetic interference.

    C. The SAE 2394 cable contains J1939 wiring for communications between the trailer ABS ECU and the truck ABS ECU.

    D. Three wires in the SAE 2394 cable are a larger wire gauge than a standard cable.

10. Which of the following is most likely to cause a no-crank situation on a truck with an Allison 4th Generation 2000 series transmission?

    A. The NSBU switch is out of adjustment.

    B. The shifter is not connected to the J1939 data link.

    C. The wait-to-start light is still illuminated.

    D. An open circuit exists between the transmission ECU park/neutral indication output terminal and the crank inhibit relay or engine ECM neutral input terminal.

11. A voltage of 0V is measured between the CAN+ and CAN– terminals at the diagnostic connector, using a DMM with the key in the run position. The multiplexed IPC is not operating and several other electrical problems exist. Of the choices listed, which of the following is most likely a cause of this problem?

    A. CAN+ and CAN– are shorted together.

    B. One terminating resistor is missing.

    C. PLC signal is not present.

    D. VIM module is defective.

12. The ABS warning light does *not* illuminate in the IPC when the key switch is placed in the ignition position for a lamp check on a truck that uses a fail-safe relay to control the ABS warning light. Which of the following would probably *not* be the cause of the problem?

    A. Open circuit between the ABS ECU and the fail-safe relay coil

    B. ABS warning lamp is burned out

    C. Missing ABS warning lamp fail-safe relay

    D. Open circuit between ABS warning lamp fail-safe relay and the ABS warning lamp

13. A particular model of truck with a J1939 multiplexed IPC uses J1939 messages transmitted by the ABS ECU to control the ABS warning lamp. What would probably be the method used to cause the ABS warning lamp to illuminate on this truck if the ABS ECU electrical connectors had been disconnected?

    A. Shorting clip in the ABS ECU power connector provides a ground for the fail-safe relay.

    B. VIM module provides a PLC signal indicating that the ABS warning lamp should be illuminated.

    C. Fail-safe relay detects the missing J1939 messages from the ABS ECU.

    D. The lack of J1939 messages being received from the ABS ECU causes the IPC to illuminate the ABS warning lamp.

14. Which of the following is true for an Eaton hybrid electric system?

   A. Orange wiring insulation is used to indicate 12V control circuits.

   B. The ECA is controlled by the HCU.

   C. The Eaton hybrid electric system is a series hybrid system; the diesel engine drives a generator to supply voltage to the electric motor.

   D. The lithium-ion batteries produce 500V AC.

15. What is the best way to troubleshoot a truck electrical problem?

   A. Use manufacturer-provided diagnostic procedures.

   B. Use circuit diagrams and other information specific to the truck you are working on to aid in troubleshooting.

   C. Use ESTs and other specialized test equipment.

   D. All of the above may be necessary to find the cause of an electrical problem, combined with a strong knowledge of the fundamentals of electricity.

Glossary

accessory air valve assembly (AAVA) Modular air solenoid valves utilized in Freightliner multiplexed electrical system.

active-high A pulled-down digital input; the input is considered active when a voltage above a specified threshold is present at the input.

active-low A pulled-up digital input; the input is considered active when a voltage below a threshold is present at the input.

aftertreatment Reduction of exhaust emissions outside of the combustion chamber.

air management unit (AMU) Air accessory module containing air solenoid valves and switches used in Freightliner multiplexed electrical system.

alternating current (AC) A continuous change in the amplitude and direction of current flow.

alternator A voltage generator that produces an alternating current that is electrically rectified to a DC voltage.

American wire gauge (AWG) A standard indicating the diameter of the conductor in wire. The smaller the gauge number, the larger the diameter of the wire.

ammeter A tool used to measure electric current. An ammeter is typically part of a digital multimeter.

ampere (amp) A unit of measure of electric current flow. One ampere is one coulomb of charge passing a stationary point per second.

analog multiplexing Use of one conductive path by several parallel-connected switches, each with a different resistance value, that are connected to a single analog input. The voltage at the analog input is used to determine the switch state.

analog to digital (A/D) converter An electronic device that transforms an analog signal into a digital signal.

analogy A comparison of a concept that is understood to explain a more difficult concept that is not understood. An analogy always breaks down as the level of detail is increased.

anode (diode) The positive material end of a diode.

anti-spin reduction (ASR) See **automatic traction control (ATC)**.

armature (motor) The rotating component in an electric motor.

armature (relay) A movable component in a relay.

automated manual transmission (AMT) An electronically controlled clutch/transmission system that uses a standard mechanical transmission.

automatic traction control (ATC) A means of minimizing wheel spin during acceleration by the ABS ECU requesting the engine ECM to reduce fueling if both wheels are spinning or through the use of differential braking if only one wheel is spinning.

backbone The main communications path in a bus-type topology. For J1939, the backbone refers to the main section of the twisted pair of wires.

backfeed A condition that exists when current is supplied to some device in the direction opposite of normal, such as in a dual-filament lamp that has an open ground connection. Also called feedback.

base (transistor) The terminal of a bipolar transistor that controls the amount of current flowing through a transistor.

BCI group number Battery Council International numbering system used to describe the dimensions of a lead acid battery.

bimetallic gauge A gauge that uses the movement of a metal strip composed of two different types of metal with different temperature expansion rates. Current flow through a heating element causes the strip to deflect.

binary A numbering system that only uses two characters: 0 and 1. Also called a base 2 system.

bit A single binary character.

blown A fuse that has been subjected to excessive current and has opened.

body controller The name of the body control module for International truck models since 2007.

breakout box A tool used to obtain closed-circuit voltage measurements for troubleshooting, typically at the connector for an electronic module.

breakout T A tool designed to obtain closed-circuit voltage measurements for a sensor or switch connector.

brush A spring-loaded sliding contact device that makes electrical contact with a rotating component.

brushed DC motor A type of DC motor that uses brushes to provide electrical connection to the armature windings.

bucking coil A winding in a three-coil magnetic-type gauge. The bucking coil's magnetic field opposes or "bucks" one of the other coil's magnetic fields, causing the gauge to move accordingly.

bulkhead module (BHM) The main electronic body and chassis control module on the Freightliner M2 truck.

The BHM is used to control several body and chassis electrical features and communicates on the J1939 data link.

bus A method of connecting the communications lines between electronic devices.

byte A group of eight bits.

Canadian Motor Vehicle Safety Standard (CMVSS) Collection of laws governing motor vehicles. CMVSS 108 addresses lighting requirements.

capacitor An electrical device that stores energy in the form of an electric field.

carbon pile A high-power variable resistor used to load a circuit for testing.

case based reasoning The concept of solving new problems based on the solution for similar problems in the past.

cathode (diode) The terminal of a diode connected to the N-type material.

cell A single voltage-producing element in a battery.

charge A fundamental property of electrons and protons. There are two types of charge, positive and negative. Electrons have a negative charge and protons have a positive charge.

charging Current flowing into a battery to restore the chemical energy stored in the battery.

chassis module (CHM) A slave electronic module on the Freightliner M2 truck used to control several chassis electrical features. The chassis module receives commands from the bulkhead module via the J1939 data link.

chuff test The self-test performed by the ABS ECU when ignition voltage is first supplied to the ABS ECU. Each modulator solenoid is actuated in a specific sequence during the test.

circuit A complete path for the flow of electric current.

circuit breaker A thermal circuit protection device. There are three SAE classifications of circuit breakers.

circuit protection device (CPD) Circuit protection device, such as a fuse, a circuit breaker, or a fusible link. The CPD is used to protect the wiring harness from excessive current flow in the event of a failure such as a shorted to ground circuit.

clearance lamps The lamps that are located at the top corners of the front and rear of the truck or trailer.

clipping A means of attaching a wire harness to the truck.

clockspring A device used to permit electrical contact between switches in the steering wheel and steering column.

closed loop control An automatic control system in which a process is regulated by feedback. The process being controlled is measured, which is compared to the desired outcome. The control system adjusts accordingly to achieve the desired outcome.

cold cranking amps (CCA) A battery rating that indicates how much current a new fully charged battery can provide continuously for 30 seconds at a temperature of 0°F and still maintain a terminal voltage greater than 1.2V per cell (7.2V terminal voltage for a typical 12V truck battery).

collector A terminal of a bipolar transistor.

combination stop/turn A red-colored turn lamp at the rear of a vehicle that also acts as the stop lamp.

common (terminal) The terminal in a multiple-throw switch or relay that is connected to the movable contact. In a DIN relay, terminal 30 is the designation for the common terminal.

commutation A mechanical means of reversing the current flow through the armature windings at just the right time to cause the motor to rotate continuously.

commutator The conductive segments that connect the motor brushes to the armature windings.

composite headlamps Headlamp assemblies that use replaceable incandescent bulbs instead of a sealed beam.

compound-wound A means of winding a motor's field windings and armature windings so that a series-parallel circuit is formed.

conductance The inverse (reciprocal) of resistance, measured in units of siemens.

conductor A material that offers very little opposition to the flow of electric current. Metals such as copper, iron, steel, and aluminum are conductors. Conductors have a large number of free electrons.

confirmed DTC An OBD diagnostic trouble code which the conditions for setting have been detected for two consecutive drive cycles. This will cause the MIL to illuminate. Also known as a MIL-on DTC

connectors Electrical components that hold terminals.

conspicuity Reflective devices required on vehicles by FMVSS and other laws.

contacts The parts of switches and relays that open to interrupt current flow and close to permit current flow.

control circuit (relay) An electrical circuit that supplies voltage to the relay coil.

controller area network (CAN) A group of electronic modules that communicate with each other without the need for a master controller.

conventional theory The assumption that the flow of current is from positive to negative. Conventional theory describes the movement of the holes or spaces, which is in the opposite direction of the electrons.

coulomb The charge associated with 6.25×10^{18} electrons.

counter emf (CEMF) Counter electromagnetic force, voltage that is produced by a motor, per Lenz's law, that acts as series-opposing to the voltage supplying the motor. CEMF is the reason that a motor draws less current as motor speed increases.

covalent bond A chemical bonding where pairs of valence electrons are shared between atoms.

current The flow rate of electrons measured in amperes.

data link The common name or slang for a twisted pair of wires used to connect a group of electronic modules.

data mining The use of telematics information for prognostics and machine usage information.

daytime running lights (DRL) Front-mounted lamps that are illuminated any time a vehicle is being operated. DRL are sometimes reduced-intensity low-beam headlamps.

delta A method of connecting three phases of an alternator stator end to end so that a triangle or delta shape is formed.

derate Intentional reduction of maximum engine power by the engine ECM to protect the engine due to a detected malfunction. Also refers to the reduction in engine power due to an exhaust emissions system failure.

desired output In control system theory, the set point established by the controller.

diagnosable switch A special type of switch typically used in a critical circuit that contains a resistor network. Switch contacts in parallel with the resistors short out specific resistors, causing the resistance measured between the terminals of the switch to change with the switch opened and closed. A diagnosable switch is connected to an analog input.

diagnostic trouble code (DTC) A number sequence assigned by an OEM or by SAE to indicate that a specific diagnostic condition has been detected.

dielectric Nonconductive insulator material placed between the plates of a capacitor.

differential braking An action of the ABS ECU that causes the brake to be applied on a wheel on one side of a drive axle that is spinning due to low traction.

digital Devices or circuits in which the output varies in distinct (whole) steps.

digital multimeter (DMM) A portable handheld diagnostic tool consisting of a voltmeter, ammeter, and ohmmeter in one tool.

diode A semiconductor device that permits electric current to flow in only one direction.

diode trio Three diodes placed in a single package, used to rectify the voltage output of the three-phase stator. The output of the diode trio is used to supply the current for the rotor field.

DIP switch Dual in-line package switch, a series of small switches that are often used in electronic gauge clusters for programming the vehicle pulses-per-mile and pulses-per-revolution values for the speedometer and tachometer.

direct current (DC) Electric current that does not change amplitude or direction.

discharging Current flowing out of a battery, the chemical energy being converted into electrical energy.

doping The addition of impurities to silicon to form a semiconductor.

drain A terminal of a field effect transistor (FET).

drive (motor) The device that contains the overrunning clutch and pinion gear on a cranking motor.

drive axle ABS event The lockup or impending lockup of a wheel on a drive axle; used to cause the engine brake to be disabled and the lockup torque converter clutch to be disengaged.

dual-filament lamp An incandescent lamp that contains two filaments. The two filaments share a common ground connection, making the lamp a three-terminal device.

dual-mode hybrid system A type of hybrid electric vehicle that has characteristics of both a series and a parallel hybrid electric vehicle.

duty cycle The ratio of the time on to the total cycle time, expressed as a percentage.

earth Another name for chassis ground, commonly used outside of the United States.

EEPROM An acronym that stands for electrically erasable programmable, read-only memory, memory that can typically be erased and rewritten in small sections.

electric field Invisible lines of electric force that exist between positively and negatively charged objects.

electric shock Electric current flowing through the body, causing involuntary muscle action.

electrical potential A quantity of electrons held at a distance from a quantity of protons. Electrical potential is also called voltage.

electrical system controller (ESC) The name given to the body control module in International high performance truck models up to 2007.

electrolyte A substance that mixes with water to form a solution or mixture that conducts electricity. In a lead acid battery, sulfuric acid and water make up the electrolyte.

electromagnet A magnet that is formed when current flows through a conductor.

electromagnetic interference (EMI) The disturbance caused by electrical and magnetic fields. EMI can cause electrical noise to be present in signals, resulting in unexpected actions.

electromotive force (EMF) The force that produces the flow of electrons in a circuit, another name for voltage.

electron A subatomic particle with a negative charge that orbits the nucleus of an atom.

electron theory The assumption that current flow is from negative to positive. Electron theory describes the actual direction of movement of electrons.

electronic control unit (ECU) The generic name of an electronic module such as ABS. The engine control module may also be called an ECU by some OEMs.

electronic gauge cluster (EGC) The name of the multiplexed instrument panel cluster on International multiplexed vehicles.

electronic on-board recorder (EOBR) Electronic log books which monitor driver hours of service information.

electronic service tool (EST) Term used to describe a broad range of diagnostic tools, including PC-based OEM diagnostic programs and handheld scan tools.

electronics The control of electrons using electricity.

emitter A terminal of a bipolar transistor. Current flows from the base to the emitter and from the collector to the emitter.

energized (relay) A relay coil that has current flow, creating an electromagnet.

energy The capacity or ability to perform work.

equivalent resistance A single resistance that is used to represent the total resistance of a group of resistors connected in series, in parallel, or in some other arrangement.

error (control system) The difference between the desired output and the measured output in an automatic control system.

excitation The supply voltage required to activate a circuit. When referenced to an alternator, a rotor that is supplied with voltage and is forming an electromagnet.

exhaust gas recirculation (EGR) The rerouting of exhaust gas back into the combustion chamber for the purpose of reducing peak combustion temperature to reduce the formation of oxides of nitrogen.

failure mode indicator (FMI) SAE J1939 and J1708 two-digit identifier used in a DTC to describe the type of failure that has been identified.

Federal Motor Vehicle Safety Standard (FMVSS) Collection of U.S. laws governing motor vehicles. FMVSS 108 addresses lighting requirements.

feedback See **backfeed**.

field effect transistor (FET) A voltage-controlled transistor.

field coil The windings that cause the motor pole shoes to become electromagnets.

field diode Another name for a diode trio in an alternator, used to self-excite the rotor field.

flash memory Digital memory that can be erased and re-written in large sections. Program memory for an electronic module is typically contained in flash memory.

floating An electronic device input terminal that is not connected to anything outside of the electronic module.

forward-biased A diode that is conducting electric current.

free electron An electron that can be easily dislodged from an atom.

freeze frame Parameters such as engine coolant temperature stored in an electronic module's non-volatile memory when a DTC became active.

frequency The number of complete cycles per second of a repeating waveform, measured in hertz.

fretting corrosion Referring to electric terminals, a layer of insulating oxide material caused by micro-movement between mated terminals. The movement wears away the terminal plating material, leaving the base metal of the terminal to corrode when exposed to oxygen.

full-field A rotor winding that is supplied with unregulated positive voltage and ground. A method of determining if a voltage regulator is defective, typically not performed on-vehicle with electronically controlled fuel systems.

fuse A circuit protection device, designed for one-time use, in which an element made of a conductive material is designed to melt to open the path for current flow when subjected to excessive current.

fusible link A circuit protection device composed of a length of wire with a special insulation material designed not to ignite (sustain a flame).

gate A digital electronic hardware device used to make logical decisions. In field effect transistors, the control terminal of the transistor.

gear-reduction A type of starter motor that utilizes a planetary gear set to reduce the motor speed and increase torque output.

ghost voltage The voltage that is dropped on the internal resistance of a digital multimeter when an open circuit measurement is obtained on a circuit with a high level of series resistance present somewhere in the circuit.

ground The metal components of the vehicle that are connected to the negative battery terminal.

ground fault circuit interrupter (GFCI) Fast-acting circuit breaker, used to help protect against electric shock by comparing the current in the hot conductor against the current in the neutral conductor.

growler A tool used to determine if motor armature windings are shorted together.

Hall effect A voltage produced in a thin layer of semiconductor material when it is exposed to a magnetic field. The voltage is proportional to the strength of the magnetic field.

halogen A noble gas used in incandescent lamps to prevent tungsten from depositing on the interior surface of the lamp glass.

hardwire The use of conventional wiring as a means to connect electrical devices.

harness A modularized section of wires designed to be connected to other harnesses or to electrical components such as switches and lights.

H-bridge A method of using four transistors to reverse the polarity of the voltage supplied to an electric motor to reverse the direction of rotation of the motor.

heavy duty on-board diagnostics (HD-OBD) U.S. EPA regulated diesel exhaust emissions; the ability of an engine system to detect that the tailpipe exhaust emissions are outside of permissible levels and provide indication to the vehicle operator of the condition.

hertz A unit of frequency, cycles per second.

high intensity discharge (HID) A type of headlamp system that uses an electric arc between tungsten electrodes as the means of illumination.

high side driver A transistor used as a switch or relay that supplies a connection to a positive voltage source. High side drivers can supply 20A or more of current.

HVAC An acronym that stands for heating, ventilation, and air conditioning.

hydraulically actuated unit injector (HEUI) An electronically controlled diesel fuel system used by Navistar and Caterpillar. High-pressure engine lubrication oil is used to produce fuel injection pressures within the unit injectors.

Hydro-Max™ A hydraulic brake booster system produced by Bosch that utilizes an electric pump motor to provide brake force amplification in the event of a stalled engine or other loss of power steering flow.

hydrometer A tool used to measure the specific gravity of battery electrolyte, engine coolant, and diesel exhaust fluid.

identification lamps Three lamps centered on the front of the truck at the top edge of the cab, at the rear on the top edge of the box of a van truck, or at the rear of the top of the trailer at the center. Identification lamps are amber at the front of a vehicle and red at the rear of the vehicle.

impedance A measure of the opposition to the flow of current in an electrical circuit. Impedance is measured in units of ohms and differs from resistance in that the AC effects of capacitive and inductive reactance are included.

idle validation switch (IVS) A switch used in conjunction with the accelerator position sensor to verify that the accelerator is being depressed by the operator.

incandescent lamps Sealed electric lighting devices that contain a thin piece of wire called a filament. Electric current passing through the filament causes it to glow and give off light.

induced The voltage that is produced from the cutting of magnetic lines of force by a conductor.

inductor A coil of wire that produces a magnetic field when current flows through the wire.

injection control pressure (ICP) HEUI fuel system high-pressure engine lubrication oil pressure. The high-pressure engine oil is used within the electronic injectors to develop the fuel injection pressures.

injection pressure regulator (IPR) HEUI fuel system control solenoid used to regulate the injection control pressure.

in-range failure A type of electric failure where the voltage supplied by a sensor to an electronic module is within the normal operational range of the sensor, but does not reflect correctly the actual physical parameter being measured. No DTC for out of range high or low would be set due to this type of failure. In-range failures can be caused by resistance in the wiring harness or failed sensors.

in-rush current A high level of current that flows when a device such as a light or motor is first supplied with voltage. The current decreases rapidly as the device reaches normal operating conditions.

instrument cluster unit (ICU) display Multiplexed instrument panel cluster used on the Freightliner M2. The ICU communicates with other electronic modules on the truck via the J1939 data link, and has hardwired inputs from the multifunction stalk switch.

instrument panel cluster (IPC) A generic term for a group of gauges and warning lamps that provide the operator with information such as vehicle speed.

instrument voltage regulator (IVR) A legacy electric component used to supply a regulated voltage for a bi-metallic gauge.

insulation The nonconductive material, such as cross-linked polyethylene, that surrounds a conductor.

insulator The materials that offer a great deal of opposition to the flow of electric current. The amount of current flowing through an insulator can typically be ignored because it is so small. Materials such as glass, plastic, and paper are insulators.

integrated circuit A circuit in which the transistors, diodes, resistors, and other components are formed during the process of manufacturing the device. Integrated circuits are commonly called chips.

interlocks The means of preventing an unsafe or unexpected action from occurring, such as the release of sliding fifth wheel lock in a vehicle traveling at high speeds.

internal resistance The total resistance between the terminals of a battery. Internal resistance increases as the battery's state of charge decreases.

ion An atom or molecule in which the total number of electrons does not equal the total number of protons, giving it a net positive or negative electrical charge.

ISO symbol International Standards Organization, a graphic symbol used to designate warning and indicator lamps such as high-beam headlamps, oil level, and coolant temperature.

J1587 An SAE standard that defines the message structure used in a serial communications network.

J1708 An SAE standard that defines the physical data link wiring and hardware used in a serial communications network. Together with J1587, J1708 is known as the ATA data link.

J1939 A series of SAE standards that defines the message structure and hardware for a high-speed serial communications network. SAE J1939 describes a controller area network (CAN).

JFET Junction field effect transistor (FET), the simplest type of FET. The JFET is controlled by the difference in voltage between the gate terminal and the source terminal.

junction The location where positive semiconductor material makes contact with negative semiconductor material to form a device such as a diode.

Kirchhoff's current law A fundamental principle of electricity that states that the algebraic sum of current flowing into a junction must be equal to the algebraic sum of current that flows out of the junction, also known as Kirchhoff's first law.

Kirchhoff's voltage law A fundamental principle of electricity that states that the sum of the voltage drops measured across devices in a closed loop must be equal to the algebraic sum of the voltage sources, also known as Kirchhoff's second law.

latching (switch) A switch that remains in the last position to which it has been moved.

leakage current A small amount of current that flows through an electrical or electronic device when the device is switched off or is otherwise not conducting.

learn cycle A calibration of an actuator, also known as zero and span calibration. The zero point is the minimum position (typically closed) to which a device controlled by an actuator can be driven (i.e., fully closed). The span is the measurement from the zero position to the maximum position (i.e., fully open).

Lenz's law A fundamental law of electromagnetism that describes the polarity of the voltage self-induced in a coil as being opposite to the polarity of the applied voltage.

license lamp White light that is located above the license plate at the rear of the vehicle, often incorporated into the tail lamp assembly.

light emitting diode (LED) A semiconductor device that gives off visible or invisible light.

linear Hall effect sensor A type of non-contact position sensor that measures the strength of a magnetic field to determine the location of the target magnet relative to the sensor.

linear power module An electronic device used to control blower motor speed by introducing a transistor-controlled series resistance to the circuit.

liquid crystal display (LCD) A type of alphanumeric display that uses a liquid to display information. A voltage applied to the liquid causes the liquid to darken.

load A device through which electric current flows, such as a resistor, that changes electrical energy into another form of energy.

load dump The dissipation of electrical energy when an electric load is suddenly disconnected from a generator. Load dump can cause the charging system voltage to increase to 50V or more and stay at a high level for 100 ms or more. The result is potential damage to electronic components. Load dump is typically caused by disconnecting battery cables with the engine running.

lookup table Stored information used by a microprocessor to determine how it should control an output based on inputs.

logic The use of correct or valid reasoning to come to a conclusion.

low side driver A transistor used as a switch that supplies a connection to ground.

magnetic gauge A type of gauge that uses the principles of electromagnetism to cause the gauge pointer to move.

magnetic lines of force The imaginary lines used to quantify the strength of a magnetic field. The arrows on the lines indicate the direction in which the needle of a compass would point if placed in the magnetic field.

malfunction indicator lamp (MIL) On-board diagnostic (OBD) regulated warning lamp that illuminates when a condition that causes the exhaust emissions to be out of compliance is detected.

marker lamps The amber and red lamps located on the sides of the truck and trailer.

material safety data sheet (MSDS) A document that contains information about the potential health hazards of exposure to potentially dangerous substances, and safe procedures to use when handling these substances.

measured output Closed-loop control system measurement of the process being controlled. In a home heating system, the thermostat setting is the set point or desired output. The measurement of the actual room temperature is the measured output.

measurement error Term describing an in-range sensor failure. An example is an indication of high coolant temperature being caused either by coolant temperature that actually is too high or by a false measurement of the coolant temperature by a sensor or sending unit. The latter is measurement error.

message identification (MID) SAE J1587/J1708 designation of the electronic module or system that has detected a failure or the device that is sending specific information.

microprocessor A device that processes information in a digital form.

microswitches The small switches that are typically contained in a sealed package designed to carry a low level current.

milling The process in which the pinion gear and ring gear remove material from each other when the pinion gear does not achieve full engagement with the ring gear.

modulator A component of an air ABS system, an electrically controlled valve designed to block the flow of air to a brake chamber and to exhaust the air in the chamber through two separate solenoids.

momentary contact A switch that is spring-loaded in a specific position. Moving the switch, such as by depressing it, causes the switch to open or close. Releasing the switch causes the switch to return to its normal state.

MOSFET Metal oxide semiconductor field effect transistor, a type of transistor that is often used as a high-current switch.

multifunction switch Turn signal stalk-mounted switch that may include switches for several other features. For Freightliner M2 models, the multifunction switch contains resistive ladder-type switches, which are used to control the windshield wipers, and a headlamp dimmer switch.

multiplexing The methods used to combine more than one channel of information into a common signal path.

negative temperature coefficient (NTC) A reduction in the resistance of a material as the temperature increases.

negative voltage spike The high-amplitude reverse-polarity voltage induced in a coil when a magnetic field surrounding the coil collapses; this happens when current flow through the coil is interrupted.

no-load voltage The voltage measured across the terminals of a battery with no current flow present in the circuit. The no-load voltage is an indication of the battery's state of charge.

nonlinear A term from mathematics that describes a system where the result (output) is not proportional to the causes (inputs). For example, many electronic devices such as thermistors are nonlinear devices. A thermistor with a negative temperature coefficient might have a resistance change of only 10Ω when the measured temperature decreases from 80°F to 70°F and a $10,000\Omega$ resistance change when the measured temperature decreases from −20°F to −30°F. The mathematical equation describing the behavior of a nonlinear system has terms that contain an exponent or power other than one. Plotting a nonlinear equation on a graph would result in a line that is not straight. By comparison, a linear equation contains only terms that have exponents of one (an exponent of one is typically not shown). Plotting a linear equation on a graph would result in a straight line.

non-volatile memory Digital memory that is maintained through power disconnection.

normal state The condition of the contacts on a relay or momentary-contact switch when the relay is not energized or the switch is in the relaxed position.

N-type Semiconductor material such as silicon that is mixed with impurities to produce a material with an excess of electrons.

ohm A unit of measurement of resistance, named after Georg Simon Ohm. One ohm is the amount of resistance necessary to cause one ampere of current to flow when one volt of potential is applied across the resistance.

ohmmeter A tool used to measure electrical resistance, typically part of a modern digital multimeter.

Ohm's law A fundamental principle of electricity that describes the relationship between voltage, current, and resistance in a mathematical form.

one-way clutch Also called an overrunning clutch, a means of permitting the pinion gear to rotate faster than the motor when the engine starts and the pinion gear is still in mesh with the ring gear.

open circuit An interruption in the path of electric current. An open circuit has an infinitely high resistance.

open loop control Non-feedback control, a controller that determines its output based on current operating state and the program instructions. The controller does not monitor the actual output of the system. Failure of important sensors in an otherwise closed loop control system can cause the system to go into open loop control as a means of still permitting the system to operate, but at a reduced level of performance.

original equipment manufacturer (OEM) Generic term for manufacturers of vehicles and major vehicle systems such as engines.

oscilloscope A diagnostic tool used to observe a changing voltage or current referenced to time.

out of range high (ORH) A condition that exists when the voltage measured at an input of an electronic module is greater than the maximum expected amount.

out of range low (ORL) A condition that exists when the voltage measured at an input of an electronic module is less than the minimum expected amount.

overrunning clutch See **one-way clutch**.

oxides of nitrogen (NOx) Regulated engine exhaust emission, includes NO and NO_2. NO_x is formed in the combustion chamber, especially at high temperatures. NO_x is controlled in-cylinder by reducing peak combustion temperatures using EGR or is converted in the exhaust system (aftertreatment) by use of selective catalytic reduction (SCR).

parallel (connection) A means of connecting two or more devices so they all have two common connection points.

parallel circuit A circuit that has more than one path for electric current to flow through. Components that are connected in parallel all have a common voltage drop.

parallel hybrid system A type of hybrid electric vehicle in which the power produced by the prime mover (i.e., diesel engine) is supplemented by an electric motor.

parameter group number (PGN) SAE J1939 numerical designation of a group of related parameters contained within one 8-byte CAN message.

parameter identifier (PID) SAE J1587/J1708 defined numerical designation of a specific measurement, such as engine coolant temperature.

parasitic loads The current draws that occur with the key switch in the off position.

parking lamps The amber or white lamps located at the front of the truck.

particulate matter (PM) U.S. EPA regulated diesel exhaust emission. PM is mostly carbon compounds (soot) that make up black smoke, but PM also includes other forms of visible exhaust smoke.

pending DTC OBD exhaust emissions related diagnostic trouble code, a DTC for which the conditions for setting the DTC have been detected for one drive cycle.

permanent DTC OBD exhaust emissions related diagnostic trouble code, indicates that a confirmed DTC had been set in the past. A permanent DTC is not actually permanent; it will be erased by the ECM after three consecutive drive cycles without the problem being detected. A permanent DTC cannot be cleared by an EST.

permanent magnet A magnet that remains magnetized without an external magnetic field present. Small motors often use permanent magnets for the pole shoes.

phase The angular relationship between two waveforms that are plotted in reference to time; also a winding of a generator or a transformer.

pinion gear A small diameter gear driven by the cranking motor that meshes with the engine ring gear.

plates Grids with pasted-on lead. Lead acid batteries have two different types of plates—positive plates and negative plates.

pole (switch) The number of independent circuits that can be controlled by a switch. This can also be thought of as the number of movable contacts contained in the switch or the number of switch input terminals.

pole shoe The magnets that set up the stationary magnetic field in an electric motor.

positive engagement A type of cranking motor that does not permit the motor to begin rotating until the pinion gear is in mesh with the ring gear.

positive temperature coefficient (PTC) A material in which an increase in temperature causes an increase in the resistance of the material.

potential The capability of doing useful work.

potential energy Stored energy measured in joules. There are several types of potential energy, including gravitational, spring, and chemical.

potentiometer A variable resistor that is often used as a position sensor. A potentiometer is a three-terminal device. Two terminals are connected to each end of a resistive element and a third terminal is connected to a wiper contact. The output is a variable voltage depending upon the position of the wiper contact.

power The rate at which work is performed or the rate at which energy is transformed from one form to another. The unit of measure of power is the watt.

power line carrier (PLC) The use of a high-frequency signal added to the power supply DC voltage for communications between devices. This is specifically used in modern trailer ABS systems to signal the truck ABS ECU of a fault detected by the trailer ABS ECU.

previous MIL-on DTC OBD exhaust emissions related diagnostic trouble code, a DTC for which the conditions for setting the DTC and illuminating the MIL had occurred in the past. Previous MIL-on DTCs are erased by the ECM after 40 engine warm-up cycles provided the problem is not detected again.

programmable parameter Rewritable non-volatile memory section of an electronic module, which contains data unique to a vehicle or engine, or owner preferences.

proportional solenoid An electric solenoid that can be driven and held in a desired location within a range of positions, not just the extreme ends of travel such as a standard electric solenoid.

proton Subatomic particles that, together with neutrons, make up the nucleus of an atom. Protons have a positive charge.

P-type A semiconductor material such as silicon that is doped with impurities to form a material with a deficiency of electrons.

pull-down resistor Part of an electronic module input circuit, a resistor that completes a path to chassis ground for the input. The pull-down resistor causes the voltage at an input to drop to near 0V with nothing external to the module connected to the input terminal.

pull-up resistor Part of an electronic module input circuit, a resistor that provides a path from a positive voltage source to the input. The pull-up resistor causes the voltage at an input to rise to some value with nothing external to the module connected to the input.

pulse width modulation (PWM) A technique in which a DC voltage source is switched off and on at a specific frequency to reduce the DC voltage level to some average value of voltage. The average voltage is regulated by controlling the time that the voltage source is switched on compared to the time that the voltage source is switched off. The time for one complete on and off cycle divided by the time that the voltage source is switched on is called the duty cycle. PWM also refers to a digital communications technique.

pyrometer A gauge used to display diesel engine exhaust temperature.

random access memory (RAM) Temporary volatile memory, used as a scratch pad to hold digital information.

readiness OBD indication of sufficient engine run time under required operating conditions for the ECM to have completed all emissions system tests for that key cycle.

recombinant Refers to a type of sealed lead-acid battery where the oxygen and hydrogen gas formed during operation recombine to form water, which is drawn back into the electrolyte solution.

rectifier A device in an alternator that converts or rectifies the negative half of a waveform into a positive waveform.

reference ground Electronic module grounding scheme where sensors are supplied with a dedicated return (ground) circuit. The reference ground is connected to chassis ground at one location within the electronic module. This reduces susceptibility to electromagnetic interference and provides greater accuracy in sensor measurements.

reference voltage A positive voltage that is supplied to a sensor or other device. The reference voltage is typically regulated to some value such as +5V.

relay An electromagnetic device with one or more sets of contacts that change position due to the magnetic attraction of an electromagnet. A small amount of current is capable of controlling a large amount of current with a relay.

reluctance The measure of a material's opposition to magnetic lines of force.

reserve capacity A battery rating indicating for how many minutes new, fully charged batteries are capable of supplying a continuous load of 25A while still maintaining a terminal voltage greater than 1.75V per cell (10.5V for a 12V battery) at a temperature of 80°F.

residual magnetism The magnetism that remains in the rotor poles with no current flowing through the rotor field.

Residual magnetism is necessary to cause the alternator initially to produce enough voltage to supply the current to the field.

resistance The opposition to the flow of electric current.

resistance temperature detector (RTD) Temperature measurement device, also known as platinum resistance thermometers (PRT).

resistive ladder network A form of analog multiplexing typically used with multiple function switches. Each switch function causes a different value of resistance across the corresponding switch terminals.

resistor An electrical component designed to have a specific value of resistance.

resistor block A high-power device that contains a stepped resistor network used to control blower motor speed.

reverse-biased A diode that is blocking the flow of electric current.

rheostat A two-terminal variable resistor typically used to adjust current flow in a circuit.

right-hand rule The method of determining the direction of the arrows of the magnetic lines of force that surround a current-carrying conductor.

ring gear A large diameter gear that is connected to the engine crankshaft, and often a part of the flywheel, flex plate, or torque converter.

ripple A low-level AC voltage that exists at the output of an alternator. The AC ripple voltage rides on top of the DC voltage level.

rotor The rotating member of an alternator that supplies the rotating magnetic field used to produce voltage in the stator windings.

routing The path followed by a wiring harness, such as along the frame rail away from the exhaust.

saturation The condition in which an increase in the input signal produces no further change in the output of the device. A transistor that is saturated is used as a switch.

sealed beam A type of headlamp or other high-power light that contains the lamp filaments, a reflector, and a lens in one sealed assembly.

selective catalytic reduction (SCR) Method of treating exhaust outside of the combustion chamber (aftertreatment) to reduce oxides of nitrogen. Selective indicates that only oxygen bonded with nitrogen is targeted. Ammonia derived from a urea solution known as diesel exhaust fluid is dosed into the exhaust stream ahead of a catalyst. The ammonia in the presence of the catalyst causes the oxides of nitrogen to be transformed into nitrogen and water.

self-discharge The process of a battery discharging without any load connected across the terminals. Self-discharge occurs more rapidly as temperature increases.

self-inductance A coil inducing a voltage within itself due to the expanding or contracting magnetic field that surrounds the coil when current flow through the coil changes.

semiconductor A material that is neither a good conductor nor a good insulator of electric current. In electronics terms, semiconductor refers to doped silicon.

sender A device that transforms liquid level, temperature, or pressure into a corresponding resistance to be used by a gauge.

sending unit See **sender**.

sensor A device that converts some physical property into an electrical signal. Sensors are also known as transducers.

separate stop/turn The stop lamps and turn signal lamps at the rear of a vehicle that do not share a common filament or lamp. Stop lamps must be red; turn signals can be red or amber in color.

series (connection) A means of connecting devices sequentially or one after the other.

series circuit A circuit that has only one path through which electric current can flow. Components that are connected in series all have the same amount of current flowing through them.

series hybrid system A type of hybrid electric vehicle system in which the vehicle is propelled only by electric motors. The internal combustion engine in a series hybrid is used to generate voltage to power the electric motors.

series-aiding Voltage sources connected so that the positive terminal of one source is connected sequentially to the negative terminal of another voltage source.

series-opposing Voltage sources connected so that the negative terminal of one source is connected sequentially to the negative terminal of another voltage source.

series-parallel circuit An electrical circuit composed of components such as resistors which have some elements in parallel with each other and other elements in series.

series-wound A method connecting the armature windings of a motor and the field coils so that all are connected in series.

shield A metal foil that is wrapped around conductors to reduce susceptibility to electromagnetic interference, specifically the effects of changing electric fields. A bare copper drain wire is often wrapped around the foil and is connected to ground through a capacitor (AC coupled) or directly to ground (DC coupled) at one point in the system.

short to ground An undesired condition in which an energized circuit has made contact with chassis ground. The short to ground has a very low resistance, causing a large amount of current to flow.

shunt A specific value of resistance placed in a circuit for measuring current flow. The voltage dropped across the shunt is used with Ohm's law to determine the current flowing through the shunt.

shunt-wound A method of connecting the armature windings of a motor in parallel with the field coils.

signal conditioning Electronic module input circuitry that prepares the input signal for the next stage of the signal process. Signal conditioning includes capacitors for filtering the signal and resistor networks to modify the voltage level at the input.

silicon An element with four valence electrons. Pure silicon forms crystals, which are nonconductive. Silicon is doped with impurities to form P-type and N-type materials to produce semiconductor devices.

sinking A device that supplies a path for current flow to ground. A low-side driver is said to sink current.

slip ring In an alternator, the surface on the rotor on which the brushes ride to supply paths for current flow through the field windings. In steering wheels, slip rings are circular conductive paths that provide a means for making electric connections between stationary and rotating contacts.

smart actuator A remote device that contains a processor and a communications interface. The smart actuator is able to control some component with minimal instructions from another controlling element, such as an engine ECM.

smart sensor A remote sensing device that contains a processor and a communications interface. Signal processing and conditioning are contained within the smart sensor. Complicated sensors such as NO_X sensors are typically smart sensors.

smart switch Specialized switches containing a resistor network used on the Freightliner M2 truck. Each switch is connected to the bulkhead module or expansion module via three analog inputs. The resistance value of two identification resistors within each type of smart switch is unique to its function.

Society of Automotive Engineers (SAE) A worldwide professional organization that develops standards used throughout the automotive industry.

solenoid An electrically controlled device designed to cause some remote mechanical movement. Solenoids may be used to control the flow of compressed air.

source (transistor) A terminal of a field effect transistor. Current flows from the drain to the source terminals as controlled by the gate terminal.

source address SAE J1939 assigned numerical designation of the device that has sent a J1939 message.

sourcing A device that supplies current flow from a positive voltage source. A high-side driver is said to source current.

specific gravity The weight of the volume of any liquid compared to an equal volume of pure water. Specific gravity has no units.

state of charge The indication of the percentage of chemical potential energy stored in a battery, compared to the maximum possible amount. A fully charged battery has a 100 percent state of charge.

stator The stationary windings that surround the rotor in an alternator. Voltage is induced in the conductive windings of the stator.

stepper motor An electric motor that can be driven in precise incremental steps or positions, used in modern truck instrumentation to drive a gauge needle.

stub A short section of twisted pair conductor that connects an electronic module to the backbone or main trunk of a J1939 data link.

suppression The use of a device in parallel with a coil (inductor) to reduce the amplitude of the negative voltage spike that occurs due to self-inductance when the magnetic field collapses when current flow through the coil is interrupted.

surface charge The condition in which gas bubbles on the plates cause the battery to appear to be at a higher state of charge than actual. Surface charge is rapidly dissipated when a load is applied to the battery.

suspect parameter number (SPN) SAE J1939 message packages which refer to a measureable factor such as SPN 110 engine coolant temperature. SPNs are also used in J1939 DTCs to define the system for which a failure has been detected.

tail lamp The red lamps at the rear of a vehicle.

telematics Convergence of telecommunications and informatics. Informatics is the science of information. Telematics can be used to predict failure before the failure occurs (prognostics).

terminals The conductive pins, sockets, and rings that terminate the ends of wires.

terminating resistor A resistor that is placed across the two conductors of a twisted pair communications network at both extreme ends of the network, such as a J1939 data link. The terminating resistor helps reduce signal reflections that interfere with communications.

thermistor A variable-resistance sensor used to measure temperature.

thermocouple A voltage-producing sensor that uses two junctions of two dissimilar metals to measure temperature. The voltage produced is proportional to the difference in temperature between the two junctions.

three-bar light Identification lamps contained in a single strip.

throw The number of positions that a switch can be moved and still complete a circuit; the number of different circuits that each of the switch's poles can control.

time-division multiplexing A multiplexing technique in which the pair of wires is time-shared by each device.

tooth abutment A condition that occurs occasionally in cranking motors in which the teeth of the pinion gear do not come into mesh with the teeth of the ring gear.

torque-speed control (TSC) An SAE J1939 message (PGN 0) that indicates a desired engine torque or speed reduction requested by some electronic module.

total resistance This represents the equivalent resistance of an entire electric circuit. The total voltage supply in a circuit divided by the total resistance yields the total current flow in the circuit.

transfer function An equation that describes the relationship between the physical parameter being measured by a sensor and the output voltage of the sensor.

transformer Electrical device consisting of two inductors, which can increase or decrease a changing electrical voltage. The supply voltage is connected to the primary windings; the output is connected to the secondary windings. The ratio of the number of turns in the secondary windings to the number of turns in the primary windings determines the voltage induced in the secondary windings.

transistor Electronic device with three terminals, used as an amplifier or a switch.

tripped The term used to describe a circuit breaker that has opened due to excessive current flow.

valence band The outermost shell of an atom, which determines the electrical conductivity of the element.

variable geometry turbocharger (VGT) Turbocharger with a variable aspect ratio. The position of the vanes is controllable to modify the exhaust manifold pressure (backpressure) and to control the air/fuel ratio.

variable reluctance A type of sensor that uses a coil of wire wrapped around a permanent magnet. A low-reluctance timing disk or tone wheel is passed in front of the sensor, causing the magnetic lines of force created by the magnet to be redirected into the low-reluctance disk.

vehicle onboard radar (VORAD) Legacy forward radar system for on-road trucks developed by Eaton, now incorporated into Bendix advanced ABS systems.

virtual fusing A means of protecting a circuit by measuring the amount of current conducted by an electronic switching device such as an FET, and switching the device off if current exceeds a predetermined value.

virtual sensor Calculated or modeled physical parameter without the use of an actual sensor.

volatile memory Digital memory that is lost when power is interrupted. RAM memory is volatile memory.

volt The unit of measurement of electrical potential or voltage. One volt is the amount of electrical potential necessary to cause one ampere of current to flow through one ohm of resistance.

voltage Electrical potential, also called electromotive force.

voltage drop The difference in electrical potential measured across a resistance. Analogous to pressure drop across a hydraulic circuit restriction.

voltage regulator A device used to maintain the output voltage of an alternator to a defined value.

voltmeter A tool used to measure the difference in electrical potential or voltage between two points. A voltmeter is part of a digital multimeter.

watt A unit of measure of power. One horsepower is equal to about 746 watts. In electric power, the product of voltage and current.

Watt's law A principle that states that electric power measured in watts is equal to the voltage multiplied by the current flowing through a device.

Wheatstone bridge An electrical circuit that can be used to determine an unknown resistance through the use of three known-value resistors.

windings Coiled conductors or wire typically used to describe components within generators and electric motors.

work Movement in the direction of an applied force.

wye (Y) A means of connecting the three phases of an alternator so that all three phases share a common connection. The shape of the windings is drawn such that the phases are in the shape of the letter "Y."

xenon A noble gas used in arc lamps.

zener diode A special diode designed to permit current flow in the reverse-bias direction after the voltage dropped across the diode is greater than the zener voltage.

zener voltage The level of reverse-biased voltage dropped across a zener diode that permits the diode to conduct in the reverse-biased direction.